THE LAW OF THE EMPEROR'S NEW CLOTHES

VOL. 1
AND DARWIN SAID
Why the Supposed Evidence of Evolution Reveals
A Universe With Design

PETER-BRIAN ANDERSSON

Cover image: Charles Robert Darwin by John Collier 1883, National Portrait Gallery, London. Wiki Commons by Dcoetzee. Citations from *The Origin* are from the 1998 Modern Library Paperback edition published by Random House, Inc. New York. Citations from Scripture are from The Holy Bible, English Standard Version® (ESV®), copyright © 2001 by Crossway, a publishing ministry of Good News Publishers. Used by permission. All rights reserved.
By The Things That Are Made Press. www.bythethingsthataremade.org. Printed in the United States of America. Copyright © 2014 Peter-Brian Andersson. All rights reserved.

ISBN-10: 0692261060
ISBN-13: 978-0692261064

BRIEF CONTENTS

THE LAW OF THE EMPEROR'S NEW CLOTHES

VOL. 1
AND DARWIN SAID
Why the Supposed Evidence of Evolution Reveals a Universe with Design

PROLOGUE: No Middle Ground 3

I. THE WONDER OF LIFE & DARWIN'S REASON: IT'S EVOLUTION
   1. "O Jehovah, Quam Ampla Sunt Opera Tua!" 15
   2. One Long Argument 69

II. THE EVIDENCE FOR EVOLUTION
   3. Evolution is "A Fact"! 103
   4. Survival of the Fittest 139
   5. Bones Of Contention 179
   6. Believing is Seeing 265
   7. Looks Are Important 291
   8. Of Molecules, Mice & Men 347
   9. Long Live "Ontogeny Recapitulates Phylogeny!" 381
   10. Making a Monkey Out of You 449
   11. Ape-Men in the Family Album 499
   12. The Origin of Species 567

About the Author 600
Index 601

VOL. 2
HAND OF GOD
Why Macroevolution is a Scientific Fact When it's Really a Lie

III. THE ALTERNATIVE REASON: IT'S DESIGN
13. In The Beginning 3
14. A Revelation of God By Force 81
15. Accepting Darwin's Challenge 131
16. Exposing a Supernatural with Science 181
17. The Devil Is Always In The Details 219

IV. WHY MACROEVOLUTION IS 'A FACT' WHEN IT'S REALLY A LIE
18. Forming Fact From Philosophy 237
19. Scientific Solutions Must Be Naturalistic! 245
20. Evolution Is A Scientific Fact! 253
21. ID Is Not Science! 263
22. ID Is An Attack On Science! 275
23. ID Is A Religion! 281
24. ID Is Unconstitutional! 293
25. Evolution Is Promoted By Propaganda! 351
26. Evolution Is Not Hostile To Christianity! 383

EPILOGUE: Exchanging the Truth of God for a Lie 409

Acknowledgements 421
About the Author 422
Index 423

FULL CONTENTS
AND DARWIN SAID

PROLOGUE: No Middle Ground

I THE WONDER OF LIFE & DARWIN'S EXPLANATION: IT'S EVOLUTION

1. "O Jehovah, Quam Ampla Sunt Opera Tua!"
- 1.1 Once Upon a Time
- 1.2 The Conundrum of Life
- 1.3 The Appearance of Design in Nature
- 1.4 Darwin's Elegantly Intuitive Solution
- 1.5 Tilting at Windmills or Jousting With Dragons?
- 1.6 What's So Special About Evolution Theory?
- 1.7 In Conclusion: Lies, Damned Lies & Science

2. One Long Argument
- 2.1 What is (Neo)-Darwinism?
- 2.2 The Theory of Evolution
- 2.3 What is the Mechanism of (Neo)-Darwinism?
- 2.4 Darwinism's First Step: The Generation of Genetic Variation
- 2.5 Darwinism's Second Step: Sorting Phenotypic Variation By Natural Selection
- 2.6 In Conclusion: So Much for Theory Let's See Fact

II THE EVIDENCE FOR EVOLUTION

3. Evolution Is "A Fact"!
- 3.1 "Simply a Fact"
- 3.2 A Catalogue of the Evidence
- 3.3 Following the 3 Step Argument for Evolution from Apparent Design
- 3.4 Step 1: The Supernatural Is Obviously Not Science
- 3.5 Step 2: Species Evolution Can Be Seen Directly
- 3.6 Step 3: Believing Is Seeing: The Macroevolution Inference
- 3.7 In Conclusion: "The Single Best Idea Anyone Has Ever Had"

4. Survival Of The Fittest
- 4.1 What's Artificial About the Artificial Selection Analogy?
- 4.2 Natural Selection: The Blind Watchmaker of Life
- 4.3 The Apparent Tautology of Natural Selection
- 4.4 Natural Selection in Black and White: Moth Melanism

4.5 Natural Selection Directional or Just Oscillating? The Example of Darwin's Finches
4.6 Other Examples of Natural Selection
4.7 Natural Selection of Weasels
4.8 Philosophy Dressed in the Robes of Science
4.9 In Conclusion: Half a "Fact" is a Half-Truth

5. Bones Of Contention
5.1 The Importance of Hard Evidence
5.2 What Are Fossils?
5.3 A Few Principles of Paleontology
   5.3.1 The Fossil Dating Game
   5.3.2 Accommodating Incompleteness
   5.3.3 Interpreting Fossil Relationships
5.4 What Is The Fossil Record Predicted By Darwinism?
5.5 What Does The Fossil Record Show?
   5.5.1 Mass Extinction
   5.5.2 Increasing Diversity and Complexity. Or Not
   5.5.3 Sudden Appearance
   5.5.4 Stasis
   5.5.5 Rare Evidence of Intermediate Forms
      5.5.5.1 Running Into Deep Water Scientifically
      5.5.5.2 Picking Bones for Wild Horses
      5.5.5.3 The Flap Over Archaeopteryx
      5.5.5.4 Intermediate Vertebrate Forms
5.6 In Conclusion: Finding Darwin's "Most Obvious & Serious Objection" Still There

6. Believing Is Seeing
6.1 Facing Up To Fossil Reality: 2 Ways to Clothe an Emperor
6.2 The First Way: The Fossil Record Is "Imperfect"
   6.2.1 Testing Fossils for Incompleteness
   6.2.2 Mirror, Mirror on the Wall
6.3 The Second Way: Punctuated Equilibrium Predicts This!
6.4 In Conclusion: How Soft Can Hard Evidence Get?

7. Looks Are Important
7.1 Homology As Biological Similarity
7.2 Homology Redefined By Evolution
7.3 Biological Similarity Does Not Result From Similar Development or Similar Genes
7.4 The Problem of Sameness in Biology: Putting Humpty Homology Together Again
   7.4.1 Phylogenetic Approaches to Homology
   7.4.2 Developmental Approaches to Homology

7.4.3. Homology Today: Now You See It Now You Don't
7.5 Homology Is the Evidence for Most of the Evidence
7.6 Evolution Evidence Lying in the Eye of the Beholder
    7.6.1 A Hands-on Look at the Evolutionary Hand
    7.6.2 An Earful of Jaw
7.7 What Exactly Is Biological Similarity?
7.8 In Conclusion: Evolution Looks That Lie In the Eye of the Beholder

8. Of Molecules, Mice & Men
8.1 Getting Down To Details: The Molecular Evidence of Evolution
8.2 The Rise Of Molecular Phylogenetics
8.3 Molecular Equidistance & Separation Not Links
8.4 The Molecular Clock Hypothesis
    8.4.1 How It Ticks And How It's Set
    8.4.2 The Neutral Theory Of Molecular Evolution
8.5 A Clock That Cannot Keep Time!
    8.5.1 Not Exactly Swiss
    8.5.2 Telling Time By Relaxin
    8.5.3 Hemoglobin Horology
8.6 Why Can't It Clock? Testing The Molecular Clock Hypothesis
    8.6.1 Flies In The Face Of Molecular Evolution
    8.6.2 Losing Time With The Albumin Gene
    8.6.3 The Paradox of the C Paradox
8.7 In Conclusion: Garbage In Garbage Out

9. Long Live "Ontogeny Recapitulates Phylogeny!"
9.1 The Significance of Embryology to Evolution
9.2 The Evidence of Evolution in Embryology: Darwin's 5 Claims
9.3 Claim 1: Are Embryos Similar Across Animal Species?
    9.3.1 The Historical Background
    9.3.2 Haeckel's Embryos and the Biogenetic Law
    9.3.3 The Phylotypic Stage Is Not Highly Conserved
    9.3.4 Von Baer's Law Is Not A Law
    9.3.5 *Hox* Gene Conservation: Phylogeny Verified or *Hox* Hoax
9.4 Claim 2: Is Ancestry Inferred By Similarity In Development?
9.5 Claim 3: Are There Parallels Between Ontogeny & Phylogeny?
9.6 Claim 4: What's The Evidence From Vestigial Structures?
9.7 Claim 5: Is Ancestry Inferred By Dissimilarity In Development?
9.8 In Conclusion: Repeat a Lie Often Enough and Everybody Will Eventually Believe It

10. Making A Monkey Out Of You
10.1 Four Arguments for the Fact of Human Evolution
10.2 The Supposition of Paleoanthropology: Evolution Is Present
    10.2.1 Inference Without Hard Evidence

      10.2.2 Hard Evidence By Inference
      10.2.3 Cladistic Analysis Is Hardly Evidence
  10.3 The Science of Paleoanthropology: Finding Evolution Present
      10.3.1 The Fact of Human Evolution or Not
      10.3.2 Fossil Trait Analysis: Measurement But From No Beginning
  10.4 Inspecting the Evidence: Before We Even Begin-Begging the Question
      10.4.1 Inspecting Argument #1: Is Evolution A Fact?
      10.4.2 Inspecting Argument #2: Is Similarity Ancestral or Typological
  10.5 In Conclusion: How to Make A Monkey Out of You
  10.6 Appendix: A Brief Overview Of The Hominid Fossil Record
      10.6.1 Hominid Taxonomy
      10.6.2 Historical Overview

11. Ape-Men In The Family Album
  11.1 Inspecting Argument #3 (Part I): Is there an Evolutionary Hominin Fossil Record?
      11.1.1 Stasis Instead Of Evolutionary Change
          11.1.1.1 Stasis in Science
          11.1.1.2 Human Brain Evolution
          11.1.1.3 Ape-Men At Work: Tool Evolution
          11.1.1.4 Are Mosaics Evolutionary? The Chad Man Ape
          11.1.1.5 Evolution Going Backwards: Reversals in the Series
      11.1.2 Sudden Appearance Instead Of Gradualism
      11.1.3 Simultaneous Existence Rather Than Anagenesis
  11.2 Inspecting Argument #3 (Part II): *Erectus* Is No Transitional Species
      11.2.1 Stasis Not Evolutionary Change
      11.2.2 Sudden Appearance
      11.2.3 Contemporaneous With Ancestor & Descendant Species
      11.2.4 Erectus Is Just An Ancient Human?
  11.3 Inspecting Argument #3 (Part III): *Habilis* Is No Transitional Species
      11.3.1 A Birth By Handy Intuition
      11.3.2 Paleontology By Personality
      11.3.3 A Bone To Pick
      11.3.4 OH 62: A Case Of Standing Tall Or Standing Down
      11.3.5 The Monkey Business Making *Homo habilis*
  11.4 Inspecting Argument #4: Fossils Of Ancient Humans?
      11.4.1 No Elbow Room: KNM-KP 271
      11.4.2 The Trail Of Evidence at Laetoli
      11.4.3 Olduvai Stones
      11.4.4 Evolution from Man-Apes or Ancient *Homo sapiens*?
  11.5 The Philosophical Implications of Human Evolution
  11.6 In Conclusion: The Science of Turning Frogs into Princes
  11.7 The Evidence of Evolution from Biogeography

12 The Origin Of Species
12.1 Back To The Future
12.2 The Origin of Biopolymers
    12.2.1 The Darwin-Haldane-Oparin Primordial Soup
    12.2.2 The "Impact" Theory
    12.2.3 The Hydrothermal Vent Theory
12.3. The Origin Of A Self Replicating System From Biopolymers
    12.3.1 The RNA World Hypothesis
    12.3.2 The Thioester World Hypothesis
    12.3.3 The Self-Replicating Clay Hypothesis
    12.3.4 Life Is Inevitable! - The Self-Organizing Hypothesis
12.4. The Origin Of All Life From A Simple Self Replicating System
    12.4.1 How Much Time Is Available?
    12.4.2 Beyond Even Chance
12.5 Life from Outer Space?
12.6 In Conclusion: How to Get Something For Nothing Scientifically

Index
About the Author

PART I

THE WONDER OF LIFE
&
DARWIN'S REASON:
IT'S EVOLUTION

Prologue: No Middle Ground

How did life in all its diversity and complexity arise? By chance and the natural forces of evolution or by some supernatural creation? One of them must be true and one of them must be false because there are no other possible answers. As far as the science establishment is concerned there's not a shadow of a doubt it was all by evolution. Harvard biologist Ernst Mayr explains why: "Evolution is not merely an idea, a theory, or a concept, but is the name of a process in nature, the occurrence of which can be documented by mountains of evidence that nobody has been able to refute."[1]

What is that evidence exactly? And if as he says, there's so much evidence that "nobody has been able to refute" evolution, why is there still so much doubt in the public mind? Less than twenty percent of Americans agree![2] Forty percent of them say the world was supernaturally created in six days, sometime in the last 10,000 years. Opinion polls have not changed on this in over a quarter century either (Fig.18.1). How come America's best academics and educators have not been able to even dent, much less dispel this obstinacy? Evolution is the most fundamental theory in all biological science they say! Evolution skepticism is not confined to America either. Albeit to lesser and varying degree it's worldwide.[3,4] Fifty two percent of Britons disagreed with the Harvard professor in a recent BBC-Ipsos-MORI poll for instance.[5] How can such opposite views about so basic a question about the world be held, and with such opposite conviction, in the twenty-first century?

What we also all know and possibly from personal experience, is the evolution-creation debate is so intractable its proponents invariably argue right past each other making no impact whatsoever. One side is usually left scratching its head, incredulous at the convictions of the other. How does that happen exactly? What are the sticking points between evolutionists and creationists, and between evolutionists and design advocates? What's the crux of the controversy? And how is it that so many Americans can reject evolution and two thirds also want creationism taught in school, yet it never happen in that democracy?[6] Actually it's illegal! This book is an investigation of these questions. It's a critical analysis of the evidence on all sides of the evolution vs. creation and intelligent design debates.

First, what is "evolution"? It means not merely "slow change" or "development", it's a comprehensive scientific theory explaining how life acquired all its diversity and complexity by natural forces alone. These forces are mutation and natural selection. Niles Eldredge at the American Museum of Natural History usefully defines evolution as "the proposition that all organisms on earth, past, present and future, are descended from a common ancestor that lived at least 3.5 billion years ago, the age of the oldest fossil bacteria yet reliably identified."[7]

What's "creationism" on the other hand? It's the view according to the Miriam-Webster Dictionary, "that matter, the various forms of life, and the world were created by God out of nothing and usually in the way described in *Genesis*". The point is creationism rests on Scripture as its evidence, not science. Philosopher of science Robert Pennock defines creationism even more simply. He says creationism is "any view that rejects evolution in favor of some personal, supernatural creator".[8] There are instead a host of creationist viewpoints however. They differ over such details as which Creator(s) did it, why and how, and some include still other elements such as when this happened and how old the Earth is. Despite their differences all creationists agree there was a supernatural creation, and that a supernatural intervention into our natural Universe is a reality.

This thinking is anathema to science now. The reason its spokespersons say is because there is no evidence of a supernatural anywhere, much less in science. Besides, evolution theory declares all biological emergence is fully explained by natural forces and chance over geologic time. True, science has no objection if someone wants to believe both in evolution and in a God. This is a view called "theistic evolution" and indeed such views are held even by some scientists, though only very few elite scientists.[9] The head of the Human Genome Project, Francis Collins is one and Brown University biology textbook author Kenneth Miller is another. Evolutionary theory is claimed fully sufficient without such supernatural notions

4

however, and even these men say their supernatural beliefs cannot be demonstrated by biology. So why believe their supernaturalism? The vast majority of elite scientists are atheist - 95% of the biologists in The National Academy of Sciences and 95% of the fellows of The Royal Society for example, and this is surely the preeminent reason why.[10,11]

While not all creationists back up their religious beliefs with data from biology, this does not mean that all creation viewpoints are necessarily empirically empty. There is a view affirming a supernatural creation of the world that claims to rest entirely on the evidence of science, not Scripture or religion. It's a conclusion made from intelligent design analysis (ID). ID concludes that patterns measurable in living things are beyond the power of natural forces and chance to construct, and so must have originated super-naturally. It declares this is directly demonstrable from the evidence of biology. This is not creationism then. It's far more outrageous. ID analysis claims that biology denies evolution, to the contrary of what all the best scientists in the world and the textbooks now say. In corollary it concludes there is an objective supernatural stamped right across our natural world, and that there must be a Creator. It's proof of God. You might want to throw up your hands at this stick-in-the-eye to everything that science says it holds dear. I once thought so too.

So why write this book? It's because that question about life's origin and complexity, and that seemingly outrageous claim of objective evidence of a supernatural in the natural world from ID theory, is a matter of rational inquiry and you should know it. This is not what the science or education establishment will tell you. In fact as far as I am aware no accredited university biology department or scientific journal, anywhere, has endorsed ID as the true solution to the origin of living things. Actually they categorically condemn it![12] This is an intriguing observation in itself. The reason is if ID really is as crackpot as they say, why have so many august personages and science organizations and journals separately voiced the same strident indignation only to merely echo each other in the protest? It creates the appearance at least, that the lady might protest too much.

Another observation to consider, is that among the advocates of ID are some impeccably well-credentialed scientists, at least by every other measure. True there are only very few, but even one would be enough to refute the claim that ID is the view only of the ignorant or misinformed. Which is what the authorities and journals say. Another observation is some ID advocates were once outspoken evolutionists! Take famous atheist philosophers Anthony Flew and James Shapiro for instance. What made them change their minds?[13,14] There's no evidence they lost them either. All this seemed worth a closer look.

Settling the question where life came from could not be more important either. The reason is easy to see. It's to ask one of the most basic questions of human existence - "Where did I come from?" It's a question we all have to answer by the way, and it's one we do all answer. The answer we each choose, whatever it is, strikes to the meaning and purpose of life and our own lives in particular because it will make your worldview. The corollaries of the answer apply to the way we think about almost everything. Thus while the question of the origin of life and its diversity is a scientific question, it's also a metaphysical question and it's also a personal question. Either there is a Creator and an overwhelming supernatural reality beyond our Universe, or else matter and natural forces are all there is. There's no middle ground. That's why getting the answer right is so important.

Perhaps you've made up your mind about this already. If so, the only question remaining is whether you're open to examining the evidence. Do you know why you answer as you do? Some answer oppositely. Do you know why they're wrong? We already know as far as academia is concerned there's nothing to debate here. As Oxford University zoologist Richard Dawkins famously says: "our own existence once presented the greatest of all mysteries, but…it is a mystery no longer because it is solved. Darwin and Wallace solved it, though we shall continue to add footnotes to their solution for a while yet."[15] So what we've learnt so far is getting the answer to the question of the origin of life could not be more important and that science, most everywhere fount and arbiter of truth today, has the answer. Is there anything left to discuss? Well, yes. Let me suggest seven things:

1. The first is that despite Dr. Dawkins' assertion this "greatest of all mysteries" is "solved", it's still challenged by an overwhelming proportion of the American public as well as by others elsewhere in the world, over a century and a half after Darwin first explained it. Why? It can't be by force of empty religious dogma because if the mystery were patently "solved", how could there be such a sustained objection for so long? Some of the most articulate objections to evolution were made back in Darwin's day, and they're still being made! In fact if the mystery were already "solved" in evolution, and all anti-evolution objections are indeed empty religious objections as he says, then evolution has proved that religion false. In the minds of many, Dr. Dawkins included, indeed it has.[16-19] So where does the truth lie? The implications, regardless of the answer, are monumental. This seemed worth investigating.

2. Objections to evolution are taken as an assault on the institution of science itself. In other words one scientific theory, evolution theory, has

come to represent all of science! In academic debate where truth is tentative and skeptical challenges to hypotheses routine, why should that be? It's because this one theory is unique in its status and intellectual reach. Dr. Dawkins explains how. "Darwinism encompasses all of life – human, animal, plant, bacterial, and, if I am right...extraterrestrial. It provides the only satisfying explanation for why we all exist, why we are the way that we are. It is the bedrock on which rest all the disciplines known as the humanities...Darwinian Evolution, as one reviewer has observed, 'is the most portentous natural truth science has yet discovered.' I'd add, 'or is likely to discover'."[20] Given such authority a detailed understanding of evolution would seem essential, regardless of whether ID has any substance. Inspecting the question of the origin of life and its different solutions seemed worthwhile therefore.

3. The debate about the origin and diversity of life is almost never cordial. This is because it's conducted across a divide as much philosophical as ever it is empirical. Why? This seemed worth exploring. From the perspective of the science community, creationism of any stripe as well as a conclusion of ID in nature is magical thinking and worse. In Richard Dawkins's words: "It is absolutely safe to say that, if you meet somebody who claims not to believe in Evolution, that person is ignorant, stupid or insane (or wicked, but I'd rather not consider that)."[21] Now we are all entitled to our own answer to this question, but if it is to be a rational one, surely not before examining all the evidence. Yet this science holds any view against evolution as scientifically speaking irrational already because it resorts to a supernatural to explain the natural world. They say a supernatural cannot be demonstrated by science. On this view ID is subjective belief, at best. Dr. Dawkins again: "Pretend as they will to scientific credentials, the anti-Evolution propagandists are always religiously motivated, even if they try to buy credibility by concealing the fact. In most cases, they know deep down what to believe because their parents recommended an ancient book that tells them what to believe. If the scientific evidence learned in adulthood contradicts the book, there must be something wrong with the scientific evidence."[22]

Well this book disagrees. You see even if the world did operate by natural forces alone, a supernatural design in nature could most definitely be detected, at least in principle. We unhesitatingly conclude ID and not natural forces when considering the origin of the heads on Mount Rushmore for instance, or if we were to come upon a valentine etched in the sand of a deserted beach. Without any need for training in physics or geology or statistics, or knowing who the designer was, or why it was done or if it was done for a religious purpose or not, we know what it would take for natural forces to produce the complexity and purposefulness of these conformations of matter, and we know it is utterly beyond their reach. Exactly the same reasoning is applied to the questions of origin in biology. What's so magical about that?

To come at the question of the origin of life and its complexity with an *a priori* decision that a supernatural is false, is already to have an answer. Who now needs evidence? There is only one certain barrier to the discovery of truth, after all. The conviction it's known already. Preconceived notions of what truth must be are certain impediments to discovering truth, and they also foster illusions. We know there has to be an illusion in the evolution-creation debate. Opposite sides are equally certain they're right! A whole lot of people are in for a mighty big shock, either way.

It's therefore ironic in light of Dr. Dawkins' accusation about bigotry above, to discover that evolution theory is a preconceived and irrefutable paradigm and the reason is simple. Science assumes anything supernatural is irrational because it can't be demonstrated empirically. But what's the evidence for that ? It's never been proved nor could it be. How could the non-existence of an unknown ever be proved? Yet it's assumed to be true in a working hypothesis for the conduct of science called "methodological naturalism", and which is believed with the faith of martyrs. Evolution could be true nonetheless, but at least we can say ID analysis in biology is a direct challenge both to evolution theory and to one of the most basic premises in twenty-first century science. We need to know where the truth sits. This book was written to that end.

A religious connotation to a conclusion arrived at by empirical means does not render that analysis automatically non-objective. This might seem self-evident, but in the evolution-creation debate it's not. In declaring itself the world's repository of objective truth "science" is rarely religiously neutral. Why? This would seem worth examining too. Ruling a line of inquiry that is the only alternative to evolution out of science simply because it might have religious connotations, specifically God, and this even before data is interpreted, is a bigotry Dr. Dawkins and the mainstream of science miss. If you feel the same you might want to read further for that reason alone. There is a still more important reason for writing however, and it has been alluded to already.

4. It's that there *are* grounds, and plenty of them, for a rational debate between evolutionists and ID advocates over hard evidence in biology here. ID theory holds the world was made by a supernatural intelligence not because of a religious belief, although it certainly leads there, but because

of patterns of complexity in nature demonstrably beyond the ability of natural forces to ever construct. This book is predicated on the obvious corollary. If indeed there is evidence for ID and for evolution as each side claims, it should be a simple matter to expose which of them is true and which of them false, and who is telling the truth and who is telling stories. Knowing the controversy and solving the Conundrum seemed worth researching and writing about. Dr. Dawkins made a false assumption above. He says science proves evolution. We're told evolution is a fact and given a plausible explanation, at least on the face of it, that is Darwinism. Against this a creation does seem magical. The question is whether those claims for macroevolution are true. Unfortunately, and to cut to the chase, the data from biology rejects them. Part II of this book lays out the reasons why while inspecting the claims evolution theory makes for itself point for point. Interestingly there is obfuscation of this bankruptcy by the science. This was another remarkable and incongruous discovery that seemed worth writing about.

5. Problems in evolutionary theory are rarely raised or allowed to reach beyond the closed circles of those involved. This is not my conspiracy theory. It happens simply because evolution is assumed to be true as the starting point. The result is the amazing situation we find today where although demonstrably false, evolution is yet scientifically "a fact" and the central paradigm of biology as well. Objections to this "fact" are aggressively resisted, even in classrooms and in lecture halls, seemingly the last places for censorship. Evolution dissenters are met with sanctions. The courts have got involved. Why? How? All this seemed worth investigating.

6. Evolution is unique for still another reason worth writing about. It's a scientific theory that's legally enforced. The teaching of creationism is banned with all the force of the law in the public schools and colleges of the United States. But if the cause of our existence has been rationally "solved" and there are "mountains of evidence nobody has been able to refute", as Drs. Dawkins and Mayr declare, then why are lawyers and judges deciding the answer by bannings and by threats? This has the appearance at least, of a fascinating dogmatism in the corridors of learning. It even recalls a mindset, albeit the opposite way around against scientists in centuries past by the Inquisition. The background and the mechanism by which this could happen today, seemed worth investigating.

7. Many devout religious leaders declare that evolution and creation are not in conflict. Other devout religious leaders are just certain they are. They

can't both be right. The fact is a plain read of the Bible declares life was brought into existence by God, from nothing, just by speaking, and that the whole Earth displays His glory. This is not the same thing, by any stretch of the imagination, as the purposeless activity of a "blind watchmaker" who created nature by the selective killing of the weakest as evolution theory says. Since the Bible declares itself true and science says evolution is true, the Bible is on trial. You might not be interested in the Bible but most evolution writers are, in case you've never read Richard Dawkins or Stephen Jay Gould. Although the Bible and Bible literacy is not the issue in the evolution-creation debate, the correct solution is, you would never think so listening to scientists like them who speak about religion as if they were experts in that also. Consider for example a recent statement condemning ID by 38 Nobel laureates:

"Differences exist between scientific and spiritual world views, but there is no need to blur the distinction between the two. Nor is there need for conflict between the theory of evolution and religious faith...Neither should feel threatened by the other".[23]

What was their scientific evidence, you ask? They didn't give any. It's called relativism and it's the prevailing view in Western Civilization now. It's also on trial here. The central claim of relativism is that there is no such thing as absolute truth. But the claim is itself one. These are another set of curious observations, the causes of which seemed worth examining and writing about.

This matter also bears on the answer to a question raised earlier. Why, and with what evidence, can people believe both in evolution and the Bible? If you are not interested in religion you should know the origin of life is necessarily a religious question. The origin of the world is necessarily a religious question too. Anyway even a passing knowledge of the history of Western Civilization declares Darwin overthrew a foundational notion of that culture, namely that the Bible is inerrant and fully informative for living. Now we're told science is true and the Bible is not, that science is synonymous with rational thought, and either implicitly or explicitly depending on the tact and temperament of the moment, that religion is not.

And so we discover it's not just the Bible that's on trial, so is science itself, or at least the materialist mindset of science by which all objective truth about the world is supposedly now to be acquired. Making opposite claims as they do, needless to say one of them must be wrong. Another observation is the scientific establishment right up to Charles Darwin had no problem conducting its business being openly Christian creationist. This thinking is most everywhere now in elite academia considered a contradiction. What caused that change and with what evidence seemed worth exploring.

You should know where this inquiry ends. Evolution theory is no more than an empty illusion best parodied in the fable of *The Emperor's New Clothes* because it's bankrupt. It's the biggest hoax since the Flat Earth, only much more dangerous. It is with pointed irony then that the science and academic establishment take themselves to be the source of objective knowledge and yet are responsible for establishing, endorsing and enforcing a myth with such conviction and capability. Unpacking that seemingly absurd claim seemed worth doing, and doing in a way that presented both sides of the argument fairly for a reader wanting to find out the truth for herself. This book lays out the evidence for you. Although I come to my own conclusion as you can see, you will have to make up your own mind. The fact is you already have decided. You already have a worldview. It rests on the answer to that question. The only question is how open you are to examining your answer and all the evidence.

This inquiry is not anti-science nor is it anti-intellectual. It's the opposite. The only thing it's anti- is unsupported conviction. The reason is because unsupported conviction is the blinker to the truth and a breeder of lies. We all depend on science to test for truth and it's essential it operate in a objective way but sometimes it doesn't. Alchemy, phlogiston, bloodletting and Flat-Earthism are examples and the history of science has plenty more if you think you can always believe what scientists say. What is "science" by the way? This inquiry takes as a given the definition Richard Dawkins proposes: "Science is the disinterested search for objective truth about the material world".[24] So where does objective truth lie in the evolution-creation debate?

This book is addressed to anyone willing to make an inspection of biology and the world in the light of the arguments made for and against evolution. Evolution or Design? One of them must be true and one of them must be a lie. Do you at least know the evidence for evolution? Not the broad brushstrokes but the details. The devil is always in the details. Perhaps you will journey with me. You might be surprised. All you need is a little biology, a little time and an open mind. There is plenty room inside my skull. Come on in.

The investigation has the following construction. Chapter 1 is a review of what it is we seek the origins of - the wonder of nature. We should at least know what nature is like before we wonder where it came from. It is the thesis of this book that the puzzle of the origin of living things can be approached as rationally as a Sherlock Holmes mystery can. The mystery of the origin of life and biological complexity this book calls the "Conundrum of Life". Chapter 1 presents an overview of all the key issues. Chapter 2 is a review of Darwin's solution to the puzzle - evolution. *Part II* (chapters 3-12) is an inspection of the twelve major lines of evidence for the answer that it's evolution. Chapter 3 provides an overview and the following 8 chapters explore its empiric foundations point-for-point. After having heard from evolution, *Part III* (chapters 13-16) is an inspection of the alternative – ID and Chapter 17 is a summary of the entire inquiry. A recommended short-cut overview is obtained by reading Chapters 1, 3 and 17. All the chapters are self-contained and internally cross-referenced to facilitate the particular interests of those who wish to jump about, rather than read the book from cover to cover. The subject matter is deep and sometimes heavy so I recommend jumping unless you like Prozac. The last section, *Part IV*, is an examination of why evolution is a scientific fact although the evidence declares it's really a lie. This is finally to answer that question earlier why such opposite views can be held on this matter with such equal sincerity, why evolution and creation advocates always talk past each other, and why they always will. A short cut overview of *Part IV* is obtained by reading chapter 18 or just looking at Fig.18.3.

I have tried wherever possible to make the discussion non-technical, but in the details are always where controversy and truth lie. I therefore make no apology for the details that you do find. I hope you agree the details deal with the answers to some of the most important questions you could ever ask. At least to Sherlock Holmes, that quintessential solver of conundrums, details were always important. He once said to Watson: "It has long been an axiom of mine that the little things are infinitely the most important". How you interpret the seeming little things is what this investigation is all about. As you go beware you don't fall into the trap Watson did. It brought this rejoinder from Holmes: "You see, but you do not observe."[25] There is only one reason for this difficulty and I submit it happens here. Sherlock Holmes again: "It is a capital mistake to theorize before one has data. Insensibly one begins to twist facts to suit theories, instead of theories to suit facts."[25] We know there is an illusion here, but which of them, evolution or design, is it? You must make up your own mind. I wish you well whatever you eventually conclude.

It is with sadness I report the deaths of several of the leading actors in this drama during my writing and most notably Sir Francis Crick, Stephen Jay Gould, Sir Fred Hoyle, Lynn Margulis, Michael Majerus, Ernst Mayr, Stanley Miller, Henry Morris, John Maynard Smith, Leslie Orgel and Phillip Tobias. We discuss weighty matters in this book. Life and death no less, as if we need a list of dead people to remind us we are some number of weekends away from a grave also. This book is about life and its meaning both before death and after. I therefore write as if they and other now deceased actors in this drama are still alive because their ideas certainly are. Their writings are crystal clear and their influence, like that of Charles Darwin (1809-1882), very much felt in the world.

Citations from *The Origin* are from the 1998 Modern Library Paperback edition published by Random House, Inc. New York. Scripture quotations are from The Holy Bible, English Standard Version® (ESV®), copyright © 2001 by Crossway, a publishing ministry of Good News Publishers. Used by permission. All rights reserved. I have taken the liberty of numbering tables, text boxes and figures without distinguishing between them so that they can be found without need of a separate index. In accord with convention in American academia, I refer to distinguished faculty as Dr. and not by their particular professorial rank with no slight intended. To the contrary.

I am indebted to many people for this book. Most of all to my wife, my sweetheart and my best friend who put up with the sacrifice it exacted over many years. It could not have been written without my teachers either. While I suspect most of them will wholly disagree with this work that is not important. I am not interested in controversy or sensationalism. I could not thank them enough. You know who you are, but I single some out because they did even more for me than the rest who were already the best of the best. None of them have anything to do with this book or its ideas. Nevertheless because of the unfortunate unpleasantness that comes to those who seem opposed to evolution, I have protected their identities so far as is reasonable while also giving them credit as they deserve. I thank you all: S.G., H.V.P., W.G., R.A.F., N.N., J.N.D., M.L.S., S.K., H.R., R.Z., W.J.P., A.M.N., Y.T.S., S.L.H, D.E.G., R.G.M., J.S.K. and Z.M.v.d.S.

I also thank my parents and my friends for countless acts of kindness. They often went un-thanked and their goodness to me was usually undeserved. I regret not being more appreciative at the time. To my neurology practice partners, my office co-workers and to the San Fernando Valley medical community who teach me best medicine every day, I thank you for your unflagging example and your friendship over many years.

This book is written in memory of Desmond V. Ducasse, Headmaster of Hilton College; Kenneth C. Gomm, Headmaster of Athlone Preparatory School; Vic Pearce, Student Advisor University of Cape Town and Kaya Charlie Ngubane, guard and guide to a little boy.

REFERENCES

1. Mayr, E. *What Evolution Is*. 275 (Basic Books, 2001).
2. Newport, F. In U.S., 42% believe creationist view of human origins. http://www.gallup.com/poll/170822/believe-creationist-view-human-origins.aspx (2014).
3. Curry, A. Creationist beliefs persist in Europe. *Science* 323, 1159 (2009).
4. Miller, J. D., Scott, E. C. & Okamoto, S. Public acceptance of evolution. *Science* 313, 765-766 (2006).
5. Ipsos MORI. BBC survey on the origins of life. 30 Jan (2006).
6. Table 17.2
7. Eldredge, N. *Macroevolutionary dynamics:species, niches and adaptive peaks.* 1 (McGraw-Hill Publishing Company, 1989).
8. Pennock, R. T. Creationism and Intelligent Design. *Annual Review of Genomics and Human Genetics* 4, 143-163 (2003).
9. Larson, E. J. & Witham, L. Scientists and Religion in America. *Scientific American*, 88-93 (1999).
10. Larson, E. J. & Witham, L. Leading Scientists Still Reject God. *Nature* 394, 313 (1998).
11. Dawkins, R. *The God Delusion*. 102 (Bantam Press, 2006).
12. National Center for Science Education. Voices For Evolution. http://ncse.com/voices (2002).
13. Flew, A. & Varghese, R. A. *There is a God: How the worlds most notorious atheist changed his mind.* (HarperOne, 2007).
14. Shapiro, J. *Evolution: A view from the 21st Century.* (FT Press Science, 2011).
15. Dawkins, R. *The Blind Watchmaker*. xiii (W.W. Norton & Company, 1996).
16. Dawkins, R. *The God Delusion*. (Houghton Mifflin Company, 2006).
17. Hitchens, C. *God is not great.* (Warner, 2007).
18. Stenger, V. *God: the failed hypothesis.* (Prometheus Books, 2007).
19. Harris, S. *Letter to a Christian nation.* (Knopf, 2006).
20. Dawkins, R. *The Blind Watchmaker*. x (W.W. Norton & Company, 1996).
21. Dawkins, R. In short: nonfiction. *The New York Times*, April 9 p34 (1989).
22. Dawkins, R. *The Blind Watchmaker*. xi (W.W. Norton & Company, 1996).
23. The Elie Wiesel Foundation for Humanity. Nobel Laureates Initiative. (2005).
24. Singh, S. *Big Bang.* 497 (4th Estate, 2004).
25. Conan Doyle, A. *A Scandal in Bohemia.* (Bantam Books, 1986 (1891)).

1. "O Jehovah, quam ampla sunt opera tua!"

§1.1 ONCE UPON A TIME

By 1735 the period now called the Baroque was waning. It would last only a little longer, to around 1750. What age would come next? Before had been the Renaissance and before that the Middle Ages and still earlier even whole empires of culture and custom had come and gone - Roman, Greek, Persian and Babylonian to name a few. How invincible they each had seemed. But like the individual human lives whose efforts they each represented, it was invincible only for a time before being swept back into the sand, their treasures left behind, their monuments torn down and their citadels rudely replaced by another. What age would come next? Another age did follow, just as it would continue to change, for this is the history of man. The central question of man has remained exactly the same throughout all of the ages however although the answer, disturbingly, has not. What is that question you ask? It calls with every beat of the human heart. If you've never heard the question of the ages, just stand still for a minute. Listen to the sound inside and you will hear its soft but unmistakable calling...............Where...did...we...come...from? Assuming we all came from the same place there can only be one answer. So what is it?

"Once upon a time" is really about all time then, but we have to start our story somewhere and which brings us back to the year 1735. It was a landmark in the history of science. It was the year the Swedish physician Carl Linnaeus published his treatise *Systema Naturae*. For the first time, the world was presented with a practical way of classifying all living things. It was a massive advance on a problem dating back at least twenty centuries to Aristotle and the Ancient Greeks. The rules of nomenclature Linnaeus presented remain the basis of taxonomy to this day. Taxonomy, by the way, is the science of the classification of life. One example of Linnaeus's legacy is the "binomial" way an organism is now defined by a two word "genus" and "species" designation. Humans for instance, are classified *Homo sapiens* and mice *Mus musculus*. Linnaeus also formalized the system into which species definitions would fit. He proposed a hierarchy of levels called "taxa" (the singular is taxon), nested one inside the other and which fit together just like Matrioshkas dolls. The Linnaean method of taxonomy applied to man and mouse for example is as follows:

Kingdom	e.g.,	Animalia	Animalia
Phylum		Chordata	Chordata
Class		Mammalia	Mammalia
Order		Primates	Rodentia
Family		Hominidae	Muridae
Genus		*Homo*	*Mus*
Species		*sapiens*	*musculus*

Given the significance of this work it is of interest to discover that the title page of Linnaeus's opus opened with the following words: *"O Jehovah, Quam ampla sunt opera Tua! Quam ea omnia sapienter fecisti; Quam plena est terra possessione tua!"* It's Psalm 104, verse 24. In English it reads: *"O LORD, how manifold are your works! In wisdom have you made them all; the earth is full of your creatures."*

The Baroque was an age of musical genius too. An age of composers like Bach and Vivaldi, Telemann and Handel. It was also an age when authorities in the arts, like authorities in the sciences, saw no contradiction between their work and its dedication to Jesus Christ, Son of God. Examples of this include Bach's *Jesu, Joy of Man's Desiring*, Vivaldi's *Gloria*, Telemann's *Passion of Jesus* and Handel's *Messiah*. Rather in-your-face religion again, you must admit. Both the sciences and the arts have an opposite mindset now. What caused so complete an inversion?

What Linnaeus and those composers were acknowledging of course was the Bible and more specifically its worldview. Although that view had been accepted as true for multiplied centuries, it would not in science for much longer. On November 24th 1859 Charles Darwin published *On the Origin of the Species by Means of Natural Selection.* Sold out on the first day, it has remained ever popular since. It is also the most influential piece of scientific literature ever written. Darwin's ideas became the central paradigm for understanding biology the world over and so they have remained.

Evolution is not only about biology however. In providing the "scientific" solution to the question of the ages, in other words the objective and true answer, it's inescapably also a philosophical view. Harvard University biologist Ernst Mayr explains why: "Evolution is the most important concept in biology. There is not a single Why? question in biology that can be answered adequately without a consideration of evolution. But the importance of this concept goes far beyond biology. The thinking of modern humans, whether we realize it or not, is profoundly affected – one is almost tempted to say determined – by evolutionary thinking."[1] He also says: "Every modern discussion of man's future, the population explosion, the struggle for existence, the purpose of man and the universe, and man's place in nature rests on Darwin."[2]

Darwin's ideas rocked Western civilization, which was at least normatively Bible-believing Judaeo-Christendom, to its very foundations. What was it *The Origin* shook exactly? It debunked a supernatural creation of the world and in corollary, the authority of the Bible. This might not sound like much but it overturned a whole world order. The reason is because before Darwin the evidence for God was obvious. It was the wonder of life all around you. The meaning and purpose of life was just as obvious. It was spelt out in the Bible, the word of God: the world was made by God for God. We read:

> *"For by him all things were created, in heaven and on earth, visible and invisible, whether thrones or dominions or rulers or authorities – all things were created through him and for him." (Col 1:16).*

God is the only source of goodness and satisfaction. Man was made to glorify God and so enjoy Him forever. Life was meaningless without the Maker! Man was doomed left to his own devices. He was created and mortal, and his innate nature corrupt to the core. He lived in rebellion against his Creator which is called sin, and so was under deserved wrath for insult upon insult against Infinite Goodness. Man could not help himself from doing things he knew were wrong either. His conscience affirmed this moral brokenness and the holy Creator's righteousness required justice for the outrage against Infinite Good. Man had a savior from this body of death however, Jesus Christ, by whom God had Himself entered human history to pay for all of the sins of all of mankind of all time instead. The Bible says:

"For God so loved the world that he gave his one and only Son, that whoever believes in him should not perish but have eternal life." (Jn 3:16)

Man had only to believe to receive this ransom but doing so meant living for God thereafter, not rebelling on by living for man. That's what Western Civilization thought before *The Origin* and that's why Linnaeus, Bach, Handel and Vivaldi lived and worked the way they did.

Scroll forward to 1859. Charles Darwin now gave a completely different explanation for the wonder of life. He called it "natural selection". He said the maker was purposeless and meaningless and acted by blind forces of nature. There was no need for a God to account for the wonder of nature so maybe God never even existed. At least Darwinism declared a Creator had no objective reality. Atheism suddenly became scientific and for the first time in history it became rational, or so it is said. The problem of explaining origins in biology, the single most important question in all of science, no longer required a supernatural, no longer required a creation and no longer even demanded God. It was all done blind by chance and the laws of nature! The answer to the question of the ages was the empirical, explanatory mechanism that Darwin had just unshackled from ancient dogma and religion. Or so it seemed.

Believers who still wanted to believe in the Bible and in God were free to believe whatever they liked and in any God they liked, so long as they kept science and academic discussion about the real world, God-free. And as for the Bible and what it says about life and death and sin and the coming judgment, and that called itself the word of God, it was misleading at very best. Absent a tangible supernatural and replete with scientific error, it was more likely a myth.

This thinking is a worldview called materialism. It's the philosophy that physical matter is the only objective reality in the Universe and that everything can be explained in terms of matter and physical forces. Astronomer Carl Sagan was a believer. He said the material Universe is: "all that is, or ever was, or ever will be".[3] Naturalism is a related system of thought that holds everything is ultimately explicable by natural laws and natural causes, that there is no evidence for a supernatural nor can there be, and that appeals to supernatural solutions for questions of the real world are irrational. What's the significance of these two philosophies? They became the foundations of science after Darwin was done. It's a reasoning paradigm called "methodological naturalism".

Since operating this way science has been phenomenally successful. Arguably the greatest technological advances, by far, have been made in the past one and a half centuries. For example in less than a century man advanced from hot air balloons to the space shuttle, from Alexander Graham Bell to smart phones, from slide rules to personal computers and from parlor quartets to iPads. We might take our dishwashers, video cameras and motorcars, medicines and the internet for granted, but it is from science that they have come. Given this performance the naturalistic paradigm by which science has operated since the days of *The Origin* would seem beyond reproach. After all to invoke an overwhelming supernatural Creator of everything, even as a rhetorical possibility, is to invoke the religious and subjective right? Religion has nothing to do with either science or objective reason. Or does it? More particularly, are the naturalistic assumptions that underlie the reasoning of methodological naturalism true?

An affirmative answer would seem obvious. It's the view of science and of academia most everywhere and the popular media too. But consider the following statement which lays out that reasoning more formally:

"There are no truths apart from scientific truths and only what is scientifically demonstrated is rational. Everything else lies in the realm of subjective belief and opinion."

So what do you think? Is it true?

It's false because it's self-refuting. It can never be true! The reason is because it's not a statement *from* science, it's a second-order statement *about* science.[4] It's a pronouncement by a philosophy about science. And therefore false by its own rules (for not being a "scientific truth"). It would seem the determination of truth lies not with science but with philosophy. It also turns out philosophy can never be divorced from science.

The reason is "science" demands a number of presuppositions before it can even get going. Presuppositions such as those declaring there is an order to the world, that it is knowable, that the laws of logic are valid, that nature and the Universe are uniform enough to permit inductive reasoning and that the mathematical expression of quantities is reasonable. This is not controversial. Tufts University philosopher Daniel Dennett is an outspoken atheist and evolution advocate who says all that, much more succinctly, like this: "There is no such thing as philosophy-free science, and there is only science whose philosophical baggage is taken on board without examination."[5] What's the point here?

The point is to recognize the significance of philosophy to science and its overwhelming consequences. The philosophical foundations of science, whatever they are, must be true before beginning the business of science because they inescapably color all of its activity. This is to confront an issue at the very heart of the evolution-creation debate. You see under the rules of science i.e. the rules of methodological naturalism, "science" can only invoke natural laws and naturalistic explanations to remain rational and objective. In the words of the National Academy of Sciences: "Science is a way of knowing about the natural world. It is limited to explaining the natural world through natural causes".[6] In the words of The National Association of Biology Teachers: "Explanations or ways of knowing that invoke non-naturalistic or supernatural events or beings, whether called "creation science", "scientific creationism", "intelligent design theory", "young earth theory" or similar designations, are outside the realm of science".[7] While this might at first blush seem reasonable, guess what it means for any possibility, ever, of a supernatural creation?

And yet as we have already seen, the statement imposing these rules on science is not only false by the same logic that rules only scientific truths objective, it is to apply an arbitrary double standard. People who say life was supernaturally created rather than evolving by natural forces by Darwinism are taken to be unscientific, and therefore objectively speaking false, even by definition. And any mention they might make of a Creator in explaining their views, which is inevitable of course (if there is design in biology there must be a Designer), is for them to be doubly misguided because that is now to mix "religion" with "science".

But who said faith and reason can't intersect? Who said religious faith can't be tested for objective truth? Let me assure you it has never been shown, at least not with evidence. And yet that notion is everywhere now in intellectual circles pronounced, and accepted, even as if it no longer even required evidence. Here for instance was the pronouncement in *National Geographic* recently: "There is, of course, no way to prove religious faith scientifically".[8] But where did that idea come from? No really. Where?

Robert Pennock is a philosopher of science at Michigan State University and an acclaimed advocate for evolution's truth and creation's falsity. He explains why the naturalistic rules of are the way they are and why they must stay the way they are:

"Of course, science is based upon a philosophical system...Science operates by empirical principles of observational testing; hypotheses must be confirmed or disconfirmed by reference to empirical data".[9]

And we would surely all agree. The problem is not this statement however. It's what he says immediately after it:

"Supernatural theories, on the other hand, can give no guidance about what follows or does not follow from their supernatural components."

The problem is he offers no evidence nor "empirical principles" nor "observational testing" which has showed this claim to be true. Yet this folk wisdom, because that's what it is otherwise, is the foundation of twenty-first century science I kid you not. The reason is it's the basis for methodological naturalism. It had better be true! We will need to see for ourselves. This is the inquiry of the book. Welcome.

§1.2 THE CONUNDRUM OF LIFE

The question of the ages that Linnaeus and Darwin answered so oppositely can also be considered a conundrum, the conundrum of the wonder of life and hereafter called just the "Conundrum". It's as much a question and a conundrum now as it was back then. The reason is while the origin of man and other living things is an intellectual question for biologists, it's as much a personal question for every individual and always will be. It's the central question of human existence. It's not just a scientific question either. It's a moral question. To see why, look at the question a little more.

The Conundrum can be stated very simply: "What's the origin of life and its complexity and diversity?" Formulated like this it might seem esoteric and not personal at all. In case that's your sense of it, the following five questions are subsumed by the Conundrum to show that to be very much mistaken: "Who am I?"..."Where did I come from?"..."Why am I here"?..."What is my relationship to other humans?"..."And other living things?"

The Conundrum is not just a question that must be personally asked and personally answered, it's also a question that cannot be ignored. Even to not have consciously formulated an answer, or to respond "I don't know" or "I don't care" while surely a display of the very height of ignorance, is still to have reached an answer. The reason is these responses are to have rejected a supernatural creation and the implications of a transcendent Creator and a simple analogy can be used to show it.

Suppose someone said to you: "There's an atom bomb hidden in the building!" Your reaction would show whether you believed it or not. Responses of "I don't know" or "I don't care" though ignorant and perhaps even irrational are still to have answered. They are to have answered: "I don't believe you". So it is when considering the reality or not of a power infinitely more devastating than an atom bomb. The solution to the Conundrum of Life is always answered even if it is not articulated, and even by those who know nothing about science. The way we live out our lives declares which answer we believe is true. Whether we believe naturalism is all there is, or whether we believe there is an Almighty Creator who knows everything and sees everything we do. The Conundrum demands, and always gets, an answer from every single one of us.

In case you haven't realized it by now, there are only two answers to the Conundrum. The order and complexity of life either came about by natural forces, in other words by evolution somehow, or the world was created supernaturally somehow. Their individual particular details are important of course and yes there are a multitude of variations, but these are secondary details on an irreconcilable dichotomy. Take even the solution that is a seeming combination of the both of them, the view that a Creator enabled evolution to proceed unguided after first setting it up by a miraculous creation. This is a very popular view today called "theistic evolution" (§1.3;§26.1). It's no hybrid though because it's creationist at heart. The reason is because it retains a supernatural Creator over evolution for reasons of personal private faith despite evolution theory declaring a Creator both undetectable and unnecessary. So too at the other extreme is the seeming non-solution that rejects both evolution and creation. This is the view holding to the hand wringing of Immanuel Kant who declared that no answer to the Conundrum is possible because there is insufficient evidence to decide (§13.1). This view always arrives at one of the two alternatives by default however. Just look at how their advocates live and think and premise the world, and you will know which one it is.

The Conundrum is the most important scientific question you could ever ask. The solution, whatever it is, is the most important principle in all of biology. In it is implied the meaning and purpose of life, and the corollaries apply to the way we think about almost everything. It makes a difference if natural forces are all there is, or instead there is a transcendent Almighty Creator who made us and sees everything we do, and knows everything we think. On the former view a closed physical universe is the only reality. If this is so the great debates of the day, in our day say over abortion or homosexuality or the definition of marriage, are merely debates of moral equivalence and personal preference because there are no absolutes and there is no absolute truth. Except of course that one that says there are no absolutes. Everyone can do what seems right in their own eyes.

On the other hand if there is a greater reality to our world than the material, then there is such a thing as absolute truth and it's not the created but the Creator who determines what that is. It also follows that if you were created no matter how hard you looked you could never discover the meaning of life or make sense of human existence, until you recognized meaning had nothing to do with you and everything to do with the Creator (§10.1). If the world was not created the opposite applies, which of course is the way the world now works.

Another observation is that despite the overwhelming gravity of the

Conundrum and its foundational relevance to living, few people make a serious effort to solve it! This is an extraordinary discovery. We spend more time looking for our car keys than where we came from. For most I submit the final solution is not even deliberated with evidence but made by *a priori* philosophical convictions. Convictions either theistic in the case of creationism (because of a particular supernatural belief system), or naturalistic in the case of evolution (because of methodological naturalism). Evolution and creation exhaust all possible solutions to the Conundrum of Life. What's your answer? What's your evidence?

The first of the answers, evolution, will be inspected in detail shortly (§2-§12). We already know what it is, at least in outline, because it is the solution of the science and the education establishment. Creationism on the other hand is not so straightforward. The word refers only to a supernatural power causing life. What power? Creationism has any number of explanations when it comes to specifics, depending on which God or Gods did the job and why. The meaning of the word creationism is even more confusing because some evolution advocates, for example Kenneth Miller at Brown University and Francis Collins, Head of the Human Genome Project and Director of the National Institutes of Health, hold that a supernatural God created living things by using naturalistic evolution. Few would publicly call them "creationist" however, at least not in the pejorative way the word is used in the evolution-creation debate. In that sense the term creationism is more than just believing in a supernatural Creator. It is: 1) rejecting evolution as sufficient to explain the origins of living things, and 2) doing so for evidence from religion not evidence from biology.

The majority of creationists in the West are Bible believing Jews and Christians who take *Genesis* at its plain meaning, but in principle a creationist can take her pick from any one of a host of other options and there are rather a lot. Here is the smorgasbord: Either pick from theism proper then choose one from monotheism or polytheism or henotheism and then pick a particular God they like from one of those categories; or choose a God from within Deism or Pantheism or Panentheism or Stoicism or Neo-Platonism. We can't stop there either because some forms of creationism have still more to say, such as how old the Earth is. In fact the type of creationism with which we are most familiar is the Bible believing young-earth creationism of "creation science", also called "scientific creationism". This is the view that life and the Universe were created in six literal days within the last ten thousand years (§21.1). "Old-earth" creationism on the other hand accepts the dates for the ages of Earth and Universe as declared by the science (around 4.5 billion and 14 billion years respectively §13.2-4). Young-earth vs. Old-earth dating is to demonstrate only one difference among creationist ideas however and one that does not begin to scratch at the variety. The point is creationism is a veritable Pandora's Box of different beliefs. It is also a demonstration of how ridiculous is the idea that all religions are equal. How can they be when they make contradictory truth claims?

Actually there are two ways they can. The first way is if their differing claims are all equally inscrutable, and the second is if their differing pronouncements are all equally false. But we are getting sidetracked. The issue for us is simply this: How are we to keep the menagerie of creationist solutions from being mere intangible conviction so that we might pin down the broader notion of a supernatural creation of the world or not, and subject that to an objective skeptical inspection? Is it possible, even in principle?

Sure it is. Contrary to popular belief a creation of the world is not necessarily a dogma. It becomes rational and refutable when formulated as entirely dependent on objective evidence, on objective evidence of "intelligent design" in nature. Intelligent design (ID) is the science that studies intelligent agency. ID theory is a theory of information rooted in statistics and mathematics. When applied to biology it's a powerful tool to study the Conundrum. It concludes there is objective and tangible evidence of a supernatural in the world from the evidence of biology alone. This is not creationism. How can ID analysis conclude there is evidence of a supernatural in the natural world?

First we need to restate the Conundrum in a way germane to a design question. It is to ask whether it was intelligent agency or natural forces that caused a particular event or object. This is the problem we face here. It's one thing to conclude design when we catch an agency in the act, or have a reliable eyewitness to the fact, but how are we to choose between intelligent agency and natural forces if all we have is the object or the event itself? If an intelligent agency intervened in nature, in other words if a supernatural creation of any biological structure happened, it happened unwitnessed, once, and a long time ago. Enter ID theory.

ID theory operates by looking for patterns in an object or event, in this case biological systems or structures in nature, that have both a complexity and purposefulness to them beyond the remotest statistical possibility of arising by natural forces. The reasoning operates as simply as it does solving the question of whether the edifice that is Mount Rushmore was designed or whether it was done by undirected natural forces and so is merely an illusion of design (Fig.1.1). We have absolutely no difficulty solving this question. We tell illusion from fact as we always do, by looking at the details, and which is the way illusions and lies are always sorted from truth. The detection of complex yet purposeful patterns with meaning in objects of nature would, if found that is, constitute the detection of a design and the rejection of an illusory appearance of design. It would then be to detect design, thus a designer of nature, which is to discover a Creator.

Fig.1.1. **The faces on Mount Rushmore.** Design vs. appearance of design on a mountainside. How many faces do you see and what's the biggest difference between them? There are at least 14 here. (Image credit: National Park Service).

The crucial difference between the conclusion of a creation as a result of a test by ID theory and that of creationism resides in the word "detect". For while ID's conclusions in biology may seem sensational, its methods are entirely biological, entirely empiric and entirely scrutable. ID is not creationism because: 1) it has no commitments *a priori* either to design or to evolution but is a conclusion after a search for both of these alternatives; 2) the conclusion it makes hangs or falls on the quality of the evidence to rule in or rule out the always more parsimonious solution of natural forces; and 3) its conclusions are not made for faith or religious reasons because ID theory has no belief system or creed, no scriptures, no religious revelations and no rituals.

But what kind of pattern in biology could be so audacious as be beyond ever being caused by natural mechanisms? One example, first proposed in 1996 by biochemist Michael Behe at Lehigh University in Pennsylvania, is called "irreducible complexity" (§15.2). Behe gives the illustration of a mousetrap to understand the concept but real-life examples include molecular machines like the flagellum, the cilium and the energy metabolism enzyme called ATP synthase (§15.3).[10] If the analogy he uses is true and we will have to see if it is, the argument against evolution and for ID boils down simply to this: "Who ever heard of half a mousetrap"? Take away any one of its constituents - platform, hammer, spring, catch or holding bar, and it will not work. A mousetrap must be constructed all at once or not at all. It could not have evolved by natural selection. Be careful before retort it could have evolved from something simpler that was not (yet) a mousetrap. Natural selection requires an object to function if it is to be selected and evolve, and half a mousetrap cannot function.

It also turns out there are a plethora of objects in biology that consist of multiple constituent parts, each performing different minimal functions, all to the same purpose. These are machines made of molecules, literally, and for which coordinate interaction of every constituent is necessary for function of the whole to happen (§15.3). These biological systems cannot be reduced to simpler components and their construction therefore implies a purposeful arrangement. Since evolution proceeds in naturalistically manageable small steps over eons, the discovery of an object that could not have come about in small steps is not only to refute evolution, it is to demonstrate design, objectively. Again the validity and any evidence for that argument, as well as its objections, will need to be inspected for substance. The only point for now is to show you there is a legitimate two-sided debate over the Conundrum that should be heard with an open mind. ID is not mystical thinking. It stands or falls on whether there is evidence or not.

The reasoning of ID theory is not magical for still another and far more obvious reason. We do it every day never concerned we are practicing magic. We did it looking up at Mount Rushmore earlier, but here are some everyday examples. Are the scratches on my car door accidental (naturalistic) or are they the work of vandals (design)? Was the gate deliberately left open (design) or was it the wind (naturalistic)? Police detectives and crime novel fans use ID even more obviously. Did she fall (naturalistic) or was she pushed (designed)? Was the fire spontaneous (naturalistic) or was it arson (design)? Fraud investigation, copyright protection, intellectual property infringement and cryptography are still more examples. Searches for alien life in space called SETI (Search for Extraterrestrial Intelligence) are probably the most sensational example, but still exactly the same thing. Namely could the pattern observed in an event or object, in the case of SETI a pattern encoded in electromagnetic waves coming from space, have arisen entirely by natural forces or was it created by an intelligence? We would conclude extraterrestrials if we discovered a purposeful complex pattern, in other words a message, in those electromagnetic waves.

An example of just that kind of discovery was provided by astronomer Carl Sagan in his book and film of the same name called *Contact*. In the story a series of electromagnetic pulses representing the sequence of prime numbers from 1 to 101 was received from space. This is to find ID. The discovery was taken by Dr. Sagan, the scientist in the film played by Jodi Foster, the viewers and the book readership, as proof of alien life. To reject this same reasoning just when it is applied to biology, assuming there were similar evidence that is, would seem contrived.

ID theory has serious limitations however. It detects design only by virtue of an "information content" statistically beyond any reasonable probability of arising by chance and the natural forces of the laws of science. What level of probability is that? The level can be set to individual discretion. What would it take for you? The most restrictive yet suggested is 1 chance in 10^{150}.[11] The enormity of that number is deceptive. To get a sense of it there are 10^{80} atoms in the Universe, the fastest meaningful interval of time is 10^{-43} seconds (the Planck time), and there have been around 10^{17} seconds, at least accepting the reckoning of the science community, since the Big Bang that they say began both time and the Universe. We will get to their details later but for here, take it that any odds greater than the product of these three numbers are odds so small they are for all practical purposes impossible. We should be able to solve the Conundrum without any shadow of a doubt. The same way we solve the question of whether there is design or not in the edifice of Mount Rushmore without a shadow of a doubt. So much for the statistics then, what about those limitations of ID theory?

It's that ID cannot determine who did the designing, how or why. These are second order questions beyond its scope which require other, non-ID methods of analysis. It's like trying to work out who sculpted Mount Rushmore. Try working that out just from the faces. It was Gutzon Borglum. The limitation has an important corollary, and it is one very relevant to the Conundrum. It's that ID theory has no religious agenda and no religious commitments. That's not what you will hear from science and the mainstream media however and why is worth pondering. We will investigate this later.

Another point is that the detection of ID in an event or object does not necessarily mean the detection of optimal design. You may have ideas about the quality of the workmanship at Mount Rushmore for instance, but this does not make it any the less designed. The methods of ID theory only distinguish the information in a particular object or event as arising entirely by natural causes or by an intelligent agency, with or without natural forces contributing in part also. An example of the combination at Mount Rushmore is wear on the sculptured faces by wind and rain.

The reasoning behind ID analysis can be explained still another way. Suppose we were walking along a seashore and came upon the inscription *"Ben loves Suzie"* in the sand. We would immediately conclude ID and exclude natural causes. We would do it even if we had no idea who wrote the words, why or when, and we would do it even if it were insisted the beach had lain deserted for millennia and that the waves and winds were particularly vigorous at this shore. Consider that writing as a crude example of ID, a creation, by an intelligent agency. It's a crude example because biological examples are incomparably more complicated than that. Actually biological things are the most complicated things we know of. To quote Oxford University zoologist Richard Dawkins, just in regard to the genetic code, "each nucleus...contains a digitally coded database larger, in information content, than all 30 volumes of the *Encyclopedia Britannica* put together".[12] Imagine all that written out in beach sand and you start to get a feel for the matter at hand.

If there were objective evidence of a supernatural design in nature there would have to be an arrangement, like the grooves of that valentine in the sand, which could not have arranged themselves in that pattern absent intervention by an intelligent agency. The exact person, methods and purposes of that agency is a second question beyond the scope of ID analysis. It requires investigation by different methods and demands different information from different sources, in this case camera surveillance records of the beach parking lot, a list of beach patrons and specimens of their handwriting.

This limitation of ID analysis would be a problem if design were to be detected in nature, but a problem not nearly as large as the Conundrum that we seek to answer, and anyway a problem which would follow it not precede it. While this might be considered a handicap, it's all we need for the purposes of accepting or refuting a creation because evolution is the only alternative. If design were demonstrated in nature it would be a monumental conclusion to be sure. But no more monumental than knowing a creation to be false and which is the view in most places now.

There is still another requisite needed before setting out to solve the Conundrum of Life and it is just as important as an open mind. It's an understanding of the question itself. We need to know what life is like to know what must be satisfied by the solution that answers where it came from. Where do we get that information? Just look at life. At nature. So join me for a whirlwind tour of the panorama of biology. Dr. Richard Dawkins

28

calls it a "baroque extravaganza"[13] and so it is.

Wait a moment. Why is biology a baroque extravaganza? Why is it so majestic, so astonishing and so wonderful? Because of its appearance of course! How? It arises from five characteristics which are exhibited by every living thing on the planet. The solution to the Conundrum will have to explain them all. For wherever we choose to look at an organism, whether on highest mountain or furthest forest or deepest ocean or buried miles under the earth's surface, and no matter what level we choose to focus in on - from the individual molecules of a cell's machinery to the coordination of whole biospheres, life always and everywhere demonstrates these features: 1) diversity, yet 2) unity, 3) complexity, 4) adaptiveness and 5) shared patterns of construction and/or function. The wonder of this appearance is what excites the readership of journals like *National Geographic, Nature* and *Scientific American* or viewers of David Attenborough and *Discovery Channel,* and it's what drives biologists to do what they do and make the many personal sacrifices they all do. As for us, we need to get a better feel for each to know what life is like before we can reasonably consider how life came to be that way. Take a look at the features of living things. There's much to marvel at! The wonder of life is above you, under you, in you and all around you, wherever you look.

§1.2.1   The Diversity of Living Things

The science of biology is already centuries old but only a fraction of the total number of species alive has even been named, much less studied in detail. About 1.7 million species are currently known but the total alive on Earth today is estimated at between 3 and 30 million with working figure of around 10 million[14-17] and a detailed estimate of 8.7 million.[18] Even this upper limit may be a gross underestimate however given recent proposals by other authorities that the total number of living species approaches 100 million, or even more.[16] Either way we can say that despite 250 years of taxonomic classification the total number of species of life on earth has not even been estimated to the nearest order of magnitude, and that even by most conservative estimates little more than 14% of living things have been described and just 1% investigated to any substantive degree. At the current rate of discovery, describing the remaining species will take another 1,200 years![18] It is an astonishing level of ignorance in the twenty-first century and the reason is simply because the diversity of life is so great.

A second observation is that as better technologies have been applied to cracking this problem, so more ecosystems and more living things have been discovered. Examples include tropical forest canopy and Deep Ocean as unique, and highly species diverse, ecosystems. As a result estimates of total species number, at least so far, have only had to be revised upwards. Six new animal phyla have been erected since 1937 and as recently as the mid 1980's, total species diversity was thought to be (only) 4.5 million.[15,19] Most of the unknown species are taken to be tropical arthropods but even among the vertebrates, which are the best known taxon, new bird and mammal species are being discovered on at least a yearly basis. In sad commentary too, every 16 years for the past four centuries one mammal species has been lost to extinction.[16] The havoc wrought by humans is causing an estimated thousand times greater rate of extinction than before civilization began.[20] It is the greatest rate of extinction since the one that killed off the dinosaurs 65 million years ago, the science says (§5.5.5). The numbers involved? Between 10,000 and 100,000 bacterial, plant and animal species are being lost, every single year.[21,22]

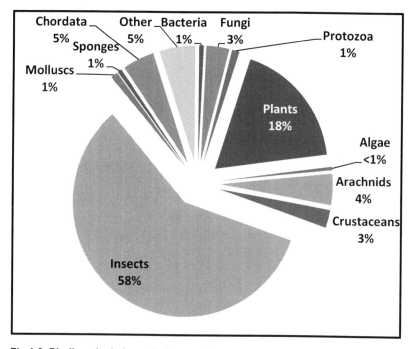

Fig.1.2. Biodiversity is irregularly spread across nature. The relative proportions of the 1.32 million known species of life listed in the *Species 2000 & ITIS Catalogue of Life*, November 2012. By far most living things are insects! The phylum Chordata, 95% of which are the vertebrates, represent only 5% of all species. (Data from Bisby et al, 2012.)[25]

A third observation is that species diversity is unevenly spread by taxonomic group (Fig.1.2). For instance 1.0 of the 1.7 million known species are animals and the arthropod phylum (the group that includes insects, spiders, centipedes and crabs) accounts for 900,000 of them, and just one family, the weevils, has over 50,000 species. The *Cycliophora* phylum on the other hand, first discovered in 1995 living on the mouthparts of Norwegian lobsters, has only 4 species.[23] Diversity is not evenly spread by geography either. Tropical forest covers less than 10% of the globe and yet it houses 90% of all species. E.O. Wilson at Harvard is among the leaders in this field. In another example he found the same species diversity up one tree in Peru as there is all the British Isles (43 different species of ant from 26 different genera).[14]

The official term to describe the variety of life is "biodiversity". It's a contraction of biological diversity. The word was coined only recently, in 1988. Though there are differences of opinion as to its exact meaning, most ecologists take "biodiversity" as being reflected at three hierarchical levels and not just in absolute species number as we have considered it so far. The levels are: (1) the number and variety of different ecosystems on the planet (for example tropical savanna, desert, freshwater, ocean, rainforest, tundra); (2) the number and variety of species within one particular ecosystem; and (3) the number and variety of genes within particular individuals of a species in an ecosystem.[16] Let's go back and focus on just the second of these levels to get a feel for the diversity of life on Earth.

At the highest dimension of Linnaean taxonomy are the three so-called "domains". Two represent single celled organisms without a nucleus called "prokaryotes" (the word is derived from the Greek words "pro" = before and "karyos" = nucleus). They are the *Archaea*, first discovered by Carl Woese at the University of Illinois in the late 1970s and the other are the *Eubacteria* (Bacteria) and therefore not needing any introduction because they veritably surround us inside and out. In regard to the former location, individuals of this domain number about 10^{12} per gram dry weight of human stool - about half its dry mass - and that's when it's normal![24] The third domain are the Eukaryotes. They do have a cell nucleus by contrast, and may be either uni-cellular (protozoan) or multi-cellular (metazoan). The Eukaryotes consist of four subdivisions – the protozoans, fungi, plants and animals. The animals are divided into vertebrates and invertebrates (i.e. those with or without a backbone), and consist of 37 phyla distinguished from each other by the presence of a unique body plan or "Bauplan". There is as much species diversity for the other three divisions of the Eukaryote domain as there is for animals. Even the protists (which include algae, amoebae and plankton) have this variety notwithstanding their tinyness. In fact they have more phyla than animals do and the differences among themselves are so impressive Indiana University embryologist Rudolf Raff says: "Protist groups are as morphologically disparate from one another as tulips and truck drivers".[14] This still does not scratch the surface of biodiversity however.

We need to include all the extinct species in the appraisal too. How many are there? From the fossil record it appears at least 99% of all species that have ever lived are extinct! Paleontologists interpret a continuous background rate of extinction throughout the fossil record onto which are superimposed six so-called "mass extinctions" where that effect was transiently catastrophic (§5.5.1). The sixth extinction is what man is doing today and the most devastating extinction was the Permian extinction, dated by the science as happening 245 million years ago. About 96% of all species living at that time were wiped out. It's hard to imagine such devastation. But what does this have to do with our investigation you ask? When the number of extinct species is included in calculations of total biodiversity, the number of species jumps to one billion and up. Conservatively. The diversity of life is stupendous. The solution to the Conundrum of Life will need to explain how at least a billion different species came to be living on the Earth.

§1.2.2 The Unity of Life

Made more striking in the light of this huge diversity is the way in which there is yet a cohesive unity of species within nature. Species are found as interrelated organizations of different species, each depending on the other in order to survive. This at least appears to be purposeful, because any one component depends on the other different components at any particular level we choose to focus in on. Check this claim out.

Zooming down to the molecular level of nature we find exquisitely controlled enzyme activities, allosterism, feedback loops and a coordination of multiple interacting biochemical pathways that is the metabolism of the cell. We see the unified yet diverse interacting members of the blood-clotting cascade, respiratory chain, photosynthesis pathway and other intracellular enzyme systems. Pulling back on the focus, we see subcellular organelles like mitochondria, ribosomes, the golgi body and endoplasmic reticulum working just like factories do. We find assembly lines coordinately operating, just like a metropolis, to produce thousands upon thousands of different proteins for multiplied different uses in the cell or for export to be used elsewhere in the body. We see a bristling network of receptors on the cell surface, literally sensors on the outposts of the cell, in a complexity far beyond anything at Heathrow or on an aircraft carrier, and coupled by complex signal transduction pathways and other molecular crosstalk to each other and to the cell interior.

Pulling back on the focus still further, and moving out from an individual cell to whole groups of like cells (called tissues), we find they are together coordinated in behavior also, a phenomenon called homeostasis. Diabetes and hypothyroidism are diseases where this control is lost and left untreated, can be fatal. Zooming out still more from tissue to organ level, we find structures like eyes and ears, we see separate components doing different things (for the eye it is visual cortex, optic nerve, photoreceptor cell, photoreceptor pigment, lens, pupil and cornea) yet all working together in an interdependent and highly coordinated way. Cells, tissues and organs make up individual organisms, individuals form co-existing species populations, and populations fit together into all kinds of different symbiotic relationships and food chains and ecosystems in a harmonious whole. This organization is not fixed either, but delicately balanced and subject to perturbation both from the physical forces of the environment and from other living things. The repercussions of environmental disasters, whether natural or by human design, and the difficulty in manipulating ecosystems in the way naturalists or property developers intend, are a reflection of this delicate fine-tuning.

One illustration of this fine-tuning gone awry, is the havoc wreaked on Australian and New Zealand ecosystems by a small European rabbit. In the same year that *The Origin* was published, one Thomas Austin released twenty-four of the little immigrants onto his property in Southern Victoria, Australia. Frisky does not begin to explain their busyness from then on. Just seven years later over fourteen thousand of their descendants had been shot on his property alone. Today this rabbit plague costs his country over a billion dollars Australian a year, it has spread to New Zealand, and it remains an ecological menace despite a whole catalogue of attempts to rid them. These efforts have included importation of myxomatosis virus but within two decades half of the rabbits had developed resistance (just in case there's any doubt microevolution is not a fact), the importation of ferrets (which only served to spread bovine tuberculosis on to livestock and deer), infecting them with rabbit hemorrhagic disease but this also infected the indigenous animals and even some plants, and most recently attempts have been made to infect them with rabbit calcivirus, which among other creatures it is feared to harm is the kiwi, New Zealand's national bird. It seems efforts to improve on the unity of life, even with foresight and planning by experts, easily backfire with even an opposite of intended effect. This is a testimony to the intricacy and wonder of the biological unity of nature.

The solution to the Conundrum of Life has to account for how the diversity of life is yet also constituted into a purposeful delicately poised, interacting and unified whole at any and every level you choose to look, from individual molecules all the way up to interactions between entire ecosystems.

§1.2.3 The Complexity of Life

A third feature of life is its intricate complexity, even in supposed "simple" organisms like bacteria. This feature of living things, like diversity and unity, can be examined at multiple different levels and exists at every one we care to zoom in on. At low magnification we find the convolutions and sophistications of whole ecosystems, food chains and the social organization of single populations like ant and bee colonies. Some of the most striking evidence of biological complexity is found in social behavior. For example bees have language, specifically dance language (very different from the human kind). Karl von Frisch at the University of Munich earned the Nobel prize in 1973 for this discovery. Forager worker bees inform the hive of food location by round, waggle and tremble dances. Another example is how some ants farm aphids like cows. They march them up trees to pasture in the forest canopy during the daytime, then herd them down into the termite nest at night where they are milked for essential nutrients. There are plenty of other examples of biological complexity. Ever looked at the construction of a bird feather? A spider's web? Considered the tensile properties of its silk which can bear a weight a 1000 times its own and yet be stretched to 4 times its length without snapping? Or considered the solution to the manifold problems required for successful animal migration, which in the case of birds is not just travel to another continent, but a particular rooftop? Bat or whale echolocation? Human visual perception? Hearing? Consciousness? But we can get even more exotic and Baroque, because life is just like that.

Consider the bombadier beetle (*Brachinus sp.*). It squirts a steaming solution of quinone at all its foes. This flame-throwing mix is formed by the (very) exothermic reaction of hydrogen peroxide and hydroquinone, which is catalyzed by a specific enzyme. You can't do this simply by mixing the two compounds together in a test tube, as Oxford zoologist Richard Dawkins has shown.[26] Success demands highly specific and highly coordinated preconditions. The most fortunate production of oxygen gas as a by-product of the reaction is usefully employed by the beetle to propel the scalding mixture in an appetite suppressing jet. One which it can aim, also very conveniently, anywhere in a 360 degree radius. Needless to say, this activity is highly complicated and its coordinately functioning yet separate and obligately necessary components appear painfully purposeful, even if only to beetle predators.

Peering into the cell and focusing down to the molecular level is where we see the complexity of life the best though. In Darwin's day the cell was considered just a "simple little lump of albuminous combination of carbon", to use the words of University of Jena zoologist Ernst Haeckel. He was a world famous biologist and author, and a contemporary of Charles Darwin we meet later (§9.3.2). No "little lump", the cell instead is a metropolis of factories and production lines with molecular machinery of bewildering complexity. Complexity in construction, complexity in function, complexity in control and complexity in coordinate regulation with other molecules in the cell, in the tissue and in the functioning of the multicellular organism considered as a whole is what is found.

Another point in considering biochemistry is this. There is no simpler reductionistic level for this complexity in biology than its constituent molecules. And yet biological complexity, like the other four traits of life's wonderful appearance, is found from the molecular level up with no hierarchy of intensity. The molecular level, the last stop of biology before physics and chemistry, is every bit as complex as an entire ecosystem. Why should that be? Every level of biology is diverse, every level has a unity, every level is complex and every level has adaptativeness. This observation may seem mundane, but think about it for a moment because it is very curious. The interactions of the very molecules that make up life are complex, "digital", non-uniform, and we can't get to any simpler level in biology, since their behavior is described by the fixed laws of physics and chemistry.

It turns out the molecules of the cell are themselves constructed from a blueprint molecule called the genetic code, and the DNA sequence that is the genetic code, is written in a language every bit a language as English is a language. The principal difference between them is that the genetic code has only 4 letters in its alphabet while English has 26. Okay, emoticons excluded. The simplest cell needs at minimum around 200 proteins to function and each protein contains at least several hundred amino acids, on average. Proteins are made of one or more chains of amino acids which are like a necklace(es). The amino acids must be in a precise order in that chain for the protein to function. Since there are twenty possible amino acid "beads" to choose from at every position along the chain, the

35

possibilities of generating any one particular protein randomly, let alone of multiple proteins working coordinately and interdependently in molecular machines and metabolic pathways, become astronomical very quickly (§12.4.2).

A further level of molecular complexity is added protein "necklaces" folding in on themselves in highly specific ways to form a protein's so-called secondary, tertiary and quaternary 3-D structure that must be assumed if it is to function at all. Enzymes are proteins and they work with at least the sophistication of the way that a key and a lock works. The difference between the white of a raw egg and a fried egg, is a crude example of the difference in 3-D folding properties of the protein albumin, and it is also the difference between life and death. In summary we need to explain how life got to be so diverse, so complex and how it appears to be so purposeful at every level we look.

§1.2.4　　　　　The Adaptativeness of Life

Life is goal directed. This is so self-evident it seems circular to say it, but wherever we see a creature, barring environmental catastrophe that is, so there we find it adapted to that particular environment. We usually don't take this any further than saying: "Organism X is adapted to permit its survival which is why it is there to be seen in the first place". But this is as superficial as it is unsatisfying as an explanation in science. The solution to the Conundrum will have to be better than that! Another observation is that almost all manner of hostile environments on Earth are yet home sweet home to the varieties of life we find there, and they are adapted to these conditions in quite wonderful ways. If the exuberant diversity, the intricate and delicate unity and the exquisite complexity don't suggest the appearance of purposeful order to you, this last feature of life - adaptations to a particular environment, are explicitly that. Purposeful. Consider the camel. Cacti. The bacterial flagellum. Thermophilic bacteria. The Venus flytrap. Bird wings. Lungs. The eye. The ear. Tapeworms. Angler fish. Humming birds. Penguins. Whales. On and on it goes...

§1.2.5　　　　　The Shared Patterns of Life

There is a fifth feature of living things that also gives them the appearance of wonder and design. It's the observation of shared patterns of similarity among otherwise dissimilar creatures. Examples are four legs or two wings or five digits in otherwise very dissimilar creatures. Like a four leg construction plan in mice and elephants and crocodiles, or two wings in bats and birds or a five digit limb in frogs and porpoises and humans (Fig.7.1). Not only structures are shared but chemistry and chemical pathways of life are common across nature and the genetic code identical no matter whether you are a microbe, a moth or a man. This fifth feature of giving the nature an appearance of design is inspected later (§7;§8).

In summary the solution to the Conundrum of Life we now go in search of must explain the diversity, the unity, the complexity, the adaptations and shared patterns of life that are synonymous with it, and do so with unambiguous empirical support. The solution should also explain the origin of life and of matter itself. We will not need the assistance of hand waving or dogma, no matter how sincerely believed. There is plenty of it by the way, so watch your head as we go. I will do my best to impartially illuminate the arguments and the primary data so you can make up your own mind. I have made my own conclusions. This is unavoidable but I defy you to find fault on empirical, objective grounds. If you find fault on philosophical, religious or emotional grounds I do not care to hear. Our search is for the objective empirical answer, reality, the truth. We will discover that it's not how you look but how you see that matters, and this is what makes all the difference. The data we review is accessible is to anybody interested to make this inquiry on their own and comes directly from the writings of the leading experts and premier journals like *Science*, *Nature* and *Scientific American*. Ready? Here we go.

§1.3   THE APPEARANCE OF DESIGN IN NATURE

What we have discovered so far is the Conundrum seems no conundrum after all. Instead the solution appears obvious. Living things are created because their every appearance declares it! This "appearance of design", as it is now called, is nowhere controversial as an appearance either because evolutionists see it too. Richard Dawkins even defines biology as "the study of complicated things that give the appearance of having been designed for a purpose".[27] It's why Nobel laureate and co-discoverer of DNA Sir Francis Crick says: "Biologists must constantly keep in mind that what they see was not designed, but rather evolved."[28] We can't rely on appearances alone of course. The earth once looked flat. So how can we tell if the appearance of design in nature is real, or just an illusion as evolutionists say? Reason demands a causal account be given. The stakes could not be higher either, because the philosophical implications of the two contending solutions are as opposite as they are immeasurably opposite. And one of them has to be false.

The next observation is that although the Conundrum reduces in the end to one of only two possible answers, evolution of some kind or supernatural creation of some kind, yet we find the evolution-creation debate to be utterly intractable! Creationists marvel at the majesty of what to them is obviously the work of the Almighty, while evolutionists say majesty got there in decidedly non-majestic steps by natural forces alone. How can one side win when the other is so entrenched? Any marvelous change can be explained by accumulative, purposeless evolutionary change as it can by a saltatory "God did it" miracle. Can the impasse be resolved? Sure it can.

It's all about what objective evidence can be mustered for either side's claim. The pretender should be obvious with a closer look (*Parts II & III*). The Conundrum question is a very simple question after all.

In the past what was obvious about the appearance of nature was taken to be obviously real. It is the reason for worship. For those living in Western Civilization it was also confirmation of what the Bible said. In case you've never read the Bible, let me assure you the answer to the Conundrum is made very clear there - God did it all. This is made clear from the first verse of the first chapter of its first book because we read: *"In the beginning God created the heavens and the earth"* (Gen 1:1). The stakes are high all right. Science is on trial. The Bible is on trial.

The latter is well known to evolution advocates, but surprisingly not to nearly as many Christians. The reason is large proportions of them, around a third in America, have no problem believing both in Jesus Christ and in Darwinism (Fig.18.1). This answer to the Conundrum is called theistic evolution. It's a most interesting idea if you like spectator sport because of the acrobatics it requires of its believers, at least if they want to show an objective basis for believing it to skeptics. Skeptics come at them from opposite sides, which is the cause of the cool acrobatics. Coming at them on one side are all the remaining Christians who say the Bible repudiates evolution. Coming them from a pole away are all the remaining evolutionists, the atheists, who say the whole point of evolution theory is it's fully explanatory without any need of a supernatural Creator. Not wanting to take sides here but merely show you the high stakes of all the answers to the Conundrum, is there any basis to such objections to theistic evolution, setting aside for now whether or not evolution is even true? I mean for a Christian, and if you are not for this question just pretend, is Christian theistic evolution in any way a contradiction? Consider two of the objections that are made.

The first problem for theistic evolution is it has to refute the plain meaning of the Bible's verses claiming a special creation. There are a rather lot and not only in *Genesis*. For other places check out Ex 20:11; Deut 4:32, 32:5; 2 Ki 19:15; 1 Chron 1:1,16:24-26; 2 Chron 2:12; Neh 9:6; Job 10;8, 12:7-9, 33:4, 38:4-39:40, 40:15-41:34; Ps 8:3, 33:6-9,15, 19:1, 74:16-17, 89:11-12, 90:2-3, 94:9, 95:5, 96:5, 100:3, 102:25, 104:2-26, 115:15, 119:73, 121:2, 124:8, 134:3, 136:5, 139:13, 146:6, 148:3-12; Prov 3:19, 8:27-8, 22:2, Eccl 11:5; Is 37:16, 40:25-26, 42:5, 43:1, 44:2,24, 45:7-12,18, 48:13, 51:13, 54:5, 66:2; Ez 28:13; Ez 31:9,18; Jer 1:5, 10:12, 27:5, 32:17, 33:2, 51:15; Hos 6:7; Joel 2:3; Jonah 1:9, Zech 12:1, Mt 13:34, 19:4; Mt 25:34; Mk 10:6, 13:19, 16:15; Jn 1:1-3, 17:24; Acts 4:24, 7:50, 14:15 17:24-26; Rom 1:19-20,25, 5:12,14; 8:22, 9:20; 1 Cor 11:8-9, 15:45; 2 Cor 11:3; Eph 1:4; Col 1:16; 1 Tim 2:13,14; Heb 1:2-3,10, 4:3, 11:3; 1 Pet 1:20, 4:19; 2 Pet 3:5; Jude 1:14; Rev 4:11,10:6,14:6-7.[29]

If this much of the Bible is merely allegory as theistic evolutionists say, what else even they call literal is really allegorical? And how would we know since we can no longer take it to mean what it says? The authority of the text that claims to be the word of God collapses for lack of clarity. Yet its words are clear. Read them. The problem is not whether they're clear or not, it's whether they're true or not.

The second problem for theistic evolution is it refutes the plain words of Jesus Christ who surely alone determines what followers must believe to fairly be called "Christian". Jesus said: *"the Scripture cannot be broken" (Jn 10:35)* and Scripture *"is truth" (Jn 17:17)*. Not that Scripture contained truth, it *is* truth. Theistic evolutionists say the Genesis creation narrative is allegorical, but Jesus affirmed the historicity of Adam and Eve (*Mt 19:4-5; Mk 10:6*). Theistic evolutionists say humans evolved after millions and millions of years, but Jesus said *"at the beginning of creation God made them male and female" (Mk 10:6)*. Jesus's family tree is listed from Adam in a genealogy beginning with a literal first man called Adam with literal descendants *(Lk 3:38)* which is repeated in 1 Chronicles 1. In Genesis 5 Adam's age at death is also given *(Gen 5:3)*. Theistic evolutionists say Adam as the first human was formed from another animal but Genesis 2 says Adam was formed "from the dust of the ground" *(Gen 2:7)*. Actually it says Adam's body was made before he was even alive: *"The LORD God formed the man from the dust of the ground and breathed into his nostrils the breath of life, and the man became a living being" (Gen 2:7)*. Genesis says Eve was made directly too, from a rib surgically removed from Adam *(Gen 2:21)*, and Jesus affirms the account precluding an understanding this too might be allegorical *(Mk 10:5-6)*. Other contradictions a belief in

Jesus Christ must reconcile with a belief in macroevolution are listed in Table 26.1. So let's not have any silliness that "his powerful Word [i.e. Scripture] is entirely compatible with the Darwinian theory of evolution", as is so often said in religious circles and as said here by Cambridge University molecular biologist and Director of the Faraday Institute for Science and Religion, Dr. Denis Alexander.[30]

This hybridization of evolution with a supernatural creation is a worldview that gives new meaning to the term "personal belief. The reason is in the final analysis theistic evolution believes neither in science (for which a supernatural is subjective and irrational), nor in evolution theory (for which a supernatural is gratuitous and intangible), nor in the Bible (for which special creation is quintessence and clear), nor even in the teaching of its Founder (whose words about a supernatural are explicit). Whether you choose to believe the Bible or not, is the issue and the question for you. At least there can be no question about its ambiguity. I told you the stakes are high regardless of what answer to the Conundrum you choose to believe.

Theistic evolutionists do not apologize about these stakes or this grounding. Francis Collins, the Director of the National Institutes of Health and Head of the Human Genome Project is among the most famous of its public advocates. He says in defense: "The theistic evolution perspective cannot, of course, prove that God is real, as no logical argument can fully achieve that. Belief in God will always require a leap of faith."[31]

There are three problems with his view, regardless of whether or not evolution is true. Of course if evolution is false then theistic evolution is poppycock as well. The first problem is if God can never be "real", why believe Dr. Collins when he says God exists? The second is he offers no evidence for his claim the existence of God cannot be empirically demonstrated. The third problem is this is not what God says about his existence! We read the opposite in the Bible: *"Let all the earth fear the LORD, let all the people of the world revere him. For he spoke, and it came to be"* (Ps 33:9). We also read *"the whole earth is full of his glory"* (Isa 6:3) as evidence of His existence, and *"The heavens declare the glory of God; the skies proclaim the work of his hands"* (Ps 19:1). The Bible says nature and the universe are a demonstration of God's presence and God's glory. I offer one more example, and it is also another example of how uncompromising the Bible is about the solution to the Conundrum.

It is that although some philosophers like Immanuel Kant have declared the Conundrum to be insoluble, and although all evolutionists - theistic and atheistic alike - say we cannot detect God in nature and never could, and although many Christian leaders say the evidence for a supernatural creation is refuted by evolution, the Bible says the evidence of God is absolutely obvious. Even to those who have never read it! It is the appearance of design in nature (§1.3). We read in *Romans*:

> "The wrath of God is being revealed from heaven against all the godlessness and wickedness of men who suppress the truth by their wickedness, since what may be known about God is plain to them, because God has made it plain to them. For since the creation of the world God's invisible qualities – his eternal power and divine nature – have been clearly seen, being understood from what has been made, so that men are without excuse." (Rom 1:18-20)

Considering the internal contradictions of theistic evolution to orthodox evolution theory and to a plain read of the Bible, it's remarkable how widely it's believed. The reason is the evidence of evolution is considered so much more compelling than the plain words of the Bible.

In a still further twist, this anti-evolution language is not what the science and education establishment see when they read the Bible! For example listen to Oxford University biologist Mark Ridley in his widely acclaimed textbook *Evolution*:

> "it is worth stressing that there need be no conflict between the theory of evolution and religious belief. It is not an "either/or" controversy, in which accepting evolution means rejecting religion. No important religious beliefs are contradicted by evolution..."[32]

No need to single him out. There are plenty of others who say exactly the same thing.[6,33-36]

And so we find something even more extraordinary. It's that although the elite of science profess a solution to the Conundrum which is entirely naturalistic, and although surveys of their own religious beliefs indicate that by far their majority are atheist, they claim both the expertise and the authority to define the limits of theistic faith - for theists (§18.1). To say this another way, those who are atheist, who are hostile to God by definition, get to define what reasonable and rational faith in God is to be. The amazing thing is how this is accepted by so many Christians without them smelling so much as the faintest whiff of rat.[37] Why evolution proponents should be so interested in religion when they tell us it's outside science and outside reality is another peculiar observation to make in passing. We return to this much later (§26.3).

Evolution or creation or both? There's sincere conviction on all sides and the stakes could not be higher. Evolution is on trial. ID is on trial. The Bible is on trial. Science as methodological naturalism is on trial and the notion that the existence of God is not empirically demonstrable to everybody is on trial. Come and take a look!

§1.4 DARWIN'S ELEGANTLY INTUITIVE SOLUTION

So how did the appearance of design in nature come to be explained by science the way it is today? To answer, we must pick up the story in 1802 when the English theologian William Paley (1743-1805) reiterated the ancient argument for design in a memorable new way. He said the appearance of nature could be likened to the order and complexity of a wristwatch. He made the case in a treatise called *Natural Theology*. His case can be condensed to simply this:

A watch has purposeful complexity and adaptation because it's designed.

Nature has purposeful complexity and adaptation.

Therefore nature is designed.

Is his argument valid? The force of his argument rests on logic that requires a watch and nature be identical or else be subject to counterargument on grounds of disanalogy. Namely that because watches are not absolutely identical to living things (e.g. they don't move, feed, grow or respond to stimuli), how can we be sure this particular characteristic, the appearance of design, is among those traits which they do share?

It was Scottish philosopher David Hume (1711-1776) who exposed this flaw in his *Dialogues Concerning Natural Religion* published in 1779. Ironically you will have noticed it was published three years after his death and twenty three years before Paley's argument was even made. Don't let that put you off. It's the way the evolution-creation debate is framed these days. The fact is, the claim Paley made for design from the appearance of nature goes back to the dawn of man. Anyway it had been made many times in science before, by among others Plato (c424-c348BC), Thomas Aquinas (1225-1274) and Isaac Newton (1643-1727). Paley's watch argument for design and Hume's objection fly about today as if counterpoints also because of the clarity by which Richard Dawkins has defined evolution as the "blind watchmaker", invoking Paley's terminology. This is what Hume said in his refutation to Paley:

"Look around the world and every part of it: You will find it to be nothing but one great machine, subdivided into an infinite number of lesser machines, which again admit of subdivisions, to a degree beyond what human senses and faculties can trace and explain. All these various machines, and even their most minute parts, are adjusted to each other with an accuracy, which ravishes into admiration all men, who have ever contemplated them. The curious adapting of means to ends, throughout all nature, resembles exactly, though it much exceeds, the productions of human contrivance; of human design, thought, wisdom, and intelligence. Since therefore the effects resemble each other, we are led to infer, by all the rules of analogy, that the causes also resemble; and that the Author of nature is somewhat similar to the mind of man; though possessed of much larger faculties, proportioned by the grandeur of the work, which he has executed"[38]

In other words Hume agrees, and rapturously, that life appears designed. Arch-evolutionist Richard Dawkins goes even further, so obvious is this appearance to say: "one thing I shall not do is belittle the wonder of the living 'watches' that so inspired Paley...When it comes to feeling awe over living 'watches', I yield to nobody."[39] But that's not where Dawkins ends of course and it's not where Hume ended either, because he continued:

"The exact similarity of the cases gives us perfect assurance of a similar event...But wherever you depart, in the least, from the similarity of the cases, you diminish proportionally the evidence; and may at last bring it to a weak analogy, which is confessedly liable to error and uncertainty."

Thus the appearance of nature is indeed wonderful but we cannot be sure the appearance is real. To ram this home, Hume gave some examples:

Humans have similarities with frogs and fishes.
Frogs and fishes have a blood circulation.
Therefore humans have a blood circulation.

The argument looks sound but only because we chanced on a correct result. The differences between supposedly similar things can instead be very significant as his next example shows:

Humans share similarities with frogs and fishes.
Frogs and fishes are cold-blooded.
Therefore humans are cold blooded.

What can we conclude from these philosophical arguments for and against design? Hume's case would seem a valid rejection of Paley's inductive argument, namely the appearance of design cannot be trusted. How valid an objection it really was, is a matter we take up again much later, but accepting it for now it is hardly satisfying (§16.9.5). The reason is while Hume had showed Paley's argument was no proof, it remained intact as an argument from analogy. Hume had not shown whether it was the dissimilarities or the similarities between watches and nature that were what was pertinent. More importantly Hume had absolutely nothing to say about the cause of that appearance of design in nature which even he said, "ravishes into admiration all men". The appearance needed explanation! Where did it come from? As a result, the Biblical notion of a supernatural

The wonder of life

creation which most definitely did explain that observation, prevailed. Enter Charles Darwin (1809-1882).

Darwin took the appearance of design head-on and scotched it as real. He did so by making recourse to a supernatural unnecessary and therefore implausible when compared to his simpler, sufficient, explanatory and solely naturalistic alternative. It was an elegant solution to the question of the ages. The Conundrum had been solved logically, and scientifically it would seem, and completely oppositely to what the solution had been before. No wonder Darwin receives the accolades he does! The diversity of species was instead caused by what he called "descent with modification". It is a succinctly descriptive term for what is now called "evolution". What is evolution? Niles Eldredge at the American Museum of Natural History defines it like this:

"Evolution is the proposition that all organisms on earth, past, present and future, are descended from a common ancestor that lived at least 3.5 billion years ago, the age of the oldest fossil bacteria yet reliably identified."[40]

Darwin's hypothesis declares all of life is descended from a single common ancestor. Life is diverse because of differential survival by natural selection over long periods of time on the random variation of individuals in populations. Individuals with new features that promote survival will inevitably increase to become whole populations with that feature emerging as a novel character trait in a species, and ultimately become a whole new species as still further beneficial changes accumulate themselves over time. And so on, up all the Linnaean taxonomic levels, to produce all of the majesty of nature. Its diversity, complexity, unity, order, adaptiveness and the shared similarities with natural selection were all inevitable! The complexity of life he said, is the logical consequence of novelty that arises by cumulative random change shaped by environmental forces. The unity of life is explained because the new species arise already within a food chain and interlinked ecosystem. They would not be there for us to see otherwise. As for the adaptiveness of life, for all their wonder and complexity, these devices are that way because they were shaped by the environment over eons. Of course creatures appear adapted to their environment. That's how they came to be there!

Even the similarities between otherwise very different creatures like the five fingered ("pentadactyl") limb shared by fish, reptiles, amphibians and mammals Darwin gave mechanistic reinterpretation (Fig.7.2). The term "homology" was already in use to refer to these similar structures across species (§7.1). Darwin upended the prevailing idea they were features of a common design plan of the Creator. It was instead because they were inherited from a common ancestor with the same trait.

Seven years later in 1866, anatomist Ernst Haeckel at the University of Jena used Darwin's idea to draw the first evolutionary tree. Among other words now common in the evolution parlance he coined "phylogeny" and "monophyletic" to convey the postulate of a single common ancestor. Like Darwin's postulate, this tree was entirely theoretical (Fig.1.3). A tree all of leaves and no branches might well be called something else but that's not what science saw, and the rest is history.

Ever since, taxonomy has had as its aim a mechanistic explanation of life rather than Linnaeus's of separating organisms by other seeming arbitrary characteristics like morphology or method of reproduction. The goal now is to uncover the ancestral evolutionary relationships between all species, and thus explain how they came to be. This new framework of classification with an evolutionary interpretation and directive is called "systematics". It has great appeal of course, because how else but for a supernatural are we to make sense of the incredible wonder of life? This classification system is also claimed not to be "so it is", but "how it is". And a mechanistic explanation is a scientific explanation.

Evolution is the reason for the origin and diversity of life says the science and education establishment and there is no doubt about it. Actually Richard Dawkins says it even more clearly than that. He says:

"Today the theory of evolution is about as much open to doubt as the theory that the earth goes around the sun...."[42]

At Harvard on the other side of the Atlantic, Stephen Jay Gould made the same point, also using his knowledge of astronomy, saying:

"evolution is as well documented as any phenomenon in science, as strongly as the earth's revolution around the sun rather than vice versa. In this sense, we can call evolution a "fact". Science does not deal in certainty, so "fact" can only mean a proposition affirmed to such a high degree that it would be perverse to withhold one's provisional assent)."[43]

The comparison they and others make (e.g.[44,45]) between the truth of heliocentricity and the truth of evolution is misleading though. Celestial mechanics is directly observable but evolution is an inference to an unwitnessed past. More importantly, understanding how something works is not the same thing as understanding how it came to be and work that way. Evolution is an answer to the question of life's origins. If an astronomical comparison is to be fairly used here, it should be the answer to the question of heliocentricity's origins. Where did the Sun and the Earth and their motions come from?

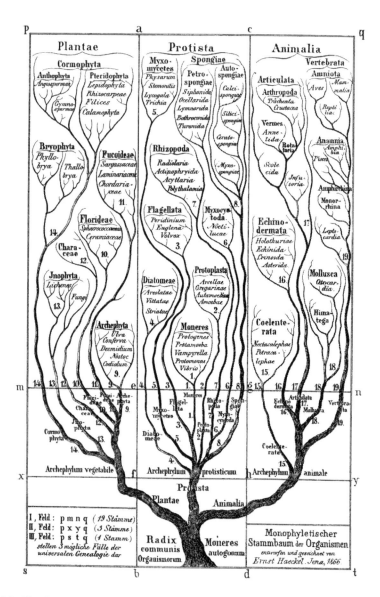

Fig.1.3. The first phylogenetic tree of life, drawn by Ernst Haeckel in 1866.[41] Note "Monophyletischer" (monophyletic) in the legend. The phylogeny was entirely theoretical.

Why exactly is evolution taken to be such a fact then? Because the evidence is so convincing of course. Editor in chief of *Scientific American* John Rennie says: "arguments that creationists use are typically specious and based on misunderstandings of (or outright lies about) evolution."[46]

But what is the evidence for evolution? Michael Ruse at Florida State University is a philosopher of science and also among the most published and most acclaimed advocates for evolution today. He gives a succinct summary of why the evidence for Darwinism is so compelling:

"If one subscribes in any honest way to the principles of empirical science then the *Origin* should convince one that organisms are the end-product of a gradual process of law-bound change, however caused. Homologies, embryological similarities, fossils, and rudimentary organs all attest to evolution. How, otherwise, can one explain the ridiculously useless isomorphisms that exist between the forelimbs of man, the horse, the bat, the porpoise, and the mole when each and every one of these limbs is used for a different purpose? And perhaps above all, the nature of the geographical distribution of organisms should make one an evolutionist. The finches and tortoises of the Galapagos just do not make sense without evolution...one often sees it said that "evolution is not a fact but a theory." Is this the essence of my claim? Not really! Indeed, I suggest that this wise sounding statement is confused to the point of falsity: it almost certainly is if, without regard for cause, one means no more by "evolution" than the claim that all organisms developed naturally from primitive beginnings. Evolution is a fact, *fact*, FACT!...We have to argue to evolution, we do not see it directly...The evidence that I have a heart is all indirect...but does anyone really believe that it is not a fact that I have a heart? Of course it is a fact, and in like manner I suggest that Darwin proved to us that evolution is a fact. The evidence he gives us is overwhelming".[47] (Table 3.1)

§1.5 TILTING AT WINDMILLS OR JOUSTING WITH DRAGONS?

It would seem there is nothing more to discuss. Darwin solved the Conundrum. But consider this: why have the critics of evolution not been silenced? It's been over one and a half centuries since *The Origin* was first published and yet opinion polls in the United States show only 19% of adults agree there was no supernatural creation (Fig.18.1). Is 89% of the country stupid? It's an extraordinary phenomenon whatever the explanation. The elite of the academic and education community, and not just *Scientific American* editor John Rennie, say objection to the truth of evolution is never on an empirical, rational basis but because of false religious belief. Richard Dawkins explains:

"Pretend as they will to scientific credentials, the anti-Evolution propagandists are always religiously motivated, even if they try to buy credibility by concealing the fact. In most cases, they know deep down what to believe because their parents recommended an ancient book that tells them what to believe. If the scientific evidence learned in adulthood contradicts the book, there must be something wrong with the scientific evidence."[48]

Certainly there are all manner of religious objections to Darwinism and indeed, they most often have come from Bible-believing Christians and orthodox Jews. It's also true that for most their objections are not because of experiments they've done in biology, but because of verses of Scripture that to their reading are as unambiguous about where life came from as it was not by evolution (§1.3). The issue of Bible literacy and Bible authority is not the question for us here though. What exactly a religion is, and whether it excludes reason or truth as he says, is not here either. What is of interest, and what is the question, is whether there are any empirical objections to evolution independent of the Bible.

And so ignoring the assertion that skeptics of evolution always have a religious agenda, and setting aside claims made by academia everywhere that evolution is a fact and there can be no rational controversy, perhaps a closer look at evolutionary theory might uncover reasons for doubt on one or more empirical grounds. It turns out to be a most worthwhile exercise.

The first discovery apparent even from a brief reading of *The Origin* is that evolutionary theory became "a fact" while it was still only a theory. The reason is because Darwin published *The Origin* claiming:

1) The basis of evolution, "descent with modification", rested on the inheritance of traits that changed during the lifecycle of an organism. This is now called Lamarckism (§2.2). While it might have a logical ring, it's false because it's based on an incorrect premise. Inheritance is by genes and an organism cannot modify its own genes. Modern concepts of genetics were added to Darwinism later. Sixty to one hundred years later. After evolution had already been accepted as true.

2) Evolution is effected by natural selection by which individuals with advantageous traits transmit their particular novel traits to the next generation and so accumulate. It sounds great in principle, but Darwin did not have even one example. He instead used the phenomenon of artificial selection known to breeders. The problem is breeders in experiments over centuries had found there were instead clear limits to the amount of novelty that could be

generated. No one had ever had produced a new species. Dog breeders only produce more dogs, and rose breeders only produce more roses. The examples Darwin gave in *The Origin* were imaginary. Literally. He said: "In order to make clear how, as I believe, Natural selection acts, I must beg permission to give one or two imaginary illustrations..." (p.120)

3) The large scale changes necessary to make complex organs like eyes and ears or new life forms like bats and whales from simpler more ancient precursors, occur by accumulation of successive small changes over time. The principle is called uniformitarianism and it is just like the changes seen by breeders over time, only multiplied over much, much more time. This extrapolation, the extrapolation of "microevolution" to "macroevolution", was without any mechanism of exactly how these large-scale changes could occur (without relying on intuitive but nevertheless completely blind faith that is). Needless to say "chance" is not a scientific mechanism if the precise details of what happened by chance are omitted. Evidence of the functional evolutionary intermediates in a supposed pathway of increasing complexity, or a plausible sequence of intermediate steps would be a minimum requirement, but he did not give any.

4) Darwin declared fossils to be the extinct intermediates of evolutionary series which had led to the species we find alive now. Thus fossils are the direct historical evidence that evolution happened. But he did not have even one example.

5) Given the title *The Origin of Species,* you would think it would say where species come from. It doesn't. While Darwin did explain how organisms become more adapted to a habitat and thereby had evolved through natural selection, he did not explain how such a continuous process could produce the discrete morphologic and genetic entities that are new species. Jerry Coyne at the University of Chicago and Allen Orr at the University of Rochester are evolutionary geneticists. They say Darwin "conflated the problem of change within a lineage [i.e. adaptation] with the problem of the origin of new lineages [i.e. speciation]",[49] and about the origin of species *The Origin* is "largely silent...and the little it does say is seen by most modern evolutionists as muddled or wrong".[50] Darwin didn't even have a coherent concept of what species were.

As Ernst Mayr says: "Darwin was fully conscious of the fact that the change from one species into another was the most fundamental problem of evolution. Indeed, evolution was, almost by definition, a change from one species into another one. In view of this central position of the problem of species and speciation...one would expect to find in the *Origin* a satisfactory and indeed authoritative treatment of the subject. This, curiously, one does not find. Indeed, the longer Darwin struggled with these concepts, the more he seemed confused. In the end, the *Origin*...was vague and contradictory both on the nature of species and the mode of speciation."[51]

So after even a cursory look we discover *The Origin* was:

1) without a valid theory of inheritance, although evolution rests on inheritance as its mechanism;

2) without any example of natural selection, although that is the mechanism of the process;

3) without any example or empirical evidence of exactly how the large-scale changes necessary for macroevolutionary change could be extrapolated from microevolution, although that what is meant by "evolution" and a block to such a thing had been the consistent observation in breeding experiments over centuries;

4) without any example of an evolutionary fossil series, although this was the best historical evidence for what is a historical process.

5) without a coherent concept of what a species was, although its thesis rested on the notion that species change and that one species changes into another.

For what was presented as a scientific solution, it seems rather a lot of unscientific faith was still required. An enormous amount of research has been done in the century and a half since these beginnings of course, and all such inconsistencies must have been resolved by now, surely. Another little known and seldom publicized fact is deficiencies in evolution theory were very apparent to Charles Darwin. Rather than claiming his theory as true as heliocentricity, he said:

"When we descend to details, we can prove that no one species has changed; nor can we prove that the supposed changes are beneficial, which is the groundwork of the theory. Nor can we explain why some species have changed and others have not."[52,53]

Darwin was even more frank when it came to explaining the all-important question of what there was before natural selection, namely how life arose in the first place:

"Though no evidence worth anything has as yet, in my opinion, been advanced in favor of a living being, being developed from inorganic matter, yet I cannot avoid believing the possibility of this will be proved some day in accordance with the law of continuity."[53,54]

As for objections from the science community to *The Origin* when it was first published, there were several. Among the dissenters was Harvard geologist and paleontologist Louis Agassiz (1807-1873). He called *The Origin* "a scientific theory untrue in its facts, unscientific in its method and mischievous in its tendencies".

Yet despite these setbacks so convincing was evolution in the theory that Darwinism was accepted in short order with the belief it was only a matter of more research to iron out all the wrinkles. Whether this has really happened we will need to see, but at least as regards the claims made in *The Origin,* there is an obvious difference between the coherency of the broad-brush strokes of the theory and its details on a closer inspection. It seems the wonder of "God Creator" might merely have been supplanted by the wonder of "Evolution". At a minimum there at least appear to be reasons to question the standing of evolution as an unequivocal truth.

Why do leaders of the science community make such emphatic claims that evolution is "a fact" then? This is one of the most interesting aspects of the controversy because if there really is a disparity from the truth, the behavior is completely antithetical to the scientific method and to academic conduct. This is the subject of *Part II*. It's an amazing story where truth is stranger than fiction and a lie as empty as The Emperor's New Clothes is uncovered as the pillar of twenty-first century science. What makes it amazing is that evolution cannot be disproved for the same philosophical reasons it is believed to be true.

The reason is because just as science excludes a supernatural by definition, so a creation demands a supernatural by definition. "Science" is in a straitjacket so tight it can never permit itself such thoughts. It will always regard those who suggest an objective supernatural as crazy, even if there was objective evidence of a creation of the natural world. Evolution and evolutionary science research is as real a reprise of the tale of The Emperor's New Clothes as you could ever hope to find. New data from biology is plugged into the preexisting Darwinian paradigm by this science despite contradictions and deficiencies, just has it always has since *The Origin* was first believed. Evolution theory morphs where necessary to accommodate inconsistencies, and so appears to get still stronger by the effort. Evolution is a fact, because it always was.

§1.6 WHAT'S SO SPECIAL ABOUT EVOLUTION THEORY?

Before looking at it's evidence some general observations about evolution are worth making. They have no bearing on whether its is true or not but they provide a backdrop to the dispute and a glimpse into the remarkable properties of evolutionary theory that make it so interesting to study. Whatever side you take in this debate, evolution science is unique and here are four of the reasons why.

§1.6.1 An Unsurpassed Intellectual Reach

Evolution theory is without peer anywhere in science in status or intellectual scope. Columbia and Rockefeller Universities geneticist Theodosius Dobzhansky famously affirmed this by his pronouncement: "Nothing in biology makes sense except in the light of evolution".[55] The National Academy of Sciences agrees, and Academy President Bruce Alberts quotes him when making the case for evolution in National Academy evolution materials provided to teachers.[7,56] Howard Hughes Medical Institute developmental biologist Sean Carroll at the University of Wisconsin says the same thing, and just as emphatically. "Biology without evolution," he says " is like physics without gravity".[57] Why does evolution theory have this special status?

Richard Dawkins explains why: "Darwinism encompasses all of life – human, animal, plant, bacterial, and, if I am right...extraterrestrial. It provides the only satisfying explanation for why we all exist, why we are the way that we are. It is the bedrock on which rest all the disciplines known as the humanities...Darwinian Evolution, as one reviewer has observed, 'is the most portentous natural truth science has yet discovered.' I'd add, 'or is likely to discover'."[48]

Harvard evolutionary biologist Ernst Mayr also explains why evolution is so special: "Every modern discussion of man's future, the population explosion, the struggle for existence, the purpose of man and the universe, and man's place in nature rests on Darwin."[2] While this power might seem surprising for a scientific theory, it's simply because the question that evolution answers is unlike most any other. It deals with ultimate origins. Most scientific theories seek to answer questions about the existing workings of the existing order of the universe. They answer such questions as "What is matter?" "How do bees fly?" "How do we see?" and "How do birds migrate?" Their answers are descriptive and predictive and data is reproducible. The question the Conundrum wants explained on the other hand, is how all that observable order got there in the first place.

The Conundrum is a question as much about history as it is about science and the answer, whatever it is, will necessarily make your worldview. The Conundrum is inescapably a "religious" question then, because it asks ultimate questions about existence. But somebody objects to say atheism is not a religion. It is. The reason is it's a worldview and a belief about ultimate origins and ultimate destiny. The courts have said so too, if only on the grounds atheism occupies in the life of a believer the same place that otherwise would filled by a religion in name (e.g. *Torcaso v Watkins* 1961; *United States v Seeger* 1965; *Kaufman v McCaughtry* 2005). Atheist unbelief is still a belief about ultimate origins and ultimate destiny and so still a religion.

§1.6.2　　　　An Unfalsifiable Scientific Fact

What "fact" means here should be clarified before going further. Although scientific theories cannot as a rule be proved to certainty (because one can never be sure that further advances in knowledge won't require an ideas' incorporation into a bigger principle or demand modification), scientific ideas can be still held with overwhelming conviction and they are. Evolution theory is one. This is what's meant by saying evolution is a scientific "fact" (§3.1).[43,45,47,58,59] That evolution theory is really "only" a "theory" does not demean this status.

When scientific theories are held with such sure conviction it is because the evidence overwhelmingly rejects all alternatives. This confidence is better called an inference to the best explanation, but that is not the point. Here is the point, in the words of French geneticist and Nobel laureate Jacques Monod (1910-1976). It's what he says about evolution being a fact. See if you agree with his point, but first some background.

Monod's much acclaimed book about evolution called *Chance and Necessity* declares that evolution fully accounts for all the wonder of living things. Chance (the effect of random events) and necessity (the necessity of natural law) are the only two naturalistic causes in the Universe of course. Evolution represents their combined action. In evolution theory "chance" is mutation and "necessity" is (the law of) natural selection (§2.4-5). But what's the evidence that evolution made the wonder of nature is "a fact"? Monod answers first by affirming what we now know (§1.6.1):

"The theory of evolution is a very curious theory. To begin with it is necessary to recall that in many respects the theory of evolution is the most important scientific theory ever formulated, because of its general implications. There is no question that no other scientific theory has had such tremendous philosophical, ideological, and political implications as has the theory of evolution."

He continues but in a most interesting way, one we wish he had told us sooner given what he just said and what he claims evolution can do:

"It is also a very curious theory in its status, which is quite different from that of physical theories. The basic aim of physical theories is to discover universal laws, laws which apply to objects in the whole of the universe, with the hope of being able, from these laws – that is to say from first principles – to derive conclusions, explain phenomena throughout the universe. When a physicist looks at a particular phenomenon it is with the hope that he will be able to show that he can deduce this phenomenon from universal laws, from first principles. The theory of evolution, by contrast, has a different aim...We can define the aim of the theory as that of accounting for the existence today on the earth of about two million animal species and about a million plant species [since revised upward §1.1.1], plus an unknown number of species of bacteria...I might say now that I do not believe it will ever be possible to do such a thing for very profound reasons that I will try to explain...The other great difficulty about the theory of evolution is that it is what one might call a second-order theory. Second-order, because it is a theory aimed at accounting for a phenomenon that has never been observed, and that never will be observed, namely evolution itself. Therefore, if you look at the structure of the theory of evolution...you will see that the discussion always goes in the following way – that one starts from the actual data, that is to say the present structure, performances, and anatomy of a given group of animals, and then one looks at the fossil record, and from the fossil record and the classical considerations of comparative anatomy, one tries to derive filiation of these forms...And then comes the second-order theory. You want to account for these facts and you introduce all sorts of further considerations, consisting of assumptions about rates of mutations and so on that might have occurred. And finally a reconstruction of the ecology of the groups that are assumed to have come before...Clearly, no reconstruction of this kind can ever be proved, and worse than that, no reconstruction of this kind can ever be disproved.[60]

Did you get his point? If not let me tell you. Evolution theory is everywhere in the science world and in education declared a scientific "fact" yet it cannot be falsified (§3.1,§19). It might be true nonetheless, but this is still extraordinary. The reason why it happens turns on the meaning of the word "scientific" and "science".

The purpose of the enterprise of science is the discovery of truth. This book takes as a given the definition Richard Dawkins uses for science: "Science is the disinterested search for objective truth about the material world".[61] To that end the signal feature of the scientific method is that for an idea to be "scientific" it must be falsifiable.[62] The signal feature of methodological naturalism is that the idea must be natural. The scientific method can't be two things at one and the same time. This a basic law of logic called the Law of Non-Contradiction, (specifically A cannot be A and non-A at the same time and in the same sense). Science as the search for "the truth" is being conflated by the search for "the natural". If methodological naturalism is false, then truth is being suppressed by naturalism. It's a simple and yet critical point and one which will be seen to be both the cause of macroevolution's strength and the cause of its illusion (§18;§19;§20). Why exactly is evolution unfalsifiable in science? We must return to philosophy and its role in science.

The argument made against design, which is evolution's only opposition after all, goes like this. Science operates as methodological naturalism today which declares the natural world is all there objectively is.[6-9] Anything supernatural is scientifically speaking to have left the objective and rational for the intangible, the subjective and the irrational. Physicist and popular science writer Paul Davies says this simplest, and unwittingly also shows us the problem so don't laugh. He says: "Science rejects true miracles".[63]

Oxford University zoologist Mark Ridley says the same thing, only you have to follow his definition out in corollary yourself to reach the absurdity this time. "Scientific arguments employ only observations anyone can make," he says, "and consider only natural causes, as distinct from supernatural origins".[64] This is unreasonable. It declares a supernatural never verifiable and never true. How do we know that? It has never been proved scientifically I assure you. How can the non-existence of an unknown ever be proved? Berkeley law professor Philip Johnson has argued against that assumption for years, as have others before him.[65-69] Suppose for argument's sake there were a supernatural. The problem is that one of the two solutions to the Conundrum IS supernatural. Think of another naturalistic solution apart from evolution? Come on! If creationism is false then life must have evolved from simpler forms somehow. Evolution is a "fact" already. There's no need for evidence. In fact all the "evidence", the observations of living and fossil organisms, in corollary of this philosophical necessity, can only reveal the complexities of the course evolution must have taken. This is why evolution is a "fact" and why it also cannot be falsified.

An example is perhaps helpful. Take Richard Dawkins' objection to the above. He disagrees and uses the same argument Oxford biologist J.B.S. Haldane (1892-1964) once made. He says evolution is disprovable, and the discovery of even one rabbit in the Precambrian would "would completely blow evolution out of the water".[70,71] But would it really?

Setting aside the obvious objections that the dating of such a rabbit fossil was wrong or the fossil merely another hoax like Piltdown man, the reason is the fossil record contradicts macroevolution already, without need of a rabbit yet evolution is "a fact" notwithstanding (§5.4-6). To wit, the pattern of the fossil record is stasis and sudden appearance not intermediate forms. Who needs a rabbit?

In another example, consider that Richard Dawkins and Steven Jay Gould have opposing interpretations of the fossil record yet both are equally certain evolution explains the origins of rabbits (§6.3). This could not happen if fossils were controlling the truth of evolution. The best intermediate fossil series which are proposed as evidence of evolution are not rabbits, and they are at best disputable, either for impenetrable interpretations or for contradictions from other evidence (§5.6). Living fossils, reworked fossils and the Cambrian explosion are contradictions as surely as a Precambrian rabbit.

We also know fossils are not controlling for the truth of evolution because Dr. Dawkins takes his rabbit offer back even before making it to declare: "The evidence for evolution would be entirely secure, even if not a single corpse had ever fossilized...We don't need fossils – the case for evolution is watertight without them".[72] Fossils don't determine the truth of macroevolution, the interpretations of the fossils as evolutionary already do. The reason is it's the only naturalistic explanation they could have and Dr. Dawkins shows you. He says: "all the fossils that we have, and there are very many indeed, occur, without a single authenticated exception, in the right temporal sequence...not a single fossil has ever been found before it could have evolved."[73] Every fossil "could have evolved". Design is not even considered (§19).

Why is there such strident resistance even to the possibility of design do you think? This is an extraordinary observation because it's not explained by the notion that to admit ID or a supernatural or "God" is to invoke no explanation at all. Some would have us think that and Richard Dawkins is one. He says if a creation were true then "we are back to miracle, which is simply a synonym for the total absence of explanation."[74] But assume the appearance of design in nature were genuine. Admitting design would not only solve the greatest question in science, it would also be what science seeks. The truth. Rejecting the possibility out of hand beforehand seems gratuitous. If there is ID in nature scientists could never know it no matter how much research they did but by the rules of methodological naturalism are doomed to continue making modifications to evolution theory to make it fit whatever fossil or molecular or other biology data is found. Scientists will never cease their pronouncements of evolution as "fact" and of creationists as crazy. This behavior follows inexorably from the artificial definitions and artificial reasoning they have set for themselves. The behavior is so inevitable it can even be considered a law, a law about science (§19.1). Evolution is first a naturalistic worldview and its acceptance has nothing to do with the evidence of biology. The intractability of the evolution-creation debate is as simple, and as scientific, as that (§3.4;§19).

§1.6.3   A Red Rag to Academic Bulls

The academic establishment defends evolution theory with a vigor which contradicts all its notions of academic freedom. This too is extraordinary. The reason for the behavior is because a challenge to evolution is considered a challenge to science itself (§22). Which of course it is, at least if methodological naturalism is the same thing as science, but it's still an extraordinary observation.

First because rather than open discussion about the data of ID in biology, and by this I mean engaging in academic debate about a legitimate question, there is instead hostility, contempt and censure for every evolution unbeliever (§25.6). Second because this one particular theory in science has come to be associated with a whole discipline in science, biology, or even all of science! Consider these two observations in turn.

In regard to the first observation, since the implications of the answer to the Conundrum are so monumental one might have expected discussion about different solutions be encouraged but this is not so. Informing students about design is resisted and teaching creationism in the public schools and universities of the United States is illegal (§24). Even disparaging evolution in school is illegal in some parts of Pennsylvania, as is merely mentioning ID (§24.1.5). Evolution is a scientific theory that is legally enforced! And yet science has to acknowledge a legitimate debate over the solution to the Conundrum, even if the alternative to evolution is considered loony. The intractable persistence of the creationist movement is one reason, but that's not the most important reason. The most important reason is because a rational argument always has two sides or else is no argument. Evidence always cuts both ways. It can both refute and support a contention and science is supposed to argue from evidence only.

In regard to the second observation, the history of scientific thought on the Conundrum of Life can be reduced to an epigram: Before Moses - other Gods. Moses - Jehovah God. Charles Darwin - no God. Darwin's answer was an inversion of all previous solutions to the ancient riddle. What was special was his reasoned removal of an Almighty by substituting an entirely materialistic mechanism. Science and society would never be the same.

Science did not always do business this way. Newton, Pascal, Babbage, Hooke, Kepler, Kelvin, Joule, Maxwell, Faraday and Galileo are some obvious Christian creationist scientists and Karolus Linnaeus who we already met is another. Today any creation solution to the Conundrum is not just non-science, it's non-sense. As Richard Dawkins says this: "It is absolutely safe to say that, if you meet somebody who claims not to believe in Evolution, that person is ignorant, stupid or insane (or wicked, but I'd rather not consider that)."[75]

While his words have the weight and assent of academia behind them, and although we know why he says them, it's still a curious pronouncement and not because of the controversy or his clarity. Dispute is common in science just as it is in academia generally, given its task of sorting truth from falsity. What is curious, is why there is hostility to those who reject evolution? There are so many other crazy ideas intellectuals brush off but about this one they get really mad. Let me suggest three reasons why. They are the three C's:

Condescension. The first is because those who conclude design from the evidence of biology or a creation from the evidence of Scripture and religious teachings, are considered ignorant and deceived by an illusion.[46,75] Evolutionists do not see design but instead an appearance of design because that appearance is really caused by the "blind watchmaker" (§1.3).[76] All of biology is framed by that mindset. "Nothing in biology makes sense"[55] otherwise, and unbelievers of this "fact" are treated accordingly. No matter how obvious is the appearance of design in nature, or how strenuously creationists point to it, claims of supernatural design in nature are regarded as the empty claims of obdurate minds deceived, and should be treated as such (§1.6.2;§25).

Contempt. The second reason is because those who conclude a supernatural design are irrational and therefore do not deserve debate. Science, at least methodological naturalism, says so (§19). Evolutionists consider Biblical creationist or any other creationist explanations to be empty metaphysical hand waving, and ID theory in biology of exactly the same ilk, "creationism in a lab coat", untestable and unfalsifiable, and therefore not only unscientific but irrational also. Evolutionists point to many phenomena in the natural world like storms and seasons and illnesses which once had supernatural explanation, and with greater or lesser religious overtones, but which are fully explained by entirely natural causes now. They miss the fact these examples have nothing to do with the Conundrum question which is not about how the world works, but how the world got to work the way it does. Regardless, those who conclude a supernatural today are no longer taken to be speaking sense. How else than with contempt as they do, do you speak to the unreasonable and non-sensible?

Choler. The third reason is because the implications of a supernatural in nature are taken as a serious affront. It's not the condescension or contempt that gets intellectuals angry, it's the assertions of an Almighty behind nature and in charge of nature. The reason is because this discovery is a demand for accountability and worship by every individual including them. To those for whom man is the measure of all things and to those who confess theism but also do only what is right in their own eyes, this causes great indignation. They get furious! Ever seen this? Open up a conversation with any evolutionist as an ID believer, even if you have to pretend to be one, and see what happens. The reaction is very curious for why not indifferent aloofness instead? Evolutionists say a creation is metaphysical claptrap and call the Bible misleading. Evolutionists say life is caused by, and turns on, purposeless undirected chance and the creative power of natural selection which means no moral accountability. They say there is no such thing as an absolute moral value and no objective thing as sin (§11.5). And so again it is a curious fact that although armed with this worldview and all the force of counter-argument from the authority and status of science, and though faced by such a puny foe - persons outside of the fellowship of elite academia - yet intellectuals still get so upset. Could this be the unmasking of a more general principle about man and living? In this light it is ironic to find Darwin himself, in the opening pages of *The Origin,* making a plea for dispassionate evidence-based analysis to prevail:

> "I am well aware that scarcely a single point is discussed in this volume [The Origin] on which facts cannot be adduced, often apparently leading to conclusions directly opposite to those at which I have arrived. A fair result can be obtained only by fully stating and balancing the facts and arguments on both sides of the question..." (p.19)

§1.6.4   A Propagation By Unabashed Misinformation

Evolution theory is distinguished in science in still another way - how misinformation about its contradictions is actively propagated. This too is such a contradiction to what the science and education establishment stand for, and it is so outrageous, that the reader might reject the idea immediately with derision. Take a look at four examples before you give up on this book and this inquiry. Still others, if you are interested, are found throughout *Part II* and in §25. Not surprisingly they bear on the problems identified earlier as unresolved by Darwin in *The Origin*:

1) Exactly one century after *The Origin* was published, Oxford University biologist and physician Bernard Kettlewell was hailed as finally providing the first experimental evidence of natural selection. Kettlewell even entitled his *Scientific American* article summarizing the work "Darwin's Missing Evidence".[77] What did he discover? He studied a population of white and peppered variants of the same moth species, and reported their relative population sizes in the wild changed according to their relative ability to camouflage themselves from bird predation on tree trunks. In the soot-polluted woods of Birmingham he found the dark variant flourished, while in the unpolluted Dorset forest the converse happened. He also produced video evidence of the birds eating moths off the tree trunks. The conclusion seemed inescapable. Here was direct evidence of evolution by natural selection.

The problem is the moths almost never rest on tree trunks in the natural condition. The experiments were also conducted during the day when moths are normally soporific, sedentary and sequestered in places of concealment. It turns out the moths were placed on tree trunks by investigators, so the moths hardly chose their own resting places. Yet leading science and biology textbooks continue to declare, as Richard Dawkins does, this work is "one of the best attested examples of natural selection in action".[78] Textbooks show students pictures of moths resting on tree trunks in daylight, staged photographs in science books no less, and fail to mention the methodological flaws of the experiments (§4.5;§25.4).

2) How changes in the relative beak size of Galapagos finches get presented as evidence for macroevolution is another example. Princeton researchers Peter and Rosemary Grant have painstakingly studied these birds for more than two decades and become famous to a lay audience, in some measure because of Jonathan Weiner's Pulitzer Prize winning book *The Beak of the Finch*. What did the Grant's discover? During times of drought the mean beak length of the finch population increased. The change was attributed to the better survival of individuals with bigger

beaks, better able to crack open the dry hard seeds that still remained as a food source. When the rains came this trend reversed again. What was shown therefore was the oscillation of a trait by natural selection caused by weather change.

This is not how the events are interpreted however. Extrapolating from the amount of change over a single drought, Peter Grant describes how it could lead to a completely new species if repeated droughts caused cumulative effects. How many droughts would it take? "The number is surprisingly small" he says, "about 20 selection events would have sufficed. If droughts occur once a decade, on average, repeated directional selection at this rate with no selection in between droughts would transform one species into another within 200 years." What is not mentioned is the reversal of this change during the intervening periods. For instance the National Academy of Sciences evolution teaching booklet describes the 5% increase in beak size in 1998, but omits to mention a decrease by the same amount the following year. Instead it calls the finches: "a particularly compelling example of speciation…if droughts occur about once every 10 years on the islands, a new species of finch might arise in only 200 years".[56] If brokers or bankers write like this they go to jail.

3) A third example concerns the claim made by Ernst Haeckel that "ontogeny recapitulates phylogeny" (§9.3). What is meant is embryos go through various stages of development representing their ancestral condition so demonstrating their evolutionary history. This thinking is what leads to the interpretation of pharyngeal arches in human embryos as being "gill slits", supposedly harking back to man's ancient aquatic ancestor and to claims "all vertebrates have similar embryos"[79] at early stages of development which become distinctive later on, in echo of their common evolutionary ancestry.

As evidence for this notion Dr. Haeckel produced a famous set of pictures in 1866 of vertebrate embryos from species as different as human and chicken and fish. His pictures showed these different species essentially identical in early development and only later becoming different. The problem is the drawings were faked (§9.3.3). This does not stop science from using them to demonstrate the greater fact of evolution in eminent textbooks until even recently, such as *Molecular Biology of the Cell* by National Academy of Sciences President Bruce Alberts and Nobel laureate James Watson (1994), *Evolutionary Biology* by Douglas Futuyma (1998) and *What Evolution Is* by Ernst Mayr (2001).

4) Fossil remnants of extinct life are now taken as among the best, if not the best evidence of evolution. Darwin predicted that since fossils represent the extinct evolutionary intermediate forms connecting the living with their

extinct ancestors, "the number of intermediate and transitional links, between all living and extinct species, must have been inconceivably great. But assuredly, if this theory be true, such have lived upon the earth."[80] The problem is despite the presence of millions of plant and animal species, it has been an inordinate struggle for science to show evolutionary fossil sequences at all. Rather than taking this obvious appearance for what it at least appears to be, what was considered more logical was to agree with Darwin the fossil evidence of missing links already existed and just needed to be found. They were just not looking hard enough. It took more than a century to even admit the fossil record evidence was even real.

Niles Eldredge and Stephen Jay Gould did so in 1973, and in another confirmation of §1.6.3, to fierce condemnation. It lead to the theory of punctuated equilibrium to explain the appearance of the fossil record (§6.3). Gould and Eldredge said: "paleontologists have worn blinders that permit them to accumulate cases in one category only: they have sought evidence of slow, steady and gradual change as the only true representation of evolution in the fossil record".[81] How could so many scientists have been in "blinders" and for so long? Gould later explained what had been going on, at least as he saw it:

> "The proponents of the synthetic theory maintain that all evolution is due to the accumulation of small genetic changes, guided by natural selection, and that transspecific evolution is nothing but an extrapolation and magnification of the events that take place within populations and species...I well remember how the synthetic theory [§2.2] beguiled me with its unifying power when I was a graduate student in the mid-1960's. Since then I have been watching it slowly unravel as a universal description of evolution. The molecular assault came first, followed quickly by renewed attention to unorthodox theories of speciation and by challenges at the level of macroevolution itself. I have been reluctant to admit it – since beguiling is often forever – but if Mayr's characterization of the synthetic theory is accurate,[82] then that theory, as a general proposition, is effectively dead, despite its persistence as textbook orthodoxy."[83]

And this is from an outspoken advocate who went to his grave certain evolution is a fact. The point, is opposite interpretations about evolution are yet taken to be equally true. Tell me what's going on? I told you evolution theory is special! You will discover an inspection of its complexities and ramifications to be a fascinating exercise. Solving the Conundrum of Life correctly could hardly be more worthwhile too. Come and look. We have lots of work to do.

§1.7   IN CONCLUSION: LIES, DAMNED LIES AND SCIENCE

The stage for our inquiry is set. Evolution and design exhaust all solutions to the Conundrum. Which of them is true? The academic establishment claims to have solved the Conundrum to such certainty that creationism of any stripe and even a creation implied by ID analysis is pigheaded falsity. The business of the science is taken to be demonstrating how evolution happened, not whether it happened. The teaching of science is taken to demand the instruction of children, including kindergarteners, about this fact (§26.2).[6] Suggesting a skeptical inspection of the evidence for evolution is to invite criticism with the contempt accorded a revisiting of the evidence for alchemy. The reason is the answer of evolution is already known to be "a scientific fact" (§3.1).

The alchemy comparison breaks down however, and not just on grounds of academic freedom by censoring inquiry. It is because the question of the Conundrum of Life is a metaphysical question much bigger than alchemy. It's bigger even than science, or at least the materialistic reasoning paradigm of science that is its present formulation. Science has mandated itself to finding solely naturalistic solutions to the Conundrum of Life. Notwithstanding that philosophical bind, evolution could still be true of course. We will need to evaluate the evidence in detail to decide. This is the subject of *Part I* of our inquiry.

Members of the science establishment universally acknowledge the wonder of life and the appearance of design in nature. For instance Harvard evolution biologist Richard Lewontin says: "Life forms are more than simply multiple and diverse, however. Organisms...have morphologies, physiologies and behaviors that appear to have been carefully and artfully designed to enable each organism to appropriate the world around it for its own life."[84] He says this appearance is an illusion though. This is the thesis of the "blind watchmaker" Richard Dawkins famously popularized already some decades ago. Here it is, in a nutshell:

> "Natural selection is the blind watchmaker, blind because it does not see ahead, does not plan consequences, has no purpose in view. Yet the living results of natural selection overwhelmingly impress us with the appearance of design as if by a master watchmaker, impress us with the illusion of design and planning."[76]

The metaphor was borrowed from William Paley, only Dawkins turns it on its head by Darwinism. He defines the entire enterprise of the investigation of life that is biology, on similar grounds too. He says: "Biology is the study of complicated things that give the appearance of having been designed for a purpose."[27] Like most in science and education he claims not fooled by the appearance of design. There must be lot to evolution then, and there must be a lot of evidence to make him so emphatic. Yet it was Darwin himself who first opened the door to refuting evolution. In rejecting design in nature he invited a challenge. He said:

> "If it could be demonstrated that any complex organ existed, which could not possibly have been formed by numerous, successive, slight modifications, my theory would absolutely break down. But I can find no such case."(p.232)

Two questions follow. First, is there such a case? Second, what's the evidence for his theory? We need to know his theory before we can review the evidence, and this is the task of the next chapter. The answer to the second question will be reviewed as lines of evidence, one by one, after that (§3-§12). And as for the first question, we tackle that much later (§15). It is to evolution theory and its wonderful blind watchmaker we must go first. Who was Charles Darwin and what, exactly, is Darwinism?

REFERENCES

1. Mayr, E. *What Evolution Is.* xiii (Basic Books, 2001).
2. Mayr, E. *One Long Argument.* 7 (Harvard University Press, 1991).
3. Sagan, C. *Cosmos.* 1 (Ballantine Books, 1993).
4. Moreland, J. P. in *The Creation Hypothesis* (ed J.P. Moreland) 41-66 (InterVarsity Press, 1994).
5. Dennett, D. *Darwin's Dangerous Idea.* 21 (Simon & Schuster, 1995).
6. National Academy of Sciences. Teaching About Evolution and the Nature of Science. http://www.nap.edu/books/0309063647/html/index.html (1998).
7. National Association of Biology Teachers. *Statement on Teaching Evolution*, <http://ncse.com/media/voices/national-association-biology-teachers-2000> (2000).
8. Lovgren, S. Evolution and Religion can co-exist say scientists. *National Geographic News*, http://news.nationalgeographic.com/news/2004/2010/1018_041018_science_religion.html (2004).
9. Pennock, R. T. *Tower of Babel.* 195 (The MIT Press, 1999).
10. Behe, M. *Darwin's Black Box.* (Free Press, 1996).
11. Dembski, W. A. *The Design Inference.* (Cambridge University Press, 1998).
12. Dawkins, R. *The Blind Watchmaker.* (W.W. Norton & Company, 1996).
13. Dawkins, R. *River Out Of Eden.* xi (Basic Books, 1995).
14. Raff, R. A. *The shape of life. Genes, development, and the evolution of animal form.*, (The University of Chicago Press, 1996).
15. Grant, V. *The evolutionary process. A critical review of evolutionary theory.* 10-12 (Columbia Unversity Press, 1985).
16. Wilson, E. O. & Perlman, D. L. *Conserving Earth's Biodiversity.* (Island Press, 2000).
17. UNEP-WCMC. *Global Biodiversity: Earth's living resources in the 21st century*.http://www.unep.org/geo/geo3/english/220.htm (World Conservation Press, 2000).
18. Mora, C., Tittenso, D. P., Adl, S., Simpson, G. G. & Worm, B. How many species are there on Earth and in the Ocean? *PloS Biology* 9, 1-8 (2011).
19. Grant, V. *The Origin of Adaptations.* (Columbia University Press, 1963).
20. Wilson, E. O. On global biodiversity estimates. *Paleobiology* 29, 14 (2002).
21. Eldredge, N. *The Triumph of Evolution.* 15 (W.H. Freeman & Co., 2001).
22. Leakey, R. E. & Lewin, R. *The Sixth Extinction.* (Anchor, 1995).
23. Funch, P. & Kristensen, R. M. Cycliophora is a new phylum with affinities to Entoprocta and Ectoproca. *Nature* 378, 711-714 (1995).
24. Stephen, A. M., Wiggins, H. S. & Cummings, J. H. Effect of changing transit time on colonic microbial metabolism in man. *Gut* 28, 601-609 (1987).
25. Bisby, F. et al. in *Species 2000 & ITIS Catalogue of Life* www.catalogueoflife.org/col/ (Reading, UK, 2012).
26. Dawkins, R. *The Blind Watchmaker.* 87 (W.W. Norton & Company, 1996).
27. Dawkins, R. *The Blind Watchmaker.* 1 (W.W. Norton & Company, 1996).
28. Crick, F. *What mad pursuit.* 148 (Penguin Books, 1990).

29 www.BibleGateway.com.
30 Alexander, D. R. *Creation or Evolution Do we have to choose?* 351 (Monarch Books, 2008).
31 Collins, F. S. *The Language of God.* 201 (Free Press, 2006).
32 Ridley, M. *Evolution.* 66 (Blackwell Science, Inc., 1996).
33 Allen, W. L. From the Editor. *National Geographic* (2004).
34 Raven, P. H. & Johnson, G. B. *Biology.* (McGraw-Hill, 2002).
35 National Science Teacher's Association. NSTA Position Statement: The Teaching of Evolution. http://www.nsta.org/positionstatement&psid=10 (2003).
36 Scott, E. C. in *Evolution: investigating the evidence* Vol. 9 (eds J. Scotchmoon & D.A. Springer) (The Paleontological Society, 1999).
37 National Center for Science Education. Statements from Religious Organizations. *http://www.ncseweb.org/resources/articles/5025_statements_from_religious_orga_12_19_2002.asp* (2002).
38 Hume, D. *Dialogues Concerning Natural Religion.* (Oxford University Press, 1993(1779)).
39 Dawkins, R. *The Blind Watchmaker.* 5 (W.W. Norton & Company, 1996).
40 Eldredge, N. *Macroevolutionary dynamics:species, niches and adaptive peaks.* (McGraw-Hill Publishing Company, 1989).
41 Haeckel, E. *Generelle Morphologie der Organismen.* (Georg Reimer, 1866).
42 Dawkins, R. *The Selfish Gene.* 1 (Oxford University Press, 1989).
43 Gould, S. J. Dorothy, Its Really Oz. *Time* 154, 59 (1999).
44 Mayr, E. *What Evolution Is.* 12-13 (Basic Books, 2001).
45 Futuyma, D. J. *Evolutionary Biology.* 15 (Sinauer Associates, 1986).
46 Rennie, J. 15 Answers to creationist nonsense. *Scientific American*, 78-85 (2002).
47 Ruse, M. *Darwinism Defended.* 57-58 (Addison-Wesley Publishing Company, 1982).
48 Dawkins, R. *The Blind Watchmaker.* ix-xi (W.W. Norton & Company, 1996).
49 Coyne, J. A. & Orr, H. A. *Speciation.* 11 (Sinauer Associates, 2004).
50 Coyne, J. A. & Orr, H. A. *Speciation.* 9 (Sinauer Associates, 2004).
51 Mayr, E. *One Long Argument.* 26 (Harvard University Press, 1991).
52 Darwin, F. *The life and letters of Charles Darwin.* Vol. 2 (D. Appleton and Co., 1899).
53 Pearcey, N. R. in *Mere Creation* (ed W.A. Dembski) 73-92 (InterVarsity Press, 1998).
54 Darwin, F. *More Letters of Charles Darwin.* (D. Appleton and Co., 1903).
55 Dobzhansky, T. Nothing in biology makes sense except in the light of evolution. *American Biology Teacher* 35, 125-129 (1973).
56 National Academy of Sciences. *Science and Creationism: A View from the National Academy of Sciences.* 2nd Ed edn, (The National Academies Press http://books.nap.edu/html/creationism/, 1999).
57 Carroll, S. B. *Endless Forms Most Beautiful* 294 (W.W. Norton, 2005).
58 Ridley, M. *Evolution.* 2nd edn, 65 (Blackwell Science, Inc., 1996).
59 Mayr, E. *What Evolution Is.* 275 (Basic Books, 2001).

60 Monod, J. L. *On the Molecular Theory of Evolution* (ed M. Ridley) 389-95 (Oxford University Press, 1997).
61 Singh, S. *Big Bang*. 497 (4th Estate, 2004).
62 Popper, K. R. *The Logic Of Scientific Discovery*. (Hutchinson, 1959).
63 Davies, P. *The Fifth Miracle*. 81 (Simon & Schuster, 1999).
64 Ridley, M. *Evolution*. 68 (Blackwell Science Ltd, 2004).
65 Johnson, P. E. *Darwin on trial*. (InterVarsity Press, 1993).
66 Johnson, P. E. *Reason in the balance*. (InterVarsity Press, 1995).
67 Johnson, P. E. *Defeating Darwinism by opening minds*. (InterVarsity Press, 1997).
68 Johnson, P. E. *Wedge of Truth*. (InterVarsity Press, 2000).
69 Gish, D. *Evolution: the fossils say No!*, (Creation-Life Publishers, 1973).
70 Wallis, C. *The Evolution Wars* in *Time* Aug 7 (2005).
71 Dawkins, R. & Coyne, J. A. in *The Guardian* (London, 2005 1st Sept).
72 Dawkins, R. *The Greatest Show on Earth*. 146 (Free Press, 2009).
73 Dawkins, R. The Greatest Show on Earth 175 (Free Press, 2009)
74 Dawkins, R. *River Out Of Eden*. 83 (Basic Books, 1995).
75 Dawkins, R. In short: nonfiction. *The New York Times*, April 9 p34 (1989).
76 Dawkins, R. *The Blind Watchmaker*. 21 (W.W. Norton & Company, 1996).
77 Kettlewell, H. B. D. Darwin's missing evidence. *Scientific American* 200, 48-53 (1959).
78 Dawkins, R. *Climbing Mount Improbable*. 87 (W.W. Norton & Company, 1996).
79 Futuyma, D. J. *Science on trial*. 206 (Sinauer Associates Inc., 1995).
80 Darwin, C. *The Origin of Species*. 408 (Random House, Inc., 1998(1859)).
81 Eldredge, N. & Gould, S. J. in *Models in Paleobiology* (ed T.J.M. Schopf) 82-115 (Freeman, Cooper, 1972).
82 Mayr, E. *Animal species and evolution*. 586 (Belknap Press of Harvard University, 1963).
83 Gould, S. J. Is a new and general theory of evolution emerging? *Paleobiology* 6, 119-130 (1980).
84 Lewontin, R. Adaptation. *Scientific American*, 213-231 (1978).

2. Darwin's "One long argument"

§2.1   WHAT IS (NEO)DARWINISM?

Charles Darwin was born in Shrewsbury, England on 12 February 1809. That same day in a one room log cabin a continent away Abraham Lincoln was born also. Charles was the fifth of six children. His mother was the daughter of the famous potter Josiah Wedgewood. She died when he was only eight and he was raised by his four sisters. His stereotypically stern father Robert Waring Darwin and his grandfather Erasmus Darwin were renowned physicians. It was therefore expected that he would follow in their footsteps like his older brother. Packed off to Edinburgh University Medical School the experience was more than enough for the young Charles. Medicine both bored and sickened him, and after two fruitless years his exasperated father finally sent him to Christ's College at Cambridge University to become a minister instead.

It was a momentous decision for although Darwin would later write that at Cambridge "my time was wasted, as far as the academical studies were concerned, as completely as at Edinburgh and at school", this was how he would become a naturalist. Natural history had always been his first love. The reason Charles Darwin could become a naturalist while training to be a vicar, was because "virtually all the naturalists in England at the time were ordained ministers, as were the professors at Cambridge who taught botany...and geology".[1] How that could be, is a telling observation of the Christian roots of science and education in Western Civilization.

After completing three years study for a B.A. degree, in 1831 Darwin was offered a job of ship naturalist and companion to Captain Robert Fitzroy on the *HMS Beagle*. The Admiralty had commissioned the ship to survey the coasts of South America and the Pacific Islands. The journey would last five years and circumnavigate the globe. It was to become the defining experience both of his life and for modern biology.

Following his return to England, Darwin spent the next twenty-three years ruminating over his ideas. Finally on November 24, 1859 he published *The Origin*. The full title was *On the Origin of Species by Means of Natural Selection or the Preservation of Favoured Races in the Struggle for Life*. The book sold out on the first day and has remained ever popular. It went through six editions, the last in 1872. This triumph followed fifteen months of feverish writing in fear of being scooped by another British naturalist, Alfred Russell Wallace.

In the spring of the previous year, Darwin had discovered to his horror that Wallace shared similar views. In fact the first announcement of what is now simply called "evolution", was made jointly by both men at the July 1, 1858 meeting of the Linnaean Society of London. Thomas Bell was the Society president that year. Likely you have not heard of him. History would not have remembered him either, except he made this assessment of the work that was presented. The year 1858 he said, had "not been marked by any of those striking discoveries which at once revolutionize, so to speak, the department of science on which they bear".[2] While the concept of biological evolution might seem obvious now it wasn't obvious to Thomas Bell, it wasn't obvious in 1858, and it wasn't obvious before Darwin and Wallace. So how did the thinking about evolution which culminated in *The Origin* develop?

It took two conceptual leaps. The first was to reject typological thinking or "essentialism". This is the idea that living things do not change into other living things in any significant way. Essentialism considers species to represent discrete separate types or "kinds". While the historical roots of essentialism are biblical (Gen 1:11,12,21,24,25) the notion was shared by secular philosophers, including Pythagoras (c569-c475 BC) and Plato (c429-c347 BC).[3]

The second leap was to discard another ancient idea, one that had originated fifteen centuries earlier with Aristotle (384-322 BC). It was the idea that living things could be arranged on a ladder of increasing complexity called the *Scala Naturae* or "Great Chain of Being", and literally "stairway of nature". The ladder began with inorganic matter on the lowest rungs followed by "simple" plants like moss higher up, then more complex plants, then more "complex" animals and eventually to man who was on the top rung. The idea was not evolutionary. It was simply an ordering of living things that had been supernaturally and separately created, with each rung occupied by a separate natural type or "species", which is a word derived from the Latin meaning "kind". The key to unlocking Darwinism was to convert Aristotle's ladder into an escalator. Presto. One species changes into another and you have evolution. But what could be the mechanism of such a thing?

The problem of finding a mechanism is obvious. Natural forces don't tend to act directionally over time but rather as variations back and forth about some mean position. Living things on the other hand are not random. They are highly ordered, highly purposeful, highly complex and highly adapted. They die otherwise (§1.2). If life evolved spontaneously by natural forces, then powerful trends toward increasing complexity, and preserving it, from simpler systems must have acted. But how could undirected natural forces accomplish such a staggering task?

Enter French naturalist Jean Baptiste Lamarck (1744-1829). It was he and not Darwin or Wallace, who proposed the first theory of evolution. Lamarck was the first to see the *Scala Naturae* not as a ladder but a biological escalator, even calling it Nature's Parade ("La Marche de la Nature"). As its mechanism, in 1809 he postulated parental traits that had either developed or atrophied during the life of an organism (depending on use or disuse respectively), are transmitted to offspring and the change is cumulative over generations. The idea held, for example, that an animal stretching to feed overhanging branches would cause its neck to lengthen over its lifetime and after multiplied generations caused the giraffe.

While this idea might have a plausible ring, it's false because it's based on an incorrect premise. Inheritance is through genes and an organism cannot modify its own genes to change a trait. Thus the inheritance of acquired physical characteristics, the essence of Lamarckian theory, is wrong. Charles Darwin was a Lamarckian.

Darwin formulated his entire theory without a valid concept of inheritance. The seminal discoveries about genes made by Gregor Mendel were published 7 years after the first edition of *The Origin*, and remained unnoticed until the end of the nineteenth century. Long after evolution theory was already accepted as true.

Lamarck's theory was not only false, it was incomplete. Darwin's contribution was to provide a mechanism for the escalator. It was "natural selection". Darwin's other seminal contribution was to explain the roles of individuals and populations within an evolving species. He said evolutionary change occurred in populations within a species, a concept today called microevolution. It's formally defined as the "change in gene content and frequency within a population" and is a directly observable indisputable phenomenon.[4] It is for instance how microbial antibiotic resistance occurs and how we got Chihuahuas and Great Danes by selective breeding. Thus Darwin recognized what others had disregarded for centuries. The reason this had happened, and for so long, was because a philosophical view had blinded observers to objective data. It was not the first time this had happened in science and it was not to be the last time either.

Darwin's observation was to point out that far from being "fixed" as essentialism declared, individuals in sexually reproducing populations of a species were unique! For instance we might all be *Homo sapiens* but you don't have to look very hard or very far to see we are also all very different. How is that observation evolutionary? If variation within a species is sorted by environmental selection. Darwinism holds that over time the small differences among individuals summate to become large differences between populations in a species, and ultimately make a whole new species when they can no longer interbreed with members of the original ancestral population. In the same way more and more change accumulates to produce new forms of life and new organ systems like eyes and ears and whale sonar and bat echolocation. This trans-species change is called "macroevolution". That Darwin got the Lamarckian mechanism by which the variation was transmitted all wrong did not matter to the evolutionary notion he proposed or to his contribution to biology. His views were the deductions from 4 observations:

1. In every generation many more individuals are born than ever survive to reproduce themselves to the next generation.
2. There is variability between individuals in a population.
3. These variations can be inherited.
4. An individual's survival and reproductive success is affected by environmental forces.

Those individuals in a population with the greatest probability of surviving to reproduce, are interacting most successfully with their environment. They are the most likely to transmit whatever advantageous genetic variation they have. The variation represents a new trait or change in "phenotype", a term meaning an observable physical or biochemical feature. The phenotype is a consequence of an organism's genes or "genotype". Thus the diversity and complexity of life lies ultimately in the genes, in the genetic code of DNA. The reason is because genes direct development from a single-celled zygote to a multicelled, multiorganed senescent adult. Embryonic development is what connects genotype with phenotype. We look at this critical link in evolutionary theory later. You might be shocked at what you find (or don't find, depending on your view §9). The process of the elimination and selection of traits by environmental forces is what is meant by natural selection. The term "adaptations" is used for phenotypic features enhancing an organism's fitness. An organism's fitness is its ability to transmit its gene, and this in turn is a function of its reproductive success.

What was Darwin's key step in formulating this grand theory? His discovery of natural selection. This was how he recalled his Eureka moment:

"[B]eing well prepared to appreciate the struggle for existence which everywhere goes on from long-continued observation of the habits of animals and plants, it at once struck me that under these circumstances favourable variations would tend to be preserved, and unfavourable ones to be destroyed. The result of this would be the formation of new species. Here, then, I had at last got a theory by which to work."[5]

What is it that evolves in Darwinism exactly? Not individuals but populations in a species. Recall the definition of microevolution earlier: "change in gene content and frequency within a population".[4]

To see this and which is also to see evolution at ground level, consider the appearance of a new genotype in a population as the starting point of evolution. It is a new gene profile in a parental sex cell caused by one or more mutations that enter the gene pool through the offspring. Next, depending on whether the phenotype that gene specifies is beneficial or deleterious, so the frequency of individuals with that genotype will increase or decrease in the population. Only the fittest survive! The relative proportion of population specified by that genotype will increase or decrease accordingly. The population evolves as more mutations accumulate over time under natural selection, and eventually this results in a population with so many trait changes that it can no longer breed with the original parental form. It has become a whole new evolved species. In the same way still more dramatic changes at genus, order, family and kingdom level are made, and the entire panorama of life is constructed. Darwinism is so logical, so simple and so plausible.

The incorporation of Mendel's data demonstrating the true basis of inheritance, and the incorporation of information from population genetics and other fields like biogeography, taxonomy, paleontology and embryology into Darwin's theory, took place between the 1920s and the late 1950s. This was chiefly the work of Ronald Fisher, J.B.S. Haldane and Sewall Wright in population genetics,[6,7] Theodosius Dobzhansky,[8] Julian Huxley[9] and Ernst Mayr[10] in systematics, George Gaylord Simpson[11,12] in paleontology and G. Ledyard Stebbins in botany.[13] We will meet them all. The end result of the amalgamated and revised theory is variously now called the Synthetic Theory, the Modern Synthesis (after the book of the same name by Julian Huxley referring to the fusion of genetics with Darwinism), the Evolutionary Synthesis or Neo-Darwinism. In a semantic quirk the latter had previously referred to something else, namely work by August Weismann in 1883 debunking claims of Lamarckian type inheritance in Darwin's theory.[14] For simplicity's sake we will refer to the current formulation of evolution synonymously as Darwinism and Neo-Darwinism, recognizing the significant historical and theoretical distinctions between them.[15,16] Getting right up to date now, what is Darwinism today?

§2.2 THE THEORY OF EVOLUTION

At the simplest level "evolution" or Darwinism has two tenets which are also its two successive steps. The first is the generation of genetic variation by mutation that results in phenotype variation of offspring. The second is the sorting or natural selection of these various traits. Condensed to its core, Darwinian evolution occurs as a result of natural selection acting on the variability among phenotypes of different individuals in a population. Phenotypic differences become adaptive changes because maladaptive ones die out and "fitter", more beneficial ones, are selected for transmission. A great deal more is subsumed within the 666 pages of *The Origin* however.[17]

Harvard University Alexander Agassiz Professor of Zoology Emeritus Ernst Mayr, who Stephen Jay Gould called "the greatest living evolutionary biologist,"[3] reviewed the development of evolutionary thought since Darwin in his masterful analysis called *One Long Argument*.

The title of his book represents the same words Darwin that used to describe the thesis he presented in *The Origin*. Mayr dissects Neo-Darwinism into five separate sub-theories: 1) Evolution, 2) Multiplication of species, 3) Gradualism, 4) Natural Selection and 5) Common Descent.[18] The first and the fifth were accepted quickly, within the first few years of *The Origin's* publishing, and the others by variable degrees until the final consensus of the Modern Synthesis.[3] The division of Neo-Darwinism into its component sub theories is a convenient way to inspect evolution in detail. We will take them in that order.

§2.2.1 Evolution

By this first sub theory of Neo-Darwinism is meant that life is changing. It might come as a surprise but the word "evolution" is nowhere found in *The Origin*. Darwin did use the word "evolved", but only in the sixth and last edition, only once and as the last word of the book. There he wrote: "from so simple a beginning endless forms most beautiful and most wonderful have been, and are being evolved." (p. 649). Darwin instead used the more descriptive formulation "descent with modification" for what we call evolution. "Evolution" merely means change, but in the context of biology it means Darwinian change. The changes of Darwinian evolution have two dimensions - a vertical and a horizontal, and two degrees of magnitude - microevolution and macro-evolution.

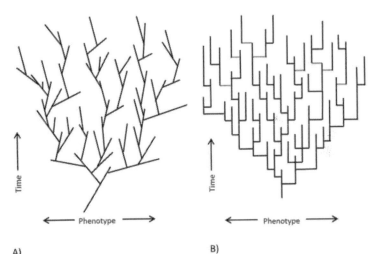

A)   B)

Fig.2.1. The two models of evolutionary change. Line slope represents rate of evolutionary change in a lineage. Branch points represent the emergence of new species. Phyletic gradualism or anagenesis (A) is the evolutionary change Darwin explained in *The Origin*, by which evolution is within a lineage with no change in tempo when speciation occurs. Punctuated equilibria (B) by contrast is evolutionary change entirely concentrated in speciation. (Drawing after Stanley, 1998).[19]

Vertical evolution is change that accumulates within a single line of descent or "lineage" and is called "anagenesis" or "phyletic evolution".

Lineages reconstructed from fossil data can show differences between them such that taxonomists consider one species to have graded into a completely different species. The two species in such a continuum are called "chronospecies" or successional species. The loss of the ancestral chronospecies as a result of change into a new descendant form is called "phyletic extinction" or "pseudoextinction" to distinguish it from the termination of a lineage (think of the dinosaurs!) called extinction.[19]

Horizontal evolution on the other hand is when a lineage splits into two or more new branches. This is called speciation or "cladogenesis". In contrast to vertical evolution the change here happens only at branches, and not linearly within established species.

These two patterns of evolutionary change can be demonstrated graphically (Fig.2.1). In anagenesis, speciation (branching) has no effect on the rates of evolution of the descendant species and no powering effect on a trend. Thus speciation in anagenesis is incidental to the rate of evolution. It just adds a new direction to the continuing process.

On the other hand horizontal evolution, also called the punctuational model of evolution (Fig.2.1B), holds that evolutionary change only happens with speciation.[16,19] While there is universal agreement in the science community that both anagenesis and cladogenesis happen, which of them is dominant is highly controversial and also highly acrimonious.

Anagenesis is dominant according to the so-called "ultra-Darwinists" like Richard Dawkins, Ernst Mayr and John Maynard Smith, while most evolutionary change is instead considered cladogenesis by such notables as Harvard's Stephen Jay Gould, Niles Eldredge at the American Museum of Natural History and Steven Stanley at Johns Hopkins University. The latter present their view in a theory called punctuated equilibrium. These different theories about evolution's mode and tempo will be taken up again when we consider the fossil record (§6.3).

Evolutionary change is also distinguished by magnitude. Specifically that occurring within species and called "microevolution", and that making a new species called "macroevolution". The changes of microevolution were appreciated long before Darwin. Examples are descriptions of ammonite variation by J.C.M. Reinecke in 1818 and Friedrich August Quenstedt in 1852. Nobody thought this capable of causing changes beyond the species level however. Darwin's theory holds that the mechanism of micro- and macroevolution is the same, and that the difference in degree is simply the difference in summated effect over much more time. Simply stated, macroevolution is accumulated microevolution, or in the words of our guide Ernst Mayr:

"The proponents of the synthetic theory [Neo-Darwinism] maintain that all evolution is due to the accumulation of small genetic changes, guided by natural selection, and that transspecific evolution [macroevolution] is nothing but an extrapolation and magnification of the events that take place within populations and species."[20]

Consider this idea further however.

While the processes involved in micro- and macroevolution are claimed to be the same, they are obviously very different and not just in degree. Macroevolution is a historical process not directly observable, and is inferred, while microevolution is a directly observable and reproducible phenomenon. Genetics and ecology are the main research methods used to investigate microevolution while paleontology, comparative morphology and molecular biology are the main methods taken to reveal macroevolution. The extrapolation from microevolution to macroevolution is the lynchpin of evolutionary theory. The validity of evolution as being the true solution to the Conundrum hangs on it. The evidence for this inference will be the focus of *Part II*.

It should be said a small minority of evolutionists has instead made a distinction between microevolution and macroevolution, including at a mechanistic level. In 1940 Richard Goldschmidt at Berkeley popularized this opinion in his book *The Material Basis of Evolution*. He instead said that no Neo-Darwinian force (namely point mutations producing changes in single alleles, the reassortment of alleles by meiosis and the action of natural selection on the genetic variability – we discuss these concepts momentarily §2.3), could generate changes of macroevolution. He said macroevolution was instead caused by more marked mutations, in other words major chromosomal rearrangements, which would happen in a single generation. While acknowledging that most would be deleterious, he postulated some as being very successful which he (tellingly) called "hopeful monsters".

The idea was ridiculed by his peers aided no doubt by his unfortunate choice of terminology. It was not until the 1970's that distinctions between micro and macroevolution were revived again. This time it was by Niles Eldredge, Stephen Jay Gould and Steven Stanley. They provided a different postulated mechanism called punctuated equilibrium (§6.3). Still another and more recent refinement distinguishing between micro and macroevolution comes from evolutionary embryology. This is the notion "[e]volution is generated by heritable changes in development."[21] In other words it's mutations in regulatory genes controlling embryonic development that are the most critical for causing macroevolutionary change. This is an important idea, both for evolution theory and for solving the Conundrum, and one we take up again later (§9.5).

§2.2.2 Multiplication of Species

By this second sub-theory of evolution is meant that species are not fixed in their character. As Darwin said it: "I am fully convinced that species are not immutable.[22] What exactly is a species? Ernst Mayr provides the most widely used definition, also called the biological species concept: "Species are groups of interbreeding natural populations that are reproductively isolated from other such groups."[23] The inability to interbreed as a result of accumulated microevolution leaves a species as an independently evolving evolutionary unit. Reproductive isolation permits new species to arise as mutations progressively accumulate within the population.

The formation of new species is caused by several potential mechanisms all of which act to prevent interbreeding. These can be usefully classified as operating either before fertilization or after fertilization (i.e. pre- or post-zygotic. For the detail-oriented the pre-zygotic mechanisms are temporal isolation, ecological isolation, behavioral isolation, mechanical isolation and gametic isolation; and the post-zygotic "mechanisms" are hybrid sterility, hybrid inviability and hybrid breakdown).[14] These details are not important to us. What's important is that it turns out defining a species is not nearly as easy as it first appears.

For instance by Mayr's definition, groups can be classified as different species even though they look exactly the same. There are four species of fruit fly *Drosophila* that are microscopically identical, for example. Another problem is Mayr's definition is no use to paleontologists trying to sort the bones of extinct creatures into separate species. Still another is that some species don't reproduce sexually at all. The problems in defining a species are acknowledged by Mayr, who concedes "no system of nomenclature and no hierarchy of systematic categories is able to represent adequately the complicated set of interrelationships and divergences found in nature."[24]

The reasons for the difficulty in defining a species are understandable but not the point. The point is the problem this creates when considering evolution and how and whether evolution works. In evolution what is that evolves exactly? A population in a species. But this is measured by the emergence of a new species (§2.2). We need to know what a species is, and how to measure it, or we will have problems measuring evolution. If evolution is a stronger concept than even the taxonomy of species, guess which one of them is horse and which is cart? We return to this later (§3).

§2.2.3   Gradualism

By this sub theory is meant that the progress of evolutionary change in Neo-Darwinism is slow. In Darwin's words: "As natural selection acts solely by accumulating slight successive favorable variations, it can produce no great or sudden modifications; it can act only by very short and slow steps." (p 626) Thomas Henry Huxley was among Darwin's most enthusiastic supporters. So much so he was also called "Darwin's bulldog"! Yet Huxley disagreed with him here. Writing to Darwin before *The Origin* was published he said: "You have loaded yourself with unnecessary difficulty in adopting *Natura non facit saltum* [Nature never makes jumps] so unreservedly."[25] Darwin was resolute. The reason is he realized what was at stake. If not by gradualism then any sudden saltational change could be interpreted as miraculous. As he wrote to another friend Charles Lyell, the most influential geologist of the day, "I would give nothing for the theory of natural selection, if it requires miraculous additions at any one stage of descent." Darwin was just as clear about gradualism in *The Origin*: "If it could be demonstrated that any complex organ existed, which could not possibly have been formed by numerous, successive, slight modifications, my theory would absolutely break down. But I can find out no such case." (p.232) He thus opens his theory to falsification. Although he could not find any, we will look for ourselves later (§15.2).

§2.2.4   Natural Selection

Natural selection is the process promoting transmission of the most adapted traits, and thereby their causative genes and mutations on to the next generation. This discovery is indisputably a great contribution to biology. Darwin explained it like this:
"As many more individuals of each species are born than can possibly survive; and as, consequently, there is a frequently recurring struggle for existence, it follows that any being, if it very however slightly in any manner profitably to itself, under the complex and sometime varying conditions of life, will have a better chance of surviving, and thus be *naturally selected*" (p.21). "[A]ssuredly individuals thus characterized will have the best chance of being preserved in the struggle for life; and

from the strong principle of inheritance, these will tend to produce offspring similarly characterized. This principle of preservation, or the survival of the fittest, I have called Natural Selection." (p.168)

There are four requirements for natural selection to operate:
(1) Sexual reproduction (because it permits genetic variation in offspring §2.4),
(2) Variation in phenotypic characteristics among individuals of a population,
(3) Heritability of different traits, and
(4) Differential ability to survive and reproduce as a result of the possession of a particular trait.

While mutation and the shuffling of genes (called "recombination") are "random" processes, natural selection is not random. Natural selection is responsible for the constructing power of evolution and is what Richard Dawkins famously calls "the blind watchmaker". In ratchet-like fashion the theory holds, natural selection builds up biological complexity over time.

There are two qualities that individuals in a species are selected for by natural selection. The first is superior survival and by implication a greater ability to reproduce, which Darwin called "natural selection". The second is a greater ability to mate in the "struggle between individuals of one sex, generally the males, for the possession of the other sex", which he called "sexual selection". With the incorporation of population genetics into Darwinism in the Modern Synthesis, Neo-Darwinism redefined Darwin's conception of natural selection in terms of reproductive success rather than greater survival of an individual.

Greater evolutionary "fitness" does not refer to physical fitness therefore, but to greater reproductive fitness . The ability to transmit genes on to the next generation. Natural selection is really about differential reproduction in a species therefore, because that is what drives the evolutionary process. To the extent there is any improved physical fitness that does not directly contribute to reproduction, it is irrelevant. Being able to fly further might seem an obvious evolutionary advantage, but not if it results in less successful matings or an increased exposure to predators.

Columbia and Harvard University paleontologist George Gaylord Simpson (1902-1984) was a co-architect of the Modern Synthesis (§2.2). He explains all this for us:

"Natural selection favors fitness only if you define fitness as leaving more descendants. In fact geneticists do define it that way, which may be confusing to others. To a geneticist fitness has nothing to do with health, strength, good looks, or anything but effectiveness in breeding."[26]

§2.2.5   Common Descent

By this fifth and final sub theory is meant that all of the diversity of life, either alive now or extinct in the fossil record, is descended from one original common ancestor. As Darwin put it: "all the organic beings which have ever lived on this earth may be descended from some one primordial form" (p.643). Niles Eldredge says it like this: "Evolution is the proposition that all organisms on earth, past, present and future, are descended from a common ancestor that lived at least 3.5 billion years ago, the age of the oldest fossil bacteria yet reliably identified."[27] The notion is only an inference, but it quickly became accepted and Darwin died a national hero. He was buried where Britain buries most of its heroes, in Westminster Abbey, and just a few reaches away from that well known physicist and creationist, Sir Isaac Newton. Ernst Mayr summarizes the status of common descent in science now: "Though there are still a number of connections among higher taxa to be established, particularly among the phyla of plants and invertebrates, there is probably no biologist left today who would question that all organisms now found on the earth have descended from a single origin of life."[28] How does evolution work exactly?

§2.3   WHAT IS THE MECHANISM OF (NEO-)DARWINISM?

The concept of biological evolution is simple and is reducible to just two steps, encapsulated by Darwin's phrase "descent with modification". The first step or "modification" is the production of genetic variation, which is done either by a mutation to form a new "allele" of a gene, or by the random reassortment of preexisting alleles in meiotic cell division. The term "allele" refers to the different variant forms of a particular gene that occupy a particular site (or "locus") on a chromosome. Alleles are caused by mutation. Examples of human alleles are brown or blue at the eye color locus or the blood group alleles A, B or O. New gene combinations are transmitted to offspring, and the consequence is a new or modified trait from the parental condition. Exactly what trait is changed will depend on the particular action of the new protein encoded by the new mutated gene or allele recombination.

The second step of the process of evolution is "descent", which is by natural selection. This is the sorting by the environment of individuals by their trait variations and more specifically the positive selection of those which promote reproductive success in some way. These individuals increase in relative proportion. The effect is that genes causing a selective advantage are more likely to be transmitted and so the number of individuals with them increases preferentially over other individuals who do not. Over generations of such "descent" the successful trait ("modification") in a population eventually spreads to all members of the species, and with further accumulation of beneficial traits from new gene mutations, ultimately becomes so different from the ancestral form that it can no longer interbreed with them. Presto, the population has evolved into a new species. Most mutations are not improvements but detrimental however, and such individuals are instead subjected to negative selection. They either die out taking those gene arrangements with them to extinction, or else diminish in the population.

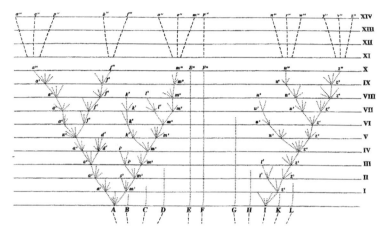

Fig.2.2. Darwin's schematic of an evolutionary tree. It was the only figure in *The Origin* and also entirely theoretical, as he had no example in all of nature for this idea.

Thus the process of evolution operates at the level of the individual to create a population effect. It's repeated in each generation leading to microevolution, and it summates over multiple generations to become macroevolution. This reasoning leads naturally to the concept of the evolutionary tree. Interestingly an evolutionary tree was the only figure in *The Origin* (Fig.2.2). Even more interestingly, it did not contain even one example from all nature. Data for the positioning of organisms on evolutionary trees, and which are the staple fare of biology textbooks as evidence for the process, was discovered after evolutionary trees had already been made, not the other way around.

To get an appreciation of the mechanism of evolution in more detail some familiarity with genetics is necessary. A crash course in three paragraphs will be more than sufficient for us. Hold on here we go:

Gametes or sex cells have only one set of chromosomes and so are called "haploid" (from the Greek *haplous* meaning single). At fertilization the haploid chromosome set of the egg cell is paired with the haploid set of the sperm cell to produce a fertilized egg or "zygote". The zygote divides repeatedly, eventually developing into a new adult organism. The non-gamete cells of the body (also called "somatic" cells) thus have a duplicate set of chromosomes, one of paternal and one of maternal origin, and so are called "diploid"(from the Greek *diplous* meaning double). Humans for example inherit 23 chromosomes from each parent making the diploid number of man 46. This is very close to the average for all nature, which is 50. The individual paternal and maternal chromosome pairs of a diploid set are called "homologous", and carry the two copies of every gene (with the exception of the X-Y chromosome pair in males). If the genes at a particular position (locus) on homologous chromosomes are identical the condition is called "homozygous". If they differ between the maternal and paternal chromosome at a particular locus it is a "heterozygous" offspring. As we already know, different forms of a gene at a particular locus are called alleles and are responsible for different traits in the population because they code for different proteins. Whichever of the different alleles is expressed in the offspring, and to what degree, depends on which one is "dominant" and which is "recessive". For instance in humans brown is dominant over blue at the eye color locus. Two more paragraphs to go.

The gene is a DNA molecule. It consists of two complementary strands of four nucleotide bases called adenine, guanine, cytosine and thymine, which are bound to deoxyribose sugar and phosphate in a double helix ladder. James Watson and Francis Crick discovered this structure of DNA to win the Nobel Prize in 1953. The total DNA in humans amounts to around 3 billion bases (3×10^9). The total number of genes this represents is unknown, but currently estimated at between 20-25,000.[29] If all the information written in this DNA were extracted from a single human cell it would fill a book of more than half a million pages.[30] A cell's chromosome number and DNA quantity, also called the "C value", varies widely across taxa without regard to their supposed more "simple" or more "advanced" evolutionary condition or not. For instance potatoes have more chromosomes than humans and ameba have 200 times more DNA than humans. This is a puzzle for an evolutionary explanation called the "C paradox" (of the evolutionary paradigm). We inspect this matter again later (§8.6.3). One more paragraph and you'll have all the genetics you need.

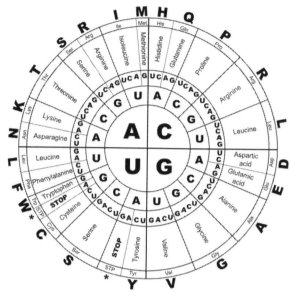

Fig.2.3 The genetic code is a language as real as English is a language. To start reading from the first letter of the codon, begin at the center circle and go outward. There are at least two codons for every amino acid. Codons that specify the same amino acid are called synonymous codons. Three "stop" codons punctuate the end of a polypeptide sequence just as a period does in English. Notice that synonymous mutations are mostly at the third position of the codon and all mutations at the second position lead to new amino acid replacements (i.e., non synonymous mutations) or to stop codons (i.e. nonsense mutations). (Image credit: J. Alves)

The chromosome represents tightly coiled DNA bound to packing proteins called histones. Without this packing the DNA molecule of the largest human chromosome would stretch out to 10 cm and if all the chromosomes were unraveled and the DNA molecules aligned, they would stretch 2 meters (1 m per haploid chromosome set). The efficiency of the packaging is an extraordinary feat when we consider that the 2 meters of DNA from the 46 chromosomes fits easily into a cell nucleus with a diameter of only 6 x 10^{-9} meters. Another way of showing the packing power is to consider that if the entire DNA from every cell of the human body (of which there are about 10^{13}) were aligned end-to-end it would stretch from here to the sun and back, 50 times. (It's 1.5 x 10^{11}m = 150 million kilometers ≈ 93 million miles to the Sun). The sequence of bases in DNA is written out in a biological language called the "genetic code". It's a language as real as English or Spanish is a language except its alphabet has 4 letters not 26 (Fig.2.3). Those letters are the 4 possible constituent bases of DNA (A, G, C, T). The genetic code is with extreme exception universally shared across all of life. Each of the 20 possible amino acids found in living things is specified by three letter combinations of the four bases called "codons".

Since there are 4x4x4 or 64 possible codons (four possible bases to choose from at each one of the three positions), there is more than one codon for each amino acid. This is called the redundancy of the genetic code. Three of the codons are used only for punctuation – they are "stop" codons which indicate the termination of a gene sequence. A gene sequence is simply a string of codons aligned end to end, and which specifies a string of amino acids aligned end to end. Done.

We can now consider the two steps of Darwinian evolution in more detail. This is a level of detail many readers may not desire, nor is it necessary to continue their search for the solution to the Conundrum. If this is you proceed directly to §2.6 and I will meet you there. Consider §2.4 and §2.5 as reference material for some rainy day.

§2.4 DARWINISM'S FIRST STEP: THE GENERATION OF GENETIC VARIATION

The generation of genetic variation means the transmission new combinations of alleles and/or entirely new alleles. This happens either by the random mixing of existing alleles called recombination, or by the generation of entirely new alleles called mutation. The end result of this change in genotype is a change in the proteins produced in the offspring compared to the parent, which in turn causes a modified or entirely new trait as specified by those proteins. How the genetic material is transmitted is by the gametes (sex cells). Meiosis is the process by which gametes are formed. It consists of two consecutive rounds of nuclear and cell division which follow a single round of DNA (gene) replication (Fig.2.4).

The first division is called meiosis I. This separates homologous pairs of chromosomes after they become aligned together on the equator of the cell. Before meiosis I can proceed, the DNA strand of the chromosome must be duplicated to form the familiar double DNA molecule X-shape we commonly associate with a chromosome. The > and < components of a chromosome are called its sister "chromatids" and bound at a waist called the "centromere". The chromatids of a duplicated chromosome are simply two separate DNA molecules that are identical ("replicated") copies of each other. A mutation is any error in the replication of a DNA strand that results in a change in base sequence. As can be seen from Fig.2.4 meiosis I generates two haploid daughter cells. Paternal and maternal homologous chromosomes are allotted randomly to each daughter cell. Before the homologous chromosomes are separated however, the chromatids of homologous pairs randomly exchange pieces of DNA in a process called "crossing over". This results in non-complementary chromatids on the homologous chromosomes of the daughter cells of the meiosis I division. The second meiotic division, called meiosis II, distributes these non-complementary chromatid DNA strands to produce 4 daughter cells each with a unique haploid genotype. In the final step a complementary DNA strand to each daughter chromatid is made, so restoring the chromosome to its usual diploid, double complementary (identical) chromatid structure.

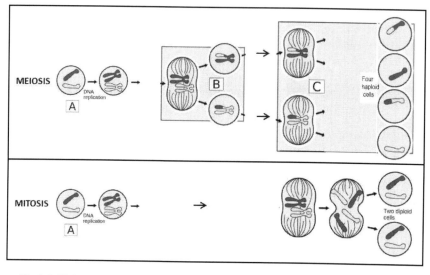

Fig.2.4. Meiosis and mitosis compared. Meiosis is how gametes are produced and mitosis is how all other cells are produced. Replicating the chromosome is the first step for both types of cell division (A). The major difference is meiosis has two cell divisions but mitosis one. In the first meiotic division homologous chromosomes are separated to form haploid daughter cells (B). This does not happen in mitosis. In the second meiotic division the paired chromatids are separated into haploid daughter cells (C). Mitosis is similar to meiosis II except daughter cells are diploid.

How does genetic variation occur exactly? There are four different ways and all happen in meiosis. The first way is by crossing over - when pieces of DNA are swapped between homologous chromosome pairs. The second is by shuffling of homologous chromosomes so that offspring get a new mixture of maternal and paternal chromosomes. The third way is by fertilization when maternal and paternal haploid chromosomes pair for the first time. The new genotype produced from the parents by any one of these three processes is called a "recombinant" because it represents "recombined genes" that were pre-existing. No mutations occurred. The phenotypic effect produced by the genetic change is dependent on the particular dominant versus recessive effects of the alleles distributed. It is also dependent on the phenomenon of "pleiotropy" (the process by which a single gene may control more than one phenotypic effect) and on "polyphenism" (by which environmental influences may produce different phenotypes).

The fourth way genetic variation is produced is the mechanism with which we are most familiar, mutation. Rather than just a new mix of genes in the offspring, a mutation is a new gene, a change to gene's DNA sequence, the formation of a new allele. Mutations can occur at any stage of meiosis or the cell cycle. Consider each of these four processes in turn to see how much genetic variation they make for microevolution. The potential for genetic diversity is very impressive indeed.

§2.4.1 Crossing Over

In meiosis I homologous chromosomes align in pairs on the equator of the cell. Having four chromatids between them, the complex is also called a tetrad. The chromatids exchange pieces of DNA with their homologous counterpart at random places along their lengths. The sites at which they attach to do are called chiasmata. The size and the location along the chromatid chosen to be swapped is random, but when it is done it is with exquisite precision nonetheless. Swapping is only between homologous chromatids, and breaks occur only in the identical relative position on the separate paternal and maternal chromatid so that there is no net change in the amount of genetic material, just a shuffling of pre-existing genes between chromosomes. The end result is two reciprocally different chromosomes from the parents.

This "crossing over" produces a new recombinant frequency which varies from zero percent (no crossing over of alleles took place) to 50% percent (in which there is independent assortment of all of the alleles). The actual frequency with which different allele combinations are found in recombinants thus depends on how far away from each other along the chromosome. The closer they are the less likely they are to swap independently. How much variety can be generated? If homologous chromosomes are heterozygous at n loci, it follows that crossovers can occur at $n-1$ intervals. Since there are two different daughter genotypes produced, there are $2(n-1)$ new genotypes from a single generation with only a single crossover. However more than one crossover can occur, multiplying this variety further each time by that factor.

§2.4.2   Random Independent Assortment of Homologous Chromosomes

Maternal and paternal chromosomes are independently shuffled in meiosis I. How much variety does this generate? For humans with 23 maternal and 23 paternal chromosomes, there are 2^{23} or 8.4 million potential combinations. In other words by chromosome reassortment alone parents can produce $2^{23} \times 2^{23}$ or over 70 trillion (7×10^{13}) different offspring! This number is representative for other species too because the average haploid number in nature is 25. The exact number of different diploid genotypes that can be generated by this sorting of alleles where n is number of genes and r the number of alleles of each gene is given by the formula $[r(r+1)/2]^n$.

From the perspective of the offspring, the paternal grandparent's genes are shuffled in the paternal gamete and the maternal grandparent's are shuffled in the maternal gamete. In the shuffling process half the grandparent's genes that were passed to the parents are deleted in the offspring. Looking at this from the genetic perspective of the grandparents, there is no gene mixing in their children, but rather in the grandchildren. In a genetic sense then, nothing is achieved until the grandchildren are born. Perhaps that's why grandparents love their grandchildren so much.

Since chromosomes occur in homologous pairs, the genes occupying a particular locus of the chromosome also occur in pairs. Any gene at a given locus may have different forms (alleles) but for any particular somatic cell there can be only two alleles at any particular locus, one from each parent. We have learnt already that when the two alleles are identical, the state is called "homozygous" and when maternal and paternal genes differ, the offspring is "heterozygous" for that particular gene. The existence of two or more alleles for a locus with different gene frequencies in the population is called "genetic polymorphism".

Heterozygosity in the parents provides greater potential for variety in the offspring. As University of California San Diego geneticist Francisco Ayala says, "the greater the amount of genetic variation in a population, the faster its rate of evolution."[14] Determining exactly how much variation exists in an individual is difficult, as frequently in the heterozygous state one allele is dominant and the other allele, being unexpressed, will go undetected. For example a person heterozygous for blue eye color has brown eyes because blue is the recessive allele and its allele goes undetected. The variability can be detected retrospectively however, in family trees or breeding experiments when the recessive genes are finally expressed in the homozygous offspring.

Different alleles are simply different sequences a particular gene and they code for proteins with different amino acid sequences and (therefore) different chemical properties and conformations. The differences between different protein products of alleles can be detected by their different electrophoretic mobilities (relying on the fact proteins move in an electric field at different rates depending on their size and electrical charge, and which in turn depends on their particular amino acid sequences). Alternatively different alleles can be determined by direct DNA sequence analysis of the gene locus in question called gene sequencing.

How much polymorphism is there in biology? Plants are heterozygous at around 17% of their genes, invertebrates 13%, and humans around 6.7%. Since there are somewhere between 10^4 and 10^5 genes in the human genome (with a current best guess of 20-25,000)[29] an individual will typically have about 6700 heterozygous genes.[14] As a result, around 2^{1500} or 10^{450} different gametes can be generated by allele sorting alone! By the way, the total number of humans on the planet is only 7×10^9 and the total number of atoms and constituent subatomic particles in Universe 10^{80}. There is a staggering individuality of life. Every sexually reproducing plant and animal that has ever lived, or ever will live, is unique. Ignoring the obvious anomaly of same zygote-derived multiple offspring, "identical twins", that is. There never has been, and there never will be, another you. You really are special. You are absolutely unique and science can prove it.

In summary most genetic variation in breeding populations results not from mutations, but reshuffling existing alleles into new combinations.[14] The potential for variety is enormous. In humans it can generate well over 10^{464} variants. This is impressive, but what exactly is its significance to evolution? It's helpful to stand back a moment to see the issues involved by focusing in on a local breeding population again, the smallest unit of microevolution.

Take for example a gene polymorphic for two alleles, call them *A* and *a*, and let their frequencies in that population be 60% and 40% respectively. What happens to those allele frequencies over time by recombination (crossing over, reassortment of homologous chromosomes and fertilization)? Do the allele frequencies change – in other words evolve by microevolution? Or do they blend out over time? Or do they stay the same? Make a choice there are no other options than these! Assuming 1) mating is random, 2) the gene pool (the population with all its alleles) is large, 3) and isolated (there is no migration of individuals in or out of the population), 4) the offspring are equally successful at reproduction, and 5) mutations do not occur to any significant extent, how much microevolution will there be? Think about it.

None. Allele frequencies under these 5 conditions are described by the Hardy-Weinberg law which states equilibrium among alleles in a population is reached after a single generation, regardless of initial genotype frequency. Using the example above, if the frequency of allele A polymorphism is p and for a is q, the genotype frequencies are given by the binomial formula $(p + q)^2$. Recalling high school calculus this is the same as $p^2 + 2pq + q^2$. Plugging in the values for A and a from into the equation, the equilibrium genotype frequencies are given by AA^2 = 0.36%, $2Aa$ =0.48 and aa^2 = 0.16%. We can test to see if this calculation is correct - the frequency of allele A is (0.36 + 0.36 + 0.48)/2 = 0.6, and of allele a is (0.48 + 0.16 + 0.16)/2 = 0.4. The principle is the same regardless of the number of alleles. For 3 alleles the Hardy-Weinberg equilibrium is $(p + q + r)^2$ and so on upwards in polynomial squares, $(p+q+...n)^2$ where p, q...n are the individual allele frequencies. The population always remains static. All of the above processes result in new combinations of genes in germ cells and tremendous genetic diversity, but they do not alter actual gene frequencies in the population which is microevolution. There is more to evolution than mixing alleles therefore. It is deviations from those five prerequisites for Hardy-Weinberg equilibrium that represents microevolution.

§2.4.3　　　　　　　　Fertilization

The mixing of paternal and maternal genes at fertilization is the most obvious cause of genetic variety, but it represents just another way of shuffling pre-existing genes in a population.

§2.4.4　　　　　　　　Mutation

Important as shuffling and recombining of alleles by the above three mechanisms is to generating genetic variability, and as powerful as it is in generating variety, it is mutations alone that are the ultimate source of genetic novelty. Novelty is essential for evolution. It is mutations that are the ultimate cause of those different alleles shuffled by recombination too of course. Nobel laureate Manfred Eigen puts it the simplest and the best: "Mutations are the source of evolutionary progress."[32]

Mutations may occur at any stage of the cell cycle, from when DNA is duplicated before cell division (called replication) or when messenger RNA is made for protein synthesis (called transcription) or when messenger RNA is read during polypeptide assembly (called translation). All have the end result of generating an altered protein product, which in turn represents an altered phenotype from the parent. As regards evolutionary processes, a mutation is only relevant when it can be inherited i.e., when it occurs to the gene and when it occurs in the gametes. Mutations in differentiated tissues such as skin are important, for instance in causing senescence or carcinoma, but they are not transmitted to offspring and have no significance to evolution unless they somehow impact reproductive success. Mutations can be classified either as single base pair changes to the DNA sequence of a gene (also called "point" mutations) or as multiple base changes (called "chromosomal" mutations). Consider the two in turn. As regards the point mutations there are three different types:

(1) Missense - by which one base of the four possible is substituted for another. There are two types of missense point mutation. When the substitution causes a different amino acid to be specified by the codon it is called a "non-synonymous" missense mutation. We already know that more that most amino acids are specified by more than one codon and thus a point mutation may not change the amino acid specified by the codon at all. Mutations which are phenotypically silent because they change the codon to one that specifies the same amino acid are called "synonymous" missense mutations.

(2) Nonsense – by which one base is substituted for another to transform a codon specifying an amino acid into one of the three stop codons, so causing premature termination of the reading frame of the gene. This results in a truncated protein product of that gene.

(3) Frameshift - by which rather than a one-for-one substitution of a nucleotide base it is the insertion or deletion of a single base from the gene. As a result, it changes the entire reading frame of the gene.

As regards the chromosomal mutations the different types are classified according to the mechanism by which they cause the multiple base changes simultaneously. These are by chromosome fusion, duplication, deletion, inversion, insertion, transposition and translocation.

Mutation rates are very low. The spontaneous mutation rate per nucleotide base ranges from 10^{-7} to 10^{-11} across life but when corrected for the effective genome of the organism, is relatively similar from bacteria to man.[33] The typical spontaneous mutation rate per gene locus in humans is around 10^{-5} per gamete formed with a range of 10^{-4}-10^{-6}, and which is a mutation rate of around 10^{-8} per nucleotide per generation.[33,34] Here are some specific examples: Hemophilia A gene 1.3×10^{-5}, albinism 3×10^{-5},

Huntington's chorea 2×10^{-6}, bacterial lactose fermentation 2×10^{-7} and bacterial streptomycin sensitivity 10^{-8} to 10^{-9}. The massive majority of mutations lower the viability from the wild type form. For instance almost 99% of experimentally induced mutations to barley (*Hordeum vulgare*) are detrimental.[35] On the other hand, experimenters found that 0.1-0.2% of the mutants were better adapted to particular environmental conditions imposed on the offspring plants. This is where microevolution comes in.

The process of changing a population by mutation alone is very slow because mutations are so rare. For instance if a mutation occurs at a frequency of 10^{-6} in a gene Y to produce allele y (assuming the population is completely homozygous for Y to begin with) after a single generation the frequency of y alleles is $1 \times 1/1000,000 = 0.000,001$ (one y in a million). The next population generation would start with one in a million less Y alleles from which to mutate to y, so the calculation of mutant frequency y in the population becomes $0.999999 \times 1/1000,000 = 0.000009$. The new y frequency is 0.0000019 and the amount of Y in the population has dropped to 0.9999981. Not only is the rate of increase by the new mutant extremely slow, the rate gets progressively slower with each generation. The reason is because the proportion of genes available to mutate becomes progressively less. After n generations of mutation, the proportion of Y is approximately $e^{-n\mu}$, where μ is the mutation rate and e is the base of natural logarithms. Plugging in actual numbers from our example, we see that after a hundred thousand generations the amount of Y is $e^{-(100000) \times (10^{-6})} = e^{-0.1} \approx 0.904$.[36] Thus even after one hundred thousand generations there are still only 10% mutant y alleles in the population, and moreover the rate of that snail's pace increase is decreasing logarithmically!

Geneticist Ronald Fisher calculated the extinction rate for a selectively neutral mutant allele to be 37% by the first generation, 60% by the third and 99% by the 127th. Even if that mutant is given a 1% selective advantage, the extinction rates differ only at the third decimal place until the 7th generation and at the 127th generation, the extinction rate is just 2% lower than when the mutant is neutral. This is why recombination is important to microevolution. By contrast with mutation, generating variety in the offspring by recombination is incomparably faster. Recombination can only occur with meiosis however, and meiosis only happens with sexual reproduction. What have we learnt so far?

Mutations are the ultimate source of phenotypic variation but they cannot produce microevolution alone. This is because 1) they are rare, 2) the overwhelming majority of them are adverse, and 3) because they are entirely random but nature is by contrast manifest order and purposeful adaptation (§1.2). We have also learnt recombination cannot produce evolutionary trends. While by contrast to mutation the forces of recombination are rapid and generate incredible variety, they do not produce genuine novelty and reach equilibrium, the Hardy-Weinberg equilibrium, after only one generation. Clearly something else is required to provide evolutionary direction to incremental change and clearly it is circumstances that produce deviations from Hardy-Weinberg equilibrium. So what is it that causes microevolution exactly?

Microevolution, you will recall, is defined as a change in the frequencies of alleles in a breeding population. The question that asks "What causes microevolution?" can be stated more formally as "What are the evolutionary forces? "Just as there are four ways to get genetic variety so there are four evolutionary forces. They are:

(1) mutation,
(2) gene flow,
(3) natural selection, and
(4) genetic drift.[37]

To see them at work, consider again a local breeding population and its gene pool. The allele types and their frequencies will change i.e. undergo microevolution in that population if:

1. New alleles are formed within the population, such as Y mutating to y, called "mutation";
2. New alleles are brought into the gene pool by immigrants into or out of the breeding population, called "gene flow";
3. Individuals with a particular allele reproduce better than those with other alleles, called "natural selection"; or the
4. Relative frequencies of alleles in the pool change by random fluctuation alone. The magnitude of this is appreciable when the population is small and called "genetic drift".

Considering the two-step process of microevolution introduced earlier, mutation and gene flow (which derives ultimately from mutation) represent the factors that provide genetic variability, while natural selection and genetic drift provide particular variant frequencies in the population. Both steps are required if microevolution is to happen. Incidentally, the force of natural selection provides a third reason why mutation alone is insufficient to cause evolution: selection determines whether or not a particular mutation will establish itself in the population or be eliminated. Mutation makes an offering, but it is natural selection that has the last word because it is natural selection that determines its fate in the population. Of the three processes only natural selection is adaptive, the others are random and operate independently of the environment.

The relative intensity of these four forces under different conditions in a population can be quantitated in theoretical models by assigning them values of between 0 and 1. Performing calculations of evolutionary trend by thought experiment is easy. Demonstrating it in biological reality is notoriously difficult. The reason is the difficulty controlling the multitude of variables in an ecosystem (§1.2.3).

The mutation rate (u) of a gene is considered to vary from immutability (zero), to mutation (one) with every generation. Gene flow (m) varies from no migration of new individuals into the population (zero), to complete replacement of the population (one). The particular value of m is given by the ratio (total number new immigrants/total number individuals in breeding population. The selection co-efficient (s) by natural selection varies from no selection (zero), to total allele substitution in a single generation (one). The particular value of s is given by 1 − [reproductive rate of unfavored allele]/[reproductive rate of favored allele]. Thus if the relative rate of reproduction per generation of our alleles above due to natural selection is 100A: 98a, then the value of s = 1 − (98/100) = 0.02, or a 2% selective advantage. Finally, the extent to which random genetic drift is operative depends on the population size being small in comparison with the other forces. Drift is controlling allele frequencies when either population size given by N is ≤ 1/2m or m ≤1/2N, while conversely gene flow m is controlling allele frequencies when N ≥ 1/4m or m ≥ 1/4N. (Both forces are operating in between these extremes (N = 1/4m to 1/2m). The same calculations apply to mutation pressure u. Drift is operating when N ≤ 1/2u and mutation pressure when N ≥ 1/4u.[38]

Biodiversity according to evolution theory is ultimately just the result of accumulated random mutations over geologic time (§1.2.1). It doesn't take a rocket science to realize random changes to a complex system more likely reduce rather than create more complexity, yet living things are the most complicated things we know of. This observation is acknowledged but refuted by evolution advocates, who point to the evolutionary significance of rare beneficial mutations that instead enhance fitness and overwhelm that expectation. The specifics of this at a theoretic level are worth a very close inspection therefore.

It was Cambridge astronomer Sir Fred Hoyle (1915-2001) who first coined the term the "Big Bang" (§13.2). An accomplished mathematician, he also provided a simple mathematical model of the evolutionary process that at least on a theoretical level, impugns its supposed obvious ability to generate complexity to anything close to the degree needed to solve the Conundrum of Life.[39] Recall that in the genetic code a three base sequence (codon) of DNA nucleotides specifies for each of the 20 different amino acids. Since there are four different bases in the alphabet of the genetic code, there are 4^4 or 64 possible codons. There are 61 codons code for the 20 possible amino acids and the remaining three are stop codons that specify termination of the message. The genetic code thus has a redundancy of 3:1. What this means is that the effective mutation rate, as assessed by a change in the amino acid sequence coded by the codon of a gene, can conveniently be approximated by ignoring the third member of the triplet of bases that represents a codon.[40]

Now if the chance of any single base pair being miscopied in a germ cell is set at around 3×10^{-9}, the chance of this DNA point mutation leading to a change in the amino acid sequence of a protein is $(3 \times 10^{-9})+(3 \times 10^{-9})$ or 6×10^{-9}. Using humans as the example, about 95% of the human genome of around 10^9 base pairs is not expressed as protein, which leaves about 10^8 base pairs as genes that code directly for proteins.

It follows that the number of DNA mutations that occur per generation and per chromosome set is $3 \times 10^{-9} \times 10^8$, or about 0.3. Since there are 2 chromosome sets, a maternal and a paternal set, there are $0.3 + 0.3 = 0.6$ mutations arising per individual per generation. Actually the mutation rate typically quoted for humans is 10^{-5}/gene/generation, and since there are about 2×10^5 genes, every human has around 2 new mutations per generation. The theoretical treatment agrees close enough with what is known of biological reality and concedes more to evolution that it deserves. So far we appear to be in good shape on the road to microevolution and thereby to seeing what is really at stake in solving the Conundrum by macroevolution.

Now consider the case of asexual reproduction as occurs in bacteria. The problem emerges when we consider that although there is a "continuing flood" of new mutations every generation, the vast majority of them are detrimental and tend to cause loss rather than gain of function. Thus the defense by evolutionists to this problem we saw above is misdirected.

The real issue for Neo-Darwinism to answer here is not the power of the rare beneficial mutation. It's how to escape an ever accumulating burden of detrimental mutations that threaten to overwhelm it. Simply removing bad mutations by inverse (so-called "revertant") mutations is useless because they occur too infrequently to help - about 6×10^{-9} per chromosome set per generation, compared to an additional 0.3 new mutations per chromosome set every generation.

Thus rather than the logical inevitability of mutations creating increasing complexity the way Neo-Darwinism tells us, we find just the reverse. A reduction in fitness by just under 0.6 in every individual, every generation. As Hoyle explained, "the empirical fact that the degeneration of the somatic cells occurs in not more than a few tens of generations emphasizes the amazing ability of species to preserve information in sex cells...with such widespread evidence of senescence in the world around us, it still seems amazing that so many people think it "obvious" that the biological system as a whole should be headed in the opposite direction, traveling from inferior to superior, traveling as it were from age to youth."[41]

Asexual reproduction, also called binary fission or "budding", can produce no positive evolution therefore since all mutations are transmitted to the next generation and the more frequent bad ones inevitably swamp the rare advantageous ones. The solution to this problem is sexual reproduction, specifically meiosis. By shuffling allelic mutations by crossover and random association of chromosomes, the onslaught of bad mutations is uncoupled from the rare beneficial mutation. This frees up natural selection to promote the advantageous mutation in the population. Natural selection cannot reverse minor deleterious mutations however, and over time they will still compound. There is still a problem. We will need to see if this problem we find in theory here is also a problem in biological practice. At least direct experience namely breeding experiments and the consequences of consanguinity, confirms the theoretical treatment here.

§2.5 DARWINISM'S SECOND STEP: SORTING PHENOTYPIC VARIATION BY NATURAL SELECTION

While step 1 is the purposeless generation of variability in the genotype, step 2 is the nonrandom differential reproduction of variants by natural selection. The non-random aspect is why it's selective of course. Natural selection sculpts the order of life. It's what gives life the appearance of being designed by cumulative selection according to evolution theory. As Richard Dawkins says:

"Natural selection is the blind watchmaker, blind because it does not see ahead, does not plan consequences, has no purpose in view. Yet the living results of natural selection overwhelmingly impress us with the appearance of design as if by a master watchmaker, impress us with the illusion of design and planning."[42]

Mount Improbable is the metaphor he uses for explaining the appearance of design in nature (§17.2). He uses a metaphor of "climbing Mount Improbable" to refer to the attaining of biological complexity (the diversity, unity, complexity and adaptiveness of life) by small steps of purposeless natural selection. Geneticist Massimo Pigliucci at the University of Tennessee explains why evolution accounts for the illusory appearance of design in nature:

> "Natural selection, combined with the basic process of mutation, makes design possible in nature without recourse to a supernatural explanation because selection is definitely nonrandom, and therefore has "creative" (albeit nonconcious) power. Creationists usually do not understand this point and think that selection can only eliminate the less fit; but Darwin's powerful insight was that selection is also a cumulative process – analogous to a ratchet – which can build things up over time, as long as the intermediate steps are also advantageous."[43]

While the generation of genetic variability is the first step of the evolutionary process, the second step of natural selection operates not on that genetic variability but indirectly through the variability in phenotype of a particular individual produced by the genetic variation. In the words of Johns Hopkins paleontologist Steven Stanley, "Genetic components have significance only in terms of phenotypic expression."[44] Differential reproduction represents the "fitness" of an individual. How well an organism survives is not the issue, it's the successful transmission of genes to the next generation that is the arbiter of fitness. What exactly is the cause of the selective discrimination will vary from situation to situation. The specifics of the cause are not defined by the term "Natural Selection", the point being it is not a specific mechanism but a generic description of a process.

A related point is that while natural selection is non-random, the particular mutations that reproduce best under particular natural selection constraints, are only seen to have been adaptive after the fact. In retrospect. Thus the details of exactly what force of natural selection is important and exactly what alleles are adaptive needs to be reconstructed historically, post hoc.

The diversity of life is ultimately the outcome of speciation – the "splitting" of one species into two new different ones, or the "budding" off of a species from an ancestral species which is left intact. In each case the two different species evolve independently thereafter. The end result is reproductive isolation of each new species, but that is the consequence of speciation. What is its cause? This is the burning question of evolution. It is also the trouble that when we descend to the details, we need to rely on more theory to explain the theory. We need to look at evidence from biology ourselves then, and we do that in the next chapter.

§2.6 IN CONCLUSION: SO MUCH FOR THEORY LET'S SEE FACT

There you have it, the Neo-Darwinian theory of evolution. It reduces to a change in allele frequency in a population of a species and it proceeds in two steps. The first step is the generation of genetic variety, and the second step is natural selection of that genetic variety. Genetic variety arises by mutation and gene flow (which is the immigration of new mutant alleles into the population) and genetic drift (the random up-and-down fluctuation in relative allele frequency seen in small populations). Novelty relies ultimately on new mutations however. Natural selection selects which mutations will be established in the population and which will die out to extinction. Those individuals with traits that permit them to survive and produce the most offspring are the "fittest", and those traits spread through the population. This is microevolution and it is not disputed, anywhere.

What is disputed is the notion that over time these microevolution changes summate to generate all of the diversity and complexity and adaptations of living things, and called macroevolution. Richard Dawkins will hear none of this disputing. He praises the extrapolatory power of Darwinism. "Never were so many facts explained by so few assumptions", he says. "Its economy in doing so has a sinewy elegance, a poetic beauty that outclasses even the most haunting of the world's origin myths...In one way or another, all my books have been devoted to expounding and exploring the almost limitless power of the Darwinian principle – power unleashed whenever and wherever there is enough time for the consequences of primordial self-replication to unfold."[45] What evidence he has for this sentiment is the question before us. There is a lot at stake, not just for biology and not just for the philosophy of science (§1.1). How can we tell if microevolution begets macroevolution? We go to that evidence next. So much for theory, let's see fact.

REFERENCES

1. Mayr, E. *One Long Argument*. 3 (Harvard University Press, 1991).
2. Jones, S. Go milk a fruit bat! *New York Review Of Books* July 17 (1997).
3. Mayr, E. *What Evolution Is*. (Basic Books, 2001).
4. Eldredge, N. *Reinventing Darwin: The great debate at the high table of evolutionary theory*. 3 (John Wiley & Sons, 1995).
5. Darwin, C. *The Autobiography of Charles Darwin*. (W.W. Norton, 1969).
6. Fisher, R. A. *The genetical theory of Natural Selection*. (Clarendon Press, 1930).
7. Wright, S. Evolution in Mendelian populations. *Genetics* 16, 97-159 (1931).
8. Dobzhansky, T. *Genetics and the Origin of Species*. (Columbia University Press, 1937).
9. Huxley, J. S. *Evolution : The Modern Synthesis*. (Allen & Unwin, 1942).
10. Mayr, E. *Systematics and the origin of species*. (Columbia University Press, 1942).
11. Simpson, G. G. *Tempo and mode in evolution*. (Columbia University Press, 1944).
12. Simpson, G. G. *The major features of evolution*. (Columbia University Press, 1953).
13. Stebbins, G. L. *Variation and Evolution in Plants*. (Columbia University Press, 1950).
14. Ayala, F. The mechanisms of evolution. *Scientific American* 239, 56-69 (1978).
15. Ruse, M. *Darwinism Defended*. (Addison-Wesley Publishing Company, 1982).
16. Gould, S. J. Is a new and general theory of evolution emerging? *Paleobiology* 6, 119-130 (1980).
17. Darwin, C. *The Origin of Species*. (Random House, Inc., 1998(1859)).
18. Mayr, E. *One Long Argument*. (Harvard University Press, 1991).
19. Stanley, S. M. *Macroevolution, Pattern and Progress*. (Johns Hopkins University Press, 1998).
20. Mayr, E. *Animal species and evolution*. 586 (Belknap Press of Harvard University, 1963).
21. Gilbert, S. F. *Developmental Biology*. 6th edn, (Sinauer Associates, Inc., 2000).
22. Darwin, C. *The Origin of Species* 23 (Random House, Inc., 1998 (1859)).
23. Mayr, E. *Populations, species and evolution*. 12 (Harvard University Press, 1970).
24. Mayr, E. *Systematics and the origin of the species*. 103 (Columbia University Press, 1942).
25. Huxley, L. *The life and letters of Thomas Henry Huxley*. (Macmillan, 1900).
26. Simpson, G. G. *This view of life*. 273 (Harcourt, Brace & World, 1964).
27. Eldredge, N. *Macroevolutionary dynamics:species, niches and adaptive peaks*. 1 (McGraw-Hill Publishing Company, 1989).
28. Mayr, E. *One Long Argument*. 24 (Harvard University Press, 1991).
29. International Human Genome Sequencing Consortium. Finishing the euchromatic sequence of the human genome. *Nature* 431, 931-945 (2004).

30 Alberts, B. *et al. Molecular biology of the cell.* 3rd edn, (Garland Publishing, 1994).
31 Alberts, B. *et al. Molecular Biology of the Cell.* 4th edn, (Garland Science, 2002).
32 Eigen, M. *Steps toward life.* 124 (Oxford University Press, 1992).
33 Drake, J. W., Charlesworth, B., Charlesworth, D. & Crow, J. F. Rates of spontaneous mutation. *Genetics* 148, 1667-1686 (1998).
34 Nachman, M. W. & Crowell, S. L. Estimate of the mutation rate per nucleotide in humans. *Genetics* 156, 297-304 (2000).
35 Gustafsson, A. Mutations, environment and evolution. *Cold Spring Harbor Symposia in Quantitative Biology* 16, 263-281 (1951).
36 Griffiths, A. J. F., Miller, J. H., Suzuki, D. T., Lewontin, R. & Gelbart, W. *An Introduction To Genetic Analysis.* 7th edn, (W.H. Freeman, 2000).
37 Grant, V. *The evolutionary process. A critical review of evolutionary theory.* (Columbia Unversity Press, 1985).
38 Grant, V. *The evolutionary process. A critical review of evolutionary theory.* (Columbia Unversity Press, 1985).
39 Hoyle, F. *Mathematics of evolution.* (Acorn Enterprises LLC, 1999).
40 Hoyle, F. *Mathematics of evolution.* (Acorn Enterprises LLC, 1999).
41 Hoyle, F. *Mathematics of evolution.* 29-30 (Acorn Enterprises LLC, 1999).
42 Dawkins, R. *The Blind Watchmaker.* 21 (W.W. Norton & Company, 1996).
43 Pigliucci, M. Design yes, intelligent no. A critique of intelligent design theory and neocreationism. *Skeptical Inquirer* 25 (2001).
44 Stanley, S. M. *Macroevolution, Pattern and Progress.* 56 (Johns Hopkins University Press, 1998).
45 Dawkins, R. *River Out Of Eden.* xi-xii (Basic Books, 1995).

PART II

THE EVIDENCE FOR EVOLUTION

3. Evolution is a fact!

§3.1   "SIMPLY A FACT"

Two points must be appreciated to have any understanding of the evolution-creation debate. The first point is the debate is not over microevolution but whether microevolution can become macroevolution. The second point is the science and education establishment has no doubt that the first point is true. Evolution is a scientific theory, but it is also scientifically "a fact". In case there's any doubt in your mind, here is a panel of experts to clear the air.

    How much evidence for evolution has accumulated in the one and a half centuries since *The Origin* was published? The verdict in *National Geographic* recently, was that "the evidence for evolution is overwhelming."[1] This was not news. Already back in 1922 the world's largest scientific society, the American Association for the Advancement of Science (AAAS) had declared: "No scientific generalization is more strongly supported by thoroughly tested evidences than is that of organic evolution...The Council of the Association affirms that the evidence in favor of the evolution of man are sufficient to convince every scientist of note in the world".[2]

    Forty years later and almost forty years before the *National Geographic* verdict above, Columbia University and Modern Synthesis co-architect Theodosius Dobzhansky summed up the evidence like this: "The business of proving evolution has reached a stage when it is futile for biologists to work merely to discover more and more evidence of evolution. Those who choose to believe that God created every biological species separately in the state we observe them, but made them in a way calculated to lead us to the conclusion that they are the products of an evolutionary development, are obviously not open to argument."[3] In 1973 Dr. Dobzhansky restated this in the premier biology education journal *The American Biology Teacher*, presumably for the benefit of any teachers who still didn't know: "Let me try to make crystal clear what is established beyond reasonable doubt and what needs further study, about evolution. Evolution as a process that has always gone on in the history of the earth can be doubted only by those who are ignorant of the evidence or are resistant to evidence, owing to emotional blocks or to plain bigotry. By contrast, the mechanisms that bring evolution about certainly need study and clarification. There are no alternatives to evolution as history that can withstand critical examination. Yet we are constantly learning new and important facts about evolutionary mechanisms."[4]

In 1991 Ernst Mayr repeated the consensus of science again: "The basic theory of evolution has been confirmed so completely that modern biologists consider evolution simply a fact."[5] In 2001 he added: "Is evolution a fact? Evolution is not merely an idea, a theory or a concept, but is the name of a process in nature, the occurrence of which can be documented by mountains of evidence that nobody has been able to refute...It is now actually misleading to refer to evolution as a theory, considering the massive evidence that has been discovered over the last 140 years documenting the existence. Evolution is no longer a theory, it is simply a fact."[6]

Evolution textbook author Douglas Futuyma explains why evolution is a scientific fact:"The theory of evolution is a body of interconnected statements about natural selection and the other processes that are thought to cause evolution, just as the atomic theory of chemistry and the Newtonian theory of mechanics are bodies of statements that describe causes of chemical and physical phenomena. In contrast, the statement that that organisms have descended with modifications from common ancestors – the historical reality of evolution – is not a theory. It is a fact, as fully as the fact of the earth's revolution about the sun. Like the heliocentric solar system, evolution began as a hypothesis, and achieved "facthood" as the evidence in its favor became so strong that no knowledgeable and unbiased person could deny its reality. No biologist would think of submitting a paper entitled "New evidence for evolution;" it simply has not been an issue for a century."[7]

In 2002 the Editor-in-chief of *Scientific American* John Rennie gave an update in an article entitled "*15 Answers To Creationist Nonsense*". It turned out to be only a repetition of what had already been repeated. He said: "When Charles Darwin introduced the theory of evolution through natural selection 143 years ago the scientists of the day argued over it fiercely, but the massing evidence from paleontology, genetics, zoology, molecular biology and other fields gradually established evolution's truth beyond reasonable doubt. Today that battle has been won everywhere – except in the public imagination."[8]

Mark Ridley at Oxford University is the author of the premier undergraduate textbook on evolution. He says: "In modern evolutionary biology the question of whether evolution happened is no longer a topic of research, because the question has been answered".[9] The last word comes from his zoology department colleague, Richard Dawkins, in his definitive review of the evidence in *The Greatest Show on Earth*: "Evolution is a fact. Beyond reasonable doubt, beyond serious doubt, beyond sane, informed, intelligent doubt,...The evidence for evolution is at least as strong as the evidence for the Holocaust, even allowing for eye witnesses to the Holocaust. It is the plain truth we are cousins of the chimpanzees, somewhat more distant cousins of monkeys, more distant cousins still of aardvarks and manatees, yet more distant cousins of bananas and turnips...continue the list as long as desired."[10]

One would never suspect there could be any problems with Darwinism at all! The truth of evolution is scientifically self-evident and the tone of the experts, like the title of Dr. Rennie's article, underlines it for any who still aren't sure (§1.6.3). Looking a little closer the view that lay public skepticism of evolution is merely witlessness, flies in the face of their sophistication in other areas of science.[11,12] Why should the public be so stupid only when it comes to evolution? It also doesn't explain why some accomplished academics also say Darwinism is bankrupt.[13-24] Their voices can be dismissed, but not on grounds of ignorance. The explanation must lie elsewhere therefore (§18).

In a phenomenon even more paradoxical, the ability of evolution to maintain its enormous intellectual reach is itself taken as confirmation of its truth. Nobel laureate Sir Peter Medawar (1915-1987) is one who says so. He says: "It is naïve to suppose that the acceptance of evolution theory depends upon the evidence of a number of so-called "proofs"; it depends rather upon the fact that the evolutionary theory permeates and supports every branch of biological science, much as the notion of the roundness of the earth underlies all geodesy and all cosmological theories on which the shape of the earth has a bearing. Thus anti-evolutionism is of the same stature as flat-earthism." So Dr. Medawar defends the paradox with a suspiciously circular argument. Evolution is true because it underlies all of biology so it is true. This looks like dogma resting on, well, nothing. If evolution indeed "supports every branch of biological science" then what supports evolution?

The way the Nobel laureate ridicules objections to evolution rather than refuting them is another paradox. The reason is flat-earthism was once believed by the science establishment with as much conviction as the public, and the illusion lasted for as long as it did exactly because of this sort of bluster and unassailable logic. In fact if the history of science declares anything, it's that flat-earthism is only one of many illusory theories given artificial life support by circular reasoning long after brain death. Like all who ignore history he is bound to repeat it. Illusions are always true until they are exposed, and when they are the only surprise is how they could have lasted for so long.

Robert Pennock at Michigan State University is a philosopher of science and another undisputed expert in this field. He agrees with Dr. Medawar: "It is the great explanatory power of evolutionary theory – that it accounts for the facts so well – that testifies to its truth," he says.[25] The point here, and the problem, is not that a theory in science is consistent with the evidence. We expect no less. It's that a theory with an intellectual sweep so great that it "permeates and supports every branch of biological science," to use Medawar's words, might be so great as to define the interpretation of its own data. And so accommodate almost any data as evidence in confirmatory, but circular logic. We are here to decide if that concern is reasonable or not. Take a look.

We already know macroevolution is a process that can only be inferred, yet it's considered as certain as heliocentricity which can be proved directly (§1.4). In this disparity perhaps you might sense the whiff of fish already. If evolution is an inference, how is it a "fact"? Because of the evidence of course. Ernst Mayr again:

> "Evolution is a historical process that cannot be proven by the same arguments and methods by which purely physical or functional phenomena can be documented. Evolution as a whole, and the explanation of particular evolutionary events, must be inferred from observations. Such inferences subsequently must be tested again and again against new observations, and the original inference is either falsified or considerably strengthened when confirmed by all of these tests. However, most inferences made by evolutionists have now been tested successfully so often that they are accepted as certainties."[26]

That's why evolution is "a fact". The evidence is so convincing. Okay, no surprise so far.

The surprise is that the evidence for evolution is so convincing it can't be falsified. How can it be if the only alternative to natural origins, a creation, is already scientifically false for invoking a supernatural? The "tests" he speaks of can only confirm evolutionary inferences in one way or another and so render evolution further a fact. He is not ashamed to say so either. Any alternative makes "no sense," he says:

> "To be sure, at first, the thought that life could have evolved was merely a speculation. Yet, beginning with Darwin in 1859, more and more facts were discovered that were compatible only with the concept of evolution. Eventually it was widely appreciated that the occurrence of evolution was supported by such an overwhelming amount of evidence that it could no longer be called a theory. Indeed, since it was as well supported by facts as was heliocentricity, evolution also had to be considered a fact, like heliocentricity...It shows how remarkably congruent are the conclusions drawn from the most diversified branches of biology, which all support evolution. Indeed, these findings would make no sense in any other explanation."[27]

Botanist Verne Grant at the University of Texas denies evolution is a dogma. The reason is because evolution rests on multiple independent lines of evidence in a phenomenon called consilience. He says: "It should be noted, finally, that the case for evolution rests, not on one or two lines of evidence alone, but on the concurrent testimony of several independent classes of facts".[28] The questions before us become clear: What are the lines of evidence for evolution? How "independent" are they? And what, exactly, is the evidence for (macro)evolution?

§3.2   A CATALOGUE OF THE EVIDENCE

The evidence for evolution is examined in *Part II*, and Table 3.1 will serve as a catalogue for the entire inquiry. The sources on that table are among the foremost authorities in biology. Wherever possible and even to tediousness they are quoted to ensure there be no chance of a misunderstanding.[10,27-35] As your guide I will make every effort to display this evidence in an impartial way, but be warned. From this same evidence that biologists say is unambiguous evidence of evolution, I will arrive at an opposite conclusion. That answer is compelled by the data and I will show you why. Why it is not obvious to these far more qualified experts in biology is a question begged, and it will be the focus of *Part IV*. You will have make up your own mind as we go but I challenge you, no I dare you, to disagree. On objective grounds that is. I do not care to hear if it is on other grounds.

The reasoning for evolution begins with a premise, here called the Naturalistic Premise. It goes like this:

The Naturalistic Premise: The existence of a supernatural is not a question science can address because supernatural events are neither measurable nor predictable. Anything to do with the supernatural is not science. Science must proceed without the possibility of a supernatural. Even if there were a supernatural, it has no detectable influence in nature or in the Universe.

Australian physicist and popular science author Paul Davies says all that much more simply, as "it is the job of science to solve mysteries without recourse to divine intervention."[36] What's the result? It restricts scientific solutions to metaphysical questions to natural causes (§1.6.2, §19). Since a supernatural is ruled scientifically impossible, all observations support the only alternative and so become confirmatory in one way or another. Evolution theory is modified when necessary to explain any seeming inconsistencies from new data, but as a whole it can never be rejected. This follows as certainly as if it were a law of science (§19). If this sounds ridiculous, that's because it is ridiculous. Empiricism has been abandoned for naturalism. It is to suppress the truth, deliberately. Abandon empiricism and sooner or later you always get absurdity though. If the origins of the Earth and all its life cannot be by a creation, then it has to be by evolution somehow (§20). Understanding what's at the core of the evolution-creation debate is not difficult. The astonishing thing is so basic reasoning flaw could lie at the heart of twenty-first century science. The question is how exactly that could happen (§18).

The Naturalistic Premise is all-important because what is not stated is what it necessarily implies: evolution of some kind. We've not even started to see the evidence for evolution but there's no surprise where we'll end up because there's only one answer left. Evolution has nothing to do with the evidence because it's already true before we get to the evidence. No wonder it is "a fact"!

There are twelve main lines of evidence (*LOE*) offered for macroevolution. The first is the most important. It's the evidence for microevolution:

LOE# 1: Species can be directly observed to undergo changes in morphologic traits over durations of relatively few years. Species are not fixed in their characteristics therefore.

When combined with the Naturalistic Premise, microevolution is taken to lead to two inferences. The first extrapolates forwards in time which we will call the Macroevolution Inference or Inference 1, and the other extrapolates backwards which we will call the Common Ancestry Inference or Inference 2, and both imply macroevolution. Here they are:

Inference 1 (The Macroevolution Inference): By inference the same processes causing the observable (microevolutionary) changes to species morphology can be extrapolated forward over time to have caused changes so great as to generate all of the diversity and complexity of life.

Inference 2 (The Common Ancestry Inference): By inference, the same processes causing the observable changes to species morphology can be extrapolated backward to show that all life is ultimately descended from a single common ancestor.

The appearance of design in nature is explained as having been generated by natural forces alone. The impressive designed appearance of objects in nature is reduced into small unimpressive cumulative steps like those directly observable in microevolution, by extrapolating them forwards and backwards in time.

This explanation Richard Dawkins calls "Climbing Mount Improbable".[37] The improbable high mountain of design is scaled by the summated small steps of microevolution (§17.2). Harvard biochemist and Nobel laureate George Wald explained the idea in *Scientific American* even more memorably: "The time with which we have to deal is of the order of two billion years…Given so much time, the "impossible" becomes possible, the possible, probable, and the probable virtually certain. One has only to wait: time itself performs miracles."[38]

With the Naturalistic Premise plus microevolution plus the two evolution inferences, the paradigm of evolution is fully erected. The remaining 11 lines of evidence (*LOE*) support that paradigm "scientifically", confirming a macroevolution inference as both a historical and biological truth, and as even "a fact". Here are the main lines of that evidence:

LOE #2: Fossils are evidence for evolution because they are found in evolutionary series and specifically as time-ordered sequences of morphologic intermediates beginning with more primitive ancestors and leading to the species found alive today.

LOE #3: Fossils are evidence for evolution because the chronologic order in which the major vertebrate groups appear in the record (fish, amphibian, reptile, bird and mammal) conforms to an expected evolutionary series.

LOE # 4: The striking morphologic similarities among totally different species called "homologies" are evidence for evolution because a common ancestry explains their presence as inherited traits. One example is the five fingered limb found in such diverse creatures as lizard, bird, whale and human and used for such unrelated activities as walking, flying, swimming and writing (Fig.7.2).

LOE #5: Just as there are striking morphologic similarities between otherwise different organisms, so there are striking similarities in their constituent molecules. Molecular similarities are evidence for evolution because a common ancestry explains them. One example is the genetic code which is essentially the same in all of life from microbes to man. Another is biochemistry itself because similar substrates, metabolic pathways and chemical reactions are used by all living things. This conservation of traits is explained by an inheritance from a single common ancestor.

LOE #6: The degree of similarity between molecular structures is explained by how closely related organisms are in their phylogenic ancestry. Molecular systematics is able to perform phylogenetic reconstructions in confirmation of the truth of evolution.

LOE #7: Morphologic similarities among embryos are evidence for evolution because embryos and larvae of different taxa have similar characteristics at early stages of development and only later show the different special characteristics of the higher taxon to which they belong. This reflects both their common ancestry and their phylogenetic relationships. For example vertebrates have very similar embryos although they are very different as adults, because they share a common evolutionary ancestor.

LOE #8: The similar genetic mechanisms employed by organisms to control development (e.g. segmentation and orientation) are evidence for evolution because they are shared by organisms as different as fruit flies and humans. This is explained by inheritance from an ancient common ancestor of both flies and humans by evolution.

LOE #9: The appearance and subsequence loss of structures during embryologic development that are retained in the adult stages of other creatures is the repetition (or "recapitulation") of ancestral traits during the development (or "ontogeny") of descendant species. An example is the gill arches found in embryos of all terrestrial vertebrates that are never used for breathing, but which reflect the evolutionary ancestry from fish of amphibians and reptiles and mammals.

LOE #10: Atrophied or non-functional ("vestigial") organs are evidence for evolution because they represent the inherited remnants after disuse of once well-developed functional structures in ancestral species with this trait.

LOE #11: Fossil, artifact and molecular data demonstrate the process of human evolution from an ape-like ancestor in confirmation of evolution.

LOE #12: The geographical distribution of plants and animals on the Earth is explained by evolution.

An overview of the evidence

Table 3.1. The evidence for evolution. The 12 main lines (*LOE#1-#12*) as advanced by leading authorities. The scientific "fact" of evolution rests on the authority of a premise of naturalism here called The Naturalistic Premise, the direct observations of microevolution in artificial selection and natural selection, and two inferences here called the Macroevolution Inference and the Common Ancestry Inference declaring microevolution accumulates to make macroevolution over time.

Line of Evidence	CITATION
THE NATURALISTIC PREMISE *Evidence leading to the conclusion of a supernatural agency can never be science.*	"Science is a way of knowing about the natural world. It is limited to explaining the natural world through natural causes."[31] *(The National Academy of Sciences)* "Science deals only with the natural world."[39] *(Kenneth Miller, Brown University).* "Explanations employing nonnaturalistic or supernatural events, whether or not explicit reference is made to a supernatural being, are outside the realm of science and not part of a valid science curriculum."[40] *(National Association of Biology Teachers)* "Scientific arguments employ only observations anyone can make, as distinct from private revelations, and consider only natural causes, as distinct from supernatural origins. Indeed, two good criteria to distinguish scientific from religious arguments are whether the theory invokes only natural causes, or needs supernatural causes as well, and whether the evidence is publicly observable or requires some sort of faith. Without these two conditions, no constraints are placed on the argument."[41] *(Mark Ridley, Oxford University)* "[A]ny 'theory' that explains phenomena by recourse to the actions of an omnipotent, omniscient supreme being, or any other supernatural omnipotent entity, is a nonscientific theory...Because such a theory cannot be challenged by any observation, it is not scientific...In dealing with questions about the natural world, scientists must act as if they can be answered without recourse to supernatural powers."[42] *(Douglas Futuyma, State University of New York)* "The Creationists believe that the world started miraculously. But miracles lie outside of science, which by definition deals only with the natural, the repeatable, that which is governed by law"[43]. "the essential characteristics of science are: (1) It is guided by natural law; (2) It has to be explanatory by reference to natural law..."[44] *(Michael Ruse, Florida State University)* "No serious scientific discussion of any topic should include supernatural explanations, since the basic assumption of science is that the world can be explained entirely in physical terms, without recourse to godlike entities."[45] *(Massimo Pigliucci, University of Tennessee)*

	"We humans can directly experience that material world only through our senses, and there is no way we can directly experience the supernatural. Thus, in the enterprise that is science, it isn't an ontological claim that a God...does not exist, but rather an epistemological recognition that even if such a God did exist, there would be no way to experience that God given the impressive, but still limited, means afforded by science. And that is true by definition...It's simply a matter of definition – of what is science and what is not...that is all science is: a set of rules and an accumulated set of ideas, some more powerfully established than others, about the nature of the material world. By its own rules, science cannot say anything about the supernatural...The argument boils down simply to this: we can invoke a naturalistic process, evolution, for which there is a great deal of evidence...Or we can say, simply, that some Creator did it and we are, after all, complex machines like watches. The analogy is as meaningless as that: it proves nothing. It could even be true, but it cannot be construed as science, it isn't biology, and in the end it amounts to nothing more than a simple assertion that naturalistic processes automatically cannot be considered as candidates for an explanation of the order and complexity that we all agree we do see in biological nature."[46] *(Niles Eldredge, American Museum of Natural History)*
THE MACROEVOLUTION INFERENCE *Microevolution can be extrapolated to macroevolution given enough time.*	"As we extend the argument from small scale observations...to the history of all life we must shift from observation to inference. It is possible to imagine, by extrapolation, that if the small-scale processes we have seen were continued over a long enough period they could have produced the modern variety of life. The principle behind this is called 'uniformitarianism'"[30] *(Mark Ridley)*
	"Thus we know from direct observation and experiment that the ingredients of evolutionary change are real and potent, just as a geologist knows that erosion is a fact of physical geology. Over the course of millions of years, it is inconceivable that erosion and other observable geologic mechanisms should fail to create great gorges and canyons; and it is just as difficult to imagine that mutation and natural selection should fail to create great changes in species over vast periods of time...It is a fact that species are related to each other by descent."[29] *(Douglas Futuyma)*
	"The extrapolation from small-scale variation [microevolution] to large-scale evolution [macroevolution] is well justified. In evolutionary parlance, Evo Devo [§9 1] reveals that macroevolution is the product or microevolution writ large."[47] *(Sean B. Carroll, University of Wisconsin)*
	"One of the major tenets of the Modern Synthesis has been that of extrapolation; the phenomena of macroevolution, the evolution of species and higher taxa, are fully explained by the microevolutionary processes

	that gives rise to varieties within species. Macroevolution can be reduced to microevolution. That is, the origins of higher taxa can be explained by population genetics."[48] *(Scott F. Gilbert, Swathmore College; John M. Opitz University of Utah; Rudolf A. Raff, University of Indiana)*
THE COMMON ANCESTRY INFERENCE *All life is descended from a universal common ancestor.*	"We cannot observe the orders of mammals diversify from mammal-like reptiles, nor the great diversity of living things flower forth from a single Precambrian ancestor. But we can test the predictions of the hypothesis that living things have a common ancestry."[29] *(Douglas Futuyma)* "Darwin's theory of common descent postulates that every group of organisms is derived from an ancestral group...In theory it should be possible to establish the ancestry of every group of fossil or still living organism...What is most gratifying is that all findings are consistent with Darwin's theory of common descent."[49] *(Ernst Mayr)*
LOE #1 *Microevolution can be directly observed.*	"Evolutionary changes within populations and some modes of species formation are established by direct observational and/or experimental evidence"[28] *(Verne Grant, University of Texas)* "We have now seen that evolution can be observed directly on a small scale: the extreme forms within in a species can be as different as two distinct species, and, in both nature and experiments, species will evolve into forms highly different from their starting point."[30] *(Mark Ridley)*
LOE #2 *Fossils are found in evolutionary series*	"The fossil record thus provides consistent evidence of systematic change through time – of descent with modification."[32] *(National Academy of Sciences)* "Many groups of organisms with good fossil records show a succession of forms through geological time. In some cases transitional forms between two different major groups are preserved as fossils."[28] *(Verne Grant)* "The most convincing evidence for the occurrence of evolution is the discovery of extinct organisms in older geological strata. Some of the remnants of the biota that lived at a given geological period in the past are embedded as fossils in the strata laid down at that period. Each earlier stratum contains the ancestors of biota fossilized in the succeeding stratum. The fossils found in the most recent strata are often very similar to still living species or, in some cases, even indistinguishable. The older the strata are in which a fossil is found – that is, the further back in time – the more different the fossil will be from living representatives."[27] *(Ernst Mayr)*

	"We predict that if evolution has occurred, old rocks will forever lack fossils of many species not yet evolved. We have never found a fossilized mammal or flowering plant in Silurian deposits, and we never will. We should and do find, on the contrary, that the early rocks contain organisms such as lungfishes and cockroaches that are believed to be ancient groups on the basis of entirely independent evidence from the anatomy and biochemistry of their living representatives. Similarly, groups such as elephants and ants which the anatomy of living species tell us are more recently evolved do not appear except in more recent geological formations. On occasion we should and do find gradual transformations from primitive ancestors to modified descendants in exceptionally good fossil deposits."[29] *(Douglas Futuyma)*
LOE #3 *The fossil record of the major vertebrate groups conforms to an evolutionary series.*	"The order of succession of major groups in the fossil record is predicted by evolution, and contradicts the separate [creationist] origin of the groups."[30] *(Mark Ridley)* "What is particularly convincing about fossil animal series is that each fossil type is found at the time level at which one ought to expect it. For instance, modern mammals began to evolve...at the beginning of the Paleocene (60 million years ago). No modern mammal, therefore, should be found in strata that are 100 or 200 million years old, and indeed none has ever been found... At the turn of the twenty-first century, the sequence of accurately dated fossils has documented evolution in the most convincing manner."[27] *(Ernst Mayr)*
LOE #4 *Morphologic similarities are explained by evolutionary ancestry.*	"It is quite remarkable how successful comparative morphology can be in the reconstruction of missing steps in an evolutionary sequence...Homology cannot be proven; it is always inferred. Homology is due to the partial inheritance of the same genotype from the common ancestor. This is the reason why homology exists not only for structural characters, but for any inheritable feature, such as behavior."[27] *(Ernst Mayr)* "If species have descended from common ancestors, homologies make sense...Homologous similarities between species (understood as similarities that do not have to exist for any pressing functional reason) suggest that the species descended from a single common ancestor...The argument from homology is inferential...You must understand some functional morphology, or molecular biology, to appreciate that tetrapods would not share the pentadactyl limb, or all species share the genetic code, if they originated independently."[30] *(Mark Ridley)*

LOE #5 *Molecular similarities are explained by evolutionary ancestry.*	"Molecular homologies, such as the genetic code, now provide the best evidence that life has a single common ancestor."[30] *(Mark Ridley)* "A modern line of evidence, not available in Darwin's time, is the close similarity of the biochemical composition and molecular structure of homologous proteins in members of different related families or orders. Good examples are the homologous forms of hemoglobin and of cytochrome c in humans, apes and monkeys."[28] *(Verne Grant)* "All species, from bacteria to mammals and trees, use the same genetic code, in which the same nucleotide sequences code for the same amino acids. All species use "left-handed" amino acids…From a chemical point of view, this universality is not necessary….The only possible reason for these chemical universalities is that living things got stuck with the first system that worked for them.[29] *(Douglas Futuyma)* "The code used to translate nucleotide sequences into amino acid sequences is essentially the same in all organisms. Moreover, proteins in all organisms are invariably composed of the same set of 20 amino acids. This unity of composition and function is a powerful argument in favor of the common descent of the most diverse organisms…In 1959 scientists…determined the three-dimensional structures of two proteins that are found in almost every multicelled animal: hemoglobin and myoglobin…During the next two decades, myoglobin and hemoglobin sequences were determined for dozens of mammals, birds, reptiles, amphibians, fish, worms and mollusks…the differences between sequences from different organisms could be used to construct a family tree of hemoglobin and myoglobin variation among organisms. This tree agreed completely with observations derived from paleontology and anatomy about the common descent of the corresponding organisms. Similar family histories have been obtained from…other proteins, such as cytochrome c (a protein engaged in energy transfer) and the digestive proteins trypsin and chymotrypsin. The examination of molecular structure offers a new and extremely powerful tool for studying evolutionary relationships."[32] *(National Academy of Sciences)*
LOE #6 *Molecular systematics can be applied to reveal phylogeny in confirmation of evolution.*	"[E]ither directly by examining the DNA, or indirectly by examining the proteins… investigations give the same answer that evolutionary taxonomists had found on the basis of morphology: pigs and cows are more similar to each other than to dogs; these together are more similar to each other than primitive mammals like kangaroos; all mammals are more similar to each other than reptiles; and so on."[29] *(Douglas Futuyma)* "[M]olecules evolve just the same as do somatic structures. On the whole, the more closely related two

EVOLUTION IS A FACT!

	organisms are, the more similar are their respective molecules…By comparing homologous genes and other homologous molecules of different organisms, one can determine the degree of their similarity…What is gratifying is the fact that when a phylogeny based on morphological or behavioral characteristics is established, it is usually found to be essentially the same as a phylogeny based exclusively on molecular characteristics."[27] *(Ernst Mayr)*
	"As the ability to sequence the nucleotides making up DNA has improved, it also has become possible to use genes to reconstruct the evolutionary history of organisms. Because of mutations, the sequence of nucleotides in a gene gradually changes over time. The more closely related two organisms are, the less different their DNA will be…The concept of a molecular clock…determines evolutionary relationships between organisms, and it indicates the time in the past when species started to diverge. The evidence for evolution from molecular biology is overwhelming."[32] *(National Academy of Sciences)*
LOE #7 Similarities in embryonic development are explained by ancestry.	"[A]ll vertebrates have similar embryos…"[29] "Embryology…reveals that related species are similar in ways that make no adaptive sense…This makes sense if they carry in their genes the imprint of their history"[29] *(Douglas Futuyma)*
	"Perceptive anatomists observed in the eighteenth century that the embryos of related kinds of animals are often far more similar to each other than are the adult forms. An early human embryo, for instance, is very similar not only to embryos of other mammals (dog, cow, mouse), but in its early stages even to those of reptiles, amphibians and fishes. The older the embryo, the more it shows the special characters of the higher taxon to which it belongs."[27] *(Ernst Mayr)*
	"If you viewed an early human embryo and an early pig embryo side by side…you'd have a hard time telling them apart. The more alike the development of two organisms is, the more closely related they are thought to be. Thus the pig and human must be closely related. As the pig and human progress in their development, their patterns become more and more different"[50] *(Raymond F. Oram, Biology textbook author)*
	"Why do we find that the embryos of organisms very different as adults – humans and chickens and dogs, for instance – have embryos that are very similar? Simply because they are descended from common ancestors. There is no other good reason."[51] *(Michael Ruse, Florida State University)*
	"The early stages, or embryos, of many animals with backbones are very similar…it is clear that the same groups of embryonic cells develop in the same order and in similar patterns to produce the tissues and organs of all

An overview of the evidence

	vertebrates. These common cells and tissues, growing in similar ways, produce the homologous structures discussed earlier".[52] *(Kenneth Miller)* "Early in development human embryos and embryos of all other vertebrates are strikingly similar"[53] *(George B. Johnson, Washington University)*
LOE #8 *Similarities in genetic control of embryogenesis are explained by evolutionary ancestry.*	"The discovery of the ancient genetic toolkit [*Hox* and related genes] is irrefutable evidence of the descent and modification of animals, including humans, from a simple common ancestor."[54] *(Sean B. Carroll)* "Embryology, the study of biological development from the time of conception, is another source of evidence for common descent...a wide variety of organisms from fruit flies to worms to mice to humans have very similar sequences of genes that are active early in development. These genes influence body segmentation or orientation in all these diverse groups. The presence of such similar genes doing similar things across such a wide range of organisms is best explained by their having been present in a very early common ancestor of all these groups."[32] *(National Academy of Sciences)* "The similarity of developmental control systems found across a wide variety of animal phyla is extraordinary...The remarkable similarities represent an array of developmental mechanisms that were assembled early in metazoan history and thus were present in the common ancestors of the phyletic lineages that diverged subsequently."[55] *(James Valentine, University of California Berkeley)* "...ancestral structures serve as embryonic "organizers" in the ensuing steps of development...the genetic developmental program has no way of eliminating ancestral stages of development and is forced to modify them during the subsequent steps of development in order to make them suitable for the new life-form of the organism. The anlage of the ancestral organ now serves as a somatic program for the ensuing development of the restructured organ."[27] *(Ernst Mayr)*
LOE#9 *Embryonic structures are recapitulated during ontogeny to reveal phylogeny.*	"[E]mbryos of different species so often resemble each other in their early stages and, as they develop, seem sometimes to replay the steps of evolution."[56] *(Bruce Alberts, President National Academy of Sciences & James Watson, Nobel laureate* "[I]n certain features, as in the gill pouches, the mammalian embryo does indeed recapitulate that ancestral condition...embryonic structures are found in thousands of cases to be indicative of their ancestry...This is same reason why all terrestrial vertebrates (tetrapods) develop gill arches at a certain stage in their ontogeny. These gill-like structures are never used for breathing, but instead are drastically

restructured during the later ontogeny and give rise to so many structures in the neck region of reptiles, birds, and mammals...What is recapitulated is always particular structures, but never the whole adult form of the ancestor."[27] *(Ernst Mayr)*

"For example all vertebrates begin development in the same way, looking rather like an embryonic fish...Embryonic stages don't look like the adult forms of their ancestors, as Haeckel claimed, but like the embryonic forms of ancestors. Human fetuses, for example, never resemble adult fish or reptiles, but in certain ways they do resemble embryonic fish and reptiles...Embryos still show a form of recapitulation: features that arose earlier in evolution often appear earlier in development...All vertebrates begin development looking like embryonic fish because we all descended from a fishlike ancestor with a fishlike embryo."[57] *(Jerry Coyne, University of Chicago)*

"Early in development, human embryos are almost indistinguishable from those of fishes, and briefly display gill slits."[58] *(Douglas Futuyma)*

"Both von Baer and Haeckel captured elements of the truth; evolution leads both to embryonic divergence and, in some lineages, to a lengthening of the ontogenetic trajectory leading to more complex adult phenotypes with greater numbers of cells, their embryos passing through simpler, quasi-ancestral forms." [59] *(Wallace Arthur, National University of Ireland)*

LOE #10
Vestigial organs are evidence of evolution.

"Many organisms have structures that are not fully functional or not functional at all... Such vestigial structures are the remnants of structures that had been fully functional in their ancestors abut are not greatly reduced owing to a change in niche utilization...They are informative by showing the previous course of evolution."[27] *(Ernst Mayr)*

"Every organism has such vestiges of structures that can only be the useless remnants of past adaptations."[60] *(Douglas Futuyma)*

"Some members of a major group often possess an organ that is atrophied or non-functional...These structures are interpreted as reduced vestiges of their well-developed homologues in other members of the same major group"[28] *(Verne Grant)*

"We humans have many vestigial features proving that we evolved...our appendix is simply the remnant of an organ that was critically important to our leaf-eating ancestors, but of no real use to us...We have a vestigial tail: the coccyx...It's what remains of the long, useful tail of our ancestors" [35] *(Jerry Coyne, University of Chicago)*

LOE #11 Fossil, artifact and genetic data demonstrate evolution of humans from an ape-like ancestor.	"The evidence is just too good...the fossil record of human evolution is one of the very best, most complete, and ironclad documented examples of evolutionary history that we have assembled in the 200 years or so of active paleontological research."[61] *(Niles Eldredge, American Museum of Natural History)* "The African record of human evolution is just one of the many cases in which transitional series from ancestors to more modern descendants have come to light."[29] *(Douglas Futuyma)* "The australopithecine ancestors of man also form a rather impressive transition from a chimpanzee anthropoid stage to that of modern man."[27] *(Ernst Mayr)* "Modern man is the culmination of a series of evolutionary trends that can be traced back to the non-human primates."[62] *(Verne Grant)* "the fossil record, comparative anatomy and development. These disciplines....define the magnitude and nature of evolutionary change in the human lineage."[63] *(Sean B. Carroll)* "We are apes descended from other apes, and our closest cousin is the chimpanzee, whose ancestors diverged from our own several million years ago in Africa. These are indisputable facts...Looking at the whole array of bones, then, what do we have? Clearly, indisputable evidence for human evolution from apelike ancestors."[64] *(Jerry Coyne)*
LOE #12 Geographic distributions of animals and plants are consistent with evolution.	"Biogeography has also contributed evidence for descent from common ancestors...Evolutionary theory explains that biological diversity results from the descendants of local or migrant predecessors becoming adapted to the diverse environments. This explanation can be tested... Wherever such tests have been carried out, these conditions have been confirmed."[32] *(National Academy of Sciences)* "Evolution also helped explain another great puzzle of biology, namely, the reasons for the geographic distribution of animals and plants...For a creationist there is no rational explanation for distributional irregularities, but they are completely compatible with a historical evolutionary explanation."[27] *(Ernst Mayr)* "Many genera, families, and other groups of medium taxonomic rank are confined to one geographical region...Meanwhile another isolated geographical region harbors a distinct organic group. Thus the hummingbirds are an American family of birds and the Hawaiian honeycreepers a Hawaiian family. The logical inference in such cases is that the species belonging to each group, or at least many of them, evolved in the region of their present abundance and diversity."[28] *(Verne Grant)*

> "[K]nowing that the continents have been drifting apart ever since the Permian, we predict that groups which supposedly evolved late in evolutionary history should be more restricted to one of a few continents than groups which supposedly evolved early. This is in fact the case. The fossils of mammal-like reptiles are broadly distributed over all the continents, and so are most of the primitive mammals such as marsupials. But orders and families of mammals such as carnivores and hoofed mammals, which are so similar to each other that they are thought to have diverged more recently, are more restricted in distribution…The effects of history are abundantly evident in the distribution of organisms. There are no native land mammals in Hawaii, not because they cannot survive there, but because they evolved on continents and couldn't cross the Pacific."[29] *(Douglas Futuyma)*

Data taken to be demonstrating the truth of the Macroevolution and Common Ancestry Inferences is either of the directly supportive kind listed above and which Douglas Futuyma calls "positive" evidence, or what he calls "negative" evidence[29] or Jerry Coyne calls "bad design".[35] "Negative" evidence is evidence that refutes a design argument and so supports evolution indirectly since ID is the only alternative to evolution. The main lines of "positive" evidence from the science are found in Table 3.1. Their claims are crystal clear as you can see and we will review the evidence for each in chapters 4-11 to see if their claims are true. But first, what is the "negative" or "dysteleology" evidence for evolution?

Dr. Futuyma tells us what it is. It's "that the natural world does not conform to our expectation of what an omnipotent, omniscient, truthful Creator would have created."[60] What that would make him judging the world, he does not say. Arguments refuting a creation on grounds it could not have happened because this or that human mind has a better way of doing it if they were Creator, or because observations in biology "make no sense" to that mind apart from evolution, are not dealt with here for the obvious reason that if one is to test and refute ID, its reasoning needs to be seriously engaged. ID is not creationism. It claims to rest on the evidence of science (§1.2). Test it. "Negative" or "bad design" arguments for the truth of evolution do not, and so are exactly that. Negative and bad and cheap. A supernatural creation can be refuted very simply. Refute a claim of some supposed evidence of ID in nature, not play make-believe at second guessing God.

Another indirect argument, also frequently used to argue for macroevolution's truth, is ID and creationism alike are necessarily false in as they are without a mechanism. Richard Dawkins' formulation of this objection was addressed earlier, but Mark Ridley says it this way: "Another powerful reason why evolutionary biologists do not take creationism seriously is that creationism offers no explanation for adaptation."9 (§1.6.2) His argument is just as specious and the reason is if there was a creation of the world, acknowledging it would be the truth. Knowing the mechanism is irrelevant to it being true. Seeking objective truth is what science is. Science does not arbitrate truth depending on whether it can first deliver a mechanism. The reasoning flaw goes a lot deeper than this of course. Reasoning by methodological naturalism as science demands every mechanism be naturalistic so they couldn't be permitted anyway. That's why this illogicality can be argued by experts with a straight face.

Something obvious has been omitted from the discussion so far and ironically, given Darwin's choice of title, it wasn't addressed in *The Origin of Species* either. It's the answer to the question of where life came from in the first place. Where did the forces of evolution come from? If evolution is a scientific fact, there ought to be a scientific answer!

And so to coherently argue a naturalistic cause for everything, and not maintain evolution was supernaturally kick-started, we need to explain how life began naturally and how the essential requirements for evolution got there naturally. Evolution itself cannot be the answer because evolution presupposes the existence of life or as Theodosius Dobzhansky once said, "prebiological natural selection is a contradiction in terms" (§1.6.1). We need to explain much more than this though. Not just the origin of life and where evolution came from, but where matter came from, where the physical anthropic properties of the Universe came from (i.e. the origins of a habitat conducive to living) and where the laws of science that describe how matter behaves in the Universe came from. So for an entirely naturalistic view that is the view of science most everywhere, two further premises in addition to the Naturalistic Premise (Premise 1) are therefore necessary (Table 3.2):

Premise 2 (The Spontaneous Generation Premise): Living things arose spontaneously from non-living things.

Scientifically speaking, life sprang into being from inorganic chemicals somehow. See Table 3.2 for other ways of saying this, but here is Douglas Futuyma's way: "We will almost certainly never have direct fossil evidence that living molecular structures evolved from nonliving precursors. Such molecules surely could not have been preserved without degradation. But a combination of geochemical evidence and laboratory experiment shows that such evolution is not only plausible but almost undeniable."[65] We need to explain the origins of the chemicals from which life came too. Thus we need still one more premise:

Premise 3 (The Everything From Nothing Premise): The universe was once formed from singularity entirely naturalistically, and the singularity ultimately from nothing.

"Nothing will come of nothing," said King Lear but you don't have to be a king to know this applies as much to a Universe as ever it did to Cordelia. Since we are here and the universe is here, there could not have been nothing before! "Something" must have pre-existed but what exactly? If you think this necessarily indicates a Creator, you would be wrong in science class. There are theories in science seeking to explain the spontaneous creation of matter out of nothing and they are taken very seriously in intellectual circles (§13.1).[68,69] Ninety-five percent of National Academy of Science biologists are atheist and a similar number of the Royal Society membership, that's how seriously (§18.1).[66,67]

Take for instance the theory held by someone we all know and love, Dr. Stephen Hawking. He is the Lucasian professor of mathematics at Cambridge University. It's the same chair that old creationist Isaac Newton once had. In 1997 Dr. Hawking told the world: "Recent developments have encouraged us to believe that it may be possible to find laws that hold even at the creation of the universe. In that case, everything in the universe would be determined by the laws of science".[70] He explained this proposal further in his widely acclaimed and immensely popular book *A Brief History of Time*. In a nutshell the claim is space and time have no boundary. The effect of that, is then the "universe would be completely self-contained and not affected by anything outside itself. It would neither be created nor destroyed. It would just BE...What place, then, for a creator". This still requires some input somewhere, as King Lear well knew. Dr. Hawking tells us where: "this idea...is just a proposal: it cannot be deduced from any other principle".

I should tell you Dr. Hawkins' idea also invokes "imaginary time" and in his model, in his words, "it is meaningless to ask: which is real, "real" or imaginary time? It is simply a matter of which is the more useful description." It would seem real time is much more useful than he gives credit. Why believe his model at all? The Everything From Nothing Premise. That's the imperative.

Another theory based on this premise comes from Russian cosmologist Andrei Linde at Stanford. It's that our universe is only one of a near infinite number of other parallel universes which eternally self-reproduce themselves.[71] But why believe in so many other universes, much less a near infinite number of them that can also reproduce themselves somehow, when there is only one Universe that we know of or can know of? This premise. Those who claim the high ground of objectivity and "science" are convinced these extraordinary claims are yet reasonable. All can be ultimately explained without need for supernatural design, even if it cannot be done right now. And yet to borrow Carl Sagan's words: "extraordinary claims require extraordinary evidence," not extraordinary dogma and certainly not by scientists. You at least keep an open mind. We have lots of looking still to do.

| THE SPONTANEOUS GENERATION PREMISE
Living things arose spontaneously from non-living things. | "[L]ife in fact is probably a universal phenomenon, bound to occur wherever in the universe conditions permit and sufficient time has elapsed."[72] "Spontaneous generation was disproved one hundred years ago, but that leads us to only one other conclusion, that of supernatural creation. We cannot accept that on philosophical grounds; therefore we choose to believe the impossible: that life arose spontaneously by chance."[38] *(George Wald, Nobel laureate, Harvard University)*

"We may furthermore conclude that the evolution of life, it is based on a derivable physical principle, must be considered and inevitable process despite its indeterminate course…it is not only inevitable in principle but also sufficiently probable in a realistic span of time."[73] *(Manfred Eigen, Nobel laureate and Chairman, Max Planck Institute)*

"…how did life originate?…As a general a priori answer to the question, it may be taken as virtually certain, unless one adopts a creationist view, that life arose through the succession of an enormous number of small steps, almost each of which, given the conditions at the time, had a very high probability of happening."[74] "It is now generally agreed that if life arose spontaneously by natural processes – a necessary assumption if we wish to remain within the realm of science – it must have arisen fairly quickly…".[75] *(Christian de Duve, Nobel laureate, Rockefeller University)*

"It is the job of science to solve mysteries without recourse to divine intervention. Just because scientists are still uncertain how life began does not mean life cannot have had a natural origin."[36] *(Paul Davies, University of Adelaide)* |

THE UNIVERSE FROM NOTHING PREMISE *The Universe arose from nothing by itself such as by a quantum fluctuation*	"I still believe the universe has a beginning in real time, at the big bang. But there's another kind of time, imaginary time, at right angles to real time, in which the universe has no beginning or end. This would mean that the way the universe began would be determined by the laws of physics. One wouldn't have to say that God chose to set the universe going in some arbitrary way that we couldn't understand."[76] "The universe would be completely self-contained and not affected by anything outside itself. It would neither be created nor destroyed. It would just BE."[77] *(Stephen Hawking, Cambridge)* "The key question is not whether the multiverse [parallel universes] exists but rather how many levels it has."[78] *(Max Tegmark, MIT)* "In it [self-reproducing inflationary theory] the universe appears to be both chaotic and homogeneous, expanding and stationary. Our cosmic home grows, fluctuates and eternally reproduces itself in all possible forms, as if adjusting itself for all possible types of life it can support."[71] *(Andrei Linde, Stanford University)*

The journey reviewing the evidence for evolution that is *Part II* of this book will assess the sufficiency of naturalism to account for life as two separate problems therefore. The first task is to assess the adequacy of evolution as a process to account for all the order and diversity of life after accepting the assumption that there indeed is an ancient common ancestor to begin the process (§4-§11). The second task will be to inspect the evidence for the assumption about ultimate origins, namely how life could arise naturalistically in the first place (§12) and where the Universe came from (§13). Before heading out it's useful to see the reasoning by which the evolution paradigm is erected because in one way or another this is not just the paradigm for biology, but for most of the reasoning in academia (§1.6.1). Are the premises and inferences of the three-step argument above reasonable?

§3.3 FOLLOWING THE 3 STEP ARGUMENT FROM APPARENT DESIGN

Among the relatively few authoritative works that lay out the evidence for evolution rigorously is Mark Ridley's textbook *Evolution*. It is also among the most acclaimed and so for two reasons affords observers like us an opportunity to see the reasoning up close.33 Follow along as he makes the case. His lesson begins with an important premise and all the reasoning that follows depends on it being true. He says: "We merely suppose it could happen by some natural mechanism". It's the Naturalistic Premise of course, which in his words is that: "Scientific arguments employ only observations anyone can make...and consider only natural, as distinct from supernatural causes." This pronouncement, for which he gives no evidence anywhere in his book, excludes any alternative to evolution. He continues thereafter with a review of the evidence (see Table 3.1 for the numbering of his arguments):

> "How can it be shown that species change through time, and that modern species share a common ancestor? We begin with direct observations of change on a small scale **[LOE#1]** and move out to more inferential evidence of larger-scale change **[The Macroevolution Inference]**. We then look at what is probably the most powerful general argument for evolution: the existence of certain kinds of similarity (called homologies) between species – similarities that would not be expected to exist if species had originated independently **[LOE#4, LOE#5]**. Homologies fall into hierarchically arranged clusters, as if they had evolved through a tree of life and not independently in each species. The order in which the main groups of animals appear in the fossil record makes sense if they arose by evolution, but would be highly improbable otherwise **[LOE#2, LOE#3]**. Finally the existence of adaptation in living things has no non-evolutionary explanation, though the exact way that adaptation can be used to suggest evolution depends on what alternative is being argued against."

What he presents then, is a three-step argument. Anything to do with the supernatural is not science and therefore not a consideration (The Naturalistic Premise) or step 1. Species can be directly observed to change over time (*LOE#1*) or step 2. This effect is magnified by time and can be extrapolated out to change beyond the species level (The Macroevolution Inference) or step 3. The reasoning for macroevolution is entirely by inference but corroborated as true by the lines of evidence ([*LOE#*2-5]). Is this reasonable, even on grounds of logic? Let's look.

§3.4   STEP 1: A SUPER-NATURAL IS OBVIOUSLY NOT SCIENCE

The first step sweeps the field of any alternative. He says: "two good criteria to distinguish scientific from religious arguments are whether the theory invokes only natural causes, or needs supernatural causes too, and whether the evidence is publicly observable or requires some sort of faith. Without these two conditions, no constraints are placed on the argument." He gives no evidence for this claim either and *Parts II and III* of this book are a formal disproof. He explains what he means:

"As a theory of adaptation, neo-Darwinism asserts that adaptations evolve only by natural selection. This assertion is a most fundamental claim, because we have no theory other than natural selection to explain adaptation; without it, we must fall back on miracles as an explanation."

So he declares creation arguments have "no constraints" and require "some sort of [private] faith" (rather than evidence) for their acceptance because they are "religious". This is misleading on two counts.

The first is that any rational claim, even if it appeals to a supernatural, has constraints even by definition. For instance the supernatural claim Jesus Christ rose bodily from the dead or the claim there was a supernatural creation of the world have constraints. They can also be tested for their validity. In regard to the former by reviewing the documents and testimony of witnesses,[79] and in regard to the latter finding evidence of ID in nature.

Second he assumes "miracles", a supernatural solution he equates as "religious", are never empirical and always objectively false. Where does that idea come from? He gives no evidence. There is none. What if a supernatural intervention did occur? It's not difficult to see this science can't go anywhere the evidence might lead. This is unblushingly admitted too. For example by Paul Davies here (§1.6.2),[80] by Niles Eldredge here (Table 3.1),[46] by Richard Lewontin here (§19.1)[81] and by Michael Ruse here (§21.1).[43] Yet solving the origin of nature is the most important question in science. Objective evidence that leads to the conclusion of a supernatural is not science and therefore scientifically false, even if it's true (§19).

§3.5   STEP 2: SPECIES EVOLUTION CAN BE DIRECTLY SEEN

The next step is to show evidence of changes in shape, size, color or the like in organisms at the species level, either in the wild or by breeding experiments. In other words microevolution. Dr. Ridley gives examples of the house sparrow (*Passer domesticus*) which has undergone striking changes in morphology on the North American continent since its introduction from Europe in 1852, melanism in the peppered moth (§4.4), and changes in beak size of the Galapagos finch (§4.5).

Artificial selection has produced far more impressive changes than what has been seen by natural selection. Darwin relied on artificial selection in *The Origin* to make his case for macroevolution (§1.5), and it's still used for that reason as Dr. Ridley explains:

> "Evolution can also be produced experimentally...Artificial selection can produce dramatic change, if continued for long enough. A kind of artificial selection, for example, has generated almost all of our agricultural crops and domestic pets."

Where is he going with this argument? He tells us:

> "The point here of these, and similar, examples is to illustrate how, on a small scale, species can be shown experimentally not to be fixed in form."

What have we learnt? Species like the house sparrow and the Galapagos finch can change in their morphology. Is "evolution" demonstrated? Well yes and no. What is demonstrated most is the imperative of being precise about terminology. As we know, there are different meanings to the word "evolution" and most importantly microevolution and macroevolution (§2.2). Evolution theorists consider them essentially the same, differing in magnitude only because of differences in duration (i.e. Darwinism = microevolution x time). Dr. Ridley does not use the word microevolution but "evolution" and so blurs the distinction. In the light of this ambiguity and the critical implications it's a most unfortunate omission. If "evolution" is to refer to the change he shows, and which is seen only at the species level then of course "evolution can be observed in nature". Evolution is surely a fact. This is microevolution which also and much more impressively generated the 420 different dog breeds now known. They vary in size from the Chihuahua at 6 inches tall to the Irish wolfhound at 32 inches, and in looks and character are as different as guacamole and Guinness®. Microevolution is a directly observable fact.

But this is not the kind of change he means nor the kind of change he is marshaling evidence for, or what we are trying to have clarified. The question is whether macroevolution is real, but the evolution he shows is merely microevolution because these sparrows have remained sparrows, moths moths and finches finches. Dr. Ridley acknowledges this, but in a curious way. He says:

> "Because the house sparrows have not reached the stage of evolving new species, the example does not distinguish between transformism and evolution."

How does he know house sparrows "have not reached the stage of evolving new species"? It seems a gratuitous evolution expectation. The question is whether change can cross the species barrier or not, and so support

128

inferences of macroevolution. Exactly what a species is becomes critical therefore (§2.2.2). He continues: "What does it mean to say a new species has evolved? This question unfortunately lacks a simple answer that would satisfy all biologists." This news comes as a little disturbing then. He explains this not the problem it appears to be though.

> "In arguing for evolution, we do not have to define a species. If someone says, what's the evidence that evolution can produce a new species, we can reply, "you tell me what you mean by species, and I'll tell you the evidence."

If you are not reassured by this, allow him to explain further. He gives two ways of defining a species. One way is "reproductive". It's the definition of Ernst Mayr that species cannot breed with each other (§2.2.2). The other way is by "phenotypic appearance". On this definition, what a species is devolves to a subjective judgment about the degree of change in size or color or shape or some other discriminator. Dr. Ridley is not perturbed by the sudden loss of objectivity and gives two reasons why.

The first is because: "The final answer often lies with an expert who has studied the forms in question for years and has acquired a good knowledge of the differences between species," and so defends it with empty appeal to authority if not a circular argument also. The second reason is because: "for relatively familiar animals we all have an intuitive phenotypic species concept. Again, humans and common chimpanzees belong to different species, and they are clearly distinct in phenotypic appearance. Common suburban birds, such as robins, mockingbirds, and starlings are separate species, and can be seen to have distinct coloration. Thus, without attempting a general and exact answer to the question of how great the differences between two species must be, we can see that phenotypic appearance might provide another species concept."

In moving with him from species definitions on reproductive grounds to species definitions on morphologic grounds we have moved from objectivity to what he admits is "intuitive" subjectivity. This is moving from clarity to ambiguity. Reproductive criteria for defining a species cannot be used for organisms that reproduce asexually nor for fossils but this weaker phenotypic definition he invokes is not necessary for the assessment of living organisms. Why does he use it then? It allows evolution to be detected now, and it can be done in two ways.

First, if morphology is used to define species then microevolution can make new species. As he says it: "The evidence from domestic animals suggests that artificial selection can produce extensive change in phenotypic appearance – enough to produce new species and even new genera – but has not produced much evidence for new reproductive species." The second way is by carefully selecting definition criteria to subdivide an otherwise single interbreeding species of different morphologies into (evolutionary) lineages. This can be called "evolution", but it does not come close to what is meant by macroevolution or the kind of evidence that evolution theory needs to produce. He has more evidence however. He introduces the concept of "ring species". This is a species population separated geographically with the extremes of its morphologic variation so diverged they cannot interbreed and so they satisfy the reproductive definition for a separate species. He explains:

> "Ring species provide important evidence for evolution, because they show that intra-specific differences can be large enough to produce an interspecies difference."

He gives the example of the salamander subspecies *E. eschscholtzii* which has unblotched coloration, and *E. klauberi* which has blotched. These are phenotypic differences not even close to the magnitude seen between different dog breeds you could round up in a single apartment complex. He also concedes there are regions in the ring of species variation at which these two salamander forms yet interbreed. His explanation for this contradiction to separate species? It's that:

> "in such locations [of interbreeding], the forms have not evolved apart enough to be separate reproductive species...When the forms reached the southern tip of California, however, the two lines of population have evolved far enough apart that when they meet...they do not interbreed because they are two normal species."

He offers no evidence for this "explanation" so it's empty too. How do we know what we are witnessing is merely microevolution in a single salamander species? Can new species as organisms clearly reproductively distinct be produced experimentally however?

Yes they can. It only happens in plants, and it requires the use of colchicine, the gout and anti-cancer drug that disrupt the normal separation of chromosomes at cell division. The process results in offspring with double the number of chromosomes, a phenomenon called polyploidy. These offspring are interfertile with other polyploidy hybrids like themselves, but not with their parental species. Polyploidy plants are also found naturally by spontaneous mutation. In fact it is estimated between 50 and 80% of flowering plant species arose that way. This is a very interesting model for research into plant hybrids then, but how polyploidy bears on macroevolution, much less on macroevolution from accumulated microevolution and what we want demonstrated, is not clear at all. This is not how Dr. Ridley sees it. He says polyploidy offspring:

> "provide clear evidence that new species in the reproductive sense can

be produced. If we add these plant examples to the examples of dogs and pigeons [above which of course are not], we have now seen evidence for the evolution of new species according to both the reproductive and the phenotypic species concepts".

There you have it. In his view you have witnessed evolution across the species barrier. This is his Step 2.

What happens in microevolution is the end result of random mutation and allele shuffling by one or more mechanisms of recombination, combined with the effect of natural selection which is differential reproductive success of individuals in a population (§2.4-5). What we need to recognize is evolutionary change in biology in the end is limited to what mutation can offer up (§2.3-5). Natural selection can only sieve what mutation has already made. We are comfortable extending a graph over time if it is linear, or a curve that can be described by an explicit formula, but the extrapolation of macroevolution from microevolution either by sieve at the theoretical level of biology or by the examples he gives from experimental is not granted by the evidence so far. A third and final step of the evolution argument is still required.

§3.6   STEP 3: BELIEVING IS SEEING: THE MACROEVOLUTION INFERENCE

Dr. Ridley summarizes the evidence and the argument so far:
"We have now seen that evolution can be observed directly on a small scale: the extreme forms within a species can be as different as two distinct species, and in nature and experiments, species will evolve into forms highly different from their starting point."

Having used the Macroevolution Inference in Step 2, his argument and which is Step 3, is the Common Ancestry Inference (Table 3.1). Watch closely.

"As we extend the argument from small-scale observations, like those described in HIV, dogs and salamanders [of microevolution], to the history of all of life we must shift from observation to inference. It is possible to imagine, by extrapolation, that if the small-scale processes we have seen were continued over a long enough period they could have produced the modern variety of life."

Is this reasonable? He is right to say extrapolation requires imagination and the reason is it's not empiric. The reason he doesn't say it that way or even see that, is because of the Naturalistic Premise from Step 1. An extension from microevolution into macroevolution by inference is demanded. It's called the principle of "uniformitarianism" or what we have called it here, the Macroevolution Inference (Table 3.1). He explains:

> "The principle of uniformitarianism is not peculiar to evolution, but is used in all historical geology. When the persistent action of river erosion is used to explain the excavation of deep canyons, the reasoning principle again is uniformitarianism."

This analogy is spurious. The process of macroevolution is biological novelty, which is new organs like eyes and ears or new life forms like bats and butterflies or new behavior like transcontinental bird migration and honeybee talk. This is nothing remotely like eroding a bigger hole in the ground by a physical process. Macroevolution represents the accumulation of additional complexity, additional biological information, and information does not accumulate linearly. A bigger hole from a smaller hole is produced by a simple iterative equation and can be extrapolated, even linearly. He counters this objection.

> "Someone who permits uniformitarian extrapolation only up to a certain point will inevitably be making an arbitrary decision. The differences immediately above and below the point will be just like the differences across the break-point."

Not so, and the reason is the extrapolation to macroevolution is without warrant. It's also because species defined as reproductively separate entities represent boundaries that are not "arbitrary" but clear. What has happened of course is the weaker subjective phenotypic species definition introduced earlier as an alternative to this clear reproductive definition has been conscripted as a measuring rod to demonstrate the fact of evolution. How does he counter the objection there are limits to evolutionary change demonstrated by the evidence of artificial selection? The fossil record.

> "The fossil record contains a continuous set of intermediates between the mammals and reptiles, and these fossils destroy the impression that "mammals" are a discrete type... *Archaeopteryx* does the same for the bird type, and there are many further examples."

If this were true this would excellent evidence. It's what Darwin said the fossil record would show. The problem is it doesn't. Don't believe me though here's Stephen Jay Gould, a paleontologist, to tell you: "The extreme rarity of transitional forms in the fossil record persists as the trade secret of paleontology. The evolutionary trees that adorn our textbooks have data only at the tips and nodes of their branches; the rest is inference, however reasonable, not the evidence of fossils. All paleontologists know that the fossil record contains precious little in the way of intermediate forms; transitions between major groups are characteristically abrupt."[83] The other problem is the rigorous species definition of inability to interbreed cannot be applied to this argument. What constitute separate fossil species is made by morphologic, and so necessarily subjective, criteria. But perhaps his evidence and his explanation has convinced you evolution is indeed a fact.

We make a curious observation in passing though. Dr. Ridley and Dr. Gould have opposite interpretations of the fossil record, so one of them must be telling tales. But that's not what's curious. What's curious is that both men are just as certain about the truth of evolution and fossils as the evidence of it. Evolution theory accommodates their opposite interpretations! How could that happen? (§20) We will inspect the fossil record later to decide for ourselves, and what it shows (§5.5;§6.3). The blind watchmaker inference from microevolution to macroevolution seems more a blind leap so far. We need to see the lines of evidence, but the entire notion is put on notice given the performance so far.

§3.8   IN CONCLUSION: "THE SINGLE BEST IDEA ANYONE HAS EVER HAD"

Tufts University philosopher of science Daniel Dennett epitomized the current consensus when he said:

"Let me lay my cards on the table. If I were to give an award for the single best idea anyone has ever had, I'd give it to Darwin, ahead of Newton and Einstein and everyone else. In a single stroke, the idea of evolution by natural selection unifies the realm of life, meaning, and purpose with the realm of space and time, cause and effect, mechanism and physical law."[84]

UCLA physiologist, ecologist, Pulitzer Prize author and National Medal of Science winner Jared Diamond agrees. He says:

"Evolution is the most profound and powerful idea to have been conceived in the last two centuries."[85]

Michael Ruse sums up the evidence for this wonderful idea for us, and numbered by the lines of evidence numbers of Table 3.1:

"If one subscribes in any honest way to the principles of empirical science then the *Origin* should convince one that organisms are the end-product of a gradual process of law-bound change, however caused. Homologies [*LOE#*4,#5], embryological similarities [*LOE#*7,#8], fossils [*LOE#*2,#3], and rudimentary organs [*LOE#*10] all attest to evolution. How, otherwise, can one explain the ridiculously useless isomorphisms that exist between the forelimbs of man, the horse, the bat, the porpoise, and the mole when each and every one of these limbs is used for a different purpose [LOE#4]? And perhaps above all, the nature of the geographical distribution of organisms should make one an evolutionist [LOE#12]. The finches and tortoises of the Galapagos just do not make sense without evolution...one often sees it said that "evolution is not a fact but a theory." Is this the essence of my claim? Not really! Indeed, I suggest that this wise sounding statement is confused to the point of falsity: it almost certainly is if, without regard for cause, one means no more by "evolution" than the claim that all organisms developed naturally from primitive beginnings. Evolution is a fact, *fact*, FACT!"[86]

It's as true as heliocentricity![7,87,88] In closing, although the Naturalistic Premise is a reasoning flaw assume it to be valid so we can test it (§1.1,§3.4,§19). Assume the evolution paradigm as constructed by the Macroevolution Inference and the Common Ancestry Inference, and assume the status of the Spontaneous Generation Premise and the Everything From Nothing Premise as true also. We now have a primitive self-replicating genetic system in a Universe that sprang into being by itself as a given. Can we find unambiguous evidence of macroevolution now, and so confirm Darwinism true like helicentricity is true? In case you're getting a little pessimistic before we start Ernst Mayr has a word for you:

"Are not the "facts" of evolutionary biology something very different from the facts of astronomy, which show that the Earth circles the sun rather than the reverse? Yes, up to a point. The movement of the planets can be observed directly. By contrast, evolution is a historical process. Past stages cannot be observed directly, but must be inferred from the context. Yet these inferences have enormous certainty because (1) the answers can very often be predicted and the actual findings then confirm them, (2) the answers can be confirmed by several different lines of evidence, and (3) in most cases no rational alternative explanation can be found."[89]

The question is what he said true? We now go to look. This inquiry should at least be interesting. French astronomer and mathematician Pierre-Simon Laplace once presented Napoleon Bonaparte with a personal copy of his treatise on celestial mechanics, the apocryphal story goes. Noticing no mention of God the Emperor asked the astronomer whether God's existence had any role in his thesis or in his reasoning. Laplace replied: "Sire, I have no need of that hypothesis." Evolutionists say the same thing about ID in biology. What's the evidence that makes them do so? Chapters 4-11 review the main lines of that evidence (*LOE*#1-12). Later in chapters 12-14 the origin of life, the origin of the Universe and then the origin of natural forces themselves are inspected, and so the validity of the Spontaneous Generation and Everything from Nothing premises assessed. Feel free to jump and skip about in the book. The inquiry is written with that proclivity in mind. I recommend you do because the subject matter can get heavy. The reason it gets heavy is because it's serious but it's serious because it's important. Finding leaves is easy. You just scrape on the surface. Finding diamonds takes work because you have to dig deep. If in the effort of digging through heaviness you find even one thing precious, this writing will have been worthwhile.

There are two forces by which evolution operates. The first is "chance" which is genetic variation by mutation and recombination and we looked at it earlier (§2.4). The second is the force of "necessity" or natural selection. This is the "blind watchmaker" taken to be the creative force of the process. We should start our search for evolution evidence there.

REFERENCES

1. Quammen, D. Was Darwin Wrong? *National Geographic*, 4-35 (2004).
2. American Association for the Advancement of Science. Present scientific status of the theory of evolution. *Science* 57, 103-104 (1922).
3. Dobzhansky, T. quoted by Futuyma, D.J. in Science on Trial 1995. p68 (1962).
4. Dobzhansky, T. Nothing in biology makes sense except in the light of evolution. *American Biology Teacher* 35, 125-129 (1973).
5. Mayr, E. *One long argument*.163-4 (Harvard University Press, 1991).
6. Mayr, E. *What Evolution Is*. 275 (Basic Books, 2001).
7. Futuyma, D. J. *Evolutionary Biology*. 15 (Sinauer Associates, 1986).
8. Rennie, J. 15 Answers to creationist nonsense. *Scientific American*, 78-85 (2002).
9. Ridley, M. *Evolution*. 2nd edn, 65 (Blackwell Science, Inc., 1996).
10. Dawkins, R. *The Greatest Show on Earth*. (Free Press, 2009).
11. Rensberger, B. The Nature of Evidence. *Science* 289, 61 (2000).
12. National Science Board. *Science and Engineering Indicators 2012*, <http://www.nsf.gov/statistics/seind12/c0/c0i.htm> (2012).
13. Macbeth, N. *Darwin retried. An appeal to reason*. (Gambit Incorporated, 1971).
14. Grasse, P.-P. *Evolution Of Living Organisms*. (Academic Press, Inc., 1977).
15. Denton, M. *Evolution: a theory in crisis*. (Adler & Adler, 1986).
16. Johnson, P. E. *Darwin on trial*. (InterVarsity Press, 1993).
17. Davis, P. & Kenyon, D. H. *Of pandas and people*. 2nd edn, (Haughton Publishing Company, 1993).
18. Spetner, L. M. *Not By Chance! Shattering The Modern Theory Of Evolution*. (The Judaica Press, Inc., 1997).
19. Hoyle, F. *Mathematics of evolution*. (Acorn Enterprises LLC, 1999).
20. Dembski, W. A. *The Design Inference*. (Cambridge University Press, 1998).
21. Wells, J. *Icons of Evolution*. (Regnery Publishing, Inc, 2000).
22. Ashton, J. F. *In six days. Why fifty scientists choose to believe in creation.*, (Master Books, 2000).
23. Gonzalez, G. & Richards, J. W. *The Priveliged Planet*. (Regnery Publishing Inc, 2004).
24. Flew, A. & Varghese, R. A. *There is a God: How the worlds most notorious atheist changed his mind*. (HarperOne, 2007).
25. Pennock, R. T. *Tower of Babel. The evidence against the new creationism*. (The MIT Press, 1999).
26. Mayr, E. *What Evolution Is*. 12-13 (Basic Books, 2001).
27. Mayr, E. *What Evolution Is*. 12-39 (Basic Books, 2001).
28. Grant, V. *The evolutionary process. A critical review of evolutionary theory*. (Columbia Unversity Press, 1985).
29. Futuyma, D. J. *Science on trial*. 197-207 (Sinauer Associates Inc., 1995).
30. Ridley, M. *Evolution*. 40-68 (Blackwell Science, Inc., 1996).

31 National Academy of Sciences. Teaching About Evolution and the Nature of Science. http://www.nap.edu/books/0309063647/html/index.html (1998).
32 National Academy of Sciences. *Science and Creationism: A View from the National Academy of Sciences.* 2nd Ed edn, (The National Academies Press http://books.nap.edu/html/creationism/, 1999).
33 Ridley, M. *Evolution.* 3rd edn, (Blackwell Publishing, 2004).
34 Shermer, M. *Why Darwin Matters.* (Holt Paperbacks, 2006).
35 Coyne, J. A. *Why evolution is true.* 81-85 (Viking, 2009).
36 Davies, P. *The Fifth Miracle.* 31 (Simon & Schuster, 1999).
37 Dawkins, R. *Climbing Mount Improbable.* (W.W. Norton & Company, 1996).
38 Wald, G. The Origin Of Life. *Scientific American* 191, 44-53 (1954).
39 Miller, K. R. & Levine, J. S. *Biology.* 3 (Prentice-Hall, 2004).
40 National Association of Biology Teachers. *Statement on Teaching Evolution,* <http://ncse.com/media/voices/national-association-biology-teachers-2000> (2000).
41 Ridley, M. *Evolution.* 66 (Blackwell Science, Inc., 1996).
42 Futuyma, D. J. *Science on trial.* 169-170 (Sinauer Associates Inc., 1995).
43 Ruse, M. *Darwinism Defended.* 322 (Addison-Wesley Publishing Company, 1982).
44 Overton, W. R. Creationism in the schools: The decision in the McClean versus the Arkansas Board of Education. *Science* 215, 934-943 (1982).
45 Pigliucci, M. Where do we come from? A humbling look at the biology of life's origin. *Skeptical Inquirer* 23, 21-27 (1999).
46 Eldredge, N. *The Triumph of Evolution.* 13,95,137,141 (W.H. Freeman & Co., 2001).
47 Carroll, S. B. *Endless Forms Most Beautiful.* 291 (W.W. Norton, 2005).
48 Gilbert, S. F., Opitz, J. M. & Raff, R. A. Resynthesizing Evolutionary and Developmental Biology. *Developmental Biology* 173, 357-372 (1996).
49 Mayr, E. *What Evolution Is.* 68-69 (Basic Books, 2001).
50 Oram, R. F. *Biology: Living Systems.* 312 (Glencoe McGraw-Hill, 2003).
51 Ruse, M. *Darwinism and its discontents.* 42 (Cambridge University Press, 2006).
52 Miller, K. R. & Levine, J. S. *Biology.* 385 (Prentice-Hall, 2004).
53 Johnson, G. B. *Biology - Visualizing Life.* 179 (Holt Rineman & Winston, 1998).
54 Carroll, S. B. *Endless Forms Most Beautiful.* 10 (W.W. Norton, 2005).
55 Valentine, J. W., Erwin, D. H. & Jablonski, D. Developmental Evolution Of Metazoan Body Plans: The Fossil Evidence. *Developmental Biology* 173, 373-381 (1996).
56 Alberts, B. *et al.* Molecular Biology of the Cell 32-33 (Garland Publishing, 1994).
57 Coyne, J. A. *Why evolution is true.* 73,78,79 (Viking, 2009).
58 Futuyma, D. J. *Evolutionary Biology.* 122 (Sinauer Associates, Inc., 1998).
59 Arthur, W. The emerging conceptual framework of evolutionary developmental biology. *Nature* 415, 757-764 (2002).
60 Futuyma, D. J. *Science on trial.* 198 (Sinauer Associates Inc., 1995).

61 Eldredge, N. *The Triumph of Evolution*. 60 (W.H. Freeman & Co., 2001).
62 Grant, V. *The Evolutionary process. A critical review of evolutionary theory*. (Columbia Unversity Press, 1985).
63 Carroll, S. B. Genetics and the making of Homo. *Nature* 422, 850-857 (2002).
64 Coyne, J. A. *Why Evolution is True*. 192,207 (Viking, 2009).
65 Futuyma, D. J. *Science on trial*. 95 (Sinauer Associates Inc., 1995).
66 Larson, E. J. & Witham, L. Leading Scientists Still Reject God. *Nature* 394, 313 (1998).
67 Dawkins, R. *The God Delusion*. 102 (Bantam Press, 2006).
68 Faber, S. M. The big bang as scientific fact. *Annals of the New York Academy of Sciences* 950, 39-53 (2001).
69 Weinberg, S. A universe with no designer. *Annals of the New York Academy of Sciences* 950, 169-174 (2001).
70 Hawking, S. L. in *Stephen Hawking's Universe: The Cosmos Explained* (ed S.L. Filken) (Basic Books, 1997).
71 Linde, A. The self-reproducing inflationary universe. *Scientific American* Nov, 48-55 (1994).
72 Wald, G. The origins of life. *Proceedings of the National Academy of Sciences USA* 52, 595-611 (1964).
73 Eigen, M. Selforganization of matter and the evolution of biological macromolecules. *Naturwissenschaften* 58, 465-523 (1971).
74 de Duve, C. *Blueprint for a cell:the nature and origin of life*. (Neil Patterson Publishers, 1991).
75 de Duve, C. The beginnings of life on earth. *American Scientist* 83, 428-437 (1995).
76 Hawking, S. J. *Black holes and baby universes and other essays*. (Bantam, 1994).
77 Hawking, S. J. *The Illustrated A Brief History of Time*. (Bantam Books, 1996).
78 Tegmark, M. Parallel Universes. *Scientific American* May, 40-51 (2003).
79 Habermas, G. R. & Licona, M. R. *The Case for the Resurrection of Jesus*. (Kregel Publications, 2004).
80 Davies, P. *The Fifth Miracle*. 81 (Simon & Schuster, 1999).
81 Lewontin, R. Billions and billions of demons. *The New York Review of Books*, 28 (1997).
82 Dembski, W. A. *No Free Lunch*. (Rowman & Littlefield, 2002).
83 Gould, S. J. *The Panda's thumb*. 181-189 (1980).
84 Dennett, D. *Darwin's Dangerous Idea*. 21 (Simon & Schuster, 1995).
85 Mayr, E. *What Evolution Is*. vii (Basic Books, 2001).
86 Ruse, M. *Darwinism Defended*. 57-58 (Addison-Wesley Publishing Company, 1982).
87 Dawkins, R. *The Selfish Gene*. 1 (Oxford University Press, 1989).
88 Gould, S. J. Dorothy, Its Really Oz. *Time* 154, 59 (Aug 23, 1999).
89 Mayr, E. *What Evolution Is*. 276 (Basic Books, 2001).

4. Survival of the fittest

§4.1 WHAT'S ARTIFICIAL ABOUT THE ARTIFICIAL SELECTION ANALOGY?

Consider the Macroevolution Inference again because this is where the rubber meets the road in evolution theory. Can we get a better handle on macroevolution although it is a historical process? First of all, where did the inference of microevolution becoming macroevolution originally come from? Not from observations of living things but from observations of rocks. It originated with Kings College geologist Sir Charles Lyell and his book *Principles of Geology* (1832).[1] Lyell's view, which is also called uniformitarianism, is that geological features of the earth are the result of minute, gradual, accumulated changes over long periods of time. Darwin appropriated his friend's idea to explain the features of living things. If you are a little put off hearing this, or if to you there would seem a difference between extrapolating processes that cause canyons and mountains and those producing the non-linear biological novelty and purposeful complexity of biology, it's not obvious to biologists. Skepticism of the Macroevolution Inference is considered "nonsense" (§3.1,6). Anyway Richard Dawkins upholds a geological metaphor for macroevolution still, as "climbing Mount Improbable" (§17.2).[2]

An exchange of correspondence between Modern Synthesis co-architect Theodosius Dobzhansky (§2.1) and a macroevolution skeptic called Frank Lewis Marsh is helpful to show the difficulty that creationists and ID advocates alike have in believing it. After hearing all the arguments for macroevolution from Dr. Dobzhansky, Marsh replied: "Alas! Inferential evidence again! Is there no real proof for this theory of evolution which we may grasp in our hands?" The frustrated professor eventually gave up trying to convert him, saying: "If you demand that biologists would demonstrate the origin of a horse from a mouse in the laboratory then you just cannot be convinced."3

Fortunately we do not demand that much and neither did Marsh. All we ask, is for the details and a mechanistic explanation of how macroevolution occurred exactly, or even could have occurred, to agree it's "a fact". This is hardly unreasonable to ask of a science.

So how can we satisfy Marsh that macroevolution is true when we can't see it directly? We make another inspection of the microevolution-that-becomes-macroevolution-if-left-long-enough inference in this chapter. We look first at the best evidence of microevolution - artificial selection. Next we look at the evidence for natural selection. What evidence has accumulated since *The Origin*? With this background we then examine the supposed inevitability of getting macroevolution from microevolution if we wait. We begin by looking at artificial selection. It's an analogy of natural selection first used by Darwin in *The Origin* and it remains elemental to arguments for evolution still (§3.5). It contains a seldom mentioned flaw however.

Microevolution (*LOE*#1) is indisputable but "evolution" in the Darwinian sense means much more. It means the inference that microevolution is extrapolated to deliver macroevolution, and that with sufficient time populations and species change so much by natural selection that entirely new tissues, organs and whole new creatures are generated from a common ancestor. Darwin had no example of natural selection to support this grand idea. He therefore resorted to artificial selection, the purposeful breeding to a directed end by an intelligence, as an analogy of the process. Which of course is neither (§2.2). His reasoning went like this:

> "Slow though the process of selection may be, if feeble man can do so much by his powers...I can see no limit to the amount of change, to the beauty and infinite complexity of the co-adaptations of life, which may be effected in the long course of time by nature's power of selection." (p.109)

Artificial selection has a drawback he never considered worth mentioning though, let alone refuting.

It's that to the contrary of his view by which he could "see no limit" to the amount of change natural selection could make if one waited, all the scientific evidence of artificial selection in centuries of experiments has consistently shown it does have limits. That's what's artificial about the artificial selection analogy. Limits and sterility. Artificial selection is not a valid analogy at all.

You should also know the changes observed by natural selection are nowhere near as impressive as those of artificial selection. Why should we expect natural selection to be so much more powerful if given more time? There have been others like Marsh who have not been persuaded by the macroevolution inference for these and other reasons, and one of them is the French zoologist Pierre Grassé.

Back in 1977 he catalogued his objections in a book called *Evolution Of Living Organisms*. Grassé was roundly criticized for his efforts and the consensus of the science establishment was that he was crazy. But if his anti-evolution views were potty, at least his scientific credentials were unimpeachable. Theodosius Dobzhansky was among those who opposed him. Although he rejected Grassé's work outright Dobzhansky made an important concession, which coming from a hostile observer at least suggests the Frenchman may have been onto something. Dobzhansky said:

> "Now one can disagree with Grassé but not ignore him. He is the most distinguished of the French zoologists, the editor of 28 volumes of *Traite de Zoologie*, author of numerous original investigations, and ex-president of the Academie des Sciences. His knowledge of the living world is encyclopedic."

What Grassé objected to was the same thing as Marsh. The lack of specifics of the micro- to macroevolution extrapolation, an inference plausible in principle but eerily quiet in detail. You see Grassé could not see microevolution accumulating, even in theory. He said:

> "What is the use of their unceasing mutations if they do not change? In sum, the mutations of bacteria and viruses are merely hereditary fluctuations around a median position; a swing to the right, a swing to the left, but no final evolutionary effect.[4]...The 'evolution in action' of J. Huxley and other biologists is simply the observation of demographic facts, local fluctuations of genotypes, geographical distributions. Often the species concerned have remained practically unchanged for centuries! Fluctuation as a result of circumstances, with prior modification of the genome, does not imply evolution, and we have tangible proof of this in many panchronic species [also called 'living fossils' §5.5.3]."[5]

Perhaps Grassé's disbelief is unjustified. We will have to see. At least the analogy of the observable effects of artificial selection for the unobservable inference of macroevolution is unpersuasive. This is not because the former is guided by an intelligence, although that is a valid objection, but because artificial selection has limits and microevolution supposedly does not.

Darwinists counter by pointing first to the power of pleiotropy (which is the ability of a single gene mutation to have multiple phenotypic effects), and second to the theoretical power of summating small change over long intervals of time using the uniformitarian thinking geologists do about rocks. The theory remains just theory if the micro- to macroevolution steps involved are absent details of how it happened or could have happened however. This is not just because explicit detail is what is requisite of a science, but because in the laboratory pleiotropic mutations never construct novelty. They only alter pre-existing elements. As a result novelty eventually exhausts rather than extrapolating onwards and upwards over time. Another problem needing explaining, with explicit details, is how the supposed novelty of accumulating microevolution escapes the burden of an overwhelmingly greater frequency of loss of function and detrimental mutations (§2.4.4).

In sum the blind watchmaker notion of additive microevolution making macroevolution is a theoretical inference at least contradicted by the known evidence if the analogy of artificial selection is used. What is the evidence to support natural selection power in the wild though? Perhaps it's different out there. What evidence for natural selection having power necessary for macroevolution has accumulated since *The Origin*?

§4.2   NATURAL SELECTION: THE BLIND WATCHMAKER OF LIFE

If gene mutation and recombination is the generator of new diversity in evolution theory, then natural selection is its sieve and storehouse. Darwin first described microevolution using the now famous phrase "survival of the fittest" in *The Origin* like this:

> "But if variations useful to any organic being ever do occur, assuredly individuals thus characterized will have the best chance of being preserved in the struggle for life; and from the strong principle of inheritance, these will tend to produce offspring similarly characterized. This principle of preservation, or the survival of the fittest, I have called Natural Selection." (p.168)

Ernst Mayr says: "The truly outstanding achievement of the principle of Natural Selection is that it makes unnecessary the invocation of "first causes" – i.e. any teleological forces leading to a particular end."[7] This is Richard Dawkin's blind watchmaker of life. He explains it so clearly yet creationists and ID advocates just don't see it. Maybe Stony Brook botanist and philosopher of science Massimo Pigliucci can get through:

> "Natural selection, combined with the basic process of mutation, makes design possible in nature without recourse to a supernatural explanation because selection is definitely nonrandom, and therefore has "creative" (albeit nonconscious) power. Creationists usually do not understand this point and think that selection can only eliminate the less fit; but Darwin's powerful insight was that selection is also a cumulative process – analogous to a ratchet – which can build things up over time, as long as the intermediate steps are also advantageous."[8]

The survival of the fittest is the effect by which new life forms and organ systems are created by selective killing. An extraordinary paradox, come to think of it. What does fitness in "survival of the fittest" actually mean? Modern Synthesis co-architect George Gaylord Simpson is here to teach us.

> "To a geneticist fitness has nothing to do with health, strength, good looks, or anything but effectiveness in breeding."[9]

Mark Ridley explains further:

> "In evolutionary theory, fitness is a technical term, meaning the probability that an individual will survive to reproduce. This condition therefore means that individuals with some characteristics must be more likely to reproduce (i.e. have higher fitness) than others."[10]

Has the claim that natural selection causes macroevolution become any clearer? Look closer because there's more here than what meets the eye. Or is it the other way around?

§4.3   THE APPARENT TAUTOLOGY OF NATURAL SELECTION

Darwin's use of "survival of the fittest" for natural selection in *The Origin* is unfortunate because in that formulation his statement is a tautology. The phrase was not coined by him anyway, but British philosopher Herbert Spencer whose reputation rivaled Darwin's at that time.

How does one know which species are the fittest? Those that survive! Natural selection predicts the fittest organisms will be those that produce the most offspring, but the fittest organisms are also defined as those producing the most offspring. The most adapted to the environment are selected and those that survive are the most adapted. Round and round in circles we go. To see this more clearly, the word "survivors" could equally well substitute for "fittest" in Darwin's original statement above. It then reads: "the survival of the survivors I have called Natural Selection." Under these conditions natural selection is an observation of demographic fact.

But rather than a tautology, natural selection is much more. It describes a phenomenon in consequence of the survival of the fittest. Namely the preferential transmission of particular gene combinations and gene variations by those organisms with the greatest reproductive success. In his book *The Selfish Gene*, Richard Dawkins goes further to say that in the final analysis it is not organisms but their genes that are in a constant struggle to transmit copies of themselves to the next generation. This gives to natural selection the role of active purposeful force rather than purposeless filter. Natural selection neither predicts nor explains the reason responsible for selection in any particular instance. That's a retrospective assessment which will vary in its details from case to case. Natural selection simply states some creatures will be more successful at gene transmission than others.

That certain individuals of a species transmit their genes to descendants better than others is hardly contentious. It's the incontrovertible demographic of every ecosystem. Some individuals survive to produce offspring in number, and others die for whatever particular reason in that particular ecosystem at that particular time. It was Darwin's contribution to biology to take this apparently mundane observation, the life cycles of individuals within a species, and postulate that with time and accumulating mutations they could have dramatic, directional population effects that to him even created new tissues, new organs and new creatures. Semantic objections to Darwinism on grounds it's tautological miss the point therefore. Nonetheless, some very prominent evolutionary theorists have wrestled with this tar baby and become stuck in all kinds of ways, including upside down. Here are three of them all caught up for you to untangle.

One response is to deny a tautology even exists: "Species are not fit because they survive, but because they have evolved superior genes," is the retort. What do we say to this? The first clause is false. Successful species most certainly are fit. They survive! The second clause is an assertion without proof - assuming species evolve their genes into new species. The "superior" genes in a population indeed are selected for by natural selection, but progressive change in a phenotype leading to a new species by evolution still needs to be demonstrated for us to believe it. Moreover the use of the term "superior" is an irrefutable post hoc description of survivors, which already by definition are better off than those now dead. Johns Hopkins University paleontologist Steven Stanley (§6.3) gives his view about the matter.

> "I tend to agree with those who have viewed natural selection as a tautology rather than a true theory. It is essentially a description of what has happened, with only weak powers of prediction, in that the kinds of individuals that are favored can only be recognized in

retrospect. The doctrine of natural selection states that the fittest succeed, but we define the fittest as those that succeed. This circularity in no way impugns the heuristic value of natural selection as a generation-by-generation description of evolutionary change."[11]

Well do you agree? Of course not. He claims natural selection has "powers of prediction...only...in retrospect"!

Harvard geneticist Richard Lewontin on the other hand, says the tautology is neutralized when natural selection is predictive. He says:

"The concept of relative adaptation removes the apparent tautology in the theory of natural selection...An analysis in which problems of design are posed and characters are understood as being design solutions breaks through this tautology by predicting in advance which individuals will be fitter."[12]

Only a few paragraphs later he acknowledges this cannot be done however:

"Hence there is no way we can predict whether a change due to natural selection will increase or decrease the adaptation in general. Nor can we argue that the population as a whole is better off in one case than in another...Unfortunately the concept of relative adaptation also requires the ceteris paribus assumption [all things being equal], so that in practice it is not easy to predict which of two forms will leave more offspring."

We went nowhere travelling with him.

Stephen Jay Gould's response to the tautology is the same as his Harvard colleague. Natural selection really is predictive and Darwin knew it. He says:

"Creationists have even been known to trot out this argument [of tautology] as a supposed disproof of evolution – as if more than a century of data could come crashing down through a schoolboy error in syllogistic logic. In fact, the supposed problem has an easy resolution, one that Darwin himself recognized and presented. Fitness – in this context, superior adaptation – cannot be defined after the fact by survival, but must be predictable before the challenge by an analysis of form, physiology, or behavior. As Darwin argued, the deer that should run faster and longer (as indicated by an analysis of bones, joints, and muscles) ought to survive better in a world of dangerous predators. Better survival is a prediction to be tested, not a definition of adaptation".[13]

There's a flaw in his argument though, and it emerges when we notice he forgets Darwin was a Lamarckian. Fitness considered as better survival by this reasoning does not necessarily apply to the Mendelian genetics of Neo-Darwinism. With the Modern Synthesis adaptation was redefined as applying only to traits that promoted breeding effectiveness and transmission of genes (§2.2). For instance a deer running faster might have no relevance to reproductive success. It may even be detrimental. Gould continues unperturbed, giving as another example the myriad extinct species observed in the Burgess fossil field of British Columbia (§5.5.2):

> "This argument applies in exactly the same way to the Burgess fauna...We must in principle, be able to identify winners by recognizing their anatomical excellence, or their competitive edge."[13]

He then makes a rather stunning about face:

> "But if we face the Burgess fauna honestly, we must admit that we have no evidence whatsoever – not a shred – that losers in the great decimation were systematically inferior in adaptive design to those that survived. Anyone can invent a plausible story after the fact."

He then informs us that Simon Conway Morris, Derek Briggs and Harry Whittington, the three paleontologists whose work forms the bulk of the current understanding of the Burgess fossils, came to this same conclusion. We are therefore left unpersuaded by Dr. Gould's original claim.

A final view on the tautology comes from University of Edinburgh geneticist Sir Conrad H. Waddington, who weighed in at the centennial celebrations of *The Origin's* publishing in 1959, hosted at the University of Chicago:

> "Natural selection, which was at first considered as though it were a hypothesis that was in need of experimental or observational confirmation, turns out on closer inspection to be a tautology, a statement of an inevitable but previously unrecognized relation. It states that the fittest individuals in a population (defined as those which leave most offspring) will leave most offspring. This in no way reduces the magnitude of Darwin's achievement; only after it was clearly formulated, could biologists realize the enormous power of the principle as a weapon of explanation."[14]

To which Berkeley law professor Phillip Johnson has replied:

> "When I want to know how a fish can become a man, I am not enlightened by being told that the organisms that leave the most offspring are the ones that leave the most offspring."[15]

We have heard a lot of opinion so far. What does the data itself say about whether natural selection is predictive or not? The conclusion by Princeton ornithologists Peter and Rosemary Grant of the most investigated example to date, the beak morphology of the Galapagos finch, is that in over 30 years of study "evolution is unpredictable because environments, which determine the directions and magnitudes of selection coefficients, fluctuate unpredictably."[16]

In summary natural selection is not prognostic of a specific mechanism of survival. It's retrospective. An important corollary is with this post-hoc view it can explain any turn of events. It claims being responsible for why the fittest survive, after all. As a forward-looking "predictive" mechanism, as a scientific law for instance in telling us how, when or why an adaptation will be selected in survivors, it's silent. There is a great temptation to use the term "survival of the fittest" as an explanation for evolution as even some experts do, but it can be nothing of the sort.

Another lawyer, Norman Macbeth, pointed this out to biologists over four decades ago and two decades before Phillip Johnson did. They did not listen to him either. He said:

"The late J.B.S. Haldane...said: 'the phrase 'survival of the fittest', is something of a tautology...There is no harm in stating the same truth in two different ways.'[17] This is extremely misleading. There is indeed no harm in stating the truth in two different ways, if one shows what one is doing by connecting the two statements with a phrase such as *in other words*. But if one connects them with *because*, which is the earmark of the tautology, one deceives either the reader or oneself or both; and there is ample harm in this."[18]

What of our task however, of getting a handle on the Macroevolution Inference to examine it? How does one test it even at a theoretical level if it is a tautology? It would seem three properties need satisfying: 1) Does selective reproductive success occur within a species? 2) If so, does it cause genes and populations in a species to change? 3) Do these changes indicate that the changing populations can evolve into whole new forms of life? Keep these questions in mind as we look at the best known examples of natural selection. What does the best evidence of natural selection declare?

§4.4   NATURAL SELECTION IN BLACK AND WHITE:
                              MOTH MELANISM

Darwin's detailed description of natural selection as the force of evolution was audacious if not science fiction, because he had no example in all nature for the idea much less show what he said it could do (§1.5). It was not until almost one century later that experiments on a moth called *Biston betularia* by Oxford University physician Bernard Kettlewell were hailed as the first experimental evidence. Kettlewell even entitled his review of the work in *Scientific American* as "Darwin's Missing Evidence".[19] Richard Dawkins considers these moth experiments as "one of the best attested examples of natural selection in action."[20] This should be very good evidence then.

The phenomenon in question is a change in the population frequencies of white *B.betularia* moths peppered with black spots called the *typica* form, and a hyperpigmented autosomal dominant allelic variant called the *carbonaria* form (Fig.4.1). *Typica* was the only type existing prior to 1848 but *carbonaria* rose rapidly thereafter to account for almost 100% of the population in the industrialized areas of Britain, but not in the countryside where a reciprocal frequency existed. This trend in the cities peaked in the late 1950s has been reversing ever since. Similar observations have been made in continental Europe and the United States.

The textbook explanation ever since Kettlewell, is that this happens because birds selectively eat the less camouflaged moths resting on tree trunks and branches. The black *carbonaria* moth populations increase in areas polluted by heavy industry because pale colored lichen on the trees dies, and soot and sulphur dioxide darkly pigment the denuded trunk. They become less conspicuous to predatory birds than the pale *typica* form that is better camouflaged by lichen thus the converse happens in rural areas. The reversal of the trend in moth population frequencies since 1950 is explained as due to reductions in atmospheric pollution in consequence of the Clean Air Act promulgated at that time.

Fig.4.1. The light and dark forms of the peppered moth Biston betularia (typica form above and carbonaria below). Kettlewell's paper in *Scientific American* shows a picture of each attached to a dark tree trunk with the legend: "Dark and light forms of the peppered moth were photographed on the trunk of an oak blackened by the polluted air of the English industrial city of Birmingham. The light form (*Biston betularia*) is clearly visible; the dark form (*carbonaria*) is well camouflaged."[19] (Image credit: Olaf Leillinger).

The experiments Dr. Kettlewell performed were as follows. He bred and or caught (surprisingly the exactly details are still unclear), then tagged and released, a mixed population of *typica* and *carbonaria* moths in both a rural and an urban forest environment. He returned to the same areas later and recaptured as many moths as he could to assess relative survivability. In the polluted Birmingham woods he recovered 52% of the *carbonaria* and 25% of the *typica*. In the rural Dorset woods he recovered an inverse ratio of 6% *carbonaria* and 13% *typica*. Kettlewell also recorded bird predation of the moths on 16mm film and demonstrated in the laboratory that *typica* preferentially rested on white surfaces and *carbonaria* on black. He seemed to have illuminated a clear-cut case of natural selection by bird predation on the basis of "crypsis" (camouflage).

University of Chicago geneticist and Modern Synthesis architect Sewall Wright called it: "the clearest case in which a conspicuous evolutionary process has been actually observed".[22] Mark Ridley calls it: "a classic example of Natural selection."[23] Richard Dawkins says: it's "one of the best attested examples of natural selection in action,"[20] Michael Ruse says: "the reason [for the phenomenon] is unequivocally a function of Natural selection,"[24] and Douglas Futuyma says: "the increase [in moth number] was due to differential predation by birds…The appearance of design is an illusion".[25] Kettlewell's own assessment was that his work was "the most striking evolutionary change ever witnessed by man".[19] There would seem nothing questionable at all. Great job, Dr. Kettlewell.

§4.5.1 The Gray Areas Of The Story

It therefore came as a surprise to discover the sentiment suddenly changed in 1998, when a story emerged that was rather different from the clear-cut textbook explanation of natural selection for over four decades. Cambridge University geneticist Michael Majerus published a book that year pointing to problems in the moth data that nobody had mentioned and this was his bottom line:

> "In the 40 years since Kettlewell's pioneering work, many evolutionary biologists, particularly in Britain, but also in other parts of Europe, the United States, and Japan, have studied melanism in this species. The findings of these scientists show that the précised description of the basic peppered moth story is wrong, inaccurate or incomplete, with respect to most of the story's component parts."[21]

Writing in *Nature*, University of Chicago biologist Jerry Coyne responded:

> "Until now, however, the prize horse in our stable of examples has been the evolution of 'industrial melanism' in the peppered moth, *Biston betularia*, presented by most teachers and textbooks as the paradigm of Natural selection and evolution occurring within a human

lifetime...Depressingly...this classic example is in bad shape, and, while not yet ready for the glue factory, needs serious attention."[26]

So what and where were the holes?

The details of the moth release experiments are eye opening for a start. The "release" was merely Kettlewell's placement of moths on exposed tree trunks. Any notion of moths flying off to find their own places of safety in the forest is false. The moths were also released during daytime and during daytime were observed to be eaten by birds, but they are normally nocturnal. The inflated densities of moths caused by moth release from too few a number of release sites could have altered normal moth and predator behavior too. Finally at least some of Kettlewell's moths were reared at remote sites, and thus in release and capture experiments could have behaved differently to the endemic wild moths he was studying.

It also turns out *B.betularia* phenotypes are not confined to the melanic *carbonaria* and speckled *typica* allelic variants. There are shades called the *insularia* phenotype in between. *Insularia* is controlled by at least 3 additional alleles which are recessive to *carbonaria* and dominant to *typica*. Thus the *typica* genotype is homozygous TT, *carbonaria* is homozygous CC or heterozygous CT or CI, and *insularia* is homozygous $I_{1-3}I_{1-3}$ or heterozygous $I_{1-3}I_{1-3}$ or $I_{1-3}T$.

In sum the moth matter is more complicated than is acknowledged. More importantly for us, the reasons for the moth population shifts may not bear on Kettlewell's conclusions. The potential problems can be reduced to these three: 1) bird predation may not be relevant to the population shifts, 2) crypsis may not be relevant, and 3) air pollution may not be relevant. We should look at each in turn.

§4.5.2 Bird Predation May Not Be Relevant

The most disturbing discovery is the claim these moths do not rest on tree trunks in the natural condition. In the words of Sir Cyril Clarke at the University of Liverpool, whose own huge data set best documents the *carbonaria* trend reversal since the 1950s: "all we have observed is where moths do not spend the day. In 25 years we have only found two *Betularia* on the tree trunks or walls adjacent to our traps."[27] This alone invalidates Kettlewell's conclusions.

Kettlewell's release was far from natural and required placing moths on tree trunks, exposing them to avian predators directly.[26] The high recovery rates of the tagged moths in the experiments are a cry of artificiality too. Their natural resting places are in non-exposed areas, namely at branch-

150

trunk junctions, under branches or on foliate twigs.[21] The classic photographs of moths resting on tree trunks as seen in biology textbooks are staged (e.g.[28-31]). Although bird predation definitely occurs when moths are placed on tree trunks, there is still no robust experimental evidence show this the mechanism for the wild moth population, and the best is Majerus's communication that he observed nine different bird species eating these moths from his back window.[21] The moth experiments are certainly an experimental model, but may not be relevant to the wild condition and natural selection in the wild is what Kettlewell's conclusions were supposedly all about. The moth experiments became still more intriguing a few years ago. It was again because of Michael Majerus, but this time for opposite reasons.

Having unwittingly pulled the moth evolution tent down in 1998, this decorated member of the British Humanist Association now announced his own moth experiments, which he conducted in his garden, and said they confirmed what Kettlewell found after all. In his 2007 address to the European Society for Evolutionary Biology he called the moths: "one of the most visually impacting and easily understood examples of Darwinian evolution...It provides after all: The Proof of Evolution". These results were not published until 2012 however, and then only as a brief report posthumously.[38] Dr. Majerus had died rather suddenly three years before, in 2009, after a brief battle with mesothelioma.

The two key findings were that he recovered fewer melanic moths than non-melanic in his release-capture experiments, consistent with a bird predation natural selection as explanation, at least at a rural unpolluted site so confirming Kettlewell. Second he reported that a third of the 135 total moths he observed over six years were indeed resting on tree trunks, vindicating Kettlewell's release method. The latter claim is at least disputable, first by the data of Sir Cyril Clarke above and second by his own data, given such small numbers in the face of the many thousands in his garden at one time or another. Majerus released over 4,800 for these experiments and this is not to speak the endemic wild moth population, so suggesting there were more he failed to see and count. The concern is also supported by his data that moths typically rest on high, less accessible branches he would not have assessed by his search methodology.

§4.5.3   Crypsis In The Manner Presumed May Not Be Relevant

There are three reasons for this conclusion. The first is while Kettlewell's results imply a correlation between the degree of moth camouflage and the degree of avian predation, the evidence remains limited just as Majerus reported in 1998 and as analysis of his own experiments afterward indicates.[21]

Second, the experimental design and interpretation of Kettlewell's results have assumed that avian vision is like human vision but the evidence at least suggests otherwise. Birds not only have greater acuity, they see in the ultraviolet light frequency range. In 1998 Majerus concluded "none of the assessments of the relative crypsis of moths as determined by humans should be applied to birds."[21] His posthumously released results to the contrary were claimed as confirming "the previously accepted conclusion that visual predation by birds is the major cause of rapid changes in the frequency of melanic peppered moths."[21]

Third, the environment in which the moths live regardless of pollution status is still heterogeneous. It is therefore reasonable to ask why moths might not find resting surfaces matching their pigmentation status nonetheless, and therefore resist visual predation, particularly in light of Kettlewell's laboratory experiment result. Another concern is the latter result could not be replicated by others.

§4.5.4   Air Pollution May Not Be Relevant

The decline in the proportion of the melanic moth form is assumed to be from falling air pollution levels. Majerus's posthumous work reported support for Kettlewell's thesis in a rural setting.[21] However in Southern Michigan, the only United States location where the melanic form of *Biston* has been recorded in frequencies similar to Britain, the area "was not conspicuously blackened before clean air legislation, nor has the habitat changed in this regard since then."[32] The recovery of the *typica* moth populations has also not been associated with changes in lichen flora in Britain[33,34] or the United States.[32] This is contrary to the crypsis prediction "that lichens should precede the recovery of the typical morph as the common form. That is, hiding places should recover before the hider."[34] Perhaps moth melanism is not an industrial disease after all. What can we conclude from the evidence? More about science than moths surely.

§4.5.6 What Do Peppered Moth Experiments Tell Us About Science?

Contrary to what the textbooks say is long settled[23,24,28-31,35,36], the exact cause of the peppered moth population changes is at least controversial if not settled at all. Although the authors of Majerus's posthumous data say Kettlewell's interpretation of bird predation is "virtually impossible to escape," the exact mechanism is likely more complex as reanalysis of Kettlewell's data by Majerus indicates.[21] The evolutionary effect, as always, is simple - a change in the proportion of alleles in a population (§2.3). Interestingly Kettlewell himself acknowledged this, writing that "Industrial melanism probably represents an example of transient polymorphism."[37]

Setting objections to the moth experiments aside and even assuming Kettlewell's conclusions to be true, what do the moth results have to do with macroevolution? This is our question and our interest after all. The moth color changes are no more striking than effects obtained by artificial selection. So what's the big deal? The moth color changes also fluctuate, reversibly, from a predominance of light forms to dark forms and back to light again. We're not talking about new species being made. The phenomenon also does not explain how the *carbonaria* or *typica* form arose in the first place, or explain the curious fact that the melanic form is autosomal dominant. In sum changes observed in peppered moth populations are microevolution but there is no evidence so far they have anything to do with macroevolution.

Why the peppered moths data is so interesting is because that's not what the science community say. To the contrary, and in Bernard Kettlewell's words:

"Industrial melanism involves no new laws of nature; it is governed by the same mechanisms which have brought about the evolution of new species in the past."[19]

Michael Majerus agrees. He says:

"It is after all, The Proof of Evolution".

Their claim that if a moth population can come to be dominated by a variant of a different color in a few years then it could change into a different creature over still more time, would seem to be a speculation requiring rather more evidence. The problem for evolution theory of course, is how to extrapolate these modest population shifts and single gene mutation population shifts into new organs and new species and ultimately all the diversity of life, but instead of even more moths we only hear crickets. Jerry Coyne offered this conclusion of the moths in *Nature*:

"for the time being we must discard *Biston* as a well-understood example of natural selection in action, although it is clearly a case of evolution. There are many studies more appropriate for use in the classroom, including the classic work of Peter and Rosemary Grant on beak-size evolution in Galapagos finches."[26]

It is to this work that we now turn.

§4.5 NATURAL SELECTION DIRECTIONAL OR JUST OSCILLATING? THE EXAMPLE OF DARWIN'S FINCHES

The most abundant birds in the Galápagos Archipelago are 14 closely related species of finch. They differ principally in beak size and shape but all look very similar and so much so they are considered to be monophyletic, originating from one ancestral species that reached the Galapagos from Central or South America then diverged to fill vacant ecological niches on the islands (Fig.4.2).[39] These birds are considered to be one of the best documented examples of adaptive radiation. This is principally due to the work of the Princeton University husband-and-wife team Peter and Rosemary Grant, who since 1973 have meticulously followed phenotypic changes in finches on one of the smaller islands in the archipelago called Daphne Major. The acclaimed and famous changes have been in beak shape, beak size and body size. The experiments are acclaimed because of their detail and duration, but they are famous because they supposedly demonstrate "evolution in real time" to use the words of Jonathan Weiner in his Pulitzer Prize winning book *The Beak of the Finch*.

The birds are collectively now called "Darwin's finches". This is itself interesting. The reason is although Darwin collected finch specimens when the *Beagle* was in the Galapagos in 1835, ascribing to him data or experiments on finches that demonstrated or even suggested natural selection has more to do with myth than truth. Like all good myths it persists though. For instance even in the journal *Science* we read: "The peculiar finches he [Darwin] collected there, each species with a distinctive beak shape, helped inspire his theory of evolution by natural selection."[41] ID advocate Jonathan Wells has collected other examples of misinformation concerning the finches from college textbooks published after 1998. One is Raven and Johnson's biology which tells us: "The correspondence between the beaks of the 13 finch species and their food source immediately suggested to Darwin that evolution had shaped them." The finches are not even mentioned in *The Origin*. What are the facts of the matter?

Darwin collected specimens of nine finch species but considered only six to be finches and in only two did he notice a difference in their diet. He never made correlations between diet and morphology or their distribution across the islands of the archipelago. He had no evidence to support any speculation of an evolutionary ancestry because he never visited the west

Fig. 4.2. **The 14 different species of Darwin's finches.** The male and female of each species are shown. The birds differ in beak and body size but are very similar. *G. fortis* is 2, *G. scandens* is 5 and *G. fuliginosa* is 3. (Reproduced from Lack, 1983 with permission.[40] Drawings by Lt.-Col. W.P.C. Tenison, D.S.O.)

coast of South America to see if the birds differed on the mainland. He made no effort to sort the finches geographically, nor is there evidence of an inquiry by him into the issue.[42] Darwin's notion the birds were an example of natural selection was not made from any quantitative data, but was a speculation he made over a decade later on the basis of their morphologic similarity. Specifically in 1845 he wrote in his *Journal of Researches*: "The most curious fact is the perfect gradation in the size of beaks of the different species. Seeing this gradation and diversity of structure in one small, intimately related group of birds, one might really fancy that from an original paucity of birds in this archipelago, one species had been taken and modified for different ends." He could not study the specimens he collected to test the idea because of his lack of systematic collecting or documentation. Darwin's finches became famous because of evolution, and not the other way around. Like fossils (§1.5;§5.1;§10.2.1). The finch data have been extrapolated to fit a theory conceived of before experiments were even done.

The person who popularized the birds as an instance of natural selection and as "Darwin's finches" was not Darwin, but Oxford University ornithologist David Lack in his highly influential book *Darwin's Finches*.[40] Lack spent one breeding season in the Galapagos in 1937 and produced maps showing the dominance of particular species with particular beak sizes to different islands in the archipelago. For instance there are two major populations of finch resident on Daphne Major, the medium ground finch (*Geospiza. fortis*) and the cactus finch (*G. scandens*), as well as a rare third population called *G. fuliginosa*. *Fortis* beak size on Daphne Major is smaller than in the same species on Santa Cruz Island only 8 km away. Lack did not know why bill size varied across different islands. Demonstrating this was caused by natural selection on the basis of seed characteristics was the work of the Grants.

There are four potential evolutionary forces (§2.4.4). The Grants showed that neither genetic drift nor a founder effect through gene flow were the cause since birds traveled between different islands. Hybridization by *G. fortis* with other finch species of different beak size i.e. "mutation" would account for beak size variation, and indeed they showed it does occur. However the effect was small, just 1.9% *fortis* and 0.9% *scandens* pairs were found to be hybrids.[43] The principal reason for the size difference was natural selection therefore, and they showed it was on the basis of seed feeding efficiency from beak size. Previous investigators had missed this, perhaps because they had only studied the birds during the wet season when food was abundant and the different species ate similarly.

The evidence for natural selection here is robust and founded on several concordant lines of experimental evidence. During the dry season when food was scarce, finch abundance was correlated with seed abundance. Beak size was correlated with size and hardness of seed that could be eaten, and there was worse survival of the smaller beaked birds during droughts when small seeds were rare. By measuring the sizes of parents and offspring, the Grants were able to show beak and body size and beak shape are inherited. Measuring the mean of each of these traits with every generation provided a measure of evolutionary change that could be tracked from year to year. The results are shown in Fig. 4.3 A-F. What was found?

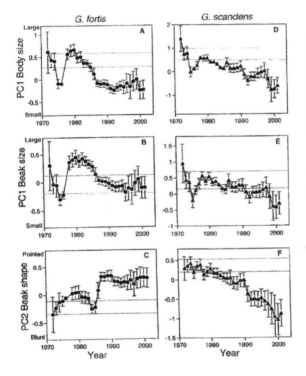

Fig. 4.3. Morphological trajectories of adult *G. fortis* (A-C) and *G. scandens* (D-F). Body size, beak size and beak shape are shown. The 95% confidence intervals of the initial 1973 observations are represented by the horizontal lines to indicate trends since. (Reproduced from Grant & Grant, 2002 with permission).[16]

Mean beak and body size in *fortis* are seen to have see-sawed in opposite directions 5 times by natural selection, in both directions, over the 30 years of the experiments. *Scandens*, a larger bird, also showed fluctuations in both directions but also a more gradual trend toward smaller size and blunter beak.[16] This represents a convergence toward the *fortis* morphology. As a result of the periodic weather disturbance La Nina in 1977-8 there was no wet season. The finch population plummeted to 15% of what it had been. Smaller birds died in greater numbers than larger and more females died than males for the same reason - size. It was found that smaller seeds were scarce from the drought, favoring the larger beaked birds which could crack larger, harder seeds. *G. fortis* born in 1978 after the drought were 4% larger than before it. The 4% size difference in beak depth, which is the width of the beak at its base, represented a distance too small to be appreciated with the naked eye but for a finch it meant the difference between life and death that year. The rapid change on the graph is caused by the massive selective population losses caused by the drought.

Four years later a reverse phenomenon was seen. As a result of exceptional rainfall caused by the weather aberration El Nino, there was an excess of small seeds. *G. fortis* born in 1985 had beaks 2.5% smaller than before the rains. Similar changes of lesser magnitude were also seen during these times in G. *scandens*. Also apparent is how the trend in scandens toward smaller size and blunter beaks (toward fortis morphology) has persisted while fortis has remained relatively stable. The reason for this is *scandens* males have been breeding with fortis females, and the hybrids only mate with *scandens* because they are imprinted on the bird songs of their *scandens* fathers. The net result is a flow of *fortis* genes into the *scandens* gene pool, called introgressive hybridization and explaining the trend toward *fortis* morphology.

The Grant experiments represent field biology at its best. What the data shows is oscillation in body size, beak size and beak shape by changes in weather patterns. This is convincing evidence natural selection is real, as hypothesized by Darwin, and this achievement by both Darwin and the Grants is worthy of great accolade. The trouble is this is not the accolade they receive nor is it the way this data showing finch microevolution is interpreted.

Returning to the three questions posed at the end of §4.3 as tests of natural selection for purposes of examining the Macroevolution Inference, the answer to the first question is in the affirmative and so is the second. Gene changes have not been demonstrated for these heritable traits but they surely can reasonably be assumed. The sticking point is over question three. We see mean population trends contributed in large part by selective death that fluctuate in both directions. They involve two bird populations that are very similar and interbreed, suggesting they are not separate species as the experts say. The point, and the question, is whether this is evidence for macroevolution.

As far as the Grants and the rest of science community are concerned it unquestionably is. In their view the experiments show the power of natural selection to create macroevolution. The beak changes observed over just a few years are extrapolated over centuries to be of sufficient magnitude to generate whole new forms of bird. Writing for *Scientific American* in 1991 Peter Grant acknowledged that "the population, subjected to Natural Selection is oscillating back and forth" but then leapt to the reasoning that if instead of fluctuation there were a cumulative effect caused by repeated episodes of selection in the same direction, then

> "about 20 selection events would have sufficed. If droughts occur once a decade, on average, repeated directional selection at this rate with no selection in between droughts would transform one species into another within 200 years. Even if the estimate is off by a factor of 10, the 2,000 years required for speciation is still very little time in relation to the hundreds of thousands of years the finches have been in the archipelago".

This macroevolution inference of finches is nowhere controversial in the science either. Mark Ridley at Oxford tells us:

> "This rough calculation is not intended to represent the exact history of the birds. Instead, it illustrates instead how we can extrapolate from natural selection operating within a species to explain the diversification of the finches from a single common ancestor about 570,000 years ago to the present 14 species...Arguments of this kind are common in the theory of evolution."[44]

The entirely extrapolationist basis of the arguments and the supposed generation of whole new bird forms over time are alone reasons for skepticism. Yet most curiously these are considered the reason for their prestige. The prestige is because they supposedly show natural selection in the wild that is extrapolatable to macroevolution. This is unreasonable for at least four reasons:

(1) Bird breeders have been producing changes to far more impressive degrees for centuries. These, and every other example of artificial selection, demonstrate there is a limit to the amount of change possible (§3.5,§4.1). There is no reason to infer the weaker effect the Grants find in the wild should be any different.

(2) Rather than plausibly supporting directional selection to macroevolution, the finch experiments are more striking for their oscillating selection. It recalls the comments made by Dr. Grassé earlier. The huge changes that formed the basis for calculations for new species were produced by population decimation, but there was recovery afterward in the reverse direction. Yet this is ignored. Take the National Academy of Sciences teaching publication *Science and Creationism: A View from the National Academy of Sciences* for instance. They make no mention of the reversal of beak size trend, nor even an oscillatory character. Instead the

Academy considers the finches a "particularly compelling example" of the origin of new species because "if droughts occur about once every 10 years on the islands, a new species of finch might arise in only about 200 years."[45] Bankers who write like this go to jail.

(3) The claim is the demonstration of a process by which a new species can be formed in about 2000 years, but it is absent a trend toward the novelty indicative of a new species. The data rather demonstrates a trend toward a merging of species – of *scandens* to *fortis* rather than divergence to something new. So what is it really, divergence or merging or both?

(4) Most problematically and also most importantly, do the finches represent different species? They interbreed. Ernst Mayr's biological species definition is explicit: "[s]pecies are...reproductively isolated."[46] (§2.2.2) To be fair nine percent of the known world total of 9,672 bird species has been reported to breed with another species while in captivity (although how this number applies to the wild is unclear),[43] but in this particular case given the evolutionary implications being made, pointing to the difficulties of the biological species definition is to miss the point. The reason is because we need to be certain we deal with separate species to accept the macroevolutionary inference. If the biological species definition is abandoned, the other bases for species classification are weak. If it is to be phenetic appearance for a new species then why should the finches today be considered different species? They look very similar. This was one of the key observations for David Lack back in 1937 yet it usually goes unmentioned. He said: "In no other birds are the differences between species so ill-defined". He did not know that they interbreed, and which was the reason to keep them as separate species. The Grants have since shown their non-interbreeding to be false and that hybrids of *fortis*, *scandens* and *fuliginosa* outperform their parents both in survival and breeding ability! This is to be contrasted with the typical offspring of an interspecies cross which has sterility and lower fitness. Think of mules and hinnies.

Thus rather than diverging into new species as the experts say, the data indicate that *scandens* and *fortis* are merging and there are no objective grounds to consider them separate species in the first place. In a more sober assessment, far removed from claims for species divergence that echo today, the Grants already said this back in 1992: "The discovery of superior hybrid fitness over several years suggests that the three study populations of Darwin's finches are fusing into a single panmictic population, and calls into question their designation as species."[43] Separate species are the crucial basis for the evolutionary inference of finch species divergence by natural selection however. In summary the finch evidence details microevolution in the wild and no more.

§4.6   OTHER EXAMPLES OF NATURAL SELECTION

Reference to textbooks will show that the peppered moths and Darwin's finches are considered the best examples of natural selection. There are many others however, and more are accumulating. Among the most impressive are the so-called "ring species" and particularly the example of the herring gull (*Larus.argentatus*) and the lesser black-backed gull (*L. fuscus*).[47]

In Northern Europe they are separate species – they do not interbreed and they also look different. The herring gull mantle is grey and its legs are pink, while the lesser black-backed gull's are black and yellow respectively. In North America only the herring gull is found and it looks very similar to the European bird. The interesting thing is that bird populations vary progressively by geographic location going from North America around the North Pole to Asia. Specifically the morphology is progressively less like the herring gull and ever more like the lesser black-backed gull. On the west side of the Bering Strait the Siberian Vega gull is found. It is classified among the same species as the herring gull, but it has a darker mantle than the European and North American bird and yellow legs. The trend continues until west of central Siberia when the morphology merges with that of the European lesser black-backed gull. Hence the term "ring species". What is the significance of this observation?

They are considered to be evidence for macroevolution by showing, to quote Mark Ridley: "that intraspecific variation can be large enough to make two species; that the differences between species are the same in kind, though not in degree, as the differences between individuals and populations, within a species."[47] The question is whether this can reasonably be called macroevolution, at least in the spirit that summated microevolution makes the novelty of macroevolution, or that it even supports the notion of a trend to that end if we were to wait longer. Dr. Ridley says yes because ring species blur the species boundaries:

"Ring species, therefore, demonstrate the existence of a continuum from interindividual to interspecies variation. Natural variation is sufficient to break down the idea of a distinct species boundary. The same argument...can be applied to larger groups than species and, by extension, to all life."

The problem is that to the contrary boundaries do not "break down...in larger groups" nor in "all life". In regard to the former the reason is simply because the phenetic differences of ring species are very modest, far less than artificial selection has wrought, and thus do not extend to higher taxon levels. Anyway morphologic similarity is overwhelmingly discontinuous across taxa, whether living or fossil or morphologic or molecular, but that is an assertion compelled by evidence we have not heard yet, so set this objection aside for now. And in regard to his latter claim, the reason is because most all the rest of nature, "all life", are not ring species. If anything ring species are freaks. There are only 24 candidates in all of nature.[48] Why should we take the exception to prove the rule? The evidence of ring species is registered however, and by the criterion of non-interbreeding they do demonstrate speciation and to that extent "macroevolution". If this is evidence of the Macroevolution Inference for you, you have to notch this one up.

Another line of evidence is that "new reproductively distinct species can be produced experimentally," says Mark Ridley.[49] Specifically crosses in the laboratory between different plant species like the primrose *Primula verticallata* with *P. floribunda* produce hybrid offspring that breed with other hybrids, but not with either of the parental species (*P. kewensis*). This is too is taken as evidence for macroevolution. While it sounds impressive, it's artificial. The hybrids cannot to do it in the wild but remain sterile, and the experiment demands use of the anti-cancer drug colchicine to disrupt chromosome sorting during cell division to double the chromosome number to produce the hybrid. The parental species both have 18 chromosomes and *P. kewensis* has 36. The many varieties of orchids, tulips and other garden plants are examples of these kinds of crosses. There is evidence this process occurs naturally in that many species are found to be polyploid (multiples of chromosome number), and Dr. Ridley estimates as many as 70-80% of angiosperm species are polyploids. These experiments provide a very good explanation for how such plants have arisen, and to that extent they represent "evolution", but multiplying an existing chromosome number is hardly biological novelty or the kind of micro into macroevolution trend being claimed. If it is good enough evidence for you nonetheless, notch this one up as well.

Insect resistance to pesticides and microbial resistance to antibiotics are dramatic and indisputable examples of natural selection, and likely examples with which we are most familiar. They don't have anything to do with our inquiry because they don't have anything to do with macroevolution though. Bacteria progress through countless generations in a matter of months much less years but none have shown macroevolution.

What can be concluded from the evidence of natural selection then? Natural selection as a process causing microevolution is not in doubt. The effect is far weaker than artificial selection though. There is no evidence to suggest the species barrier will be crossed to support an inference of macroevolution, except under the peculiar circumstances of ring species or the artificial circumstance of inducing polyploidy. Why then are these examples of microevolution considered demonstrative evidence of macroevolution? Mark Ridley explains. We just need to be patient because

> "evolution can be observed directly on a small scale [microevolution]: the extreme forms within a species can be as different as two distinct species, and in nature and in experiments species will evolve into forms highly different from their starting point. It would be impossible, however, to observe in the same direct way the whole evolution of life from its common, single-celled ancestor a few billion years ago. Human experience is too brief. As we extend the argument from small-scale observations, like those described in moths, dogs, and gulls, to the history of all life we must shift from observation to inference. It is possible to imagine, by extrapolation, that if the small-scale processes [microevolution] we have seen were continued over a long enough period they could have produced all the modern variety of life."[50]

And there you have it. It comes down to a logic that compels us first to rely on imagination and second to ignore existing evidence of the natural and artificial selection experiments. We inspected the reasoning earlier (§3.6). But is the macroevolution inference even reasonable in theory if we cannot see the change ourselves?

Paul Davies explains that once a primitive life form arose with the ability to self-replicate, there is no difficulty explaining how the fundamental properties of life, diversity, unity, complexity and adaptation could develop. He says:

> "Random mutations plus natural selection are one surefire way to generate biological information, extending a short random genome over time into a long random genome. Chance in the guise of mutations and law in the guise of selection form just the right combination of randomness and order needed to create "the impossible object". The necessary information comes, as we have seen, from the environment."[51]

So the power of the blind watchmaker, the sculpting of the cogs and wheels that are mutation into the multitudinous different watches that are complex life, is done by the "environment," by natural physical forces like weather, and by the natural biological forces of other individuals and other species in an ecosystem. We have heard this before and are left with intangibles we cannot test. Can we get closer to the reasoning of the microevolution to macroevolution leap some other way? If you ask Richard Dawkins this question, he points to the natural selection of weasels. We should look at them then.

§4.7 NATURAL SELECTION OF WEASELS

Dr. Dawkins writes so well his explanations can easily be followed. This is what he teaches in regard to the micro- to macro inference in his enormously influential book *The Blind Watchmaker*. He first sets the scene....
> "We have seen that living things are too improbable and too beautifully 'designed' to have come into existence by chance [§1.3]. How, then, did they come into existence? The answer, Darwin's answer, is by gradual, step-by-step transformations from simple beginnings, from primordial entities sufficiently simple to have come into existence by chance. Each successive change in the gradual evolutionary process was simple enough, *relative to its predecessor*, to have arisen by chance. But the whole sequence of cumulative steps constitutes anything but a chance process, when you consider the complexity of the final end-product relative to the original starting point. The cumulative process is directed by non random survival."[52]

He explains the concept in a later book as akin to climbing a high mountain in small steps.[2] Mount Improbable he calls it, and the mountain is metaphorical for macroevolution and the illusory appearance of design in nature (§17.2). In order to show the powerful effect of cumulative selection and how it can reasonably generate the illusory appearance of apparent design, he gives the following example:

How long do you think it would take a monkey typing at random on a keyboard with 26 capital letters and a spacebar to produce the following sentence from Hamlet: METHINKS IT IS LIKE A WEASEL?

Since there are 27 possible keys for each of the 28 positions of this sentence, and each is equally likely to be selected, the odds of getting it at first try is $(1/27)^{28}$ which is around 1 in 10^{40} or one chance in ten thousand, trillion, trillion, trillion. The number for this so-called "blind search" is massive and so a random undirected natural process by our monkey to construct this complexity is hopeless, even for this short sentence much less the entire genome of an organism which in his words is more "than all 30 volumes of the *Encyclopedia Britannica* put together". Mount Improbable seems impossibly high without a Designer.

He then makes a modification to illustrate the power of cumulative selection to resolve the impasse and show how evolution really works. You employ an evolutionary algorithm. There is a multitude variety and while they vary in sophistication they all have two components. One part generates variation by some chance process and another sorts it by some fixed law-like process. Evolutionary algorithms model evolution because evolution also consists of two such steps (§2.3). Evolutionary algorithms like this one have persuaded many by supposedly illustrating how macroevolution can inevitably follow from microevolution without design. The argument goes like this:

1. Start with a random sequence string of 28 characters and spaces, say WDLMNLT_DTJBKWIRZREZLMQCO_P,
2. Randomly alter (i.e. "mutate") all of the positions that do not agree with the target sequence.
3. Choose (i.e. "select") the sequence that most closely matches the target.
4. Repeat the procedure again and again by leaving ("selecting") letters at positions that match the target but randomly altering those at positions that do not ("accumulate selected mutation").

How long does it take to type the sentence now?
After 1 generation it's WDLTMNLT_DTJBSWIRZREZLMQCO_P
after 10 it's   MDLDMNLS_ITJISWHRZREZ_MECS_P
after 20 it's   MELDINLS_IT_ISWPRKE_Z_WECSEL
after 30 it's   METHINGS_IT_ISWLIKE_B_WECSEL
after 40 it's   METHINKS_IT_IS_LIKE_I_WEASEL
after 43 it's   METHINKS_IT_IS_LIKE_A_WEASEL

Instead of 10^{40} the computer gets it in only 43 generations! Cumulative selection is impressive all right. The hopelessness of the impasse has been traversed and "Mount Improbable" scaled easily. Generating the entire works of Shakespeare now appears feasible and so does generating the different genomes of life by cumulative selection. Evolution really is amazing. Or is it?

What has he shown us? Not natural selection. Instead of being blind and purposeless and undirected this computer simulation was, well, the opposite. We knew the result before we even started. In fact there was no alternative outcome but that target verse from Act III Scene II of *Hamlet*. Far from generating complexity the cumulative selection generated none whatever. Information theory considers complexity to be proportional to the reciprocal of the probability of its occurrence. The more improbable an event so the more information it contains and the more complex it is. We look at this in detail later (§16). For now in this example however, we see the probability of getting to the target was 100%. The information complexity is zero. What we want to know is how biological complexity got to be so high and how it purposelessly accumulates by microevolution through mutation and natural selection. He gives us no answer here.

Weasel selection is even weaker than a artificial selection example. Neither address the issue at hand which is to demonstrate how we are to get new biological information that is macroevolution from microevolution. Having shown us this dramatic example of "evolution" by cumulative selection Dr. Dawkins does concede its drawbacks:

> "Although the monkey/Shakespeare model is useful for explaining the distinction between single-step selection and cumulative selection, it is misleading in important ways. One of these is that, in each generation of selective 'breeding', the mutant 'progeny' phrases were judged according to a resemblance to a *distant ideal* target, the phrase METHINKS IT IS LIKE A WEASEL. Life isn't like that. Evolution has no long-term goal."

Why mention it then? The model is irrelevant by his own admission. More importantly, he has nothing better to offer. What happens in real life i.e. with natural selection? He explains:

> "In real life, the criterion for selection is always short-term, either simple survival or, more generally, reproductive success." He tells us no more about how this could generate complexity over time when weasel selection and artificial selection could not, nor does he do any computer simulations with letters or numbers that at least model what he claims.

Instead he switches tack, to a computer simulation of a recursive branching algorithm, which is subjected to random changes and selection to generate stick figures with variant morphologies that he calls "biomorphs". This produces all manner of line drawings of fantastic fuzzy imaginary creatures. Though he gets wistful about their diversity and beauty, if they are intended as a metaphor for the appearance of the diversity of life these cartoons are not even close enough to be a miss. He produces no numbers to confirm the activity is generating true novelty, nor does he return to the weasel example to show how it would happen in real life. It's all taken as a brute given. Without any mathematics the appearance of the power of natural selection at making biomorphs is kept immune from scrutiny. Pop goes the weasel.

The power of evolutionary algorithms, and I mean all of them, is no better than performing a blind search. Their supposed power at generating complexity is always tucked into the "fitness function" somewhere, the second law-like "natural selection" of its two components, and the complexity they make is always present from the start, as seen in weasel natural selection. This has been formally demonstrated by the aptly named no-free-lunch theorems, the first of which was proven by David Wolpert and William Macready in 1996.[53,54] What the no-free-lunch theorems do is provide an accounting of the subtle powers of an evolutionary algorithm's fitness function which can be beguiling. They appear powerful complexity generators when in fact complexity can all be extracted from the source code of the original algorithm. When it comes to getting biological complexity for free for the most complicated things we know of, life forms, it should not take mathematics to convince us. All the same these theorems are there for those who need it. There would seem no escape.

Stuart Kauffman now at the University of Calgary proposes an algorithm nonetheless. His reasoning is instructive because it shows the straightjacket imposed by the Naturalistic Premise (§3.2). Evolution must be true, somehow. Kauffman says:

"The no-free-lunch theorem says that, averaged over all possible fitness landscapes, no search procedure outperforms any other...In the absence of any knowledge, or constraint, on the fitness landscape, on average, any search procedure is as good as any other. But life uses mutation, recombination, and selection. These search procedures seem to be working quite well. Your typical bat or butterfly has managed to get itself evolved and seems a rather impressive entity. The no-free-lunch theorem brings into high relief the puzzle. If mutation, recombination, and selection only work well on certain kinds of fitness landscapes, yet most organisms are sexual, and hence use recombination, and all organisms use mutation as a search mechanism, where did these well-wrought fitness landscapes come from, such that evolution manages to produce the fancy stuff around us?"

In other words so irrational is the notion that the appearance of design is real that wrestling on with this problem, though mathematically demonstrably futile, is not only rational it's fruitful. He does not have an solution at present though. Actually he admits: "No one knows" and so opens himself to being wrong twice over, and for different reasons. While he does not have a solution, he does point in the direction where he thinks the solution is to be found.

It's the proposal that in evolution the fitness function also evolves (they are called "co-evolving fitness landscapes") and is caused by co-evolving fitness functions. ID theorist William Dembski has brought that idea firmly back to earth again (§16.4). He points out Kauffman's notion merely multiplies the original problem with each co-evolving step. How did the fitness functions get to be so clever at evolving so well? Instead of providing relief, Kauffman's efforts merely underline the harsh realities of the original crushing problem he sought to escape. There's no such thing as a free lunch. Surely we know that.

§4.8 PHILOSOPHY DRESSED IN THE ROBES OF SCIENCE

We've made no headway in getting our question answered. Niles Eldredge restates our question again and our problem:

"How do ultra-Darwinists handle macroevolutionary adaptive change? Dawkins' (1986) *The Blind Watchmaker* considers the problem of adaptation in detail...Dawkins is at pains to reaffirm the plausibility of the Darwinian vision of incremental change accumulating under natural selection to shape the adaptations of organisms – the design in nature seen by biologist and creationist alike. Though elaborately explicit on *how* selection induces incremental change, Dawkins is almost silent on the ecological and evolutionary contexts of adaptive change, especially long term adaptive change; that factors that promote and retard adaptive change through evolutionary time...there remains a critical gap between a Darwinian-based understanding of how selection works to modify phenotypes on a generation-by-generation basis and a theory that addresses actual patterns of such change in the history of life."[55]

Our problem is seeing in the detail how macroevolution comes from microevolution, and which by the way Dr. Eldredge has not solved either (§6.3). The question he raises is not how much time is required for the process. It's whether microevolution could generate macroevolution at all. This is because of the need of an input to generate entirely new genetic information and an entirely new order, not simply manipulate pre-existing traits. The literary analogy is we are being asked to accept is that with enough time the text of the genetic code of one ancestral organism can be modified in small steps to an entirely different code and thus to an entirely different organism. This is the idea that the genius of Hamlet can be gradually modified, letter-by-letter, into the genius of Othello. And with every change along the way, spelling, syntax, grammar and an intelligible plot must of course be preserved, because the intermediate organisms in a lineage must remain viable. Any interruption is otherwise an extinction that ends the effort. This is only another verbal argument however, and it is one that is countered. At best then, it speaks to the stalemate in the evolution-creation debate. The reason for the stalemate is because this aspect of the debate is conducted with eloquent description and impenetrable reasoning, not mathematics.

In this regard consider the challenge to macromutation made by Richard Goldschmidt in *The Material Basis of Evolution* back in 1939 (§2.2). Although he was convinced about the truth of evolution, he questioned what it takes for granted. Namely that new gene mutations arise somehow, are selected for somehow, and the process is repeated over and over to somehow to eventually account for all the complexity and diversity of life. This is what he said:

"At this point in our discussion I may challenge the adherents of the strictly Darwinian view...to try to explain the evolution of the following features by accumulation and selection of small mutants: hair in mammals, feathers in birds, segmentation of arthropods and vertebrates, the transformation of the gill arches in phylogeny including the aortic arches, muscles, nerves, etc.; further, teeth, shells of mollusks, ectoskeletons, compound eyes, blood circulation, alternation of generations, statocysts, ambulacral system of echinoderms, pedicellaria of the same, cnidocysts, poison apparatus of snakes, whalebone, and, finally, primary chemical differences like hemoglobin vs. hemocyanin, etc. Corresponding examples from plants could be given."[56]

His challenge was made over seven decades ago and it remains unanswered. There are plenty of theories but further theories to support evolutionary theory are not in the spirit of his challenge. He sought explanations from direct evidence, not more narrative.

As example of what constitutes the current quality of how Goldschmidt's challenge is met, this is Richard Dawkins' reply using the example of the eye. Further examples are the quintessence of his writing and to which the reader is recommended.[57 60] He says: "Richard Goldschmidt's problem...turns out to be no problem at all." He argues by claiming, without independent evidence, that the order we see in nature can be subdivided into smaller (evolutionary) steps, so working backwards from what appears to be designed to the illusory appearance of design. "The important thing about light intensity...distance of image from the center of retina, and similar variables, is that they are all *continuous* variables". But this is all in his mind because it is not seen in nature – half an eye is not a reality. It comes as no surprise then to see him arguing that the extrapolation inference that is Step 3 in the Argument From Apparent Design is really linear after all (§3.6):

> "Every one of us knows from personal experience, for example on dark nights, that there is an insensibly graded continuous series running all the way from total blindness up to perfect vision, and that every step along this series confers significant benefits. By looking at the world through progressively defocused and focused binoculars, we can quickly convince ourselves that there is a graded series of focusing quality, each step in the series being an improvement over the previous one. By progressively turning the colour-balance knob of a color television set, we can convince ourselves that there is a graded series of progressive improvement from black and white to full color vision...Considering each member of the series of Xs connecting the human eye to no eye at all...Not only is it clear that part of an eye is better than no eye at all. We also can find a plausible series of intermediates among modern animals...Some single-celled animals have a light sensitive spot with a little pigment screen behind it. The screen shields it from light coming from one direction, which gives it some 'idea' of where the light is coming from. Among many-celled animals, various types of worm and some shellfish have a similar arrangement, but the pigment-backed light-sensitive cells are set in a little cup...Carry ON.[61]

The trouble is his explanation is the only independent evidence evolution happened that way. The problem is that arranging modern day creatures with marvelous and marvelously different kinds of eyes into an evolutionary series without a demonstrable ancestry is an artificial construct which may have no basis in reality, no matter how plausible it may sound in print. It bears pointing out that not even one species among all the roughly 1.7 million now known can be shown in an unambiguous way to have come about by an accumulation of mutations. The theory is well ahead of the data then, at least on this point.

Lynn Margulis at the University of Massachusetts is the author of the "symbiont" theory of mitochondria. This theory holds that mitochondria, the structures that produce energy for the cell, were once independent bacteria that took up permanent residence in a host bacterium then evolved into the organelles that we see now. She has no doubts about the truth of evolution either. But as one of the foremost experts in microbiology, this is how she feels about the macroevolution of bacteria into eukaryotes: "I have seen no evidence whatsoever that these [evolutionary] changes can occur through the accumulation of gradual mutations."[62] Back at Berkeley, geneticist Richard Goldschmidt was not satisfied by reassurances like Dawkins' either. He said:

> "Microevolution by accumulation of micromutations – we may also say neo-Darwinian evolution – is a process which leads to diversification strictly within the species, usually, if not exclusively, for the sake of adaptation of the species to specific conditions within the area which it is able to occupy...Below this level, microevolution has even less significance for evolution (local mutants, polymorphism, etc.). Subspecies are actually, therefore, neither incipient species nor models for the origin of species. They are more or less diversified blind alleys within the species. The decisive step in evolution, the first step toward macroevolution, the step from one species to another, requires another evolutionary method than that of sheer accumulation of micromutations."[63]

Another demonstration of the stalemate over the supposed inevitability of macroevolution is posed by "living fossils". They are exactly what the word implies in an evolutionary paradigm of biology, and examples include horseshoe crabs, coelacanths, cycads, cockroaches, sturgeons, alligators and aardvarks. Take for example the cockroach. Assuming dating of the Earth and the fossil record to be what science says, the fossil record demonstrates those alive today very similar to fossils dated 280 million years old. Thus they appear not to have evolved. At least in these cases mutations do not equal evolution. Pierre Grassé used that observation to point out a plausibility problem for macroevolution:

> "Schizophytes and others of the same type have existed since extremely remote times, as far back as the Precambrian period. The main historical fact to be remembered is that, in spite of their infinite number, the have only *once* given rise to a cellular organism, following a process about which we know nothing...In nature, at present, schizophytes and among them especially bacteria – because of their great numbers of mutations – vary extensively and in the greatest disorder; but they "go around in circles" revolving around their specific form, and in the end change very little, that is to say, do not evolve. Bacteria have remained bacteria for the past two or three billion years."[64]

Evolutionary theorists counter by arguing "stabilizing selection". Touché. In another lunge Goldschmidt points out a most curious slowing down of the evolutionary rate. Something special happened once back in time to cause the wonderful panorama of life, but since then the evolutionary change has decreased:

"The genesis of phyla stopped in the Ordovician; of the classes, in the ; of the orders, in the Paleocene-Eocene. After the Eocene...mammals and birds continued to specialize in various directions. The extent of evolutionary novelties gradually changed. They no longer affected the structural plan but only involved details. The only form which evolution took was speciation: in insects since the Oligocene, in mollusks since the Miocene, in birds and simians since the Pliocene, and in some glirines and hominids since the Holocene; *Homo sapiens*, the last in line is probably 100,000 years old. Evolution has not only slowed down, it has also decreased in scope and extent." [65]

This too can be countered by evolutionists. We are going in circles because evolutionary theorists have theories to explain these and other supposed difficulties and they can be drawn as quick as a six gun. There would seem no way to refute the eloquent claims about what happened in a history we are unable to witness. How can we know? How can this controversy over apparent or real design, the solution to the Conundrum of life, the reality of micro- to macroevolution, ever progress to objective analysis, to empirical science?

First we need to know the data for the macroevolutionary inference. There are 11 more lines of evidence and microevolution is only the first (*LOE#1*). We will look there next. Second it is to employ mathematics not rhetoric and specifically to deploy information theory. It may have struck you already as odd that for a scientific debate there has been a paucity of numbers in the discussion so far, when science is numbers. It was Leonardo da Vinci who said: "No human investigation can be called real science if it cannot be demonstrated mathematically." If evolution is such a fact why is this so? The lively debate in words from both sides of the evolution-creation debate, one side of which has to be completely wrong, would evaporate with numbers and it would also resolve the matter definitively. "Mathematics is the science of what is clear by itself" Prussian mathematician Carl Jacobi once said. Obviously we need to apply mathematics to the evolution-design question. If we do what do we find?

A meeting between mathematicians and evolutionists with this goal in mind took place in 1966, 1969, 1980 and again in 1984. Interestingly rather than a happy accord over a "fact" as certain as heliocentricity, this meeting of minds was hugely acrimonious. An exchange between the leading lights of evolutionary theory and mathematics was recorded at the first of these meetings, held at the Wistar Institute in Philadelphia in April 1966. The report is worth reading and not just because it is funny. It's because evolution theory was pronounced dead by mathematics that day.

The cause of death was the advent of a new generation of digital computer. At last the evolutionary scenarios so talked about could be formally tested by crunching the numbers. In the end the meeting proceedings spoke more about the unshakable faith of biologists in the evolution construct than they did about macroevolution mathematics. When Manhattan Project participant Stanislaw Ulam declared that at least mathematically there was not nearly enough time for the eye to have been formed by the accumulation of successive mutations, he was told by Sir Peter Medawar and Sir C.H. Waddington that the problem lay with his numbers and not with Darwinism. The fact is, they said, the eye had already evolved and he was conducting his science back to front.

Next Ernst Mayr spoke up to protest, only to say the same thing: "Somehow or other by adjusting these figures we will come out all right. We are comforted by the fact that evolution has occurred." MIT mathematician Murray Eden then told the audience life could never begin by random selection, which of course only leaves design (§1.2.). He produced computer simulations by which if even the entire planet were covered with bacteria to a depth of an inch for five billion years, it was still mathematically impossible for even one pair of genes to be produced by mutation. We look at this problem later (§12.4.2). The folding of proteins he also declared to be mathematically impossible. But it was computer scientist Marcel-Paul Schutzenberger from the University of Paris who really stirred up the hornet's nest.

He concluded that by his numbers, "[t]here is a considerable gap in the neo-Darwinian theory of evolution, and we believe this gap to be of such a nature that it cannot be bridged with the current conception of biology." Sir C.H. Waddington demonstrated how any alternative to evolution must always be biologically false as the Naturalist Premise demands, because he snapped back: "Your argument is simply that life must have come about by special creation." (§3.2,§3.4,§19) It was time for another surprise, because it turned out the mathematicians were just as opposed to the implications of their own calculations as were the biologists. Although they had just razed Darwinism, they refused to accept they had just demonstrated ID. Schutzenberger bristled at Waddington's charge, and in chorus with his colleagues in the audience, shouted back: "No!" The mathematicians offered no alternative however.

At a subsequent meeting held at Chicago's Field Museum of Natural history in 1980, it was decided there would be no record of the proceedings. Considering the interest the three previous meetings had generated, that decision is extraordinary and begs the question why. It appears searches for scientific truth can be made in smoky backrooms after all. I find no mention of the heavyweight mathematicians being invited this time either, although the debate was again over the feasibility of extrapolating macroevolution from microevolution.[66] From the information which was disclosed, it seems the proceedings were as rancorous as before though. In his report to the journal *Science*, Roger Lewin summarized what happened. The conclusion of the meeting was extraordinary alright, almost as extraordinary as it was for *Science* to publish it: "The central question of the Chicago conference was whether the mechanisms underlying microevolution can be extrapolated to explain the phenomena of macroevolution. At the risk of doing violence to the positions of some of the people at the meeting, the answer can be given as a clear, No."[66]

§4.9   IN CONCLUSION: HALF A "FACT" IS A HALF-TRUTH

What can we say so far? For a theory we're told is a "fact" there seems an awful lot of disagreement over its certain fundamentals, and an application of mathematics to the science is more incendiary than salve. Perhaps the macroevolution inference is becoming clearer nonetheless. We've covered a lot of ground and find no mathematics, but instead lot's of opinion and narrative theory. Microevolution becoming macroevolution after studying natural selection is as intangible as when we started - we're not making headway by studying natural selection. Can we ever condense the macroevolutionary inference into something objective we can get our hands on and test?

Of course we can. Macroevolution demands two assumptions. The first is that biological novelty will inexorably accumulate, and the second is that biological complexity is the result of cumulative selection. In other words biological information is reducible to a multitude non-complex, non-saltationary, cumulative, microevolutionary steps. Mark Ridley calls it the "principle of uniformitarianism". Richard Dawkins calls them "continuous variables" (§3.6;§4.10). The notion of micro- to macroevolution would be rejected if biology was found to be digital or quantized then. We look at that later. To present that analysis now and mathematically, is to shortcut the purpose of our journey which has scarcely begun. We need to proceed by a systematic inspection of all of the evidence for evolution, from the perspective of science, because we have two tasks to perform. While we need to solve the Conundrum, we also need to explain why the controversy is intractable on both sides of this debate. To that end we need to tour each of the main lines of evidence (*LOE#*2-12), one non-saltatory step at a time.

In summary so far though evolutionists overstate their case. Microevolution is indeed a "fact", but macroevolution is just an inference. Perhaps it is true, but how do we know? Nature at least looks designed and the evidence of microevolution rejects an automatic extrapolation into macroevolution. The science establishment does not see it that way and Verne Grant explains why:

> "Evolution was indeed a theory at one time, but a bit of water has flowed over the dam since 1859. Much of evolution is now in the category of verified fact. Evolutionary changes at the levels of microevolution and speciation have been observed by competent biologists and can now be regarded as proven facts. Evolutionary changes at the level of macroevolution are in a somewhat different position, in that they are historical phenomena and consequently were not observed directly by any human observer. Lack of eyewitness testimony is an inherent problem in any attempt to reconstruct past events. The biblical account of creation, which was written long after the event, suffers from this same difficulty. The simple dichotomy of theory vs. fact does not cover the situation in macroevolution. Evolutionary history cannot be observed directly, but can be inferred from the fossil record. The fossils themselves are facts. Macroevolution is a necessary inference from those facts."[67]

So let's look at the facts. We begin by an inspection of the fossil record.

REFERENCES

1. Mayr, E. One Long Argument (Harvard University Press, Cambridge, MA, 1991),4.
2. Dawkins, R. Climbing Mount Improbable (W.W. Norton & Company, New York, 1996.
3. Pennock, R. T. Tower of Babel (The MIT Press, Cambridge, MA, 1999),55.
4. Grasse, P.-P. Evolution of living organisms (Academic Press, Inc., New York, NY, 1977),87.
5. Grasse, P.-P. 130 (Academic Press, Inc., New York, NY, 1977) 130.
6. Dawkins, R. The Blind Watchmaker (W.W. Norton & Company, New York, 1996),21.
7. Mayr, E. Darwin's influence on modern thought. Scientific American 283, 79-83 (2000).
8. Pigliucci, M. Design yes, intelligent no. A critique of intelligent design theory and neocreationism. Skeptical Inquirer 25 (2001).
9. Simpson, G. G. This view of life 273 (Harcourt, Brace & World, 1964.
10. Ridley, M. Evolution (Blackwell Science, Inc., Cambridge, 1996),72.
11. Stanley, S. M. Macroevolution, pattern and progress (Johns Hopkins University Press, Baltimore, 1998),192-3.
12. Lewontin, R. Adaptation. Scientific American, 213-31 (1978).
13. Gould, S. J. Wonderful Life (W.W. Norton & Company, Inc., New York, 1989),236.
14. Waddington, C. H. in Evolution after Darwin (ed. Tax, S.) 381-402 (1960) 381-402.
15. Johnson, P. E. Darwin on trial (InterVarsity Press, Downers Grove, IL, 1993),22.
16. Grant, P. R. & Grant, B. R. Unpredictable evolution in a 30-year study of Darwin's finches. Science 296, 707-11 (2002).
17. Haldane, J. B. S. Darwinism under revision. Rationalist Annual, 24 (1935).
18. Macbeth, N. Darwin retried. An appeal to reason (Gambit Incorporated, Boston, MA, 1971),63.
19. Kettlewell, H. B. D. Darwin's missing evidence. Scientific American 200, 48-53 (1959).
20. Dawkins, R. Climbing Mount Improbable (W.W. Norton & Company, New York, 1996),87.
21. Majerus, M. E. N. Melanism. Evolution in action (Oxford University Press, New York, 1998),97-156.
22. Wright, S. Variability within and among natural populations (Chicago University Press, Chicago, 1978.
23. Ridley, M. Evolution (Blackwell Science, Inc., Cambridge, 1996),109.
24. Ruse, M. Darwinism Defended (Addison-Wesley Publishing Company, Reading, MA, 1982),101.
25. Futuyma, D. J. Science on trial (Sinauer Associates Inc., Sunderland, 1995),125-6.
26. Coyne, J. A. Not black and white. Nature 396, 35-6 (1998).

27. Clarke, C. A., Mani, G. S. & Wynne, G. Evolution in reverse: clean air and the peppered moth. Biological Journal of the Linnean Society 26, 189-99 (1985).
28. Miller, K. R. & Levine, J. Biology (Prentice-Hall, Upper Saddle River, NJ, 2000.
29. Raven, P. H. & Johnson, G. B. Biology (McGraw-Hill, New York, 2002.
30. Guttman, B. S. Biology (McGraw-Hill, Boston, 1999.
31. Patterson, C. Evolution (Cornell University Press, Ithaca, NY, 1999.
32. Grant, B. R., Owen, D. F. & Clarke, C. A. Decline of melanic moths. Nature 373, 565 (1995).
33. Lees, D. R., Creed, E. R. & Duckett, L. G. Atmospheric pollution and industrial melanism. Heredity 30, 227-32 (1973).
34. Grant, B. S. & Howlett, R. J. Background selection by the peppered moth (Biston betularia Linn).: individual differences. British Journal of the Linnean Society 33, 217-32 (1988).
35. Futuyma, D. J. Evolutionary Biology (Sinauer Associates, Sunderland, 1998.
36. Schraer, W. D. & Stoltze, H. J. Biology: The Study of Life (Prentice Hall, Upper Saddle River, N.J., 1999.
37. Kettlewell, H. B. D. in Evolution (ed. Ridley, M.) 62-6 (Oxford University Press, Oxford, 1997) 62-6.
38. Cook, L. M., Grant, B.S., Sacceri IJ & Mallett, J. Selective bird predation on the peppered moth: the last experiment of Michael Majerus. Biology Letters 8, 609 (2012).
39. Sato, A. et al. On the origin of Darwin's finches. Molecular Biology and Evolution 18, 299-311 (2001).
40. Lack, D. Darwin's finches (Cambridge University Press, Cambridge, 1983 (1947).
41. Zimmer, C. Darwin's avian muses continue to evolve. Science 296 (2002).
42. Wells, J. 159-75 (Regnery Publishing, Inc, Washington, 2000) 159-75.
43. Grant, P. R. & Grant, B. R. Hybridization of bird species. Science 256, 193-7 (1992).
44. Ridley, M. Evolution (Blackwell Science, Inc., Cambridge, 1996.
45. Wells, J. Icons of Evolution (Regnery Publishing, Inc, Washington, 2000),174-5.
46. Mayr, E. Populations, species and evolution (Harvard University Press, Cambridge, MA, 1970),12.
47. Ridley, M. Evolution (Blackwell Science, Inc., Cambridge, 1993),40-2.
48. Irwin, D. E., Irwin, J. H. & Price, T. D. Ring species as bridges between microevolution and speciation. Genetica 112-113, 223-43 (2001).
49. Ridley, M. Evolution (Blackwell Science Ltd, Malden, MA, 2004),53.
50. Ridley, M. Evolution (Blackwell Science, Inc., Cambridge, 1993),43.
51. Davies, P. The Fifth Miracle (Simon & Schuster, New York, 1999),120.
52. Dawkins, R. The Blind Watchmaker (W.W. Norton & Company, New York, 1996),43
53. Wolpert, D. H. & Macready, W. G. No Free Lunch theorems for optimization. IEEE Transactions on Evolutionary Computing 1, 67-82 (1997).
54. Dembski, W. A. No Free Lunch (Rowman & Littlefield, Lanham, MD, 2002.
55. Eldredge, N. Macroevolutionary dynamics (McGraw-Hill Publishing Company, New York, NY, 1989),60-1.

56. Goldschmidt, R. B. The material basis of evolution (Yale University Press, New Haven, 1940),6-7.
57. Dawkins, R. The Blind Watchmaker (W.W. Norton & Company, New York, NY, 1996.
58. Dawkins, R. River Out Of Eden (Basic Books, New York, 1995).
59. Dawkins, R. Climbing Mount Improbable (W.W. Norton & Company, New York, 1996),73-91, 73-107 (ch).
60. Dawkins, R. The Selfish Gene (Oxford University Press, Oxford, 1989),13.
61. Dawkins, R. The Blind Watchmaker (W.W. Norton & Company, New York, NY, 1996),84-6.
62. Mann, C. M. Lynn Margulis: Science's Unruly Earth Mother. Science 252, 378-81 (1991).
63. Goldschmidt, R. B. The material basis of evolution (Yale University Press, New Haven, 1940),183.
64. Grasse, P.-P. Evolution of living organisms (Academic Press, Inc., New York, 1977),59-60.
65. Grasse, P.-P. Evolution of living organisms (Academic Press, Inc., New York, 1977),70-1.
66. Lewin, R. Evolutionary Theory Under Fire. Science 210, 883-7 (1980).
67. Grant, V. The evolutionary process. A critical review of evolutionary theory (Columbia Unversity Press, New York, 1985),13-4.

5. Bones of contention

§5.1 THE IMPORTANCE OF HARD EVIDENCE

Why are fossils so important to evolution? Because macroevolution is about the history of life and fossils *are* historical life. Fossils are the most important of all of the lines of evidence for evolution because they are as direct as the evidence could get! Paleontologist Steven Stanley says it this way: "While many inferences about evolution are derived from living organisms, we must look to the fossil record for the ultimate documentation of large-scale change. In the absence of a fossil record, the credibility of evolutionists would be severely weakened. We might wonder whether the doctrine of evolution would qualify as anything more than an outrageous hypothesis."[1] The fossil record is where macroevolution's truth most needs to be confirmed. Well is it?

If we take the word of science there's absolutely no doubt about it. As Harvard's Modern Synthesis co-architect George Gaylord Simpson famously said, "Fossils demonstrate that evolution is a fact".[2] Why? Because fossils can be reconstructed into organisms ancestrally related to other reconstructed organisms, and when aligned chronologically they show progressive morphologic change in accord with the predictions of Darwinism confirming the macroevolution hypothesis. This is *LOE#2* and *LOE#3* (Table 3.1). If so, this line of evidence is very impressive indeed.

Still, it could never be more than an inference. The reason is fossils are variably incomplete remnants of never seen and now extinct organisms that first must be reassembled from what remnants remain, then interpreted, and then an assumption made that the one is the ancestor of another. The only thing that can be "a fact" about a fossil is its morphology, its constituents and where it was found! Numbers can be assigned to these properties, but what they mean biologically and historically is the question paleontology must answer and that answer is always, at best, an inference. Even its age is an inference (§5.3.1). Henry Gee is Senior Editor for the journal *Nature*. He is also a paleontologist. He wrote a whole book about this matter called *In Search of Deep Time*. He says:

"If we can never know for certain that any fossil we unearth is our direct ancestor, it is similarly invalid to pluck a string of fossils...arrange these fossils in chronological order, and assert that this arrangement represents a sequence of evolutionary ancestry and descent...To take a line of fossils and claim that they represent a lineage is not a scientific hypothesis that can be tested, but an assertion that carries the same validity as a bedtime story".[3]

This paleontologist is in no doubt that macroevolution is true however, and so begs the question why. This is also when studying fossils starts to get really interesting. When his views were seized upon by anti-evolutionists as an admission of defeat, Dr. Gee responded. He never denied what he said though. He just rephrased himself as if a whole book hadn't made his position clear already. He said: "Unfortunately many paleontologists believe that ancestor/descendent lineages can be traced from the fossil record, and my book is intended to debunk this view".[4] In this you might already see why a study of fossils and their handlers could be full of surprises.

How do paleontologists study fossils exactly? Niles Eldredge at the American Museum of Natural History is our guide. He asks what a paleontologist does, and then answers:

"[C]ollecting fossils is very much like going to the beach. If you sort the different kinds of shells from the high-tide line, they will fall into a number of discrete piles. If you then go under the waves and retrieve living specimens, replete with all the soft anatomical parts, and with reproductive functions intact, you will establish that your piles of shells on the beach correspond to discrete reproductive communities. A paleontologist cannot check living samples beneath the waves. But the inference is that our samples of different shells, drawn from a single layer [of geologic strata], conform to biological species as any similar sampling of beach shells would. Then our paleontologist collects samples of shells from older and younger strata, and finds much the same sorts of shells, equally sortable into different piles. It is inference, to be sure."[5]

Unable to "look-beneath-the-waves" means, among other things, that a paleontologist invariably does not have the "soft anatomical parts" of the internal organs when making reconstructions of exactly what kind of animal the fossil represents, and of course she can never know which fossil creature so constructed could breed with any other. Verne Grant spells out the science for us: "Evolutionary history cannot be observed directly, but can be inferred from the fossil record. The fossils themselves are facts. Macroevolution is a necessary inference from these facts."[6] This is *LOE#2* again but the question now is how and why "macroevolution" is so compelling from fossils if it requires such inferences? Perhaps fossils overwhelmingly demonstrate macroevolution as he says nonetheless.

There are some obvious hurdles that will have been satisfied then. Arrangements of the bones of selected extinct organisms into chronological sequences are demonstrative of ancestral relationships, we're told. We need to be satisfied such a series is not artificial, but how can we tell? There are multitudes of fossil species. Billions of them (§1.2.1). And it's only evolution that declares such relationships exist. Dry bones can't declare the notion of macroevolution a fact, only support the claim. At an even more fundamental level, how can we tell a particular fossil organism is a separate species from some other fossil organism? Ernst Mayr's definition of species as reproductively independent organisms won't do here (§2.2.2). And even accepting a selected arrangement of fossil organisms as an ancestrally related series and their morphological changes as caused by macroevolution, the next question is how, exactly, did evolution do it? The answers to all of these questions rely on inferences. Reasonable they might be, but they are a compounding of inferences upon inferences all the same.

As a result an alignment of a particular selection of fossil specimens defined on morphologic grounds as being separate species by inference, arranged in a particular sequence by inference, and considered ancestrally related to another by inference, is at best no more than supportive of an *a priori* inference of macroevolution. This is still not evidence to be sneezed at, assuming it's true, but it's rather removed from claims fossils declare evolution "a fact".[2] Especially when that "fact" is declared equivalent to the "fact" the earth moves around the sun and when Stephen Jay Gould, a paleontologist, says so also.[7-10] (§1.4) Such claims rather seem to speak to something else - the convicting power of macroevolution and where that might come from. At least we know it can't be from fossils, and this is my only point. We have two more immediate tasks before considering that question about the origin of the conviction however.

The first is whether fossils really are compelling inferential evidence of macroevolution. In other words whether the fossil evidence is coherent with evolutionary theory, whether its working inferences and hypotheses are reasonable, and whether they withstand a skeptical scrutiny. Our second task is not so much an inspection of the fossil evidence as an inspection of their handlers. How does this science do business gathering and interpreting fossil evidence? What exactly is the reasoning of the seashell-collecting-but-unable-to-check-under-the-waves enterprise that is evolutionary paleontology? This will also answer that other question of why and in what sense macroevolution is today a fossil "fact".

Knowing now how important the fossil evidence is, it might come as surprise to discover that Darwin proposed his theory without any fossils. He noted fossils showed biological turnover, and more "complex" species (like mammals) appeared more recently in the fossil record than less "complex" (like worms), but these were observations equally compatible with creationism and indeed that was their prevailing explanation. What he needed to show was a succession of transitional fossil forms consistent with his postulate of evolution, but he could not show even one (§1.5).[11] "In fact Darwin's evidence for his theory," says Oxford University anatomist Sir Wilfred Le Gros Clark, "was derived almost entirely from his observations on living organisms – their variation in nature and under domestication, the tendency of their populations to increase rapidly in numbers and the inference that they are necessarily exposed to what he termed the "struggle for existence", their geographical distribution and so forth. All this kind of evidence...was no more than circumstantial".[12] Fossil evidence was still required. Although evolution theory was constructed without fossils, we expect confirmatory fossil evidence has emerged since then of course.

The change in paleontological opinion is ironic either way, because rather than using fossils to prove evolution, Darwin was forced to defend evolution from fossil critics! He used an entire chapter in *The Origin* to do it. Darwin said the absence of intermediates between species in the fossil record that ought to be abundant by his theory of gradualism, was the "most obvious and serious objection which can be urged against the theory".[13] He could account for their absence though. How?

He said "the explanation lies, as I believe, in the extreme imperfection of the geological record"[13]. By this he meant the fossil evidence was an illusion, not a real reflection of the true fossil record, because all the intermediate fossils or "missing links" he was postulating were not yet discovered. We will assess how reasonable this hope was later but either way his claim was a presumption, it was his only defense and it was made in spite of the fossil evidence not because of it (§6.2). You will also notice he was claiming the fossils the most missing were the ones that were really the most abundant. At least in his time then, evolution had a greater truth than fossil data. We expect that's all changed now too, of course.

Another surprise is Darwin's evolutionary interpretations of fossils took a rather long time to take hold. Over eight decades. The surprise is not the duration; the surprise is that it was caused by the skepticism of paleontologists! The two most influential in Darwin's day were Louis Agassiz at Harvard and Sir Richard Owen at the Natural History Museum in London, and both were outspoken critics of Darwinism. They interpreted fossils as instead the remnants of a succession of supernatural creations and extinctions.

What can we conclude so far? There is a curious back to front reasoning at work and we've scarcely begun. Fossils became evidence of evolution after evolution theory was accepted not before. It was not until 85 years after *The Origin* was published to be exact. It was with the publication of *Tempo and Mode in Evolution* in 1944 and *The Major Features of Evolution* in 1953 by George Gaylord Simpson that paleontology finally entered the evolution fold. As if to make up for lost time, the union now was hot and heavy. Simpson was insistent: "Fossils demonstrate that evolution is a fact,"[2] and with rare exception paleontologists have been as insistent ever since.[3] Take Richard Cowen at the University of California at Davis. He says: "Evolution might have been accepted without fossil evidence but fossils now seem inexplicable without evolution".[14] The curious thing is that when we read *The Origin* we find Darwin making apologies for fossils. What changed? Perhaps new fossil evidence has been found since then. Before we can answer these questions we need to know what fossils are, how they are researched and how they are interpreted being the variably complete remains of organisms never seen before. The next two sections will provide this background. The inquiry resumes at §5.4 and I will meet you there.

§5.2 WHAT ARE FOSSILS?

Fossils are the remains of ancient plants and animals. The term generally refers to specimens older than the Holocene period dated as beginning 10,000 years ago (Fig.5.1). The word "fossil" comes from the Latin *fossilis* meaning "something dug up". Fossils can be formed in four principal ways: (1) by preservation of the whole organism intact as when trapped in plant resin that turns to amber; (2) by encrustation of an organism's outer surface with a mineral coating of pyrite, carbonates, phosphates and/or silicates; (3) by petrification of the organism by quartz (SiO_2) or calcite ($CaCO_3$) mineralization; or (4) by molding which is when the organism leaves an impression or "cast" of its shape in rock sediment.[15] Animals can also leave evidence of their behavior, and such remains are called "trace fossils". These include fossil footprints, trails, burrows and feces.

The study of fossils is called paleontology, a word coined from the Greek root for "science of ancient life" by the British geologist Charles Lyell in 1838 (1797-1875). Lyell was a friend and intellectual father figure to Charles Darwin who had a profound influence on him, particularly in regard to his ideas of geological "uniformitarianism".[16] In short, Darwin applied Lyell's ideas about rocks to living things (§2.2;§3.6).

Geological uniformitarianism is the concept that the laws governing the physical, chemical and geologic processes which we observe today are identical to those that have always operated in the past. This inference permits extrapolation back in time, which together with evidence used to declare the earth of great age, is a cornerstone for the claim Darwinian gradualism caused all the diversity of life. Microevolution observable becomes extrapolatable to macroevolution never observable because "[t]ime is the great enabler", to quote University of California Irvine geneticist Francisco Ayala.[16] Although Lyell coined the name for the science of paleontology, he was not the founder of this science however.

That person was the Frenchman Georges Cuvier, or more correctly Baron Georges Léopold Chrétien Frédéric Dagobert Cuvier (1769-1832). Cuvier made the key observation that while some fossils are very similar to living creatures others are very different, and represent the remains of entirely novel forms of life. He offered three possible explanations:

(1) Fossils represent the remains of life that is now extinct;
(2) Fossils represent the remains of extant life still undiscovered;
(3) Fossils are the extinct precursor forms of creatures that evolved into other creatures some of which we see alive today.[17]

These explanations are not exclusionary. Cuvier argued the first of them was the correct one, a concept also called "catastrophism". Evidence for the notion is incontrovertible, or as McGill paleontologist Robert Carroll says it, fossils are "irrefutable evidence of extinction" (§5.5.1).[18] This is one fossil fact for sure.

For the record, skeptics of evolution in Darwin's day like Agassiz and Owen thought so too, but under Lyell's influence Charles Darwin did not. He instead declared: "[t]he old notion of all the inhabitants of the earth having been swept away by catastrophes at successive periods is very generally given up" (p.449). Darwin rather chose Cuvier's third hypothesis, namely that evolutionary transformation, not environmental catastrophe was the reason why creatures went extinct. This view accepts fossils as extinct life but goes a lot further to claim competition by natural selection as the cause and macroevolution into totally new species as the consequence.

Which only leaves Cuvier's second explanation. This was the choice of a contemporary of Cuvier also living in France at that time. It was the ambassador from America who later became its third president, Thomas Jefferson. We meet this great man much later (§24.1-3). Jefferson believed Cuvier's second explanation was true not for reasons of science or biology though, but because he believed his God would not permit any of His creation to go extinct. There are a few creatures that satisfy the second Cuvier hypothesis and one is the coelacanth. It is a fish first caught swimming in waters off the Southern African coast in 1937. Its most recent fossils are dated at 70 million years old. As a result the fish had long been given up by the experts as extinct. Another example is the Mountain Pygmy possum (*Burramys parvus*) which has fossils also dated as very ancient, only for it to be found alive and well and living in Australia in 1966. Such cases are rare oddities however, and Jefferson's claim as a general explanation is false. Anyway the God of the Bible makes no such claim, at least not in print (*Gen 6:7; 7:21-23; Isa 65:17; 2 Pet 3:7,11*).

In summary Cuvier's first two explanations can be verified by direct evidence and although the third may appear intuitively satisfying, it is an inference. The question is if the inference is true? To tell, we need to know how fossils are studied. The next section provides an overview of three of the most basic principles of fossil interpretation. They address the questions raised earlier about paleontology. Namely its reliability, its objectivity and its ability to determine the reality of a particular fossil interpretation.

§5.3 A FEW PRINCIPLES OF PALEONTOLOGY

Take in your hand an ancient bone or shell because that's what fossils usually are. Look at its shape and feel its contours. How are we to make sense of this relic? What do these signify about its biology and its history? The first problem is to date it. The second is to interpret it as a living creature knowing that fossilization is rare, that fossils may be damaged over time, that a fossil typically represents only a part of an organism never seen alive, and that the fossil record as a whole may a biased reflection of what creatures were living at the time this one died. The third problem is even more important. How we are to go about relating our reconstructed fossil creature to others? How we are going to interpret its morphologic features, its ridges, grooves, holes and protuberances so as to classify and interpret what they mean in the context of other fossils in the record that look similar and different and which date younger and older than it. Are any ancestrally related or not? The task of paleontology is to reconstruct the original plant or animal to which the fossil(s) belong, reconstruct the context of that organism in the ecosystem of other organisms living at the time, and reconstruct the biological history of that creature in relation to the rest of the fossil record. To do this we at least need to 1) date the fossil, 2) resolve that inherent bias of the fossil record, and 3) determine fossil relationships in the record. Consider these three tasks in turn.

§5.3.1 The Fossil Dating Game

A fossil is nothing but a relic absent an interpretation from paleontology of its biological meaning. The first step to that end is an accurate dating. The methods used to do that in this science are either "relative" by which one fossil is dated against another, or "absolute" in which the fossil or the rock strata in which it is found or found nearby is dated.

Relative fossil dating is founded on the principle of "superposition". This is the notion that the Earth's strata are deposited in successive layers in a geologic column over time with the oldest lying deepest and the youngest at the surface. Telling the chronological sequence of geologic strata is not as simple as it seems though. Movement of the earth's crust can disrupt or even completely invert the original order of geologic strata deposition in a phenomenon known as "overthrust". There is also no known geologic column with a complete set of strata of all geologic time to serve as the reference standard. Even the most extensive geologic formations have discrete layers which are considered "absent". For example the Grand Canyon extends down for over a mile, but this represents is only one percent of the entire geologic column of the earth and more than half of the strata that "should" be present are not.[19]

As a result paleontologists piece together partial sequences of overlapping layers from different areas. In general, marine deposits are more complete than non-marine. The extent of stratigraphic completeness (or incompleteness depending on your view) has been estimated. An average 30 million year section of marine geological deposit is considered around 33% complete for one million year resolution intervals, and variably less than this for non-marine deposits.[20]

Absolute fossil age determination is done by radioisotope dating. Though called absolute, the answer is in the end also made by inference.

The relative proportions of parent and daughter radioisotope pairs in material, generally rock close to where the fossil was found, are measured. Radioisotopes spontaneously decay at stochastically constant rates. The assumption and the inference, is the laws of physics and chemistry have not changed since the universe began and the constants of nature have not changed either. If these assumptions are wrong the dating enterprise collapses. This inquiry will take the dates accorded the Earth by the authorities in science as a given. If we find reasons to doubt the authorities however, say if we were to find macroevolution a hoax (§17.3), or the big bang demonstrably false (§13.4-5), or find something equally outrageous to a Big Bang old earth worldview - say discover that the entire Universe has an axis aligned with our planet (§13.5.2.2), or discover that both nature and the inanimate Universe are demonstrably patterned through and through with ID (§13.4-5;§14.1-4;§15.2-3,5;§16.6-7;§17.3), then I take it these claims about the age of the Earth and the age of fossils will need a skeptical relook also.

Making the assumptions of the science authorities the age of a rock can be calculated very simply by knowing the half-life ($t_{1/2}$) or decay rate of a particular radioisotope and the relative amounts of the parent and daughter isotope in the sample tested. This requires still further assumption and inferences, that no daughter isotope was produced at the time of the rock's formation and no parent or daughter isotope was added or removed since. Willard Libby at the University of Chicago won the Nobel Prize in 1960 for the development of this technique. The most commonly used radioisotope pairs are the conversion of uranium ^{235}U to lead ^{207}Pb ($t_{1/2}$ 7.13 x 10^8 years), potassium ^{40}K to argon ^{40}Ar ($t_{1/2}$ 1.3 x 10^9 years) and uranium ^{238}U to lead ^{206}Pb ($t_{1/2}$ 4.5 x 10^9 years).

While extremely accurate in principle, this is not so simple to perform in practice. The radioactive elements required for the process are typically not located in the fossil itself or in the rocks that fossils are found called sedimentary rock, but in igneous rock. Thus igneous rock needs to be found close to the fossil bed and ideally bracketing in the strata above and below.

The geological time scale used to date fossils is shown in Fig.5.1. You will see the scale is a hierarchy of eras, eons, periods and epochs of non-uniform and non-linear durations and bearing names without consistent etymological basis. The terms are summarized in Fig.5.1 in an attempt to make them more digestible for our journey. They are worth more than passing familiarity, because we will refer back to this chart often before we're done.

EON	ERA	PERIOD		EPOCH	Ma
Phanerozoic	Cenozoic	Quaternary		Holocene	0.011
				Pleistocene	0.8
		Tertiary	Neogene	Pliocene	2.4 / 3.6 / 5.3
				Miocene	11.2 / 16.4
			Paleogene	Oligocene	23.0 / 28.5 / 34.0
				Eocene	41.3 / 49.0 / 55.8
				Paleocene	61.0 / 65.5
	Mesozoic	Cretaceous			99.6 / 145 / 161
		Jurassic			176 / 200
		Triassic			228 / 245 / 251
	Paleozoic	Permian			260 / 271 / 299
		Pennsylvanian			306 / 311 / 318
		Mississippian			326 / 345 / 359
		Devonian			385 / 397 / 416
		Silurian			419 / 423 / 428
		Ordovician			444 / 488
		Cambrian			501 / 513 / 542
Precambrian	Proterozoic	Late	Neoproterozoic (Z)		1000
		Middle	Mesoproterozoic (Y)		1600
		Early	Paleoproterozoic (X)		2500
	Archean	Late			3200
		Early			4000
	Hadean				

Fig.5.1. The geologic time scale taught by science. The subdivisions of geologic time from greatest to smallest are eons, eras, periods and epochs (top row). There are 2 eons, the Precambrian extending over 90% of Earth's history up to 542 Myr ago, and the Phanerozoic (meaning "abundant life") for 10% of total Earth history. Note the time scale in million years ago (Ma) is logarithmic (far right column). Concerning the eras of the Phanerozoic ("-zoic" means "life"), the terms from past to present mean "ancient life" or Paleozoic, "middle life" or Mesozoic and "recent life" or Cenozoic. There are 12 periods. Their names are typically taken from the region where a rock of that dating was first found. Cambrian is from the Roman name for Wales, Ordovices and Silurian were Welsh tribes during the Roman Conquest, Devonian is from Devon, Mississippian is after the Mississippi River Valley, Pennsylvanian similarly, Perm was the name of a Russian province, the first Triassic rocks studied were considered to be of a 'triad' of components, Jurassic is named for the Jura mountains between France and Switzerland, and Cretaceous means 'chalky' because rocks of this age include the prominent chalk formations of English and French coasts. The names for the epochs of the Cenozoic era are based on additions to the suffix "-cene" which means "recent". The successive terms are thus literally "ancient-recent" or Paleocene, "dawn-of-recent" or Eocene, "few-recent" or Oligocene, "fewer-recent" or Miocene, "more-recent" or Pliocene, 'most-recent' Pleistocene and 'entirely-recent' Holocene. (Image credit: U.S. Geological Survey 2009 with minor adaptation).

§5.3.2   Accommodating Incompleteness

Once our fossil is dated the next problem to reckon with is its rarity. Rather than the fossil record being a representative sample of all creatures that have ever lived, consensus thinking is that it's biased. We will need to take account of this in interpreting our specimen then. The several reasons offered are not difficult to understand.

The first is that fossilization is rare. David Raup at the University of Chicago has estimated that only one percent of all the species that have ever lived are trapped in the fossil record.[21] Fossilization is favored by conditions permitting rapid burial in low oxygen concentrations. Thus organisms living in or near water are more likely to be preserved than land dwellers. The geographic locality of the fossil is also important. For instance plants and creatures fossilized in deep ocean floor are less likely to be discovered, and arid rather than tropical conditions favor fossil recovery because of more limited weathering and exposure. Fossilization also occurs more readily in organisms with hard tissues, like shells, bones, teeth and cuticle. Deposits in which hard as well as soft tissues are well preserved are called konservat Lagerstätten ("lode places"), but they are rare. Examples include the Burgess Shale in British Columbia discovered by Charles Doolittle Walcott in 1909, the Ediacara fauna in Australia discovered by R.C. Sprigg in 1947 and the Chengjiang fossils in China discovered by Xian-Guang Hou in 1984. We will inspect these deposits ourselves shortly.

While soft tissues can be preserved, the fossil record is with these rare exceptions essentially confined to hard tissues. This is another source of bias. The reason is because much if not most of the characteristics of an organism reside in its soft tissues. How would we know what a mussel was if all we had was its shell? To borrow Niles Eldredge's analogy of paleontology earlier as shell collecting, we "cannot check living samples beneath the waves".[5] Most animals have no hard parts at all however, and on this basis J. William Schopf at UCLA has calculated only around forty percent of genera could be captured by the fossil record at all. Another source of potential bias is that some fossil species are named on the basis of no more than jaw fragments, or even just teeth! Actually Stephen Jay Gould informs us: "The majority of fossil mammals are known only by their teeth."[22] I told you there's a lot more to the business of studying than first meets the eye.

A living example of the so-called "incompleteness" of the fossil record was alluded to earlier – the coelacanth. The last fossil crossopterygian fish is found in rocks dated 70 million years old and so was taken as long extinct. It had also been taken to be an evolutionary intermediate between fish and amphibians. However in 1937 a living crossopterygian, the coelacanth, was caught off the South African coast and more have been found since. It has even been filmed swimming in the wild. Two lessons can be learned from this fish story.

The first is that a gap of 70 million years was found in the fossil record underlining concerns at least for this fossil, of fossil record reliability. The second is that when the coelacanth was dissected the fossils were discovered to have been over-interpreted from an evolutionary mindset. The skull, teeth and spine fossils of a living coelacanth had been predicted to show amphibian traits in keeping with it being an evolutionary intermediate between fish and amphibians. Even before 1937 there were reasons for skepticism of this evolution. No fin fossils had been found connected to the spine (vertebral column), yet all amphibian limbs are connected to the spine. No fin fossils had been found with a structure suggestive of an "elbow-knee" joint to support gait as amphibians limbs do, and there was no crossopterygian soft tissue fossil to corroborate the notion of fish-to-amphibian internal organ features either. There was no way to refute the hypothesis however and the evolutionary notion was consensus. Until a live coelacanth was found, that is.

When the coelacanth was cut open it was found to be a fish and a rather typical one . For instance it had a spiral-valved intestine, a two-chambered heart and the gut drained to the sinus venosus. No evolutionary intermediate. Today coelacanths have the opposite interpretation from paleontology, namely that they are "living fossils". Living fossils are the same today as their fossils dated millions of years ago (§5.5.3). You will have noticed the coelacanth remains as much evidence of evolution as it was as when interpreted as a fossilized evolutionary intermediate.

The problem here is not that the fossil record was incomplete. The coelacanth was as much a fish as its fossil was a fish. The problem is fish-amphibian intermediate characteristics, an evolutionary prediction, were interpreted from fish fossils. Determining progressive change in the fossil record over geologic time is the cornerstone to proving evolution. The lesson is that without sufficient fossil structural evidence, interpretations of what a fossil truly represents must be made with caution. An evolutionary inference brought to the fossil record and by which it must be interpreted, can be misleading rather than edifying. This might have relevance to that paradox pointed to earlier about what fossils declared in Darwin's day and what they oppositely declare now. We will need to see for ourselves. To cut to the chase though, in the end the concern about paleontology has nothing to do with the record but with the way fossils are being investigated. Ever since Darwin the truth of evolution has lead the truth of the fossil evidence but you will have to see that yourself to believe it (§5.4-5). For now, here is McGill paleontologist Robert L. Carroll again. He tells us about the single most important feature of the fossil record we need to know here: "It would seem a simple matter to establish whether species change progressively or remain stable for long periods of time. In fact, only rarely are geological processes sufficiently regular and continuous to lead to the preservation of numerous fossils that are evenly distributed throughout a long time range of species. Most fossiliferous horizons record only an extremely short period of time and are separated by long gaps from other productive zones. Because of this problem, few paleontologists have been concerned with change within species and most simply assumed that it followed the pattern hypothesized by Darwin." [23] If Carroll's charge is true, and I submit it is, why have fossils ever since George Gaylord Simpson been unequivocal evidence of evolution? We will need to see the fossil evidence and how it is interpreted to answer, but we still need one more section of background first.

§5.3.3　　　　　Interpreting Fossil Relationships

The third problem to resolve before interpreting our fossil, and it is another potential pitfall to overcome, is that evolutionary paleontology rests on inferences. If you think that because paleontology is a science then it is fully objective, this section is for you. There are several reasons why subjectivity can enter fossil analysis. There are the obvious ones already mentioned, like the bias of the fossil record but these are problems that can be accommodated because they are obvious. Others are more serious because they can slip in and slip by unnoticed.

The first is arranging extinct organisms or their fossilized remnants by morphologic criteria is inevitably subjective. At very least an arbitrary decision needs to be made regarding which of a fossil's features are to take precedence over which of its other features. Measuring fossils, their shapes, sizes, lengths and age, is an exact and an exacting science, but how one interprets these numbers afterward is to introduce subjectivity. What constitutes a different species in the fossil record, for instance? Arbitrary decisions made on morphologic grounds. Specifically if fossils are very differently shaped then they are different species or "paleospecies", to use the vernacular. This is by inference and the classification is by subjective criteria.

The second reason subjectivity enters fossil analysis is because the classification of fossil species is conducted according to an evolutionary paradigm and mindset. Where is the independent truth for macroevolution to validate the idea? If macroevolution is true this is all well and good, but what if it is not? If this is not clear consider how Berkeley paleontologist James Valentine explains the Linnaean system employed for fossil classification: "Linnaean taxa at any level have arisen from a species that branched from an ancestral taxon of the same level; indeed they must have done. Therefore, the only Linnaean taxa that are not paraphyletic [i.e. do not contain all the descendants of the most recent common ancestor] are those that have never given rise to daughter taxa of the same rank. For Linnaean families in the fossil record, between about the third and one half seem to be paraphyletic, the others becoming extinct without daughters. Paraphyly provides a nested aggregational hierarchy of taxa with similar ranks. Probably no system of classification has as much organizing power as a properly nested hierarchy."[24]

If evolution is true then his statements and his conclusion are true and the method has the additional advantage of providing an explanatory classification of life. But suppose the macroevolution he assumes is false. All he does then is to impose it on the fossil record. Evolution is found very quickly that way even if it wasn't there before. Fossils so labeled are hardly independent evidence of evolution either. Yet they are taken as independent evidence. They're *LOE#2, LOE#3* (Table 3.1). What's going on? Come and take a look.

The third problem is where this rubber meets the road. How are fossils ordered and classified? By looking at their features of course, as is done for living organisms. This too is reasonable but difficulties come in when we need to determine relationships between fossil specimens at the same time, an earlier time, and at later times in the record. How can we be certain that different fossil forms represent different paleospecies? Consider the fossils of a Chihuahua and a St. Bernard if all we had were their bones and both were extinct and neither had been seen before. Can the morphologic species concept be relied on as a taxonomic foundation on which to base crucial inferences with evolutionary implications? We will have to examine the evidence to tell, but let's grant the paleospecies assumption as valid for now (§5.4-5). The real problems lie elsewhere anyway. To put it simply, how are we to relate one organism to another when all we have are bones? How can we know if one creature is ancestrally related to another?

Inferences. The point is simply that if fossils are to be used as evidence for evolution, they should not be interpreted with an evolutionary mindset and evolutionary inferences to begin with. This is obvious. The bones should declare evolution, not the other way around. Yet perhaps that's not so obvious. In this regard hearing how Stephen Jay Gould describes the process of fossil interpretation is instructive. Imagine yourself a paleontologist again. How do you go about making sense of the fossil you hold in your hand? Learn from the expert.

"Evolutionary and genealogical inferences rest upon the study and meaning of similarities and differences, and the basic task is neither simple nor obvious. If we could just compile a long list of features, count the likenesses and unlikenesses, gin up a number to express an overall level of resemblances, and then equate evolutionary relationship with measured similarity, we could almost switch to automatic pilot and entrust our basic job to a computer" [25]

If so paleontology would be pleasantly objective. Instead one sees how the paradigm of evolution has become the basis for interpretation. The potential pitfall is that all the "evidence" collected under the constraints of an evolution paradigm could unwittingly be used to prove evolution in circular reasoning. This would confirm concerns raised earlier by the curious case of the coelacanth. He continues:

"What do scientists "do" with something like the Burgess shale [fossils], once they have been fortunate enough to make such an outstanding discovery? They must first perform some basic chores to establish context – geological setting (age, environment, geography), mode of preservation, inventory of content. Beyond these preliminaries, since diversity is nature's principal theme, anatomical description and taxonomic placement become the primary tasks of paleontology."[26]

And there lies the rub. How is "taxonomic placement" done? The fossil specimen sits in your hand indifferent to your world view. How do you go from this cold bone stone to a contextual paleontological explanation, by science? Dr. Gould tells us:

"Evolution produces a branching array organized as a tree of life, and our classifications reflect this genealogical order. Taxonomy is therefore the expression of evolutionary arrangement... Classifications are theories about the basis of natural order, not dull catalogues compiled only to avoid chaos. Since evolution is the source of order and relationship among organisms, we want our classifications to embody the cause that makes them necessary."[26]

He explains how a fossil's features are interpreted and compared with others in the record;

"Similarities come in many forms: some are guides to genealogical inferences, others are pitfalls and dangers. As a basic distinction, we must rigidly separate similarities arising by separate evolution for the same function. The first kind of similarity, called homology, is the proper guide to descent. I have the same number of neck vertebrae as a giraffe, a mole, and a bat...because seven is the ancestral number in mammals, and has been retained by descent in nearly all modern groups (sloths and their relatives excepted)."[25]

And there you have it. The fossils are interpreted to fit a theory already assumed to be true, rather than used to confirm that inference first. What's the evidence for this evolution if the fossils are not the evidence? Comparative anatomy and the inference of homology. It's not fossils. This is to see one line of evidence and one that demands evolution be true for it to be even rational, namely homology (*LOE#4-6*), being used to defend another line of evidence as demonstrating the truth of evolution that are fossils (*LOE#2-3*). This would seem circular.

We will inspect homology next, but in essence it's biological similarity in different species taken to be caused by inheritance from a shared evolutionary ancestor (§7). Here is Ernst Mayr's definition: "Homology is due to the partial inheritance of the same genotype from the common ancestor". See what I mean? The similarity you see is an evolutionary concept by definition already. We can't use genotype in the assessment of fossils. Mayr continues "Homology cannot be proven; it is always inferred." Which is another inference. Evolution makes homology true which makes evolution true. How is homology inferred? He explains:

"The claim that certain characteristics in rather distantly related taxa are homologous is at first merely conjecture. The validity of such an inference must be tested by a set of criteria (Mayr and Ashlock 1991)[27], such as position in relation to neighboring organs, the presence of intermediate stages in related taxa, similarity of ontogeny, existence of intermediate conditions in fossil ancestors, and agreement with evidence provided by other homologies."

More disturbing facts about fossils, which are the most important of the lines of evidence for evolution, emerge then. Supposedly independent lines of evidence that confirm evolution, fossils (*LOE#2-3*), homology (*LOE#4-6*) and embryology (*LOE#7-9*), are not so independent after all. Actually they're intimately related and one of them, homology, is entirely dependent on the truth of evolution to begin with. The effect? Round and round we go. Fossils are evolutionary because of homology which is true because of evolution. Fossils are evolutionary and therefore evolution is scientifically true.

This is seen more clearly when Dr. Gould continues from where he left off in contrasting homology with "analogy", the term for biological similarity not caused by inheritance but taken as having evolved independently in unrelated organisms. Think of it as similarity caused by coincidence, not ancestry (§7.1). He shows how the distinction between

194

these two lies in the eye of the beholder, how that eye is an evolutionary inference, and how that inference can be entirely preconceived:

> "The second kind of similarity, called analogy, is the most treacherous obstacle to the search for genealogy. The wings of birds, bats, and pterosaurs share some basic aerodynamic features, but each evolved independently [IO]; for no common ancestor of any pair had wings [AI]. Distinguishing homology from analogy is the basic activity of genealogical inference. We use a simple rule: rigidly exclude analogies and base genealogies on homology alone [CR]. But the basic distinction between homology and analogy will not carry us far enough."[25]

The abbreviations above are inference only (IO), argument from ignorance (AI) and circular reasoning (CR). The truth of homology as reflecting an evolutionary relationship on the basis of similar morphology is assumed, not established independently. This is circular again and yet it is the view of paleontology, everywhere.

In his book *What Evolution Is* Ernst Mayr explains how fossils are the "most convincing" of all the evidence for evolution because they document phylogeny in the specimens. How is that done? It is to reiterate what Dr. Gould just declared, in all its dizzy cicularity:

> "The study of phylogeny is really a study of homologous characters. Since all members of a taxon must consist of the descendants of the nearest common ancestor, this common descent can be inferred only by the study of their homologous character. But how do we determine whether or not the characters of two species or higher taxa are homologous? We say that they are if they conform to the definition of homologous: *A feature in two or more taxa is homologous when it is derived from the same (or a corresponding) feature of their nearest common* ancestor."[28]

The next step toward interpreting our fossil specimen and relating it to others in the record, is to decide which of the characters of our fossil are homologous and which are analogous (§7.4). It is thus to choose which of its traits are "shared-but-primitive", meaning characteristics shared with an ancient common ancestor in the evolutionary tree, and which of them are "shared-and-derived", meaning those shared with a recent common ancestor arising at a distal branch point of the tree and so looking similar just by coincidence (Fig 5.2).

Consider the example of mice and men. Hair is considered a shared and derived character but the vertebral column a common "primitive" character as it's found in all vertebrates. This is supportive evidence for evolution but can hardly be called definitive, and it's done firmly within an evolution paradigm already taken to be a reality. This form of classification is also

195

called "cladistics", "phylogenetic systematics" or "Hennigian systematics" after German biologist Willi Hennig (1913-1976) who pioneered the method. It represents an explicit searching for shared characters as defining evolutionary relationships, and so enables evolutionary reconstructions to be made post hoc. If macroevolution is true it makes good sense but if macroevolution is false the whole dippy business is bunk. There is a jargon associated with the methodology worth knowing for our analysis of the raw data. This is reviewed in three short paragraphs next. Don't panic if it's too heavy. Jump ahead for now.

Hennig coined the terms "plesiomorphy" for a primitive trait present in the common ancestor of a subsequently diversified group of organisms (also called a "clade") and "apomorphy" for a derived trait for those traits representing further specialized change thereafter (Fig.5.2A). "Symplesiomorphy" refers to a shared primitive trait, "synapomorphy" a shared specialized trait, and "autapomorphy" when a trait is unique to one group. The purpose of cladistic analysis is to uncover species with a common ancestry, also called "monophyletic species".

The reasoning of cladistic analysis is easy to understand. Evolution causes the modification of heritable characters on organisms over generations. Thus a new character in an organism and its descendants not seen in that organism's ancestor or relatives, can be inferred as a new evolutionary feature. Those characters shared by organisms from their common ancestor, or shared derived characters, are called "synapomorphies". A "monophyletic" group is one that consists of the common ancestor and all of the descendants, and "sister groups" are two monophyletic groups linked by synapomorphies. For instance hair, the three inner ear bones and mammary glands are only found in mammals. Determining which biological similarities are because of a shared ancestor (called homology and indicative of evolutionary phylogeny) and which instead evolved independently from unrelated species as a result of evolutionary adaptations to environmental conditions (called analogy and not indicative of phylogeny), is the critical step for uncovering true evolutionary trees. An example of homology is the tetrapod limbs found in vertebrates as different as amphibians, reptiles and mammals (§7.1) and an example of analogy are the wings of bees, birds and bats which we are told, evolved independently instead.

In summary we can say three things about fossil interpretation. Darwin's theory was formulated and accepted without and in spite of, fossil evidence, fossil evidence was discovered or accepted as evolutionary eighty-five years later by the Modern Synthesis (§2.1), and the reasoning employed in paleontology has the potential for an evolutionary fossil interpretation bias by circular reasoning.

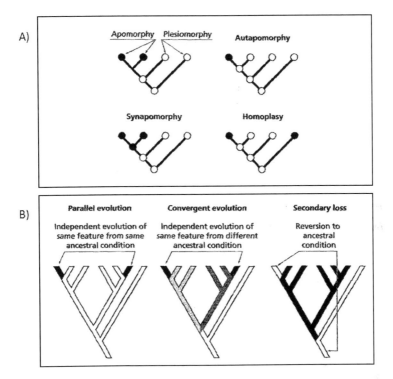

Fig. 5.2. The terminology of cladistic analysis. A) Trees show which characters are considered ancestral (or "primitive" or "plesiomorphic") in which case they have the same tree root or shared common ancestor, and which characters are descendant (also called "derived" or "apomorphic"), in which case they do not. Unshaded nodes represent ancestral character traits and shaded nodes the derived traits. Unique, derived character traits are called "autapomorphies" and shared derived traits are called "synapomorphies". Traits that are the same can either be inherited from an ancestor with the same trait (this similarity is called "homology"), or the similarity evolved independently in two separate lineages coincidentally (not because of a shared ancestor and this similarity is called analogy or "homoplasy"). Sorting out whether a similarity is homology and which therefore indicates phylogeny, and which is homoplasy and so which does not, is critical for determining evolutionary relationships. B) The three different kinds of homoplasy. (Reproduced from Page and Holmes, 1996 with permission).[29]

Philosopher of science Michael Ruse has addressed the vulnerability of paleontology to being misled, although he used it to say evolution in the fossil record is a fact! How he could come away with that conclusion is another question begged, and one I will leave you to ponder until much later (§20). For now consider why our examination is as much of fossil handling as it is of the fossils themselves because this was his observation: "The paleontologist is not in a position to watch and experiment with his organisms, where necessary performing direct checks on the genotypes. He frequently has to reconstruct the very phenomena he intends to explain."[30] Morphology and comparative anatomy are used to classify fossils. This is reasonable, but does it follow that if there is morphologic difference among fossils it means that one creature morphed into another? Does it follow that morphologic difference implies taxonomic difference at and above the species level? Science says yes. These notions are the basis and the directive of paleontological inquiry. We need to look at the fossil evidence for ourselves to see if we agree.

§5.4 THE FOSSIL RECORD PREDICTED BY DARWINISM

What does Darwinism predict the fossil record should show? The evidence of its five key features, of course. Ernst Mayr spelt them out earlier: evolution, gradualism, multiplication of species, natural selection and common descent (§2.2). Charles Darwin was very clear about what fossils should show "if this theory be true" and we need go no further than read *The Origin*. He said:

"By the theory of natural selection all living species have been connected with parent-species of each genus by differences not greater than we see between the natural and domestic varieties of the same species at the present day; and these parent-species, now generally extinct, have in their turn been similarly connected with more ancient forms; and so on backwards, always converging to the common ancestor of each great class. So that the number of intermediate and transitional links, between all living and extinct species, must have been inconceivably great. But assuredly, if this theory be true, such have lived upon the earth" (p.408).

Taking the five features in turn we should therefore see:

1. *Evolution*, because intermediate morphological fossil forms dominate the fossil record. As Darwin put it above, "the number of intermediate and transitional links, between all living and extinct species, must have been inconceivably great." This would be indicative of evolution by...

2. *Gradualism*, because descendant species in the fossil record are seen to undergo incremental morphologic change from ancestral forms by

"differences not greater than we see between the natural and domestic varieties of the same species at the present day". He said: "If it could be demonstrated that any complex organ existed, which could not possibly have been formed by numerous, successive, slight modifications, my theory would absolutely break down" (p.232) He explained: "On the theory of natural selection we can clearly understand the meaning of that old canon in natural history "Natura non facit saltum [Nature does not make jumps]". This canon, if we look to the present inhabitants of the world, is not strictly correct; but if we include all those of past times, whether known or unknown, it must on this theory be strictly true." (p.261) There could be no jumps because saltation would to return to creation of course (§2.2).

Little appreciated is that Darwin declared extinction in the fossil record is gradual. The reason is evolution and extinction are opposite sides of the same process. The survival of the fittest requires the extinction of the less fit, competition by natural selection is the reason, and this is by gradualism (§2.2). He said:

"On the theory of natural selection, the extinction of old forms and the production of new and improved forms are intimately connected together. The old notion of all the inhabitants of the earth having been swept away by catastrophes at successive periods is very generally given up...On the contrary, we have every reason to believe, from the study of the tertiary formations, that species and groups of species gradually disappear, one after another, first from one spot, then from another, and finally from the world." (p.449-450) He said: "There is reason to believe that the complete extinction of the species of a group is generally a slower process that their production: if the appearance and disappearance of a group of species be represented, as before, by a vertical line of varying thickness, the line is found to taper more gradually at its upper end, which marks the first appearance and increase in numbers of the species." (p.450). He also said: "With respect to the apparently sudden extermination of whole families of orders, as of the Trilobites at the close of the palaeozoic period and of the Ammonites at the close of the secondary period [the K-T extinction at the Cretaceous-Tertiary boundary Fig.5.3], we must remember what has already been said on the probable wide intervals of time between our consecutive formations; and in these intervals there may have been much slower extermination." (p.454)

There should also be...

3. *Multiplication of fossil species* and progressive diversification as organisms move up the geologic time scale to more recent times. Not sudden saltatory appearance because "these parent-species, now generally

extinct, have in their turn been similarly connected with more ancient forms; and so on backwards." The force of evolution effecting the microevolution summating to macroevolution is the blind watchmaker of...

4. *Natural selection,* or as he said: "As natural selection acts solely by accumulating slight successive favorable variations, it can produce no great or sudden modifications; it can act only by very short and slow steps." (p.626) No saltations, but accumulated microevolution, by gradualism, which produces multiplication of fossil species by natural selection, by...

5. *Common descent.* Darwin postulated a common ancestor whose existence would be implied by the pattern of the rest of the record. "By the theory of natural selection all living species have been connected with parent-species of each genus ...and these parent-species, now generally extinct, have in their turn been similarly connected with more ancient forms; and so on backwards, always converging to the common ancestor of each great class. (p.408) So much for the theory, what does the evidence say?

§5.5 WHAT DOES THE FOSSIL RECORD SHOW?

The fossil record pattern has five key features: 1) Multiple abrupt and devastating extinctions; 2) An increase in global biodiversity and "complexity" of life forms in general, ascending through the geologic column; 3) Sudden appearance of new fossil forms; 4) Overwhelming evidence for stasis; and 5) Scant evidence for intermediate (transitional) forms. The second feature would seem to confirm Darwinism but the others if anything the opposite. Things are not so straightforward. We will need to look much more closely.

§5.5.1 Mass Extinction

Contrary to Darwin's orthodoxy that extinction is gradual, there have instead been recurrent extinctions of colossal magnitude (Fig.5.3). The largest five are called the "mass extinctions". How colossal? At very least half the earth was affected at one time and the largest of them, the late Permian, wiped out 61% of all families and 96% of all species.[31-33] Breathtaking is the devastation, but the combined effect of other smaller extinction events is even more impressive. The total species "kill" of the five mass extinctions is only 4% of the total number of species that has gone extinct in the record.[34] There has been abrupt catastrophic destruction of life recorded in the layers of the geologic column.

Darwin rejected this pattern as being real and despite the contradiction from the fossil record, gave it only passing mention. His "evidence" was no more than a conviction about the truth of evolution theory. Not fossils. More significantly, the contradiction continued to be dismissed by the science and research into extinctions and what could have caused them only really began in the 1950's. The cause or causes remains unknown too. Sea level variations, volcanic activity, climatic change and meteorite strikes are the most favored current postulates. In regard to testing Darwin's ideas about extinction, University of Chicago paleontologist David Raup has pointed out "there are relatively few cases (if any) of widespread species becoming completely extinct in historic times without human influence,[35] and "its verification from actual field data is negligible".[36] Darwin's theorizations about extinction and the fossil evidence to document it remain theory rejected by the evidence. Two matters are raised.

The first is the empirical foundations of the science that produced Darwinism. On the one hand we have the fossil evidence Darwin ignored to instead declare: "Species are produced and exterminated by slowly acting causes...and the most important of all causes of organic change is one which is almost independent of altered...physical conditions, namely the mutual relation of organism to organism – the improvement of one organism entailing the improvement or extermination of others".

The second is this claim of gradual obsolescence by competition as the preeminent cause of extinction is an obvious contradiction, and yet it went unchallenged for another century. University of Birmingham geologist Anthony Hallam explains: "This Darwinian view has been accepted uncritically by generations of evolutionary biologists, but it is not in accord with the Phanerozoic record of extinctions. Simultaneous extinctions of groups of different biology and habitat, on a global or regional scale, clearly imply adverse changes in the physical environment, on a scale that could not have been adapted for by classic Darwinian microevolution during more normal times. In other words, to use the laconic language of Raup (1991), their extinctions could have been a matter of bad luck rather than bad genes."[37]

What do we conclude from "Mass Extinction" as one of the key characteristics of the fossil record? Three things. The first is that Darwin's convictions about extinction were not formulated from fossil evidence, but in spite of it. The second is that his view was "uncritically"[37] accepted by the science for another century. How could that happen if not that contradictory evidence was subordinate to another more believable truth:

what Darwin said. At least we know it wasn't the fossil evidence. The third is that it shows the science is capable of wholesale misinterpretation of supposed fossil facts, which begs the question of whether there are more.

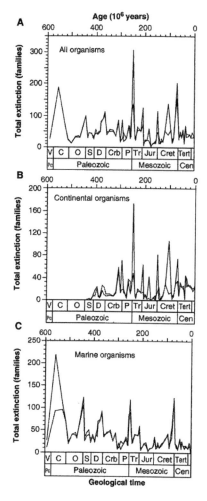

Fig.5.3. The pattern of extinctions of families in the fossil record. Plot is for all (A), continental (B), and marine (C) organisms expressed as number of families that went extinct at each stratigraphic stage of the geologic column with the lowest on the left and the highest on the right of the y axis. The rates of extinction show no sign of slowing to suggest any evolution to resist this force(es) over the entire duration of geologic column. Maximum and maximum curves are shown. Abbreviations: C Cambrian, Cen Cenozoic, Crb Carboniferous, Cret, Cretaceous, D Devonian Jur Jurassic, O Ordovician, P Permian, Pc Precambrian, S Silurian, Tert Tertiary, Tr Triassic, V Vendian. (Reproduced from Benton, 1995 with permission).[31]

§5.5.2   Increasing Diversity and Complexity. Or Not

The pattern of biodiversity through time is among the most basic of the information we need to determine whether microevolution became macroevolution. Considering macroevolution from fossil evidence is supposed "a fact,"[2,7-10] it might come as surprise to discover the pattern of the fossil record is still not even settled in the science. Experts line up on two sides of a dispute, with one side saying life followed a pattern of increasing diversity over time (the "expansionist" pattern) and the other saying global biodiversity reached an early plateau which has remained relatively constant since (the "equilibrial" pattern). If these sound like contradictory observations and yet both sides are just as certain macroevolution by fossils is a fact, you are correct. We should look at the mind-bending details.

According to the dating methods and assumptions discussed earlier, the Earth is held to be 4.5 billion years old (4,500 Myr) (§5.3.1;§13.3). The oldest sedimentary rocks are from Isua in Greenland dated at 3,800 Myr. There is a sequential appearance of life in the fossil record, in that prokaryotes appear first then eukaryotes and then multicellular organisms. This would seem a confirmation of macroevolution, but look some more at the pattern.

The oldest fossils are mats of fossilized bacteria called stromatolites recovered at Warrawoona, Australia and Fig Tree in South Africa that have been dated at 3,500 and 3,400 Myr respectively. Interpreting these fossils is easy because living stromatolites are found today, for instance at Shark Bay in Western Australia and on the north shore of the Red Sea. Two observations about fossils can be made already. The first is very soon after conditions on Earth became conducive to supporting life there were organisms able sustain highly coordinated sequences of complicated chemical reactions like photosynthesis (§12.4;§15.3.6). The second is despite the supposed 3,500 Myr of evolution since - the entire duration of life in the fossil record - living stromatolites today are not significantly different from their fossils. If we are to explain how macroevolution happened so spectacularly, we will also need to explain how it did not happen at all.

When exactly eukaryotes enter the fossil record is hard to determine because their distinction from prokaryotes is by the presence of a nucleus. We need a fossilized cell ultrastructure to tell. Proxy estimates that a fossil cell of millimeters size is too large to be a single cell are taken to indicate the oldest eukaryote is an alga called *Grypania spiralis*, dated at 2,100 Myr old. The first appearance of multicellular life in the fossil record is not until much more recently, around 570 Myr ago. How the fossils of multicellular organisms appear is a surprise given our gradualist evolutionary expectations, because they enter the record abruptly.

The oldest are called the Ediacaran fauna, a name that comes from Ediacara Gorge near Adelaide in Australia where specimens of this assemblage were first found. These fossils have since been found worldwide, and are dated by the science at between 575 and 545 Myr corresponding to the end of the late Neoproterozoic, also called the Ediacaran or Vendian Ediacaran period (Fig.5.1).[39] Ediacaran fauna consist of increasing diversities of frond and disc shaped soft-bodied fossils and simple track and burrow trace fossils of worm-like organisms. The organisms that caused these trace fossils are unknown.[40]

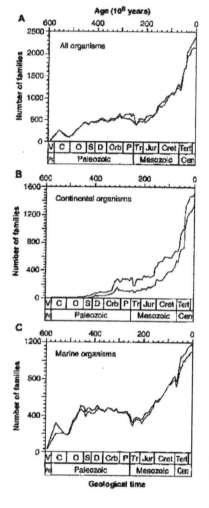

Fig.5.4. The pattern of fossil record global biodiversity over time. Plots of the maximum and minimum of families extant at each stratigraphic stage over geologic time are shown for all organisms (A), for marine organisms (B) and for continental organisms (C). Abbreviations as per Fig 5.3. (Reproduced with permission from Benton 1997).[38]

After the Ediacaran comes the Cambrian, which extends from around 545 Myr to around 490 Myr. It's a period of extraordinary fossil profusion, so much so it's called the "Cambrian explosion" even in evolution science. To use Berkeley paleontologist James Valentine's words, the Cambrian explosion is "[t]he single most spectacular phenomenon evident in the fossil record".[41] What is it? In short, and to continue to hear him speak: "The paleontological data are consistent with the view that all the currently recognized phyla had evolved by about 535 Ma[Myr]...Despite half a billion years of evolutionary exploration by the clades generated in Cambrian time, no new phylum-level designs have appeared since then."[42] Well this is a surprise! At least it was to Charles Darwin. He thought this pattern denied evolution. Being such a prominent feature of the record we should look at it more closely.

The Cambrian period is subdivided into early, middle and late (Fig.5.5). The early Cambrian period is further subdivided into four stages by names taken from fossil fields in Siberia. These are successively the Manykayaian or Nemakit-Daldynian, the Tommotian, the Atdabanian and the Botomian. The first 15 Myr of the Cambrian is the Manykayaian, which consists of burrow and surface trace fossils of a creature given the name *Treptichnus pedum*. The traces are more complex than those of the Ediacaran assemblages. During the last 3 to 6 million years of the Manykayaian another assemblage appears consisting of mineralized skeletons of small tubes, blades and cups called the "small shelly fauna", and for that reason. In the Tommotian, Atdabanian and Botomian stages that follow, a total duration of only 9 or 10 million years,[40,43-47] (i.e. 530-520 Myr although some argue for a still shorter interval from 530-525 Myr, so 5 to 6 million years)[48] but a literal instant in geological time either way, every known phylum with a fossilizable skeleton bar one appears in the fossil record.

If this news comes as a shock and you need to hear it from the experts, here is MIT's Samuel Bowring: "Numbers of phyla, classes, orders, families and genera all reached or approached their Cambrian peaks peaks during the short Tommotian-Atdabanian interval".[48] Stephen Jay Gould says: "the Cambrian explosion, the remarkable episode which lasted only 10 million years (from 530-520 million years ago) and featured the first appearance in the fossil record of effectively all modern animal phyla."[43] Berkeley's James Valentine declares the "origination of all living animal phyla by the close of this 10 million year interval."[49] No new phyla appear during the middle or late Cambrian, or at any time subsequent. The number decreases in fact. Every known body plan of life, living or extinct, appears over a period of no more than 1.7% of the entire duration of the record and if we accept the main pulse as only 5-6 million years,[48] only 0.1% of the fossil record. No links are found between phyla to suggest a common ancestor either (Figs.5.5;5.7b).

The Ediacaran fauna can't be taken as evolutionary precursors and so serve as a Pre-Cambrian evolutionary "slow-fuse", because their body plans are different. Nor are they found in profusion to supply a reservoir source of potential diversity for the Cambrian, and anyway what kind of fuse? They appear suddenly also. They are an explosion in their own right! What are we to do?

For his part Darwin said the Cambrian explosion "is the most obvious and serious objection which can be urged against the theory" (p.406). What was his answer? "The case at present must remain inexplicable; and may be truly urged as a valid argument against the views here entertained." He predicted the true gradualist pattern of the fossil record was still to be found, because "if the theory be true, it is indisputable that before the lowest Cambrian stratum was deposited long periods elapsed...the world

Evidence of fossils I

swarmed with living creatures"(p.438). The pattern of the Cambrian explosion is so contradictory it was taken to be an artifact, an illusion. Like the appearance of design. In other words what Darwin said trumped all the fossil evidence.

Evolution theory has had to come to terms with the evidence however, because stasis and sudden appearance as the pattern of the fossil record is taken as real. This is how Stephen Jay Gould sees the Cambrian explosion for instance:

"Thus instead of Darwin's gradual rise to mounting complexity, the 100 million years from Ediacara to Burgess may have witnessed three radically different faunas – the large pancake soft-bodied Ediacara creatures, the tiny cups and caps of the Tommotian, and finally the modern fauna, culminating in the maximal anatomical range of the Burgess. Nearly 2.5 billion years of prokaryotic cells and nothing else – two-thirds of life's history in stasis at the lowest level of recorded complexity. Another 700 million years of the larger and much more intricate eukaryotic cells, but no aggregation to multicellular life. Then, in the 100-million-year wink of a geological eye, three outstandingly different faunas – from Ediacara, to Tommotian, to Burgess. Since then, more than 500 million years...but not a single new phylum, or basic anatomical design added to the Burgess complement." [50]

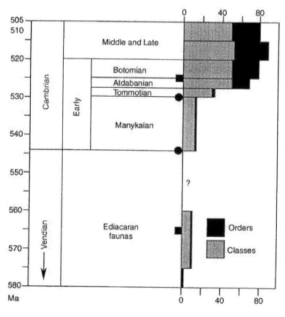

Fig.5.5. The fossil record of the Cambrian explosion. Number of new classes and orders by stratigraphic layer. The key features are sudden appearance of Ediacaran fauna, sudden appearance of Cambrian fauna, no links between the Ediacaran and Cambrian faunas, and since then in Stephen Jay Gould's words, "not a single new phylum, or basic anatomical design added".[50] The bulk of the Cambrian explosion is in the Tommoian-Adtabanian. (From Bowring et al, 1993 with permission).[48]

While the pattern of the Cambrian explosion is accepted as real, the interesting thing is evolution solutions to what Darwin called a contradiction to evolution remain hypothetical but macroevolution from fossil evidence is now "a fact". Simon Conway Morris at Cambridge calls a solution to the Cambrian explosion "surprisingly elusive".[39] The question then is not why he says it's elusive, it's why he's surprised? (§19) The fact is "the single most spectacular phenomenon evident in the fossil record"[41] is without an evolutionary solution with objective evidence. And what is it after all? A record of death not new life. Every phylum suddenly appears in the fossil record because of a sudden, catastrophic killing. Science dismisses the Flood as that cause.

Speaking of the pattern of fossil biodiversity, how many different fossil species were living at any particular time in the record? This kind of assessment has been facilitated by compilations of pooled databases like the *Paleobiology Database* and *The Fossil Record 2*. The latter was published in 1992 under the editorship of Michael Benton at the University of Bristol.[52] The bulk of the thinking about diversity trends to date has been the work of just one man however, John Sepkoski at the University of Chicago. Sadly he died suddenly in 1999, at just fifty.

Fossil biodiversity studies have mostly been done at family and genus taxonomic level, so numerous are the species (not to mention uncertainties about species identification).[31,53-55] Sepkoski documented an "expansionist" pattern in his database that recorded the times of first and last appearance of around 4,000 marine fossil families. This view is shared by James Valentine at Berkeley, Michael Benton at Bristol University, Douglas Erwin at the Smithsonian and David Jablonski at the University of Chicago (Fig.5.4A).

The countering "equilibrial" fossil record pattern argument comes from Michael Foote at the University of Chicago,[56] John Alroy at the University of California Santa Barbara,[57] and amongst others, John Sepkoski again who was a posthumous co-author on Alroy's paper. This view contends any pattern of increase (as seen by the rising slopes on Fig.5.4) is just artifact from sampling bias. The artifact has a name called the "pull of the recent" by which is meant an artificial inflating of the fossil record for most recent biota because younger rocks are more likely to be preserved and recovered, and therefore more likely to show these organisms and so more biodiversity.

Recent work has given strong support to the expansionist model by showing that at least for marine bivalves, this "pull" mechanism cannot explain the observed pattern of increasing diversity during the Cenozoic. The reason is only five percent of the around nine hundred and fifty living bivalve genera and subgenera arise in the most recent time intervals (the Pliocene and the Pleistocene).[58] For now at least, the pendulum has shifted to an expansionist fossil record pattern though not all agree.[59,60] The crux of the arguments in play? The former contend that limits in ecospace have not been reached and so diversification continues, while the latter say diversification is inhibited because niches are finite and become occupied. Try untangling that knot. Evolution can equally well explain them both!

Setting aside the Cambrian explosion, what is the pattern after that? Do we see evolution now? Consider the expansionist pattern of the fossil record as real. Look again at Fig.5.4A. The change in biodiversity for all life at the family level is exponential, rising rapidly during the Cambrian explosion followed by a further rise in the Ordovician called the "Ordovician radiation" (an increase in diversity is called a "radiation", and decrease "extinction"). During the remainder of the Paleozoic, diversity remained relatively stable followed by further exponential rises in the Mesozoic and Cenozoic and this pattern is interrupted by brief declines particularly in the Late Permian and Late Triassic which correlate with periods of mass extinction (§5.5.1). An exponential increase is just what a branching model of evolution predicts when speciation and extinction remain constant (Fig.2.1).

Separate analysis of the continental and marine fossil records shows them to have different patterns (Fig.5.4.B,C). The continental data is exponential, being dominated by the exponential radiations of land plants, insects and tetrapods. The marine diversity curve in contrast shows a brief plateau in the Cambrian, and a longer second one extending from the Ordovician to Permian, followed by an exponential rise that appears to be tapering over the last 25 My[31,38]. If the origination with the Cambrian explosion is demonstrably not Darwinian, at least afterward this fossil data seems convincing. Or does it? Look again. It has to be reconciled with Fig.5.6 and Fig 5.7b, and both reject the idea.

The problem appears when the Linnaean taxa higher than family and species are inspected. If species are descended from one common ancestor by gradualism we expect diversification at all taxonomic levels, starting with change at the species level then progressively to genus, class, order and phylum as microevolution summates to become macroevolution. The exponential curves of diversity we saw for families disappear in plots of higher taxonomic levels however. Instead there is the opposite. New phyla appear before new species and the phyla all appear at once. There is extremely rapid diversification in the Cambrian explosion followed by decline and plateau (Fig.5.5). In other words the diversity pattern is decoupled from taxonomic level. This is a contradiction to claims that microevolution becomes macroevolution.

This fossil pattern has been known for decades, but has not been a conceptual problem for evolution either then or now.[24,54,61] Most consider it an unexplained artifact.[31,54,62] But why should we accept it as an "artifact" if it is "unexplained" and also a contradiction? Why can't it at least be what it appears to be? A disproof?

Philip Signor writes of the observation: "These dissimilarities [between species and all higher orders] should not be trusted to reveal trends in species richness...higher taxa are insensitive indicators of species level trends in time."[61] His declaration what is to be "trusted" is just his opinion. The question is where the paleontologist gets the evidence for that opinion. At least we know can't be from the fossils. Michael Benton's explanation is not satisfying either. "Perhaps real diversification patterns are damped as one ascends the Linnaean hierarchy."[38] David Raup has explained the observation as due to evolution establishing the higher taxonomic ranks first, followed by an increase in the lower ranks afterward. Only this explanation is also the observation, and so no more than a logical fallacy called simplicity. The problem for evolution having reality in the fossil record has to do with a mechanism. You see the "top-down" pattern in the fossil record is a disproof of gradualism. You explain it by evolution with objective evidence. Not more evolution theory to make the fossils fit evolution theory. So far no one else has.

§5.5.3　　　　　　　　Sudden Appearance

Darwin conceptualized the pattern of macroevolution as a tree (Fig.2.2). It was his only illustration in *The Origin* and it was entirely hypothetical although it supposedly represented every organism in all of nature! Rather than the progressive change of Darwinian gradualism, sudden appearance of novel features is the rule of the fossil record however (Fig.5.6). A comparison of this drawing with the evolutionary trees found in textbooks today is even more telling, because the links between the major taxon levels remain just as hypothetical today despite a century and a half of searching (Fig.5.7b). This is not common knowledge outside of the closed circle of evolution science and it continues to remain concealed, unwittingly or not, by every author who uses linking lines on evolutionary trees (Fig.5.11;§25).

Stephen Jay Gould addressed this over two decades ago, when he pointed out what had been obvious all along:

"The extreme rarity of transitional forms in the fossil record persists as the trade secret of paleontology. The evolutionary trees that adorn our textbooks have data only at the tips and nodes of their branches; the rest is inference, however reasonable, not the evidence of fossils. All paleontologists know that the fossil record contains precious little in the way of intermediate forms; transitions between major groups are characteristically abrupt."[63]

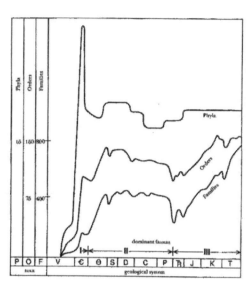

Fig.5.6. Biodiversity in the fossil record. Appearance of new phyla, orders and families of marine invertebrates by geologic time (V is Vendian or Ediacaran and € is the Cambrian). Two contradictions to Darwinism appear. The first is the rate of macroevolution is too fast and then no more. Specifically the highest taxonomic orders appear within a fraction of the duration of the record, variably estimated at 0.1%-1.7% of its entire duration. Depending on the expert either 13[51] or 16[48] phyla and 30 classes appear in the Cambrian in 6 million years. The second is the "top down" appearance of fossil diversity which is macroevolution before microevolution. The top taxonomic orders show novelty before the lowest! Compare this with Fig.5.4 & Fig.5.7a with b). (Reproduced from Valentine et al, 1990 with permission).[24]

Niles Eldredge at the American Museum of Natural History explains further:

"[I]f we are to heed geneticists (and Darwin), we simply take that natural selection model of generation-by-generation change and extrapolate it through geological time. And that, to my paleontological eyes, is just not good enough...I found out back in the 1960s as I tried in vain to document examples of the kind of slow, steady directional change we all thought ought to be there, ever since Darwin told us that natural selection should leave precisely such a telltale signal as we collect our fossils up cliff faces. I found instead, that once species appear in the fossil record, they tend not to change very much at all. Species remain imperturbably, implacably resistant to change as a matter of course – often for millions of years."[64]

This was all old news, though you would never have guessed it from the fanfare. The reason it was old news is because Darwin saw it in *The Origin*. The problem is this pattern and its implications were suppressed until

Gould and his colleagues pointed it out again, over a century later. How could this happen? The only way is that a greater truth, what Darwin said, had trumped it. George Gaylord Simpson saw saltation in the fossil record too, and this was four decades before Gould and Eldredge. He reported:

> "This is true of all thirty-two orders of mammals...The earliest and most primitive known members of every order already have the basic ordinal characters, and in no case is an approximately continuous sequence from one order to another known. In most cases the break is so sharp and the gap so large that the origin of the order is speculative and much disputed...This regular absence of transitional forms is not confined to mammals, but is an almost universal phenomenon, as has long been noted by paleontologist. It is true of almost all classes of animals, both vertebrate and invertebrate...it is true of the classes, and the major animal phyla, and it is apparently also true of analogous categories of plants."[65] Simpson could not explain this pattern, and yet it did not temper his conviction evolution was a fossil "fact".[2] What was the evidence of macroevolution for him then? We know it can't be fossils. Gould and Eldredge proposed a theory to explain the pattern as evolutionary or at least that is what they said it did, but this was three decades later (§6.3).

The fact is the signal feature of the fossil record was ignored and denied for more than a century after *The Origin* because of the greater authority of evolution theory over fossil evidence, but don't take my word for it. Here are five of the world's most eminent paleontologists who say so too, and they're all evolutionists:

The first is Chris Paul at the University of Liverpool: "With the benefit of hindsight, it is amazing that paleontologists could have accepted gradual evolution as a universal patter non the basis of a handful of supposedly well-documented lineages (e.g. *Gryphaea, Micraster, Zaphrentis*) none of which actually withstands close scrutiny...The evidence that the vast majority of species appeared suddenly, had well-defined periods of existence, and then disappeared equally suddenly, was just ignored."[66]

Here is James Valentine at Berkeley: "It has turned out that many patterns have been different from the predictions, sometimes strikingly so, and that these patterns cannot be argued away as artifacts of an incomplete fossil record. The gradualistic pattern of evolution predicted by Darwin is simply not found – simply did not occur – in many cases. The circumstances acting to produce the patterns we do find – the punctuational pattern of species lineages and the sudden introductions of major new body plans – are still not understood."[67]

Here is Niles Eldredge at the American Museum of Natural History: "Most families, orders, classes and phyla appear rather suddenly in the fossil record, often without anatomically intermediate forms smoothly interlinking evolutionarily deprived descendant taxa with their presumed ancestors."[68]

Here is Steven Stanley at the University of Hawaii: "[T]he fossil record itself provided no documentation of continuity – of gradual transitions from one kind of animal or plant to another of quite different form."[69]

And finally here is Robert L. Carroll at McGill. He also gives an explanation. It's the fossil's fault! "It would seem a simple matter to establish whether species change progressively or remain stable for long periods of time. In fact, only rarely are geological processes sufficiently regular and continuous to lead to the preservation of numerous fossils that are evenly distributed throughout the time range of species. Most fossiliferous horizons record only an extremely short period of time and are separated by long gaps from other productive zones. Because of this problem, few paleontologists have been concerned with change within species and most simply assumed that it followed the pattern hypothesized by Darwin."[70] His last sentence is a indisputable fact of the history of this science, and an indictment of it.

Carroll also says links between fossil species are rare because transitional fossils are rare. This is at best a tautology. It also requires believing the fossils most absent are really the ones most abundant. Even if transitional fossils are for some reason rare, if macroevolution is a scientific "fact" at least we can say it cannot be because of fossil evidence.

A comparison of the supposed evolutionary tree of life with the fossil record is perhaps the simplest way of convincing the reader of the point and the problem here (Fig. 5.7a vs b). Consider some specific examples that the fossil pattern of sudden appearance poses.

As regards plants, angiosperms are first seen in the Cretaceous period without fossils linking them to any other plant.[71] This is not a new observation either. Darwin saw it. He said: "Nothing is more extraordinary in the history of the Vegetable Kingdom, as it seems to me, than the apparently very sudden or abrupt development of the higher plants".[72] He also called it an "abominable mystery".[73] Why? Did I mention that Darwinism was formulated in spite of the fossil evidence?

As regards insects, there are more than 10,000 fossil insect species and more than 30,000 fossil spider species, but the transition from wingless to winged insect is not seen in the fossil record nor is there a transition between the two main types of winged insects, the Paleoptera and Neoptera. Chordates appear suddenly in the Cambrian and they are found

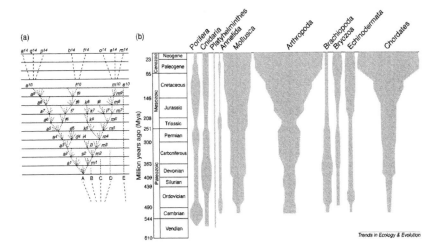

Fig.5.7 What Darwinism says about evolutionary links and what the fossils say. Left panel (a) shows an evolutionary tree exactly as first drawn by Charles Darwin in *The Origin*. Right panel (b) shows the fossil record of the main metazoan phyla. Column width (x axis) represents the number of families at each geological period (y axis) (See Fig.5.1). Note sudden appearance, separate appearance, absence of a common ancestor and how body plans persist through the record and remain separate through the record without (evolutionary) links. (Reproduced from Carroll, 2000 with permission).[45]

for every subphylum too: the Urochordates, Craniates and the Cephalochordates. As regards fish, agnathan fossils are found in the low Cambrian (555 Myr) with brains and gills and hearts,[74,75] other fish are found around 400 Myr also fully formed, and the cartilaginous fish (sharks and rays) 50 Myr later without links and distinctly different again.

As regards amphibians, they are thought to have evolved from the fish order Rhipidistia, but the earliest amphibians have four legs capable of terrestrial life with no intermediate fossils linking the fin and the leg. "The closest fossils connecting a fin and a leg come from a fish with a fin and an amphibian with a leg."[76] Turning the evolutionary argument of transitional status upside down, the mudskippers (*Periophthalmus* and *Boleophthalmus*) are unequivocally fish and closely replicate the supposed evolutionary scenario by walking in mud and standing on pectoral fins like legs as they have done for millions of years, at least according to fossil dating - yet they have not developed legs.[77]

As regards reptiles, the supposed transition from amphibian to reptile is associated with the sudden appearance of a hard egg shell.

As regards mammals, while their soft tissues are different (hair, mammary glands, body temperature) reptile and mammalian skeletons can be superficially similar. However all mammals have one lower jaw bone and 3 inner ear bones called the hammer, anvil and stapes but all living and fossil reptiles have at least 4 jaw bones and 1 inner ear bone (the stapes) with no transition. Mammals all have an Organ of Corti (the inner ear) with which to hear and reptiles have none. There are 32 orders of mammals for which there is also no record of transition. George Gaylord Simpson summarizes, mystified, the abrupt change of mammal morphology in the fossil record:

> "The most puzzling event in the history of life on earth is the change from the Mesozoic, Age of Reptiles, to the Age of Mammals. It is as if the curtain were rung down suddenly on a stage where all the leading roles were taken by reptiles, especially dinosaurs in great numbers and bewildering variety, and rose again immediately to reveal the same setting but an entirely new cast, a cast in which the dinosaurs do not appear at all, other reptiles are supernumeries, and all the leading parts are played by mammals of sorts barely hinted at in the preceding acts."[78]

Bats have no link either. The interesting thing is these contradictions to evolution have been apparent for a long time yet skepticism of evolution has been extremely rare in the supposedly skeptical science community. There are exceptions however, and one is French zoologist Pierre Grassé (1895-1985), the vitalist evolutionist we met earlier. Consider his resume before you dismiss him as a crank (§4.1). His book *Evolution of Living Organisms* summarized the state of the art, and while many decades old, it remains a stinging rebuke that is still not refuted.

> "What were the evolutionary steps from the "schizophyte" structure (that of bacteria and Cyanophyceae) to the true cell structure (i.e., a nucleus enclosed in a membrane and containing chromosomes with a well-defined structure...)? We have no idea. Paleontology does not reveal anything on this matter. Hypotheses are plentiful, but they all lack substantiation. On the animal side, the link between the uni- and multicellular organisms is still missing. In spite of extensive study, the origin of the Metazoa is still unknown...Our ignorance is so great that we dare not even assign with any accuracy an ancestral stock to the phyla Protozoa, Arthropoda, Mollusca, and Vertebrata."[79]

Sudden appearance is a consistent feature of the fossil evidence, throughout the fossil record. The Cambrian explosion discussed earlier is the most striking example. In the words of Richard Dawkins: "It is as though they were just planted there, without any evolutionary history". Indeed. Nobody can claim to be ignorant about the appearance of nature.

Debate about the Cambrian explosion has traditionally focused on why no new phyla have appeared since then. Why so many appeared and so fast would seem much more important. Discovery Institute director of Science and Culture Steven Meyer's book *Darwin's Doubt* published in 2013 is an excellent exposition of the disproof to evolution the Cambrian explosion is.[81] What was the response from the science, since it provided a 2013 review and on the state of the theory?

The response was swift and harsh. Actually it was so swift one review of over 9,400 words was published within 24 hours of Meyer's book being available to be read. The interesting thing is although misquoting and mischaracterizing where not missing the basic thesis Meyer was at pains to point to, this particular review written by a graduate student was extolled by undisputed expert Jerry Coyne and its arguments recycled in other reviews, such as in *The New Yorker* and *National Review*, so representing readership on the liberal and conservative political spectrum. Dr. Meyer should not have been shocked. His previous book *Signature in the Cell* was panned by University of California Irvine biologist Francisco Ayala without him demonstrating he had even read it at all.

While it was no surprise to hear the response the Cambrian explosion poses no contradiction to macroevolution, the arguments Meyer presented in his book were curiously nowhere rebutted and for the most part not even engaged. The objections mostly were that he had his basic facts wrong about the fossil record appearance. His critics sidestepped his arguments to discredit him before they were allowed to get off the ground.

For instance Berkeley paleontologist Charles Marshall said in *Science* that Meyer "omits mention of the Early Cambrian small shelly fossils and misunderstands the nuances of molecular phylogenetics, both of which cause him to exaggerate the apparent suddenness of the Cambrian explosion."[82] The Cambrian explosion is sudden. Marshall just calls it "apparent suddenness". Meyer did mention these fossils too, anyway they do not soften the suddenness of the explosion. This is the consensus view of the field too, including papers written by Dr. Marshall! For instance he calls the small shelly fossils "largely problematic" and "hard to diagnose, even at phylum level"[83] and excluded them from the ten million year "geologically abrupt appearance of fossils representing disparate body plans" that he too calls "the Cambrian explosion."[83,84]

The molecular nuance Marshall refers to is Meyer's disbelief that developmental gene networks evolved themselves to generate every phyla body plan in the Cambrian. The notion is absent evidence how this could be done exactly, even in theory, and it requires macroevolution be true in circular reasoning to be entertained when Meyer is pointing to the disproof of the notion by the fossil record. Marshall refused to seriously engage, much less dismantle or refute Meyer's thesis.

Los Angeles County Natural History Museum paleobiologist Donald Prothero rejected Meyer's presentation of the fossil record too, declaring "we now know that the "explosion" took place over an 80 m.y. time frame. Paleontologists are gradually abandoning the misleading and outdated term 'Cambrian explosion' for a more accurate one, "Cambrian slow fuse" or "Cambrian diversification".[85] He presented no evidence for this claim. Prothero also declared the sudden appearance of soft-bodied animals "is an artifact of preservation" and there is a gradualistic evolutionary tree in the Cambrian after all. He said:

> "The entire diversification of life is now known to have gone through a number of distinct steps, from the first fossils of simple bacterial life 3.5 billion years old, to the first multicellular animals 700 m.y. ago (the Ediacara fauna), to the first evidence of skeletonized fossils (tiny fragments of small shells, nicknamed small shellies) at the beginning of the Cambrian, 545 m.y. ago (the Nemakit-Daldynian and Tommotian stages of the Cambrian), to the third stage of the Cambrian (Atdabanian, 530 m.y. ago, when you find the first fossils of the larger animals with hard shells, such as trilobites."[85]

In *The New Yorker* the Cambrian explosion was pronounced: "not, in fact, an explosion. It took place over tens of millions of years...if one looks at a family tree based on current science, it looks nothing like Meyer's, and precisely like what Darwinian theory would predict."[86] It was if the fossil record had whole other appearance than what Meyer had said.

The second objection to Meyer's book followed from the first. Meyer's credibility was removed by the first and this one removed his believability. Dr. Prothero declared: "Meyer deliberately and dishonestly distorts the story...he deliberately ignores", Charles Marshall entitled his review: "When prior belief trumps scholarship" and other reviewers declared "he simply lifts quotes"[87] and "misleading...Meyer does not understand the field's keys statistical techniques (among other things).[86]

The third objection was that Meyer's argument was an argument from ignorance. Namely that since we don't know what caused the Cambrian explosion then God must have.[82,85-87] We discuss these matters later. The only point for now, is this too was to seriously misrepresent Meyer's argument. Hear the objection from Dr. Marshall:

> "Meyer's scientific approach is negative. He argues that paleontologists are unable to explain the Cambrian explosion, thus opening the door to the possibility of a designer's intervention. This...is a (sophisticated) "god of the gaps approach, an approach that is problematic in part because future developments often provide solutions to once apparently difficult problems."[82]

Not only is evolution absent any mechanism to produce sudden appearance, and so is really an argument from a knowledge of inadequacy rather than ignorance, the sudden biological novelty resides ultimately in new genetic material, and so represents new complex specified information. Complex specified information is what intelligent agency causes. We know this from direct experience and we know of no other cause than this cause (§1.2;§16). The ID argument is affirmative not negative then, and it is a reasoning method that comports with our intuitions (§1.2;§16.8.1).

In the end, two matters are raised by Meyer's critics. The first is whether sudden appearance is really the pattern of the fossil record after all. Figs.5.5, 5.6 and 5.7b should lay that to rest but for those who dislike figures, James Valentine and Douglas Erwin laid out a definition back in 1999: "The Cambrian explosion is named for the geologically sudden appearance of numerous metazoan body plans (many of living phyla) between 530 and 520 million years ago, only 1.7% of the duration of the fossil record of animals."

The second is the shoddy science of the critics of Meyer's science, and the absence of any attempt by any onlookers to at least set the facts straight. Meyer's arguments were never dismantled but ignored where they were not misrepresented, even as the fossil record evidence was misrepresented. We leave looking at this bad behavior by science later because I have yet to convince you such a thing could happen and be happening, much less how or why it could, in science (§28). In summary, sudden appearance is a fundamental pattern of the fossil record and although most obvious in the Cambrian, is seen throughout the geologic column now, as in Darwin's day.

§5.5.4   Stasis

Rather than showing progressive change, as a rule fossils show prolonged periods of no change. More interesting, because it is a measure of the hold Darwinism has, is how this obvious and at least on its face contrary characteristic of the fossil record, has been ignored or denied. Stasis is the fossil record norm. Waving away the absence of intermediates because they are still undiscovered will not do here yet it took over a century after *The Origin* for stasis to even be acknowledged as the true fossil record pattern!

It finally happened in 1971 when Niles Eldredge, Stephen Jay Gould and Steven Stanley turned the paleontology community upside down with the obvious, though for evolution theory absolutely revolutionary, and amazingly still controversial, notion that for fossils, "stasis is data".[88] In Eldredge's words: "back in the 1960s as I tried in vain to document examples of the kind of slow, steady directional change we all thought ought to be there, ever since Darwin told us that natural selection should leave precisely such a telltale signal as we collect our fossils...I found instead, that once species appear in the fossil record, they tend not to change very much at all. Species remain imperturbably, implacably resistant to change as a matter of course – often for millions of years."[64] Here is Steven Stanley, also spoke out in contradiction to what Darwin and Modern Synthesis paleontology was asserting: "The known fossil record fails to document a single example of phyletic evolution accomplishing a major morphologic transition and hence offers no evidence that the gradualistic model can be valid."[89] Robert Carroll does not share Stanley's interpretation of the fossil record pattern (§6.3), but concedes the observation and which is the point and the problem here: "Most of the evidence provided by the fossil record does *not* support a strictly gradualistic interpretation, as pointed out by Eldredge and Gould (1972), Gould and Eldredge (1977), Gould (1985) and Stanley (1979, 1982)".[90]

Some of the best evidence of stasis is provided by the so-called "panchronic" species or what Darwin called "living fossils" in oxymoron, a word unblushingly employed throughout science since for creatures like the coelacanth that are unchanged from their fossils. Other examples and their declared dates of appearance in the fossil record include horseshoe crabs (230 Myr), Bowfin fishes (105 Myr), sturgeons (80 Myr), snapping turtles (57 Myr), alligators (35 Myr) and aardvarks (20 Myr).[91]

§5.5.4　　　　Rare Evidence Of Intermediate Forms

While a sudden appearance of new forms out of a background of stasis is the pattern that characterizes the fossil record, there are fossil series held up by paleontologists as demonstrating the progressive change Neo-Darwinism always claimed. They are *LOE#2* (See Table 3.1). No matter how robust they may be though, and we will need to see how much, as exceptions they emphasize the overwhelming contrary character of the fossil record that usually goes unmentioned. The interesting thing is the evolution hypothesis has not been impugned a whit, either by those who still see the fossil record in the way Darwin does or in the way Stephen Jay Gould and his colleagues do. Both believe macroevolution (Table 3.1). In fact these opposite features, fossil stasis by some fossils and fossil gradualism by the others, are today both taken as confirming evolution's truth! How that seeming paradox can be accommodated scientifically, is the subject of the following chapter.

Before Gould and his colleagues pointed out the contradiction of the fossil record pattern, the explanation for the absence of intermediates was what Darwin said. Fossilization is rare and the transitional series are all still out there waiting to be discovered. Some like Michael Ruse at Florida State University still have this hope. He says: "Although it is conceded that there are many gaps in the record, it is countered, nevertheless, that there is a sizable number of well-established gradual changes to be found in the record. Hence, given all the factors making fossilization improbable, Darwinism remains totally plausible. To ask for more than this from the fossil record is unreasonable."[92] We should consider his claims. Declaring Darwinism remains "totally plausible" and what is or is not "unreasonable" to ask of "fossils" is just his opinion, but the problem is not that. It's that he offers no evidence for it. His term "sizable number" is also a relative term with a private meaning because intermediate forms are a minute fraction of the record, to the extent such alignments could be a coincidence. This is not even considered. What then is Dr. Ruse's evidence for evolution, is the question we are left with. At least we know it can't be fossils.

While Dr. Ruse's explanation for the contradictions to evolution theory by the fossil evidence is the same as Darwin's, there is a difference between them, and it is the difference between hope and faith. Hope by Darwin as although the fossil record denied gradualism, his science was arguably young and more looking could be done; but faith by Ruse because the pattern persists despite a century and a half of hopeful looking (§6.2).

The examples of transitional series found since Darwin are in science everywhere evidence of macroevolution. Ruse gives the example of the microfossil *Pseudocubus vema* , claimed to show a progressive increase in size over two million years.[93] The change is disturbingly unidirectional as Steven Stanley has pointed out: "for many invertebrates, size is indeterminate and size at death varies with growth rate and external mortality factors...Although size change often has ecological consequences, it does not produce adaptive innovations or distinctive new higher taxa. In short, it is not where the real action is."[94]

What is required instead, is a naturalistic explanation for complicated creatures like bats and whales or organs like ears, eyes and echolocation. This is macroevolution. A mere size change over millions of years does not even begin to satisfy that. Four of the most celebrated examples of intermediate fossil series, the ones that make it to all the biology textbooks, are the intermediate fossils of whales, horses, birds and humans. We will inspect the details of the first three next and humans in Chapters 10 and 11.

Evidence of fossils I

§5.5.4.1 Running Into Deep Water Scientifically - The Evolution Of Whales

The order Cetacea is the taxonomic group of whales, porpoises and dolphins. Fossil evidence for their origins by evolution is now so robust as to be called a "poster child for macroevolution."[95] Whale paleontologist J.G.M (Hans) Thewissen at the Northeastern Ohio Universities College of Medicine is among the world's leading authorities. He will be our guide. He summarizes the evidence: "We can now show that whales are in fact hoofed mammals that took to sea."[95] Don't laugh.

The evidence is a sequence of transitional fossils that begins with a four-legged, wolf-like animal called *Pakicetus* around 50 million years ago and ends with creatures similar to modern day cetaceans called the *Dorudontidae* at the Eocene-Oligocene boundary around 37 million years ago (Fig.5.8). Considering we're told ancestral whales evolved from inorganic chemicals into bacteria-like organisms into fish-like organisms into four legged mammals with hooves that ran about on dry land, and then walked back into the sea and over the next 13 million years changed into fluked, finned and streamlined deep sea diving animals with sonar, it's an extraordinary series no matter how convincing macroevolution may be.

The experts say so too. Stephen Jay Gould called the fossil whale series: "the sweetest series of transitional fossils an evolutionist could ever hope to find...The sequential discovery of picture-perfect intermediacy in the evolution of whales stands as a triumph in the history of paleontology"[97]. A review in the journal *Nature* declares these fossils "superbly document the link between modern whales and their land-based forebears, and should take their place among other famous 'intermediates', such as the most primitive bird, *Archaeopteryx* [§5.5.4.1], and the early hominid *Australopithecus* [§11.1],"[98] and commentary in the journal *Science* agrees.[99] The National Academy of Sciences calls the whale fossils: "a particularly dramatic example" of macroevolution and even shows students illustrations of what whale intermediates looked like.[100-102] Poster child indeed.

This prestige status is helpful to us because it becomes an opportunity to discover a lot more about evolution than just whale origins. For in being among the best evidence for macroevolution, whale fossils at least become an indication of how robust the rest of the fossil evidence is, and how robust the methods of analysis and audit are. Most importantly, it should show us exactly why fossils demonstrate the fact of macroevolution. We could not be better engaged.

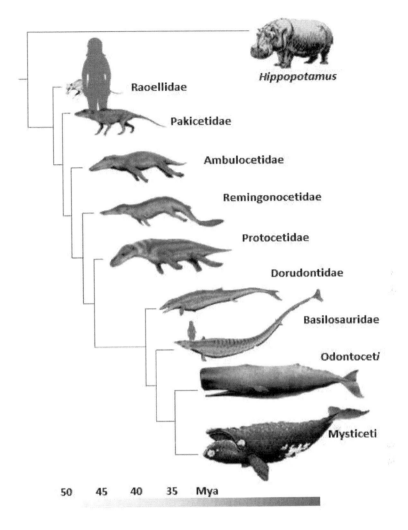

Fig.5.8. The whale evolutionary tree. Ancestral artiodactyls (even-toed ungulates) evolved into diving whales in 13 million years and their closest relative today is the hippopotamus! The transitional series is of six fossil families that start from the long-tailed, long-legged, hoofed and wolf-like Pakicetidae at 50 Mya (e.g. *Pakicetus*) which are yet called whales because of bony features in the inner ear. The raccoon-like Raoellidae (e.g. *Indohyus*) are considered a sister group. The Pakicetidae are followed by the Ambulocetidae (e.g. *Ambulocetus*) at 49 Mya with crocodile-like features, then the Remingonocetidae (e.g. *Kutchicetus*) with more aquatic features, then the Protocetidae (e.g. *Rodhocetus* at 47-38 Mya) even more aquatic, though still with toes and hooves, and then the Dorudontidae (e.g. *Dorudon*) and Basilosauridae (e.g. *Basilosaurus*) around 37 Mya with skeletons fully recognizably as cetacean including having flukes. *Indohyus*, *Pakicetus* and *Basilosaurus* scaled to human. (Image credit: Archaecetes by Nobu Tamura, phylogeny from Spaulding et al. 2009).[96]

The living cetaceans consist of two suborders, which divide themselves by mode of feeding into the Odontoceti suborder that are the toothed whales, porpoises and dolphins, and the Mysteceti suborder that are the baleen whales (Fig.5.8 bottom). The Odontoceti are predators that use echolocation to find prey, while Mysteceti filter plankton and krill through baleen plates suspended from their upper jaws. All fossil cetaceans are classified into a third suborder called the Archaeocetes, and all of them date to the Eocene epoch (about 35-55 Myr ago). The Archeocetes are the focus of our inquiry then, but since the two other suborders are taken to be their evolutionary descendants it is as well to know what extant cetaceans are like before deciding which Archaeocetes evolved into them, when and how.

There are 85 different cetacean species alive. They differ in size from Hector's dolphin which is among the smallest at 1.4 meters long and weighing less than 40 kg, to the largest animal on earth - the mighty Blue Whale, around 30 meters long and weighing up to 19,000 kg (33 elephants) and with a heart the size of small car.

Cetaceans are found widely, from fresh water to sea water and from sea abyss to sea shelf. Living cetaceans are alike in having a streamlined shape, flippers for forelimbs, no hind legs (at least in the way we define legs in all other animals), a fluke for a tail and their propulsion is by fluke and spine movement in the vertical plane. This is the opposite to fishes and seals, which use side-to-side motion in the horizontal plane. Cetaceans are warm blooded, breathe with lungs, give birth to live young, suckle, have cusped teeth and have a four-chambered heart. They are firmly classified as mammals therefore.

Yet they differ from all other mammals. They are obligatorily aquatic, have a dorsal fin, are born tail first, they lack hind limbs, tribosphenic molars, hair (vibrissae excepted) and sebaceous glands. Cetaceans are very unusual mammals. They are so different, and so obviously adapted to mammalian life in water, that George Gaylord Simpson made the following assessment: "Because of their perfected adaptation to a completely aquatic life, with all its attendant conditions of respiration, circulation, dentition, locomotion, etc., the cetaceans are on the whole the most peculiar and aberrant of mammals. Their place in the sequence of cohorts and orders is open to question and indeed quite impossible to determine in any purely objective way. There is no proper place for them in a scala naturae."[103]

What he was conceding, at least back in 1945, is cetaceans frustrate attempts at classification no matter whether criteria are anatomical (because they differ from other creatures so much), or behavioral (ditto) or of inferred evolutionary ancestry (because they have so many derived characters). To see why Simpson felt this way, consider a few examples of the latter. They are no more than variations on a theme that is consummate adaptation to the difficulties of living in water, and they are the same difficulties faced by that other water mammal, the city snorkeler now on vacation. Cetaceans have an elastic lens that permits them to focus both in air and in water including in deep water under high pressure (forget goggles). Cetaceans have mucus glands to protect the eye from salt (no more painful red eyes). Cetacean locomotion is the most energy efficient of any creature on earth (yes, really). The orca has been clocked at 55 km/hr (even Olympic swimmers at work only manage 8 km/hr). When it comes to prowess at breath holding, sperm whales can stay underwater for two hours, dive two kilometers deep and as far as is known, never get the bends (okay stop).

Heat exchange in water is twenty five times faster than in air, and temperatures fall with increasing depth under water. Whales are endothermic like humans, and need to maintain their body heat just as much as vacationing snorkelers do. Cetaceans are able to thwart heat loss by blubber and by ingenious counter current heat exchange systems of parallel alignments of veins and arteries called rete mirabilis. By the juxtaposition of vessels, heat is conducted from the warm core by arterial blood across to the returning venous blood without getting to the periphery. A temperature gradient is thus created over the animal from the warm core to the cold periphery that can be maintained even in icy seas.

Another problem cetaceans confront is how to breathe air when living in water, but needing to sleep. This solved by each half of the cetacean brain taking turns. The gray whale is able to find its way every year over a distance of around twelve thousand miles in a journey that begins from breeding grounds off Baja California, and ends at the Bering Sea in the Arctic Circle. Odontocete echolocation uses an organ system still poorly understood but at least in the case of the bottlenose dolphin, capable of detecting a ping-pong ball a whole football field away. Whales make calls under the water that can be heard for thousands of miles, literally, over an ocean. Humpbacks sing evocatively for many minutes before repeating themselves and while the biological significance of this communication is unknown, these songs are complex and structured and so suggestive of purpose. Whales are so well adapted to mammalian life in water they live over an extent of the earth from tropic to poles unmatched by any other creature, apart from man.

In sum, if whales are descended from four-legged hooved terrestrials, highly-evolved hardly begins to speak to the wonder and the power of evolution.

There are those who scoff at this wonder. Since it bears on the plausibility of a naturalistic mechanism for the origin of whales and whether this has anything to do with philosophy rather than reality, not to mention how anyone could ever think such a thing, we should hear that argument before proceeding. Douglas Futuyma at the University of Michigan is representative. As editor of the *Annual Review of Ecology and Systematics*, Emeritus Editor of the journal *Evolution*, past President of the American Society of Naturalists and author of such acclaimed textbooks as *Evolutionary Biology* and *Science on Trial*, he could hardly be more qualified to speak. This is what he says:

"If you ask, "What would I have to do to transform a primitive mammal into a bat or a whale?", the answer is, "Nothing very drastic."...Most whales, such as porpoises, are rather small. Their muscles and a thickened layer of fat give them a streamlined shape. The hind legs are reduced to vestigial pelvic bones. The front legs are flattened into paddles, with five digits (like primitive mammals); but the number of joints per digit is increased. The teeth are partly (in fossils) or entirely (in most modern whales) dedifferentiated, so they all have the same shape; and in modern (but not early fossil) species are increased in number – or else entirely lost, as in the blue whale. The most radical difference from other mammals consists largely of a forward extension of the jawbones out from under the nostrils, which are therefore situated on top of the head. In species such as the blue whale, the skin on the roof of the mouth is cornified like our calluses, and folded into sheets of baleen...The *only* characteristics that are not mere modifications of primitive mammalian features are the baleen and the dorsal and tail fins, which are rigid folds of skin and fibrous tissue, like our ears. One of the most amazing aspects of evolution is how easy it is to account for major transformations through rather simple changes in developmental processes. Most of the differences among different kinds of mammals are simply accounted for by changes in the relative rates of growth of different parts of the body."[104]

All fossil whales are called Archaeocetes and all are extinct. Six families have been interpreted from the fossil record and are taken to show a morphologic series from a fully terrestrial quadruped to a fully marine whale that diverged in the late Eocene or early Oligocene epoch into the Odontocete and Mystecete suborders alive today (Fig.5.8). No whales are known prior to the Eocene epoch. What came before the pakicetids at the base of the whale series, you ask? Dr. Thewissen replies:

"Although the cast of the early cetaceans has been filled in the past 20 years with new discoveries, the cetacean sister group has remained a hotly contested issue until recently...there were a number of viable candidates, including the extinct ungulates, the mesonychians [extinct wolf-like hoofed mammal], or one of the many artiodactyls [even toed hoofed mammal e.g. hippo, pig]. Whereas historically morphologists fell in the camp of a mesonychian sister relationship, molecular evidence suggests overwhelmingly that hippopotamids are the modern sister group of the cetaceans. Recent morphological studies agree with the hippopotamid hypothesis, whereas another suggests sister group relations between Cetacea and Artiodactyla as a whole. These theories converge if the hippopotamid lineage is ancient and includes forms that are distinctly morphologically unlike modern hippos. Both cetaceans and known artiodactyls could then be derived from a form that resembled the most archaic artiodactyls known: an animal the skeletally resembled a small, stocky mousedeer."

This is whale evolution from the science in a nutshell. We will deal with the controversies he alludes to momentarily, but know that overall plan has consensus.[98,101,105]

New discoveries of whale evolution get prominent attention in the scientific literature and the reason is obvious. Cetaceans are fascinating creatures and as a poster child for macroevolution, they show how powerful evolution is. This attention includes cover page articles and reviews in journals like *Science* and *Nature*,[98,106-110] and popular science magazines like *Scientific American* and *National Geographic*.[105,111,112] Whale macroevolution is a fossil fact everywhere. So what's the evidence? Get ready to get wet. In light of the manifold adaptations of modern Cetaceans to underwater living, unmatched by any other aquatic vertebrate, the notion that "the first whales were fully terrestrial and were even efficient runners"[98] would seem at least paradoxical, if not preposterous. Where did that idea come from?

Charles Darwin.[95,97] He wrote about whale evolution in the first edition of *The Origin*. There we read:

"In North America the black bear was seen by Hearne swimming for hours with widely open mouth, thus catching, like a whale, insects in the water. Even in so extreme a case as this, if the supply of insects were constant, and if better adapted competitors did not already exist in the country, I can see no difficulty in a race of bears being rendered, by natural selection, more aquatic in their structure and habits, with larger and larger mouths, till a creature was produced as monstrous as a whale."[97]

In response to criticism at this empty speculation his wording was muted in the subsequent editions but the substance was unchanged, even his calling a bear "almost like a whale" (p. 224). Thus the overall plan of whale evolution was conceived of without any fossil evidence. We've seen this before which is both the problem and the point. The problem is fossils require interpretation, and the point is whale fossils were sought, found and interpreted with this evolutionary pre-expectation in mind (§5.5.3). Without that presupposition there's no way a fully terrestrial hoofed quadruped could remotely be called "a whale" but in an evolutionary mind the idea of whales running on land isn't daft, it's demanded. Why? Because whales are mammals and mammalian fossils are terrestrial. If evolution is true, then whales must have evolved from terrestrial mammals somehow. The course of whale macroevolution was already known in that mind before a single fossil was found.

The science makes no apology for this pedigree either but you have to hear it to believe it. Here is our guide: "Whales indisputably are mammals...This implies that whales evolved from other mammals and, because ancestral mammals were land animals, that whales had land ancestors."[95] Here it is from *Scientific American*: "But to 19th-century naturalists such as Charles Darwin, these air-breathing, warm-blooded animals that nurse their young with milk distinctly grouped with mammals. And because ancestral mammals lived on land, it stood to reason that whales ultimately descended from a terrestrial ancestor."[105] And here it is from *Nature*: "All of the mammals that existed in the early Tertiary – some 65-50 million years ago – lived on land. So it has always been clear that aquatic cetaceans must have evolved from terrestrial mammals and returned to the water."[98]

What can we say so far? We are told the fossil record declares evolution a scientific fact. We search for objective evidence of whale macroevolution and discover this conclusion was reached before a fossil was found. The problem for paleontology after the Modern Synthesis was not whether whale macroevolution happened, only how it happened. The mission of paleontology since Darwin is to discover fossil species that document evolutionary transitions. This is worth seeing still more in black and white from experts with no intent of impugning either evolution or the business of fossil interpretation. Maureen O'Leary at Stony Brook University and John Gatesy at the University of California Riverside pronounce the aims of whale paleontology for us. Watch for the gap between what the science believes and what it knows:

"The mammalian order Cetacea made this challenging passage [from land to sea] early in the Eocene, >50 million years ago...Modern whales are clearly derived relative to the primitive mammalian form, but reconstructing the series of transformational steps that led to their origin is not trivial [i.e. whale evolution is a fact already and the task is to reconstruct it]. Did whales lose their hair before their hindlimbs, or vice versa? Did bio-acoustical modifications pre-date or postdate the move to an aquatic environment. How long did this transition take to occur? Where did the first whales evolve?"[113]

The research questions are all about how whale evolution happened not whether it happened. How do paleontologists resolve these questions then, given whale evolution was unwitnessed millions of years ago? They tell us:

"To answer such questions, a well-corroborated phylogenetic hypothesis is required to order, in space and time, the various evolutionary innovations of whales."

Evolutionary theory drives the evidence which is fossil interpretation, not the other way around. In a further refinement of this research, cladistic analysis is performed on fossils and their supposed living descendants to construct phylogenetic trees. We know cladistics uses the truth of evolution to construct an edifice that's evolutionary, which then becomes evolution evidence (§5.5.3). Where's the independent evidence whale evolution is true and the assumptions of cladistics real? If you respond it is in other transitional fossil series as *LOE#2* claims, you might be disappointed (§5.5.4.2-4). Besides, whales are the poster child and here we are.

Stephen Jay Gould will have none of this. He says the case for whale evolution is conclusive. Specifically, "Open and shut. Verdict: sustained in spades, wine and roses".[97] Before you rush off to get a glass, consider whether his conclusion was reached because a Darwinian prediction was indeed demonstrated true, or whether a subjective interpretation rendered selected fossils from multitudes as evidence in reverse. How can we tell which of them it is?

The imposter is exposed by two separate lines of inquiry. The first is by the history of the science of whale paleontology since Darwin. In other words, does history demonstrate subjective fossil interpretations from preconceived notions of whale evolution in circular reasoning? The second is by testing how well whale fossils stand up to the claims being made of them. Do they demonstrate macroevolution that can be objectively verified? Are they at least internally consistent with macroevolution, even as a hypothesis?

In light of Darwin's ideas about whale origins, the first line of inquiry has rendered the Archaeocetes fishy already but appearances could be deceptive. We will need to look. As we do, wherever possible attempts will be made to quote the leading authorities for two purposes. The first is to determine the exact merit of the whale fossil evidence and the second is to understand how the science thinks and why it is adamant whale macroevolution is a fossil fact. So, how were the pieces of the puzzle of whale origins put together?

The first archaeocete to be found was *Basilosaurus* in 1841 by the first Director of the British Natural History Museum, Sir Richard Owen (Fig.5.8). We meet him again later. Before this interpretation *Basilosaurus* had been regarded as a reptile, indeed its name means "king lizard". *Basilosaurus* is dated to the late Eocene epoch. An inspection of its skeleton shows it was a serpentine creature, similar to modern cetaceans in bone structure, around 25 meters in length and whale-sized. The *Basilosaurus* was regarded by Owen as what obviously appeared to be. An extinct cetacean. No evolution here, or at least so it would seem.

Further fossil discoveries were made in the years following but even by 1936 there were only three genera and species in the entire whale series, namely *Basilosaurus cetoides* and *Zygorhiza kochii* from the late Eocene, and *Protocetus atavus* from the middle Eocene.[114] Like *Basilosaurus* the latter two skeletons are similar to cetaceans we find alive today and even the oldest of them, *Protocetus atavus*, which we are told dates to the middle Eocene, Dr. Thewissen says is "unambiguously a cetacean".[115] His choice of words is curious, and suggests it's possible for a whale to be ambiguous, but we race ahead. First, what was the evidence of evolution found in *Protocetus*?

The teeth were not interpreted as a mosaic of characters in an extinct whale, but as evolutionary intermediate traits on the way to modern whales. The teeth were interpreted as primitive characters informative of ancestry. They were similar to those of creodonts (a group of extinct carnivores with a long dog-like skull) and thus creodonts were taken as ancestrally related to whales. The understanding of whale evolution by 1936 was reviewed by Remington Kellogg, the Smithsonian Curator then:

"In summation, it would appear that the evidence seems to point toward the concept that the archaeocetes are related to if not descended from some primitive insectivore-creodont stock, but that they branched off from that stock before the several orders of mammals that reached the flood tide of their evolutionary advance during the Cenozoic era were sufficiently differentiated to be recognized as such. Morphologically the archaeocetes seem to stand relatively near to the typical Mysticeti and Odontoceti, although all three suborders were separated from each other during a long interval of geologic time. It is not necessary to assume that any known archaeocete is ancestral to some particular kind of whale, for the archaeocete skull in its general structure appears to be divergent from rather than antecedent to the line of development that led to the telescoped condition of the braincase seen in skulls of typical cetaceans. On the contrary it is more probable that the archaeocetes are collateral derivatives of the same blood-related stock from which the Mysteceti and the Odontoceti sprang."[116]

In 1928 he had written:

"Whales at one time, geologically speaking, were land mammals and, although highly modified in some respects, they still retain all the typical mammalian features."

He had no fossil evidence for either claim. The whale fossils were bones of creatures that looked like modern whales. This evidence did not stop him from developing a detailed evolutionary hypothesis of whale origins however. Kellogg continues:

"With the passage of time their fore limbs have been modified to function as pectoral flippers, and their tail has been provided with caudal flukes to function as an organ of propulsion. No traces of hind limbs have been found in any of the living toothed whales with the exception of the sperm whale (*Physeter catodon*), but nevertheless we are fairly certain that these were present in the progenitors of the Cetacea, since the whale bone whales have one or two vestigial limb bones buried deep in the flesh of the pelvic region...We observe that the motive power was transferred from the limbs, which originally served as ambulatory organs, to the hinder end of the body, resulting in the development of caudal flukes, which became the propelling mechanism; the hind limbs disappeared; and the forelimbs were modified into fin-like organs, which served as rudders."[117]

The total fossil evidence for this whale evolution from land was...? Some teeth. The reason for whale evolution was not fossil evidence then, but supposition going all the way back to Charles Darwin. Whale paleontology, like the paleontology of every organism, was an evolutionary hypothesis in *The Origin* only needing evidence (§1.5.4;§5.3.3;§10.2-4).

Nine years later George Gaylord Simpson bluntly summarized the continued absence of fossil evidence for the idea:

"Throughout the order Cetacea there is a noteworthy absence of antecedent types, and nothing approaching a unified structural phylogeny can be suggested at present."[103]

Having made this observation he was happy to engage in some free-wheeling evolution speculation himself. He never considered this absence of evidence or Cetacean "perfected adaptation" to use his words, reasons to inspect a design hypothesis. The reason is because whale evolution was true already and you heard him say so (§5.1).[2] He continued:

"They probably arose very early and from a relatively undifferentiated eutherian ancestral stock. Thus the Archaeoceti...are definitely the most primitive of cetaceans, but they can hardly have given rise to the other suborders."

He concludes:

"Their place in the sequence of cohorts and orders is...quite impossible to determine in any objective way".

Although evolutionary whale origins were only a hypothesis, it was not doubted despite an impasse so great he called it "impossible". Why was he yet so certain? More importantly for us, what was the way forward for the science?

In 1950 Alan Boyden and Douglas Gemeroy entered the picture to tackle the problem with new techniques from immunology. They compared the immunological cross reactivity of serum proteins of extant Cetacea with other extant mammals. The experimental rationale rested on the assumption that relative antigenicity between modern day creatures bears a direct relationship to their evolutionary ancestry. Stated another way, it's the notion that relative immunogenicity (i.e. how antibodies see proteins and bind to them) is proportional to natural selection of random mutations over millions of years. If morphological difference was anything to rely on, that premise would seem a blind shot into blackness. They found the reactions were the greatest between the artiodactyls (even-hoofed mammals) and cetaceans, concluded they must be evolutionarily related therefore, and this was regarded as scientific progress.

The next leap came in 1966. Leigh Van Valen at the American Museum of Natural History pointed out the three cusped teeth of the archaeocetes were similar to those of a group of condylarth mammals called mesonychids. He also said their skulls were "remarkably similar". The latter was a contention disputed even by those who agreed with his hypothesis though. Take a look at their shapes yourself if you feel he was hard done by.[114]

Mesonychids are large, four-legged terrestrial mammals with sharp carnivore-like teeth and dated in the fossil record between 63 to 33 Mya, and thus stratigraphically exactly where ancestral whale fossil species were expected to be found. So far so good. They also had hooves. Nobody said mesonychids with sharp teeth and hooves were the intermediates between ungulates and carnivores, but rather mesonychids were the terrestrial, ungulate ancestors of whales, and the reason was their teeth shared similarities. Without any additional evidence, he said mesonychids had essentially run back into the sea to become whales in the mid to late

Paleocene epoch.[114] This is how fossils declare evolution to be a fact. They are interpreted to reconstruct the truth of evolution. He said:
> "Only two known families need be considered seriously as possibly ancestral to the archaeocetes and therefore to recent whales. These are the Mesonychidae and Hyaenodontidae..."

But why?
> "No group that differentiated in the Eocene or later need be considered, since the earliest known archaeocete, *Protocetus atavus*, is from the early middle Eocene and is so specialized in the archaeocete direction that it is markedly dissimilar to any Eocene or earlier terrestrial mammal. It is also improbable that any strongly herbivorous taxon was ancestral to the highly predaceous archaeocetes...Diverse and apparently equally valid objections exist for the various groups of Paleocene insectivores, one common to all being their small size. All marine mammals are large or rather large mammals."[118]

In case you missed it, he says *Protocetus* (the oldest whale then known) is so obviously a whale it bears no morphologic relationship to any terrestrial mammal. In other words he says no evolutionary trend is present. This was not taken as evidence and reason to question whale evolution. Instead two premises are seen operating to the contrary.

The first is a resolute conviction about the truth of whale evolution from a terrestrial ancestor. Whale fossil evolution was already true even though the only evidence was an inference from tooth similarities. The terrestrial ancestor to whales just needed to be discovered. A skeptic could consider this science contrived.

Second we see the power of the evolutionary mindset by which fossils are interpreted. Rather than seeing the complete absence of evolution because *Protocetus* was "markedly dissimilar to any Eocene or earlier terrestrial mammal", he saw evolution latent and what's more, was sure about it. The whale ancestor was from one of two families he said, and it was just a matter of finding fossils of the correct age with evolutionary features of modern whales and this terrestrial group to confirm the hypothesis. If this seems flimsy to you, the science disagrees. Van Valen's hypothesis quickly became, and remained, the accepted thinking about the origins of the archaeocetes until as recently as a few years ago. For instance the National Academy of Sciences declared in its 1998 evolution document to teachers: "The fossil record shows that these cetaceans evolved from a primitive group of hoofed mammals called Mesonychids," and as evidence presented drawings of an evolutionary series from mesonychid to *Basilosaurus*.[102] Now we are told just as certainly that the fossil record shows they evolved from arteriodactyls. It was not until 1979 that whale origins research really took off however.

This was the year University of Michigan paleontologist Philip Gingerich made a key discovery in northern Pakistan. He found the missing terrestrial ancestor of whales. He called it *Pakicetus inachus* (Fig.5.9).[119] This discovery followed hot on the heels of a reassignment of some isolated teeth found back in 1958 originally ascribed to a mesonychid called *Ichthyolestes*, but reclassified now an ancient cetacean instead.[120] The isolated teeth of some extinct, unwitnessed and otherwise unknown creature was the evidence to declare this thin legged, hooved terrestrial an ancient whale. The discovery of *Pakicetus* remains the most significant advance of whale paleontology to date because while specimens discovered since then have been more complete, they have only served to fill in the missing details of an evolutionary series between the poles that are *Pakicetus* at the base and living whales at the top. Here was the first fossil that confirmed whales were once terrestrials, that what was an ancient wolf with hooves was really a whale, and that whale evolution was indeed a fossil fact. As Gingerich and his team announced in *Science*:

> "Discovery of *Pakicetus* strengthens earlier inferences that whales originated from terrestrial carnivorous mammals and suggests that whales made a gradual transition from land to sea in the early Eocene, spending progressively more time feeding on planktivorous fishes in shallow, highly productive seas"[121]

This is not a new discovery of course because we had heard it long before. Darwin got it right again, incredibly.

In fact not only did this fossil for the first time in over a century of paleontology confirm Darwin's hypothesis about terrestrial whales, albeit a hooved wolf not a bear, it validated the evolutionary reasoning of paleontologists who accepted it as the working hypothesis. Finding additional whale intermediates suddenly got a whole lot easier with the two ends of the series now established. Sure enough further whale fossil discoveries followed in short order.

What was the *Pakicetus* fossil exactly? A partial skull. How partial can be seen from Fig.5.9A. It was wolf-like in shape with carnivorous teeth resembling those of the mesonychids. Without other data than this posterior skull fragment, the authors had no hesitation declaring it a whale. The specimen, like other pakecetid fossils found since, was lying in river sediments with land mammal fauna. Thus it was either a terrestrial or a freshwater animal. Confirmatory evidence for the latter was found by demonstrating oxygen isotope ratios (^{16}O and ^{18}O) in the bones were enriched in the ^{16}O isotope, which is the reverse of marine fossils since sea water has a relative accumulation of the heavier ^{18}O isotope (less escapes to evaporation).[122] Pakecetids are dated at 50 million years old.

Where is the whale or the whale evolution in *Pakicetus* you ask? Fossils can't speak, they need to be interpreted (§5.3.3). In that regard, Pakecetid dentition being similar to mesonychids became highly significant because as the authors pointed out, mesonychids were "the group of terrestrial mammals from which whales are thought to have evolved." There was no fossil evidence for this ancestry, only as they said, "thought". Stephen Jay Gould explains it was an accepted fact before this discovery that: "Whales must have evolved during the Eocene epoch some 50 million years ago, because Late Eocene and Oligocene rocks already contain fully marine cetaceans, well past any point of intermediacy."[97]

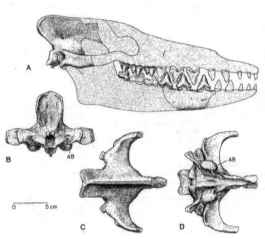

Fig.5.9. The running whale *Pakicetus*. The posterior portion of a skull seen from lateral (A), posterior (B), dorsal (C) and ventral (D) views. All the stippled rest of the skull is paleontologic "reconstruction", as is the generation of E), a whole swimming whale. (A-D from Gingerich et al 1983[121] with permission. Botttom panel from Gingerich, 1983 © National Association of Geoscience Teachers. Reproduced under Creative Commons license).[124]

The teeth of *Pakicetus* having mesonychid similarities on the one hand, were also considered to be similar to middle Eocene whales like *Protocetus* on the other. Presto. A fossil evolutionary intermediate appears. The ear bones of *Pakicetus* were also considered to have features shared by cetaceans. The conclusion was: "The basicranium of *Pakicetus* is unequivocally that of a primitive cetacean".[121] The carnivore-appearing partial skull was an ancient whale after all! Dr. Gingerich wrote of his wonderful discovery: "In time and in its morphology, *Pakicetus* is perfectly intermediate, a missing link between earlier land mammals and later, full-fledged whales."[123].

One of the consequences was that other fossil material previously considered mesonychid became reassigned as cetacean, and three pakecetid genera have since been erected (*Pakicetus*, *Ichthyolestes* and *Nalacetus*). The sum total of the evidence for the latter two genera? Some teeth.

As regards the robustness of the evolutionary evidence of *Pakicetus*, although it was of an extinct creature and consisting only of a partial skull and teeth, this was considered sufficient evidence to be a whale and sufficient for Dr. Gingerich to even publish an artist's reconstruction of what the whole creature looked like in the *Journal of Geology Education*, thus explicitly for teaching purposes (Fig.5.9E).[124] In order to understand how paleontology works it is helpful to see how he viewed the discovery that is everywhere now a landmark in the science of whale origins:

"In 1977 I organized an expedition to search for new Eocene mammal sites in Pakistan…We returned to the locality in 1978 and found, among other fossils, a partial cranium with a surprisingly small braincase (even for an Eocene mammal). Continued work produced a piece of large lower jaw with three pointed premolar teeth and some large sharply cusped molars belonging to a carnivorous mammal of some kind. Most of the fossils we found belonged to artiodactyls, rodents, a creodont, a primate, and other land mammals, but the cranium, the large lower jaw, and the sharply cusped carnivorous molars proved to represent a new genus of whale that we named *Pakicetus* after its country of origin…The age, environment of deposition, associated fauna, and functional morphology of Pakicetus indicate not only that it is the oldest and most primitive whale yet discovered, but that it is an important transitional form linking Paleocene carnivorous land mammals and later, more advanced marine whales. Whales apparently made the transition from land to sea in the early Eocene when protocetids like Pakicetus entered shallow epicontinental seas to feed on abundant planktivorous fishes living there."

There is the evidence and there is the conclusion, from the expert. But what is the evidence really? Here is Zhexi Luo at the Carnegie Museum of Natural History in Pittsburgh: "Cetaceans including pakicetids have only one unambiguous bullar synapomorphy that is absent in all noncetacean mammals – the involucrum or the pachyosteosclerosis of the medial margin of the bulla."[125] That's the evidence. That this creature differs from whales in all other respects is not relevant. That an arcane similarity in a bone feature between an ancient extinct carnivore and living whales might instead be a mosaic rather than an ancestral character is not a possibility. In the same article he admits that other features of the ear bone he says confirms *Pakicetus* a whale are not whale-like:

> "the auditory bulla...is relatively small by comparison with *Protocetus* and other archaeocetes...The bulla differs from that in other archaeocetes in that it has a much more distinct notch for the Eustachian tube...The presence of a well developed fossa for the tensor tympani indicates that Pakicetus almost certainly retained a functional tympanic membrane...there are no fenestrae or sinuses isolating the left and right auditory regions from each other [like modern cetaceans]...There is no indication of vascularization of the middle ear to maintain pressure during diving...In terms of function, the auditory mechanism of *Pakicetus* appears more similar to that of land mammals than it is to any group of extant marine mammals."

The auditory bulla of *Pakicetus* is attached to the squamosal, basioccipital and paraoccipital but in modern whales it is attached only to the periotic. This is not considered a contradiction either, for just as mosaic similarities can be interpreted as evolutionary transitions, so such differences can be interpreted as evolutionary loss of structures and intermediate evolutionary status stoutly defended. As he explains: "Bullar processes present on the squamosal and occipital of *Pakicetus* have been lost in *Protocetus* and all later cetaceans for which the skull is known". There was no doubt these creatures were whales, and so they have remained.

Writing in *Nature* in 2001 after additional non-skull Pakicetid fossils were found showing it was a terrestrial quadruped, Dr. Thewissen reiterated the conclusion *Pakicetus* was a whale even if it did not remotely look like one. The reason was simply its evolution and evolutionary adaptations:

> "Aquatic postcranial adaptations are pronounced in late Eocene basilosaurids and dorudontids, the oldest obligate aquatic cetaceans for which the entire skeleton is known, and therefore can be used to

evaluate pakecetid morphology. Aquatic adaptations include: presence of short neck vertebrae; thoracic and lumbar vertebrae that are similar in length; unfused sacral vertebrae; lack of a sacro-iliac joint; presence of a short tail with a ball-vertebra (a vertebra at the base of the fluke, with convex articular surfaces); broad fan-shaped scapula with anterior acromion and small supraspinous fossa; an ulna with a large and transversely flat olecranon; a wrist and distal forearm flattened in the plane of the hand; and tiny hind limbs. Pakecetids display none of these features...The nasal opening of pakicetids was at the tip of the snout, as in land mammals...but unlike late Eocene cetaceans...The lacrimal foramen is present in pakicetids...but is usually absent in aquatic mammals (modern cetaceans, sirenians and pinnipeds). The orbits of pakicetids are close together and are frontated (face dorsally) but are not at the most dorsal point of the head. This is unlike any other cetacean....Unlike any other cetacean, the pakecetid outer ear was unspecialized and similar to that of land mammals..."[109]

The possibility this four-legged running carnivore had no relationship to whales whatever given the differences he had just detailed, is not a consideration. Robert Carroll in his authoritative text on vertebrate paleontology explains the contradictory observations away too. He too uses evolution to explain why they are the evidence of evolution:

"The presence of a well-developed fossa for the tensor tympani indicates that Pakicetus retained a functional tympanic membrane, which is lost in advanced whales...The molars retain the cusp pattern of mesonychids. The upper molars are triple rooted, with a distinct protocone, paracone, and metacone. The lowers retain a distinct protoconid and hypoconid, exactly as in the mesonychids."[126]

Similarities are evidence of evolutionary ancestry and differences are evidence of evolutionary loss. Evolution remains true no matter what (§19;§20). And yet a problem for evolution still was able to surface. Opinion on the mesonychids shifted. Now they are not considered ancestral to whales at all. This would seem to seriously undercut Stephen Jay Gould's accolades and his assessment that:

"In terms of intermediacy, one could hardly have hoped for more from the limited material available, for only the skull of *Pakicetus* had been found. The teeth strongly resemble those of terrestrial mesonychids, as anticipated, but the skull, in feature after feature, clearly belongs to the developing lineage of whales."[97]

And Robert Carroll's assessment that: "The skulls of Eocene whales bear unmistakable resemblances to those of primitive terrestrial mammals of the early Cenozoic."[126] What can we say by way of summary?

The posterior portion of the skull of an ancient terrestrial with wolf-like features was called a whale because of a prominence on one of its ear bones and because its teeth were similar to both mesonychids and middle Eocene whales. Mesonychids were considered ancestral to whales because of similarities in tooth structure. Evolution is rather easy to demonstrate if this is all it takes. A terrestrial whale ancestor had to be found. The only question was which particular feature on which particular bone would be taken as a synapomorphy, and which would be dismissed as convergent evolution or reversal. The only surprise is that it took so long.

Fossil species between *Protocetus* and *Pakicetus* were discovered a transitional series established, and these fossils are now taken to declare whale evolution a fact. In 1993 Dr. Thewissen suggested that middle ear features on the specimen represented adaptations to underwater hearing.[127] Some years later the consensus was the creature was fully terrestrial and so they were taken as not.[109] Still further information about Pakicetus and artiodactyls would emerge and other fossil species were found to close the gap between the extremes but we need to follow the story in the order it happened.

The next advance was made at the upper end of the fossil series, with *Basilosaurus*. It was the discovery that *Basilosaurus* was an evolutionary intermediate on the way to modern whales. This was to give legs to what was then still just a fishy tale about wolf-into-whale evolution. The question posed by *Pakicetus* was this: Was there a plausible fossil link between four legged terrestrials and whales? It was already established thinking in biology that modern whales had vestigial "legs" not attached to the vertebral column that were (apparently) functionless, and thus a vestigial evolutionary remnant from the quadruped ancestor (§9.6). Was there evidence for a hind limb transition in the fossil series? The developments taken as providing the answers to these questions happened rather fortuitously, as discoveries in science often do.

In 1983 Dr. Gingerich had to move his team from Pakistan because of hostilities resulting from the Soviet invasion of Afghanistan. He decided to look for whale fossils southwest of Cairo in the Western Desert. There his team found the legs on the whale *Basilosaurus* (Fig.5.8). In 1990 they reported the discovery in *Science*, explaining the significance of the discovery for whale evolution:

"These are important in corroborating the intermediate evolutionary position of archaeocetes between generalized Paleocene land mammals that used hind limbs in locomotion and Oligocene-to-Recent whales that lack functional pelvic limbs. The foot is paraxonic [the axis of symmetry of the foot in weight bearing is through digits III and IV] consistent with derivation from mesonychid Condylarthra...The retention of well-formed pelvic limbs in Basilosaurus corroborates the morphological primitiveness of archaeocetes. Temporal and morphological intermediates are direct and important evidence of transition in evolution: an Eocene whale with functional hind limbs narrows the gap considerably between generalized Paleocene land mammals that used hind limbs in locomotion and Oligocene-Recent modern whales that lack pelvic limbs."[128]

Stand back a moment, we need to take this in. The structure in question was 3 percent of the animal's total length. It had no sacrum (the bone above the coccyx) and it was not attached to the spine. To quote the authors it "floated in muscle in the ventral body wall."[114] Clearly not used for ambulation, much less any kind of locomotion, these appendages were surely not legs. Question: How could this be an intermediate whale with hind legs? Answer: Evolutionary reasoning which renders it a foot by homology, just as it renders every structure descendant in some way from some other. This is a philosophical view. It may also yet be true and we look for ourselves in chapter 7, but this kind of data can't be used to tell. It is at best confirmatory of an evolutionary paradigm and hardly an advance on what was already interpreted about this structure in extant whales. Yet because of the significance for the coherence of whale evolution, the discovery was a report worthy for publication in a journal as celebrated as *Science*. Stephen Jay Gould provided an analysis of the work. It was too much with too little, even for him:

> "The authors strive bravely to invent some potential function for these miniscule limbs, and end up speculating that they may have served as "guides during copulation, which may otherwise have been difficult in a serpentine aquatic mammal." (I regard that guesswork as unnecessary, if not ill conceived. We need not justify the existence of a structure by inventing some putative Darwinian function. All bodies contain vestigial features of little, if any, utility)."

So Gould corrects the authors for speculating in a journal like *Science*, and then trumps them with a wilder one of his own. The objective evidence for his claim is reviewed in §9.6 and rejected. The only point for now is seeing how speculations fly and how stepping outside the evolutionary paradigm by which all the reasoning is conducted is to be left asking where the beef is, anywhere, in the whale evolution story.

It is helpful to see the power of the evolutionary mindset driving whale origins research from another perspective though. From that of the investigators. This was Dr. Gingerich's opinion of the discovery when interviewed by *Scientific American*: "I immediately thought, we're 10 million years after Pakicetus," Gingerich recounts excitedly. "If these things still have feet and toes, we've got 10 million years of history to look at." Suddenly, the walking whales they had scoffed at in Pakistan seemed entirely plausible."[105] We know the notion of walking whales was not just plausible, it was demanded. The only question was what it would look like.

The next breakthrough was to find whales with legs between modern whales and *Basilosaurus*. In 1994 Dr. Thewissen's group reported specimens of a new genus and species from northern Pakistan called *Ambulocetus natans* which means "the walking and swimming whale" (Fig.5.8). It had the skeleton of a terrestrial quadruped, the size of a sea lion. The significance of the find for whale origins can hardly be underestimated. Here for the first time was the direct fossil evidence for a quadruped that was also a whale. While that had been the accepted thinking since *Pakicetus*, it was hypothetical since the objective evidence of *Pakicetus* was only a partial skull fragment and some teeth. The investigators pointed out the significance of their find. They began by sketching the context of the discovery:

"Recent members of the order Cetacea (whales, dolphins and porpoises) move in the water by vertical tail beats and cannot locomote on land. Their hindlimbs are not visible externally and the bones are reduced to one or a few splints that commonly lack joints. However, cetaceans originated from four-legged land animals that used their limbs for locomotion and were probably apt runners."

This wild whale of a speculation is directly from the pages of *Science*. They continue using the evolutionary hypothesis to drive the reasoning and the interpretation of what they found:

"Because there are no relatively complete limbs for archaic archaeocete cetaceans, it is not known how the transition in locomotor organs from land to water occurred. Recovery of a skeleton of an early fossil cetacean from the Kuldana Formation, Pakistan, documents transitional modes of locomotion, and allows hypotheses concerning swimming in early cetaceans to be tested. The fossil indicates that archaic whales swam by undulating their vertebral column, thus forcing their feet up and down in a way similar to modern otters. Their movements on land probably resembled those of sea lions to some degree, and involved protraction and retraction of the abducted limbs."[129]

A few obvious questions follow. Why was this four-legged terrestrial so certainly a whale? The authors gave the reasons. There was no doubt about it either:

"*Ambulocetus* is clearly a cetacean: it has an inflated ectotympanic that is poorly attached to the skull and bears a sigmoid process, reduced zygomatic arch, long narrow muzzle, broad supraorbital process, and teeth that resemble those of other archaeocetes, the paraphyletic stem group of cetaceans"[129]

In an accompanying commentary to the paper written by Annalisa Berta at San Diego State University, she too questioned what a whale was. Understandably, because you didn't need a degree in biology to notice this "whale" was a running terrestrial with four legs. She argued that character based definitions were problematic, and gave a reason why. "For example, how can whales be defined as lacking hindlimbs since some whales (for example, several archaeocetes) possess them?" She means the "legs" of middle Eocene (*Basilosaurus*) and modern whales, of course. If the definition of body parts as basic as legs is to be as ambiguous as that in biology, her search for clarity is hopeless already. Still, she saw a way out.

"A more reasonable solution is to use a phylogenetic definition, that is, one based on common ancestry...because archaeocetes are more closely related to modern whales than they are to mesonychids, *Ambulocetus* is a whale by virtue of its inclusion in that lineage...Although its relationship to other whales is uncertain, *Ambulocetus* is a whale, using a definition based on ancestry."

So evolution says what's real or not, and specifically what evolutionary inferences will declare what is a whale or what is a wolf. This is to descend from the stability of character definitions, which for all their arbitrariness were at least rigid and had served well for multiplied centuries, to the speculations and quicksand of inferred ancestry which is all inferential and in permanent flux. Now the nomenclature of biology itself is evolutionary. Evidence from paleontology and biology that demonstrate evolution a fact is found in short order searching in biology this way. But Dr. Berta made two other points which were just as stunning.

The first appeared to impugn the entire premise of *Ambulocetus* as a whale. The characters used by the authors to declare *Ambulocetus* a cetacean had not been verified. She said: "Before these purported whale characters can be used in a phylogenetic definition of whales, however, the possibility that some of them may have a broader distribution (for example, in mesonychids) needs to be examined."[108] It was a dramatic admission. One might ask why the paper had any business getting the forum it had but you know why.

The second point made by Dr. Berta, also in thunderous understatement, was the authors had no pelvis to support their hypothesis of locomotion. It's significance lies in the way that the science was being conducted. The most obvious difference between the mid Eocene whales like *Basilosaurus* and modern whales was that the "hind limb" did not articulate with the vertebral column. Sure *Ambulocetus* looked like a terrestrial quadruped, but as an extinct creature - if the authors were seriously to engage the whale evolutionary paradigm, showing a leg attached to a pelvis attached to the vertebral column was what was required. Yet this omission did not even get a passing mention:

> "Thewissen et al. provide some solid comparative data to support their conclusions regarding the evolution of locomotion in whales; however, a well-corroborated phylogenetic context with which to interpret these character transformations would greatly enhance its utility. For example, since the pelvic girdle is not preserved, there is no direct evidence in Ambulocetus for a connection between the hindlimb and the axial skeleton. This hinders interpretations of locomotion in this animal, since many of the muscles that support and move the hindlimb originate on the pelvis."

In the paper the authors addressed an interesting conundrum. All fish swim using vertical tails that beat from side to side but whales swim using flukes, which are horizontal tails that beat up and down. So does another mammalian order, the Sirenia (e.g. manatees). Stephen Jay Gould explains the difficulty:

> "I don't mean to sound jaded or dogmatic, but *Ambulocetus* is so close to our expectation for a transitional form that its discovery could not provide a professional paleontologist with the greatest of all pleasures in science – surprise. As a public illustration and sociopolitical victory, transitional whales may provide the story of the decade, but paleontologists didn't doubt their existence or feel that a central theory would collapse if their absence continued. We love to place flesh upon our expectations (or put bones under them, to be more precise), but this kind of delight takes second place to the intellectual jolting of surprise. I therefore find myself far more intrigued by another aspect of *Ambulocetus*...For the anatomy of this transitional form illustrates a vital principal in evolutionary theory – one rarely discussed, or even explicitly formulated, but central to any understanding of nature's fascinating historical complexity...To give the cardinal example from seagoing mammals: The two fully marine orders, Sirenia and Cetacea, both swim by beating horizontal tail flukes up and down. Since these two orders arose separately from terrestrial ancestors, the horizontal tail fluke evolved twice independently."

You can count on fossils being found to support this notion. It's just a matter of dipping into the fossil box again. Gould continues:

> "Fishes swim in a truly opposite manner...Both systems work equally well; both may be "optimal". But why should ancestral fishes favor one

system, and returning mammals the orthogonal alternative?...Either way will do, and the manner chosen by evolution is effectively random in any individual case. "Random" is a deep and profound concept of great positive utility and value, but some vernacular meanings amount to pure cop-out, as in this case...This subject, when discussed at all in evolutionary theory, goes by the name of "multiple adaptive peaks." We have developed some standard examples, but few with any real documentation; most are hypothetical, with no paleontological backup."
This is reads rather like a confession, yet from one certain evolution is a fact. There was a purpose to this contrition though, and it was where *Ambulocetus* came in. He said:

"I love the story of *Ambulocetus* because this transitional whale has provided hard data on reasons for a chosen pathway in one of our best examples of multiple adaptive peaks...Thewissen and colleagues draw the proper evolutionary conclusion from these facts, thus supplying beautiful evidence to nail down a classic case of multiple adaptive peaks with paleontological data."

What did the authors claim? "*Ambulocetus* shows that spinal undulation evolved before the tail fluke...Cetaceans have gone through a stage that combined hindlimb paddling and spinal undulation, resembling the aquatic locomotion of fast swimming otters". Gould summed up: "The horizontal tail fluke, in other words, evolved because whales carried their terrestrial system of spinal motion to the water." This is how paleontology does business. Evolutionary reasoning drives fossil interpretation that proves evolution a fossil fact.

In 1998 Dr. Gingerich reported a new pakecetid from northern India called *Himalyacetus* which was not only the oldest whale known at 53.5 Myr and some 3.5 Myr older than *Pakicetus*, analysis of tooth enamel oxygen isotope levels demonstrated intermediate ratios of ^{16}O and ^{18}O between marine and fresh water. This is interpreted as indicative of an existence in both environments. The same analyses on *Pakicetus*, it will be recalled, showed non marine levels. If this is correct what does *Pakicetus* have to do with whale origins? Thewissen rejects both Gingerich's methodology to date the fossils and his assignment of the bones as being pakicetid.[95] Problems remain for both parties whatever way the dispute is eventually resolved. Either the date obtained by Gingerich is valid and the evolutionary series is in disarray with *Himalayacetus* swimming in sea water before *Pakicetus*, or the methods used to perform fossil dating and fossil diagnosis are so subjective that disagreement over such supposed absolutes can continue rationally even among experts. By the way, what was the sum total of the evidence for *Himalayacetus*? A lower jaw fragment bearing three molar teeth. What was the evidence that it was a whale? Here is the evidence: "Himalayacetus has the general molar form characteristic of Archaeoceti. It is clearly a pakecetid because it lacks the enlarged mandibular canal seen in all of the more advanced archaeocetes."[130]

The same year Gingerich reviewed the prevailing opinion of whale origins that "most authors now accept as a working hypothesis Van Valen's idea that the Mesonychia gave rise to the Archaeoceti."[114]. Gingerich showed impressive similarities in skeletal morphologies between the mesonychid *Pachyaena* and the extant wolf *Canis lupus* and said:

"The overall skeletal similarity of early Eocene Pachyaena to extant Canis shown...is interpreted as indicating similar behavior in life...Mesonychians are usually interpreted as solitary carrion feeders and scavengers that spent many of their waking hours trotting in search of dead animals...This is plausibly the kind of animal from which archaeocetes evolved."

Where is the comparison with whales you ask? That's the business of finding intermediate fossils which close the gap between this four legged running ancestor and a deep diving fluked whale with sonar. The similarity of this creature with a wolf was an opportunity to define the evolutionary trends that must have taken place between them. McGill paleontologist Robert Carroll writes from the perspective that mesonychia are closely related to ancestral whales, and uses morphology as evidence of the whale's evolutionary ancestry:

"The skulls of Eocene whales bear unmistakable resemblances to those of primitive terrestrial mammals of the early Cenozoic...Although the snout is elongate, the skull shape resembles that of mesonychids, especially *Hapalodectes*...Also like the early whales, *Hapalodectes* has vascularized areas between the medial portions of the upper molars...The oldest whales have been described from latest early Eocene deposits in Pakistan [i.e. Pakicetus]...The cranial remains are intermediate between those of well-known late Eocene whales and mesonychids".[126]

Plugging in additional archaeocetes following Van Valen's postulate made an evolutionary series appear and whale macroevolutionary change from mesonychids was confirmed. As Gingerich concluded:

"The most important thing that can be said about Eocene Archaeoceti is that they are beginning to fill the temporal, geographic, and morphological gap between Paleocene land mammals and Oligocene and later whales. The temporal and geographic distributions of Mesonychia and Archaeoceti...support this. Size-adjusted comparisons of morphological characteristics...show the general pattern of change from a wolflike mesonychid model ancestor through primitive remingtonocetids and protocetid archaeocetes to more advanced dorudontids and basilosaurids archaeocetes, to modern mysticetes and odontocetes...we are beginning to have enough intermediates to say with confidence what happened in the transition of whales from land to sea."[114]

This was the same year the National Academy of Sciences declared: "The fossil record shows that these cetaceans evolved from a primitive group of hoofed mammals called Mesonychids". Not much doubt there either.

In the 1990s several nearly complete mesonychid specimens had been recovered, including creatures called *Sinonyx, Dissacus and Pachyaena* and taken as advancing understanding of these extinct animals. By then the techniques of cladistic analysis could also be applied to the bones too (§5.3.3). The conclusion? Van Valen's contention that mesonychids were ancestral to whales was not supported, but instead "Mesonychians" from the Paleocene and the Eocene were considered the closest terrestrial relatives of Cetacea",[131] a clade called the Cete. These were demonstrated to be a sister group of cetaceans and the closest terrestrial relatives of cetaceans because of morphologic similarities especially of the ear region and teeth. This is how the evidence of whale evolution comes from comparative anatomy of fossil species with other fossil species, and from comparisons between the bones of extant whales with other extant creatures.

The matter of the mesonychia being sister group to the cetaceans and so confirming the truth of evolution remained true when it appeared they were not. Beginning with Van Valen, archaeocetes were considered to share teeth characteristics with the mesonychia (tall protoconids). The mesonychia had hooves. The discovery of *Ambulocetus* as ancient cetacean with "[t]oes terminated by a short phalanx carrying a convex hoof" and the Basilosaurus "foot" were both taken as striking confirmation cetaceans were ungulate descendants and that among those still living, their closest relatives were artiodactyls (like hippos, camels, pigs) as all were paraxonic. Thus the fossils had scientifically confirmed the truth of evolution.

Trouble emerged with this tidy scenario in 1994 when cladistic analysis was done on molecular characters for the first time. The problem was that the molecular analysis did not agree with results of analyses of morphologic characters. Actually this is to understate the differences. While there was agreement that artiodactyls and cetaceans were closely related, the molecular cladistic analysis demonstrated Artiodactyls are polyphyletic with the cetaceans nested within the Artiodactyls, and the hippopotamids their closest living relatives.[113] The implications of this phylogeny are that a cow is more closely related to a whale than it is to a horse and a hippo is more closely related to a dolphin than it is to a pig. This too is not taken as being so ridiculous as to prompt wholesale reexamination of the paradigm. As our guide Dr. Thewissen pointed out, the discovery "is counter-intuitive...yet it is one of the best examples of congruence between morphological and molecular estimates of mammalian phylogeny." Common sense declares something is terribly wrong. And this is supposed to be one of the best cases of macroevolution.

At first the molecular data was ignored by the paleontological community, but in 1999 Norihiro Okada at the Tokyo Institute of Technology published a report in *Nature* that forced everybody, including those with heads in sand, to engage his data. Using short interspersed elements (SINEs) which are retrotransposons integrated into the genome by reverse transcription of RNA, he found two SINE families exclusively in the genomes of whales, hippos and ruminants but not in pigs and camels. He argued that the former was a monophyletic group as a result. Following the mindset of the experts is helpful. *Scientific American* interviewed Philip Gingerich at the time for his thoughts:

"The whale-hippo connection did not sit well with paleontologists. "I thought they were nuts," Gingerich recollects. "Everything we'd found was consistent with a mesonychid origin. I was happy with that and happy with a connection through mesonychids to artiodactyls". Whereas mesonychids appeared at the right time, in the right place and in the right form to be considered whale progenitors, the fossil record did not seem to contain a temporally, geographically and morphologically plausible artiodactyls ancestor for whales, never mind one linking whales and hippos specifically."[105]

There are three morphologic features that define all artiodactyls: (1) paraxony, an anatomic term referring to a limb where the axis of symmetry runs between digits III and IV, (2) three cusps on the lower jaw deciduous premolar 4 (abbreviated in the vernacular DP/4), and (3) they have a trochleated astralagus (heel bone) which makes it very mobile whilst in other mammals it is not.[110] Since paraxony is seen in the primitive whales like *Pakicetus* and in all artiodactyls it is considered plesiomorphic and therefore not helpful in determining phylogeny. The DP/4 character is not found in the archaeocetes nor in the mesonychia. This fact alone impugns the reliability of the evolutionary inferences of morphologic characters for if it is true, cetaceans cannot be artiodactyls. But evolution can accommodate anomalies. The solution?

"If the molecular data are correct, the morphology of this tooth has, like the trochleated astralagus, a complicated phylogenetic history that includes reversals, convergences or both".[110]

Anything goes and it's all explicable by evolution. The trochleated astralagus is found only in artiodactyls and not in other mammals and therefore is taken to be synapomorphic of artiodactyls. The mesonychians do not have a trochleated astralagus. Accepting the molecular data provides two potential solutions. Either the mesonychians are not closely related to cetaceans after all (and all the similarities in teeth characters which had pushed fossil interpretation along were instead spurious convergent evolution), or the trochleated astralagus is not really the defining feature of Artiodactyls. Either way it's a huge climb down for the supposed reliability of comparative anatomy and evolutionary morphology. If the DP/4 data could be dismissed there was only one way to resolve the dilemma. Find the astralagus of the most basal whale *Pakicetus*. Could it be found?

Four years later Thewissen's and Gingerich's groups independently produced fossils of ancestral whales each bearing a trochleated astralagus. Whales were really artiodactyl ungulates and if the ankle morphology was anything to go by, great runners too. Thewissen performed cladistic analysis of the morphological characters from the bones and teeth of multiple species of artiodactyl, mesonychians, condylarths, and cetaceans. The parsimonious solutions rejected the mesonychians as being a sister group of cetaceans. They were not ancestral. All the mesonychian similarity was dismissed as simply convergence. And so the pronouncement by the National Academy of Sciences that: "The fossil record shows that these cetaceans evolved from a primitive group of hoofed mammals called Mesonychids,"[101] was the opposite only three years later. How was this handled? Dr. Gingerich:

"Although there is a general resemblance of the teeth of archaeocetes to those of mesonychids, such resemblance is sometimes overstated and evidently represents evolutionary convergence."[106]

Only by knowing the history of whale paleontology can one appreciate the impact of these words by a science that otherwise is a casual dismissal. If paleontology can be so wrong about whale ancestry and yet still be right about evolution and evolutionary origins, one might well ask how much science there is in this science.

The second question is how this can happen, and the third is what makes it yet still seem rational to all the experts? (§18;§19;§20) The mesonychid hypothesis that dated back to Van Valen died pathetically but the demise received no more attention than a drive by glance. Why, if not because there was no more objectivity to its death as a whale than there had been to its life as a whale.

The rejection of the mesonychia still has some protesters, notably Maureen O'Leary at Stony Brook. If she is right this would mean a trochleated astralagus was lost by the mesonychids, and therefore that artiodactyls cannot be defined by a trochleated astralagus after all. This upends the apple cart too, only this time from the other side. It bears mentioning a cladistic analysis of a data matrix of 8258 characters still failed to resolve a whale phylogeny. Something is terribly awry.

The most recent development directly confronted this cladistics rug pull. It was the announcement in 2007 by Dr. Thewissen of *Indohyus*, a member of an extinct group of artiodactyls called the raoellids as the link between cetaceans and the artiodactyls (Fig. 5.8).[132] The problem cladistics poses is not just the contradiction of morphologic character analysis with the molecular, but in declaring hippopotamids the closest extant relatives of cetaceans the problem is the fossils of the former are only 15 Myr old while cetaceans are 50 Myr, and the first hippopotamid fossils in Asia where the whales evolved, are only 6 Myr old. Indohyus as link and sister group to whales dates around 48 Myr so neatly solving that date problem and though a raccoon sized mammal and an artiodactyl, is classified a whale because of that one ear bone again, the involucrum. Only before *Indohyus* this ear bone had defined, at least in the whale paleontologic community, what it was to be a cetacean. Now here was an artiodactyl with a cetacean synapomorphy. Even Dr. Thewissen admitted "[i]dentification of the involucrum in Indohyus calls into question what it is to be a cetacean", but this was a question raised already long before. What can we say by way of summary about whale evolution from the fossil evidence? Five things:

First, whale evolution was a fact long before fossils were said to say so. Fossil specimens were selected by evolutionary hypotheses to confirm the notion of whale evolution from a terrestrial ancestor demanded by a paradigm first suggested by Charles Darwin and so became the evidence. How do we know the bones assigned as separated species are in any way related to one another? If this skepticism is thought to be unreasonable, consider the ease with which mesonychids, once pivotal to evolutionary notions and fossil interpretation were dismissed without ever endangering either the truth of whale evolution or even the reliability of fossil evidence. Consider how arbitrary interpretive similarities of selected characters on ear bones and teeth are enough evidence to render a partial skull of an unwitnessed extinct creature a whale, even when all the other characters of those bones say the opposite, and when Mysticetes don't even have teeth.

Consider how differences between intermediate species in an evolutionary series are accepted because they are really "pronounced adaptations" and how similarities can be either synapomorphies, or convergences, or reversals. Evolution theory can explain any possible fossil observation. This is how fossils declare evolution "a fact".

Second, the bankruptcy of the whale fossil series is supported by evidence from another bankrupt quarter. Incongruent results of cladistic analysis of molecular characters versus morphologic characters. Not only can we not be certain that fossils in the whale series are really related ancestrally, but the whale evolutionary hypothesis lacks internal coherency.

Third, the stratigraphical relationships of the intermediates are uncertain if *Himalayacetus* appears before *Pakicetus* living in salt water.

Fourth, a four-legged wolf-like land living running whale became an unequivocal obligately diving whale over the space of 15 million years or less. This is warp speed evolution, no matter how many intermediates are collected and will be aligned in the years ahead. And yet instead of unease this speed is yet championed by science as evidence of another evolutionary fact, and unblushingly taught to schoolchildren. In Dr. Thewissen's words:

"Whales underwent the most dramatic and complete transformation of any mammal. The early stages were so poorly known 15 years ago that creationists held up whales as proof that species couldn't possibly have come about through natural selection. Now whales are one of the better examples of evolution." [111]

Fifth, what is the mechanism for this top speed whale evolution? Natural selection without a mechanism is not a scientific explanation. *National Geographic* asked this question and also gave the answer: "At what stage did they [ancestral whales] come up with extra myoglobin in their muscles for storing oxygen on longer dives? When did humpbacks start singing the most elaborate, evocative songs ever heard? Because changes in physiology and behavior aren't always associated with obvious shifts in anatomy, they can be harder to track. We only know that when modern whales emerged they continued to refine their adaptations and prosper." In other words evolution is itself the answer. Fossils and other evidence will always show it if we just keep looking and also keep that in mind. Emperor's New Clothes.

§5.5.4.2. Picking Bones For Wild Horses

Very soon after Darwin's ideas were published in *The Origin*, a Yale paleontologist called Othniel C. Marsh discovered the horse transitional fossil series, and already by 1874 had declared that as regard the evolution of horses: "the line of descent appears to have been direct and the remains now known supply every important form". Although this was a series of progressive morphologies back then and today is an arborized bush, the sequence to the modern day horse has changed little since.

The series begins 55 million years ago with *Hyracotherium*, a dog-like and terrier sized creature first described by Richard Owen who never considered it equine. Marsh renamed it *Eohippus* or "dawn horse". The animal had 4 toes on the forefeet and 3 toes on the hind feet, walked on pads and had the teeth of an omnivorous browser. Afterward are sequentially found *Orohippus*, *Epihippus* (both four toed in front and three-toed behind), then *Mesohippus* around 40 million years ago, then *Miohippus*, *Parahippus* and *Merychippus* (each with three-toes on all feet), then a one-toed animal called *Dinohippus* around 12 million years ago, and the series ends around 4 million years ago with the modern horse *Equus*. At the same time as these morphologic transitions, the teeth change from browser to grazer and the body size increases (although these trends are not seen in all the lines of the evolutionary side tree branches).

The series is very plausible. The problem again is how to be certain the creatures represented by these fossils really are ancestrally related. It's only the truth of evolution that allows species of different age to be aligned and literally, morph one into another. This is at least consistent with macroevolution and so must chalk up this series on the plus side for evolution evidence. There are 3 aspects about the horse series, ignoring the absence of any ancestors for *Hyracotherium*, I leave for you to consider.

The first is a point already made. The evidence of transition in the fossil record is the extreme exception. Why should these and not all the other fossils without transition refute this process even by their overwhelming number? Here is Chris Paul at the University of Liverpool: "Examples of gradual evolutionary trends in the fossil record can be counted on our fingers (*Gryphaea*, *Micraster*, *Zaphrentis*, the horse) and we simply ignored the countless examples that do not show the expected pattern."[133] The disproportion is so great the question becomes: How do we know this series was extracted and aligned merely by chance?

Second it's one thing to arrange fossils in a series that demonstrate plausible morphing, but is the mechanism? This is what distinguishes naturalistic science. How exactly could mutations do this? How would they result in a progressive loss of toes on separate limbs? How did the marvel of engineering that is the equine hoof happen? This question has already been asked by Pierre Grasse decades ago:

'In his theoretical discussion [in *Horses* (1951)], Simpson does not linger over the structure of the hoof; yet it is the result of very innovative and precise evolution. Such a hoof, which is fitted to the limb like a die protecting the third phalanx, can without rubber or springs buffer impacts sometimes exceed one ton. It could not have formed by mere chance: a close examination of the structure of the hoof reveals that it is a storehouse of coaptations and of organic novelties. The horny wall, by its vertical keratophyl laminae, is fused with the podophyl laminae of the keratogenous layer. The respective lengths of the bones, their mode of articulation, the curves and shapes of the articular surfaces, the structure of bones (orientation, arrangement of bony layers), the presence of ligaments, tendons sliding with sheaths, buffer cushions, navicular bone, synovial membranes with their serous lubrication fluid, all imply a continuity in the construction which random events, necessarily chaotic and incomplete, could not have produced and maintained. This description does not go into the detail of the ultrastructure where the adaptations are even more remarkable; they provide solutions to the problems of mechanics involved in rapid locomotion on monodactyl limbs."[134]

Third, there is the question again of evolution in the fossil record finding itself rather than discovery being the other way around. Recalling that the series is a composite of specimens from the New and the Old World and already established by the 1870s, consider evolution scientist Othniel C. Marsh. Leonard Huxley's biography of his father Thomas Henry Huxley has the following account of the remarkable prowess of this paleontologist:

"As to each enquiry whether he [Marsh] had a specimen to illustrate such and such a point or exemplify a transition from earlier and less specialized forms to later and more specialized ones, Professor Marsh would simply turn to an assistant and bid him fetch box number so and so, until Huxley turned on him and said, 'I believe you are a magician; whatever I want, you just conjure it up.'...Thus, thanks to these important researches, it has become evident that, so far as our present knowledge extends, the history of the horse-type is exactly and precisely that which could have been predicted from a knowledge of the principles of evolution. And the knowledge we now possess justifies us completely in the anticipation that when the still lower Eocene deposits, and those which belong to the Cretaceous epoch, have yielded up their remains of ancestral equine animals, we shall find, first, a form with four complete toes...while, in still older forms...we come to the five toed animals, in which, if the doctrine of evolution is well-founded, the whole series must have taken its origin...Seldom has prophecy been sooner fulfilled. Within two months, Professor Marsh had discovered a new genus of equine mammals, Eohippus, from the lowest Eocene deposits of the West, which corresponds very nearly to the description given above." 135

§5.5.4.3 The Flap Over *Archeopteryx*

The name *Archaeopteryx* comes from Greek meaning "ancient wing". Its German name is *Urvogel* or "original bird". A total of ten *Archaeopteryx* specimens and an isolated feather have now been found since the first discovery in 1861, two years after *The Origin* was published. All are from a limestone quarry in Solnhofen, Bavaria and all have been dated to the late Jurassic period (150 Myr).

Archaeopteryx is a pigeon-sized creature with the skeleton of a therapod dinosaur (a group that includes *Tyrannosaurus*), very similar to *Compsognathus*, that chicken-sized dinosaur you may recall with dramatic cameo roles in *Jurassic Park 2*. Cladistic analysis is taken to confirm that *Archaeopteryx* was a dinosaur in having teeth, unaltered ulnar and radius arm bones, three fully developed digits with claws, no sternum, thick-walled bones without pneumatic ducts and a long bony tail.136 On the other hand *Archaeopteryx* has wings and feathers arranged in the same way as modern flying birds and there is evidence it could fly with them.137,138

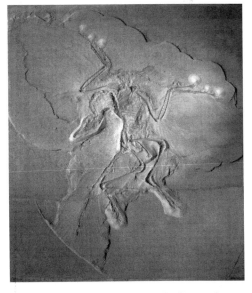

Fig. 5.10 **Archaeopteryx**. The Berlin specimen. (Image credit: H. Raab).

With these combined reptile-bird features, *Archaeopteryx* is hailed as an evolutionary intermediate just as Darwin predicted. Ernst Mayr declared it: "the almost perfect link between reptiles and birds".139 Niles Eldredge calls it: "beautifully intermediate between advanced archosaurian reptiles and birds",140 and Michael Ruse says *Archaopteryx* is the "greatest of all evolutionary links".141 Is it really?

The problem is both the ancestors and the descendants of Archaeopteryx are unknown. There is no ancestor-descendant series to invoke evolution is the point. Another problem is bird wings develop from digits two to four of the five fingered ("pentadactyl") limb, but those of *Archaeopteryx* and therapods are constructed from digits one to three (Fig.7.2). Cladistic analysis of the fossil morphology strongly indicates a therapod ancestry for the creature however. Suggestions there are undiscovered therapods in the ground with wings on digits two to four, or that a homeotic digital frameshift occurred to change digits one to three into two to four are still no more than that, suggestions.[142,143]

University of North Carolina avian paleontologist Alan Feduccia gives them short shrift because they "involve special pleading and the necessity of 'evolutionary origami.'" He summarized the status of bird evolution like this: "The riddle of bird origins will persist until discovery of new Triassic-early Jurassic fossils."[144] In other words there's still no fossil evidence but bird macroevolution is a fact and the fossil evidence just needs finding. Ignoring all the existing evidence would seem more contrived than the thinking he just scorned (§6.2.1). So there *Archaeopteryx* sits, isolated in the fossil record with no ancestor or descendant, and both dinosaur and bird features fully formed yet mixed together. How can it be called a missing link as the experts say?

McGill paleontologist Dr. Robert Carroll is another of the experts in this field. He says how it can.

"Archaeopteryx...remains the prime example of a genus that occupies the position of a "missing link" uniting two major vertebrae groups...Were it not for these feathers, Archaeopteryx would not have been recognized as a bird, as is demonstrated by the fact that one nearly complete skeleton in which the feathers were not recognized was initially identified as a dinosaur. In fact, there are no features of the bony skeleton of Archaeopteryx that are uniquely avian. All have been described in genera that are classified among the dinosaurs...In contrast with the close skeletal similarities between Archaeopteryx and its therapod ancestors, a very large morphological gap separates this genus from all other known birds. For this reason, Archaeopteryx is placed in a subclass on its own, the Archaeornithes, with all other birds classified as Neornithes." [137]

He says *Archaeopteryx* is a "missing link" in the same breath he says there's no evidence to link it, either to modern birds or an ancestor. Believe it's a link if you must. At least we know it can't be because of fossil evidence.

What then is *Archaeopteryx* if the evolutionary imperative is set aside? A mosaic creature with dinosaur features and feathered wings, of course. As a mosaic *Archaeopteryx* is reminiscent of the duck-billed platypus, another creature that confounds evolutionary explanation. On the one hand the platypus has bird-like features having a bill, webbed feet, laying eggs, nest building, its embryonic cleavage is highly meroblastic confined to a small region of the surface and its embryos form a blastoderm like reptiles and birds, not like placental mammals. On the other hand it has mammalian features too - mammary glands, four legs, fur and they have teeth when young. Why the platypus is not also accorded an evolutionary "missing link" explanation given its "intermediate" features is because, according to evolution theory, reptile evolution bifurcated separately into mammals and birds. It's theory first. Fossil interpretation comes after.

In summary *Archaeopteryx* displays combined morphologic traits fully reptilian and fully avian rather than transitional traits between them. How as a flying therapod dinosaur this bears on the supposed evolution of reptiles into birds, or even on the evolution of flight, still is conjectural after a century and a half of evolution research and theorizing. *Archaeopteryx* can't be a fossil link because it's without a fossil ancestor or descendant. There's no evolutionary precursor to the sudden appearance of its feather either and its most impressive feature after all.

The remainder of fossil record does not support bird evolution either by the way, at least in an objective sense. Reading the literature you could be forgiven for not knowing that.[145] Here's Richard Dawkin's solution to flight for instance. It's all very simple by evolution:

"How did wings get their start? Many animals leap from bough to bough, and sometimes fall to the ground. Especially in a small animal, the whole body surface catches the air and assists the leap, or breaks the fall, by acting as a crude aerofoil. Any tendency to increase the ratio of surface area to weight would help, for example flaps of skin growing out in the angles of joints...doesn't matter how small and unwinglike the first flaps were. There must be some height, call it h, such that an animal would break its neck if it fell from that height, but would survive if it fell from a slightly lower height. In this critical zone, any improvement in the body surface's ability to catch the air and break the fall, however slight the improvement, can make the difference between life and death. Natural selection will then favor slight, prototype wing flaps...And so on, until we have proper wings."[146]

Look at the metabolic adaptations to avian flight. The construction of a bird wing. A feather. Even this fossil feather will do.

Evidence of fossils I

§5.5.4.4 The Transitional Nature of the Fossil Vertebrate Classes

There's one final evolutionist position on fossils to discuss in regard to evolutionary series. It's that the order the vertebrate fossil classes appear in the fossil record chronologically, confirms the macroevolution hypothesis. In other words entire vertebrate classes are transitional series confirming the truth of evolution. This is *LOE#3* in our catalogue (§3.1). It's an argument made by experts George Gaylord Simpson and Mark Ridley amongst others, and has two parts.[147,148] Dr. Ridley explains it to us.

The first part is that "[a]natomic analysis of modern forms [extant creatures] indicates that amphibians and reptiles are evolutionarily intermediate between fish and mammals". It's the hypothesis a transitional series of evolution among the vertebrates most plausibly would be fish evolving into amphibians into reptiles into mammals and not, say, fish into mammals into amphibians into reptiles.

The second part of the argument is that "[t]his order fits with…the geologic succession of the major vertebrate groups", confirming the macroevolution inference they were eyed with in the "anatomic analysis" of the first part of the argument. It's the same argument J.B.S. Haldane offered as a proof and a potential disproof of evolution. He said he would stop believing in evolution if a fossil rabbit was found in the Precambrian. Mark Ridley extends this to finding fossil humans at the time of dinosaurs. The reason he says is because:

"[i]f evolution is correct, humans could not have existed before the main radiation of mammals and primates, and these took place after the dinosaurs had gone extinct. The fact that no such human fossils have been found – that the order of appearance of the main fossil groups matches their evolutionary order – is the way in which the fossil record provides good evidence for evolution".

He shows as evidence for *LOE#3* the spindle diagram redrawn in Fig.5.9. Are you a believer?

What he is referring to is clear, but why should evolution be concluded? The pre-eminent observation is sudden appearance of completely different body plans which is saltation. The second most obvious observation is they arise separately in the record. The links between the groups are missing (though they are joined on the figure! §25) How can this be evolution evidence?

In summary *LOE#3* is falsely premised, it ignores the saltation that is the Cambrian explosion (as the fossil record of vertebrates can be traced to the Cambrian)[150,151], the sudden appearance of classes, stasis and the extreme rarity of intermediate forms. Dr. Ridley's book *Evolution* is nearly an inch and a half thick and in its third edition, yet this is the main fossil evidence he musters to support the theory he writes so much about.[152,153] It's also an argument based more on homology than on paleontology, since the paleontological data is being conveniently ignored to fit that of homology, a line of evidence we will inspect later, and later still the evidence for transitional forms in the human evolution fossil series.

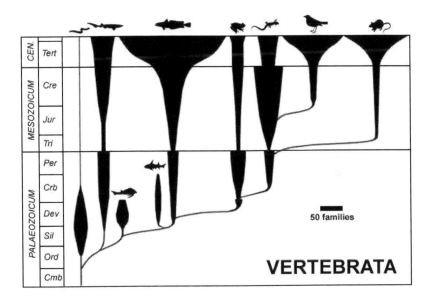

Fig. 5.11. **The order of the main groups in the fossil record indicates evolution!** Spindle width is the number of fossil families as an indication of diversity (See key bottom right). The silhouettes are vertebrate classes from left Agnatha, Chondrichthyes, Oseteichythyes, Amphibia, Reptilia, Aves and Mammalia. The two extinct classes are Placodermi and Acanthodii. Note the lines linking the vertebrate classes that demonstrate the claim of *LOE#3*, only they are artistic additions to the figure; fictional lines to show evolutionary linkage. Mark Ridley shows them too.[147] Compare with Fig.5.6b. (Image credit: Petter Bockman. Based on Benton, 1998).[149]

§5.6   IN CONCLUSION: FINDING DARWIN'S "MOST OBVIOUS AND SERIOUS OBJECTION", STILL THERE

Michael Ruse at the University of Florida says "one can see that the fossil record almost begs for an evolutionary interpretation"[154] and his view is most everywhere in the science and education establishment axiomatic. It's *LOE#2-3*. In reviewing the data for ourselves we find a surprising lack of evidence though. The rule of the fossil record is sudden appearance and stasis, and intermediate series are extremely rare. How can "[f]ossils demonstrate that evolution is a fact?"[2] If anything fossils demonstrate the opposite.

There is another discovery inescapable to any reader of *The Origin* and also seldom acknowledged. It's that Darwin was unable to reconcile his theory with fossils and freely admitted it. He had to defend evolution from the fossil evidence and all of Chapter X was written to that end. Darwin acknowledged there was biological turnover in the fossil record but saw no evidence of the evolution he claimed, concluding: "The several difficulties here discussed, namely – that…in our geological formations…we do not find infinitely numerous fine transitional forms closely joining them all together; - the sudden manner in which several groups of species first appear in our European formations; - the almost entire absence…of formations rich in fossils beneath the Cambrian strata, - are all undoubtedly of the most serious nature."(p.442-3)

Of the absence of fossilized intermediates between species predicted by gradualism he said: "this, perhaps, is the most obvious and serious objection which can be urged against this theory."(p.406) He also said: "Why then is not every geological formation and every stratum full of…intermediate links? Geology assuredly does not reveal any such finely-graduated organic chain; and this, perhaps, is the most obvious and serious objection which can be urged against the theory."(p.406)

His response to the Cambrian explosion was that Precambrian fossil life in diversity and abundance would be found one day. "Consequently, if my theory be true, it is indisputable that before the lowest Silurian stratum was deposited, long periods elapsed, as long as, or probably far longer than, the whole interval from the Silurian age to the present day; and that during these vast, yet quite unknown, periods of time, the world swarmed with living creatures."(p.438) Pressed on the absence of fossil evidence, he said that he had no answer. "To the question of why we do not find records of these vast primordial periods, I can give no satisfactory answer…The case at present must remain inexplicable; and may be truly urged as a valid argument against the views here entertained." (p.438-9)

Today unflinching paleontological opinion is that macroevolution is a fact and the fossils show it. What has changed? Not the fossils because the pattern is the same. Radioactive dating has since been developed and the number of fossils has increased by orders of magnitude but stasis, sudden appearance and rare intermediate series remain the rule. We inspected four of the best examples of the latter to see how robust they are, and they are not. How are problems with the fossil data resolved to show macroevolution then?

There are only two solutions. Either the pattern of the fossil record data is an illusion masking the true macroevolution there, or else Darwinism is false. Gone is the supposed "obviously true" status that fossils declare evolution a fact. Now we are engaged in an upside-down analysis with the aim to reject the contradiction-to-evolution appearance of the fossil evidence, only to retain the macroevolution opinion conceived of despite fossil evidence in the first place. We will attempt to sort this out in the next chapter.

REFERENCES

1. Stanley, S. M. *Macroevolution Pattern and Process.* 2 (Johns Hopkins University Press, 1998).
2. Simpson, G. *Fossils and the History of Life.* 125 (Scientific American Books, Inc, 1983).
3. Gee, H. *In Search of Deep Time.* 5, 117 (The Free Press, 1999).
4. Gee, H. Gee responds to Discovery Institute Use of Quotations. *National Center for Science Education* (2001).
5. Eldredge, N. *Reinventing Darwin: The great debate at the high table of evolutionary theory.* 112 (John Wiley & Sons, 1995).
6. Grant, V. *The Evolutionary process. A critical review of evolutionary theory.* 14 (Columbia Unversity Press, 1985).
7. Mayr, E. *What Evolution Is.* 12-13 (Basic Books, 2001).
8. Dawkins, R. *The Selfish Gene.* 1 (Oxford University Press, 1989).
9. Futuyma, D. J. *Evolutionary Biology.* 15 (Sinauer Associates, 1986).
10. Gould, S. J. Dorothy, Its Really Oz. *Time* 154, 59 (1999).
11. Pearson, P. Nature debates: The glorious fossil record (ed www.nature.com/nature/debates/fossil/fossil-1.html) (1998).
12. Le Gros Clark, W. E. The crucial evidence for human evolution. *Proceedings of the American Philosophical Society* 2, 159-172 (1959).
13. Darwin, C. *The Origin of Species.* 406 (Random House, Inc, 1998(1859)).
14. Cowen, R. *History of Life.* (Blackwell Science, Inc., 1995).
15. Doyle, P. *Understanding Fossils. An Introduction to Invertebrate Paleontology.* (John Wiley and Sons Ltd, 1996).
16. Ayala, F. Tempo and mode in the macroevolutionary reconstruction of Darwinism. *Proceedings of the American Philosophical Society* 91, 6764-6771 (1994).
17. Simpson, G. G. *Fossils and the History of Life.* (Scientific American Books, Inc, 1983).
18. Carroll, R. L. *Vertebrate Paleontology and Evolution.* (W.H. Freeman and Company, 1988).
19. Booher, H. R. *Origins, icons and illusions.* (Warren H. Green, Inc., 1998).
20. Sadler, P. M. Sediment accumulation rates and the completeness of stratigraphic sections. *Journal of Geology* 86, 569-584 (1981).
21. Raup, D. M. in *Tempo and mode in evolution:genetics and paleontology 50 years after Simpson* (eds W.M. Fitch & F.J. Ayala) (National Academy Press, 1995).
22. Gould, S. J. *Wonderful Life.* 60 (W.W. Norton & Company, Inc., 1989).
23. Carroll, R. L. *Vertebrate Paleontology and Evolution.* 570-571 (W.H. Freeman and Company, 1988).
24. Valentine, J. W. The fossil record: a sampler of life's diversity. *Philosophical Transactions of the Royal Society of London* 330, 261-268 (1990).
25. Gould, S. J. *Wonderful Life.* 213-4 (W.W. Norton & Company, Inc., 1989).
26. Gould, S. J. *Wonderful Life.* 97-8 (W.W. Norton & Company, Inc., 1989).

27 Mayr, E. & Ashlock, P. *Principles of Systematic Zoology*. (McGraw-Hill, 1991).
28 Mayr, E. *What Evolution Is*. (Basic Books, 2001).
29 Page, R. D. M. & Holmes, E. C. *Molecular Evolution*. (Blackwell Science Ltd., 1998).
30 Ruse, M. *Darwinism Defended*. 140 (Addison-Wesley Publishing Company, 1982).
31 Benton, M. J. Diversification and extinction in the history of life. *Science* 268, 52-58 (1995).
32 Raup, D. M. Size of the Permo-Triassic bottleneck and its evolutionary implications. *Science* 206, 217-218 (1979).
33 Raup, D. M. Large body impact and extinction in the Phanerozoic. *Paleobiology* 18, 80-88 (1992).
34 Raup, D. M. The role of extinction in evolution. *Proceedings of the National Academy of Sciences USA* 91, 6758-6763 (1994).
35 Raup, D. M. Extinction: bad genes or bad luck. *Acta Geologica Hispanic* 16, 25-33 (1981).
36 Raup, D. M. *Extinction: Bad genes or bad luck?*, 185 (W.W. Norton & Co, 1992).
37 Hallam, A. & Wignall, P. B. *Mass extinctions and their aftermath*. (Oxford University Press, 1997).
38 Benton, M. J. Models for the diversification of life. *Trends in Ecology and Evolution* 12, 490-495 (1997).
39 Morris, S. C. The Cambrian "explosion": Slow-fuse or megatonnage? *Proceedings of the American Philosophical Society* 97, 4426-4429 (2000).
40 Valentine, J. W., Jablonski, D. & Erwin, D. H. Fossils, molecules and embryos: new perspectives on the Cambrian explosion. *Development* 126, 851-859 (1999).
41 Valentine, J. W., Awramik, S. W., Signor, P. W. & Sadler, P. M. The biological explosion at the Precambrian-Cambrian boundary. *Evolutionary Biology* 5, 279-356 (1991).
42 Valentine, J. W., Erwin, D. H. & Jablonski, D. Developmental Evolution Of Metazoan Body Plans: The Fossil Evidence. *Developmental Biology* 173, 373-381 (1996).
43 Gould, S. J. Of it, not above it. *Nature* 377, 681-682 (1995).
44 Bell, M. A. Origin of the metazoan phyla: Cambrian explosion or proterozoic slow burn. *Trends in Ecology and Evolution* 12, 1-2 (1997).
45 Carroll, R. L. Towards a new evolutionary synthesis. *Trends in Ecology and Evolution* 15, 27-32 (2000).
46 McMenamin, M. A. S. & McMenamin, D. L. S. *The Emergence of Animals*. (Columbia University Press, 1990).
47 Fortey, R. The Cambrian Explosion Exploded? *Science* 293, 438-439 (2001).
48 Bowring, S. A. et al. Calibrating rates of early Cambrian evolution. *Science* 261, 1293-1298 (1993).
49 Valentine, J. W., Jablonski, D. & Erwin, D. H. Perspectives on the Cambrian Explosion. *Development* 126, 851-859 (1999).
50 Gould, S. J. 60 (W.W. Norton & Company, Inc., New York, NY, 1989).

51 Erwin D.H. *et al.* The Cambrian Conundrum: Early divergence and later ecological success in the early history of animals. *Science* 334, 1091-1097 (2011).
52 Benton, M. J. *The Fossil Record 2*. (Chapman & Hall, 1993).
53 Miller, A. I. *Diversity of life through time*. (Macmillan Publishers, Ltd., Nature Publishing Group, 2002).
54 Benton, M. J. Biodiversity on land and in the sea. *Geological Journal* 36, 211-230 (2001).
55 Seposki, J. J. Ten years in the library: new data confirm paleontological patterns. *Paleobiology* 19, 43-51 (1993).
56 Peters, S. E. & Foote, M. Biodiversity in the Phanerozoic: a reinterpretation. *Paleobiology* 27, 583-601 (2001).
57 Alroy, J. *et al.* Effects of sampling standardization on estimates of Phanerozoic marine diversification. *Proceedings of the National Academy of Sciences USA* 98, 6261-6266 (2001).
58 Jablonski, D., Roy, K., Valentine, J. W., Price, R. M. & Anderson, P. S. The impact of the Pull of the Recent on the history of marine diversity. *Science* 300, 1133-1135 (2003).
59 Kerr, R. A. Life's diversity may truly have leaped since the dinosaurs. *Science* 300, 1067-1068 (2003).
60 Newman, M. A new picture of life's history on Earth. *Proceedings of the National Academy of Sciences USA* 98, 5955-5956 (2001).
61 Signor, P. W. I. in *Phanerozoic diversity patterns:profiles in macroevolution* (ed J.W. Valentine) 129-150 (Princeton University Press, 1985).
62 Benton, M. J. in *Numerical palaeobiology: Computer-based modellling and analysis of fossils and their distributions.* (ed D.A.T. Harper) 249-283 (John Wiley & Sons, 1999).
63 Gould, S. J. *The Panda's thumb*. 181-9 (W.W Norton and Co. 1980).
64 Eldredge, N. *Reinventing Darwin: The great debate at the high table of evolutionary theory*. 3 (John Wiley & Sons, 1995).
65 Simpson, G. G. *Tempo and mode in evolution*. 105,107 (Columbia University Press, 1944).
66 Paul, C. R. C. in *Evolution and the fossil record* (eds K.C. Allen & D.E.G. Briggs) 99-121 (Belhaven Press, 1989).
67 Valentine, J. W. in *Phanerozoic diversity patterns* (ed J.W. Valentine) 3-8 (Princeton University Press, 1985).
68 Eldredge, N. *Macroevolutionary dynamics:species, niches and adaptive peaks*. 22 (McGraw-Hill Publishing Company, 1989).
69 Stanley, S. M. *The New Evolutionary Timetable: Fossils, Genes and the Origin of Species*. 40 (Basic Books, Inc., 1981).
70 Carroll, R. L. *Vertebrate Paleontology and Evolution*. 571 (W.H. Freeman and Company, 1988).
71 Axelrod, D. in *The Evolution of life* (ed S. Tax) 227-306 (University of Chicago Press, 1960).
72 Darwin, F. *The Life and Letters of Charles Darwin*. (John Murray, 1888).
73 Stanley, S. M. *Macroevolution, Pattern and Process*. 69 (Johns Hopkins University Press, 1998).

74 Shu D-G. et al. Lower Cambrian vertebrates from south China. *Nature* 402, 42-46 (1999).
75 Chen J.Y., Huang D.Y. & Li, C. W. An early Cambrian craniate-like chordate. *Nature* 402, 518-522 (1999).
76 Booher, H. R. *Origins, icons and illusions*. 89 (Warren H. Green, Inc., 1998).
77 Grasse, P.-P. *Evolution of living organisms*. 181 (Academic Press, Inc., 1977).
78 Simpson, G. G. *Life before Man*. 42 (Time-Life Books, 1972).
79 Grasse, P.-P. *Evolution of living organisms*. 12,13,17 (Academic Press, Inc., 1977).
80 Grasse, P.-P. *Evolution of living organisms*. (Academic Press, Inc., 1977).
81 Meyer, S. C. *Darwins doubt*. (HarperOne, 2013).
82 Marshall, C. R. When prior belief trumps scholarship. *Science* 341, 1344 (2013).
83 Marshall, C. R. Explaining the Cambrian 'Explosion' of animals. *Annual Reviews of Earth and Planetary Sciences* 34, 355-384 (2006).
84 Marshall, C. R. & Valentine, J. W. The importance of preadapted genomes in the origin of the animal body plans and the Cambrian Explosion. *Evolution* 64-5 (2010).
85 Prothero, D. *Stephen Meyer's fumbling bumbling Cambrian amateur follies*, <http://www.amazon.com/review/R2HNOHERF138DU/> (2013).
86 Cook, G. Doubting Darwin's Doubt. *New Yorker* (July 2 2013).
87 Farell, J. How nature works. *National Review* (Sept 2 2013).
88 Gould, S. J. in *Perspectives On Evolution* (ed R. Milkman) 83-104 (Sinauer Associates, Inc., 1982).
89 Stanley, S. M. *Macroevolution Pattern and Process*. 39 (Johns Hopkins University Press, 1998).
90 Carroll, R. L. *Vertebrate Paleontology and Evolution*. (W.H. Freeman and Company, 1988).
91 Stanley, S. M. 124-126 (Johns Hopkins University Press, 1998).
92 Ruse, M. *Darwinism Defended*. 218 (Addison-Wesley Publishing Company, 1982).
93 Kellog, D. E. The role of phyletic change in the evolution of Pseudocubus vema (Radiolaria). *Paleobiology* 1, 359-370 (1975).
94 Stanley, S. M. *Macroevolution, Pattern and Process*. xxii (Johns Hopkins University Press, 1998).
95 Thewissen, J. G. M. & Bajpai, S. Whale Origins as a Poster Child for Macroevolution. *BioScience* 12, 1037-1049 (2001).
96 Spaulding, M., O'Leary, M. A. & Gatesy, J. Relationships of Cetacea (Arteriodacyla) among mammals; increased taxon samply alters interpretations of key fossils and character evolution. *PloS ONE* Sep 23 4, e7062 (2009).
97 Gould, S. J. in *Dinosaur In a Haystack: Reflections In Natural History* 359-377 (Crown Trade Paperbacks, 1997(1994)).
98 de Muizon, C. Walking With Whales. *Nature* 413, 259-260 (2001).
99 Goldsmith, T. H. Everyday impacts of a most influential theory. *Science* 293, 2209-2210 (2001).

100 Working Group on Teaching Evolution. *Teaching about Evolution and the Nature of Science*. (National Academy Press, 1998).
101 National Academy of Sciences. *Science and Creationism: A View from the National Academy of Sciences*. 2nd Ed edn, (The National Academies Press http://books.nap.edu/html/creationism/, 1999).
102 National Academy of Sciences. Teaching About Evolution and the Nature of Science. http://www.nap.edu/books/0309063647/html/index.html (1998).
103 Simpson, G. G. The principles of classification and a classification of mammals. *Bull. Am. Mus. Nat. Hist* 85, 1-350 (1945).
104 Futuyma, D. J. *Science on trial*. (Sinauer Associates Inc., 1995).
105 Wong, K. The Mammals That Conquered The Seas. *Scientific American*, 70-79 (2002).
106 Gingerich, P. D., ul Haq, M., Zalmout, I. S., Khan, I. H. & Malkani, M. S. Origin of whales from early artiodactyls: hands and feet of Eocene Protocetidae from Pakistan. *Science* 293, 2239-2242 (2001).
107 Rose, K. D. The ancestry of whales. *Science* 293, 2216-2217 (2001).
108 Berta, A. What Is A Whale? *Science* 263, 180-181 (1994).
109 Thewissen, J. G. M., Williams, E. M., Roe, L. J. & Hussain, S. T. Skeletons of terrestrial cetaceans and the relationshipo of whales to artiodactyls. *Nature* 413, 277-281 (2001).
110 Milinkovitch, M. C. & Thewissen, J. G. M. Evolutionary Biology: Even-toed fingerprints on whale ancestry. *Nature* 388, 622-624 (1997).
111 Chadwick, D. H. Evolution of Whales. *National Geographic* 200, 64-77 (2001).
112 Quammen, D. Was Darwin Wrong? *National Geographic*, 4-35 (2004).
113 Gatesy, J. & O'Leary, M. Deciphering whale origins with molecules and fossils. *Trends in Ecology and Evolution* 16, 562-570 (2001).
114 Gingerich, P. D. in *The Emergence of Whales. Evolutionary Patterns in the Origin of Cetacea*   (ed J.G.M. Thewissen) 423-450 (Plenum Press, 1998).
115 Thewissen, J. G. M. in *The Emergence of Whales. Evolutionary Patterns in the Origin of Cetacea*   (ed J.G.M. Thewissen) 451-464 (Plenum Press, 1998).
116 Kellogg, R. A Review of the Archaeoceti. *Carnegie Institute of Washington Publ.* 482, 1-366 (1936).
117 Kellogg, R. The history of whales. *Quarterly Review of Biology* 3, 29-76, 174-208 (1928).
118 Van Valen, L. M. Deltatheridia, a new order of mammals. *Bull. Am. Mus. Nat. Hist* 132, 1-126 (1966).
119 Gingerich, P. D. & Russell, D. E. Pakicetus inachus, a new archaeocete (Mammalia, Cetacea) from the early-middle Eocene, Kuldana Formation of Kohay (Pakistan). *Contrib. Mus. Paleontol. Univ. Mich.* 25, 235-246 (1981).
120 West, R. M. Middle Eocene large mammal assemblage with Tethyan affinities, Ganda Kas Region, Pakistan. *Journal of Paleontology* 54, 508-533 (1980).
121 Gingerich, P. D., Wells, N. A., Russell, D. E. & Shah, S. M. I. Origin of Whales in Epicontinental Remnant Seas: New Evidence from the Early Eocene of Pakistan. *Science* 220, 403-406 (1983).
122 Thewissen, J. G. M. et al. *Nature* 381, 379-380 (1996).
123 Gingerich, P. D. The Whales of Tethys. *Natural History*, 86 (1994).

124 Gingerich, P. D. Evidence for evolution from the fossil record. *Journal of Geological Education* 31, 140-144 (1983).
125 Luo, Z. in *The Emergence of Whales. Evolutionary Patterns in the Origin of Cetacea* (ed J.G.M. Thewissen) 269-301 (Plenum Press).
126 Carroll, R. L. *Vertebrate Paleontology and Evolution.* (W.H. Freeman and Company, 1988).
127 Thewissen, J. G. M. & Hussain, S. T. Origin of underwater hearing in whales. *Nature* 361, 444-445 (1993).
128 Gingerich, P. D. Hind Limbs of Eocene Basilosaurus: Evidence of Feet in Whales. *Science* 249, 154-157 (1990).
129 Thewissen, J. G. M., Hussain, S. T. & Arif, M. Fossil Evidence for the Origin of Aquatic Locomotion in Archaeocete Whales. *Science* 263, 210-212 (1994).
130 Gingerich, P. D. & Bajpai, S. A new Eocene archaeocete (Mammalia, Cetacea) from India and the time of origin of whales. *Proceedings of the National Academy of Sciences USA* 95, 15464-15468 (1998).
131 Gatesy, J. in *The Emergence of Whales. Evolutionary Patterns in the Origin of Cetacea* (ed J.G.M. Thewissen) 63-111 (Plenum Press, 1998).
132 Thewissen, J. G. M., Cooper, L. N., Clementz, M. T., Bajpai, S. & Tiwari, B. N. Whales originated from aquatic artiodactyls in the Eocene epoch of India. *Nature* 450, 1190-1194 (2007).
133 Paul, C. R. C. in *The adequacy of the fossil record* (eds S.K. Donovan & C.R.C. Paul) 1-22 (John Wiley & Sons, 1998).
134 Grasse, P.-P. *Evolution of living organisms.* 51-52 (Academic Press, Inc., 1977).
135 Huxley, L. *The life and letters of Thomas Henry Huxley.* Vol. 2 (Macmillan, 1903).
136 Cowen, R. *History of Life.* (Blackwell Science, Inc., 1995).
137 Carroll, R. L. *Vertebrate Paleontology and Evolution.* (W.H. Freeman and Company, 1988).
138 Olson, L. & Feduccia, A. Flight capability and the pectoral girdle of Archaeopteryx. *Nature* 278, 247-248 (1979).
139 Mayr, E. *The Growth of Biological Thought.* 430 (Harvard University Press, 1982).
140 Eldredge, N. *The Triumph of Evolution.* 125 (W.H. Freeman, 2001).
141 Ruse, M. *Darwinism Defended.* 208 (Addison-Wesley Publishing Company, 1982).
142 Galis, F., Kundrat, M. & Sinervo, B. An old controversy solved: bird embryos have five fingers. *Trends in Ecology and Evolution* 18, 7-9 (2003).
143 Wagner, G. P. & Gauthier, J. A. 1,2,3 = 2,3,4: A solution to the problem of the digits in the avian hand. *Proceedings of the National Academy of Sciences USA* 96, 5111-5116 (1999).
144 Feduccia, A. Bird origins: problem solved, but the debate continues... *Trends in Ecology and Evolution* 18, 9-10 (2003).
145 Padian, K. & Chiappe, L. M. The Origin of Birds and Their Flight. *Scientific American*, 38-47 (1998).

146 Dawkins, R. *The Blind Watchmaker*. 89-90 (W.W. Norton & Company, 1996).
147 Ridley, M. *Evolution*. (Blackwell, 2004).
148 Simpson G.G. *The meaning of evolution*. (Yale University Press, 1949).
149 Benton, M. J. in *The adequacy of the fossil record* (eds S.K. Donovan & C.R.C. Paul) 269-303 (John Wiley & Sons, 1998).
150 Janvier, P. Catching the first fish. *Nature* 402, 21-22 (1999).
151 Shu, D.-G. *et al*. Lower Cambrian vertebrates from south China. *Nature* 402, 42-46 (1999).
152 Ridley, M. *Evolution*. 3 edn, 3 (Blackwell, 2004).
153 Ridley, M. *Evolution*. (Blackwell Science, Inc., 1996).
154 Ruse, M. *Darwinism Defended*. 207 (Addison-Wesley Publishing Company, 1982).

6. Believing is seeing

§6.1   FACING UP TO THE FOSSIL RECORD REALITY –
       TWO WAYS TO CLOTHE AN EMPEROR

Reviewing the fossil record for ourselves we discover it contradicts Darwin's predictions of what it should show "if my theory be true". Why this is interesting is near universal opinion in this science since the Modern Synthesis is that fossils declare evolution "a fact"(§5.1) What changed? Not the fossils. The pattern of the record is the same. There are only two explanations then. Either the fossil record data is false or else Darwin's theory of phyletic gradualism as explanation of the fossil record is false. Both alternatives have their advocates and although they are contradictory their advocates are equally certain evolution is true! It's another curious observation we make in passing because it's nowhere considered disturbing in the science. How can that be rational? (§19;§20). The two alternative evolutionary explanations of the fossil evidence will be inspected next.

§6.2   THE FIRST WAY: THE FOSSIL RECORD IS "IMPERFECT"

The contradiction between what the fossil evidence shows - sudden appearance, stasis and rare intermediate forms and what Darwinism predicts - the reverse, is on this view an artifact (§2.2;§5.4). Sudden appearance and stasis become "gaps" or "missing links", and the fossil record an unfaithful register of the true, macroevolutionary history of life to be trusted. In other words the appearance of the fossil record, just like the appearance of design, is declared to be an illusion (§1.3). This was Darwin's explanation. Writing in *The Origin* he at least acknowledged what it would mean for his theory if the fossil evidence pattern all could see was real:

> "We have seen in the last chapter that whole groups of species sometimes falsely appear to have been abruptly developed; and I have attempted to give an explanation of this fact, which if true would be fatal to my views."(p.448)

His argument that fossils "falsely appear", in other words that the fossil evidence was actually false, was made on two grounds. The fossils were at once "incomplete" and "imperfect".

The fossil record was "incomplete" because paleontology was still in its infancy and all the fossils indicating the pattern he declared were still in the ground awaiting discovery. He said:

> "only a small portion of the globe has been geologically explored with care...[and] that the number both of specimens and of species preserved in our museums, is absolutely as nothing compared with the number of generations that must have passed away even during a single formation..."(p.477).

This was just his conjecture. He did not even offer an estimate of how incomplete the record was. Generations of biologists have believed him nonetheless, even though the intermediate forms and the Pre-Cambrian fossils he declared are the most abundant, are the fossils that are the most missing! (§5.3.2)

His second reason was that fossils are "imperfect" or unreliable at leaving a true record of life's history anyway. Thus even with the most exhaustive search the fossil evidence is still going to be tainted. In other words there was a biological reality, a "perfect" fossil record which demonstrated macroevolution as he said, but which differed from the objective fossil evidence. Evolution theory on this view is a greater truth than fossil evidence. Darwin gave a list of reasons why the fossil record was "imperfect" and why it could not be trusted:

> "I have attempted to show that the geological record is extremely imperfect:...[1] that only certain classes of organic beings have been largely preserved in a fossil stage...[2] that, owing to subsidence being almost necessary for the accumulation of deposits rich in fossil species of many kinds, and thick enough to outlast future degradation, great intervals of time must have elapsed between most of our successive formations; [3] that there has probably been more extinction during the periods of subsidence, and more variation during the periods of

elevation, and during the latter the record will have been less perfectly kept; [4] that each single formation has not been continuously deposited; [5] that the duration of each formation is probably short compared with the average duration of specific forms; [6] that migration has played as important part in the first appearance of new forms in any one area and formation; that widely ranging species are those which have varied most frequently, and have oftenest given rise to new species; [7] that varieties have at first been local; and [8] lastly, although each species must have passed through numerous transitional stages, it is probable that the periods, during which each underwent modification, though many and long as measured by years, have been short in comparison with the periods during which each remained in an unchanged condition. These causes, taken conjointly, will to a large extent explain why – though we do find many links [also denied by the record §5.5.4] – we do not find interminable varieties, connecting together all extinct and existing forms by the finest graduated steps."(p.477-8)

With characteristic eloquence and missing mathematics, he made his thinking even more plain by means of a metaphor:

"The explanation lies, as I believe, in the extreme imperfection of the geological record (p.406)...I look at the geological record as a history of the world imperfectly kept, and written in changing dialect; of this history we possess the last volume alone, relating to only two or three countries. Of this volume, only here and there a short chapter has been preserved; and of each page, only here and there a few lines. Each word of the slowly-changing language, more or less different in the subsequent chapters, may represent forms of life, which are entombed in our consecutive formations, and which falsely appear to have been abruptly introduced. On this view, the difficulties above discussed are greatly diminished, or even disappear (p.443)...As we possess only the last volume of the geological record, and that in a very broken condition, we have no right to expect, except in rare cases, to fill up the wide intervals in the natural system, and thus to unite distinct families or orders."(p.466)

This view remains the opinion of the majority of scientists despite almost a century and a half of fossil searching that has only confirmed the pattern that existed in Darwin's day (§5.5). That the record is "incomplete" is not the important question. It is not even the relevant question. The reason is because incompleteness is not in dispute (§5.3.2). The relevant question and the important question is whether the fossil record is too incomplete to reliably interpret. Fortunately for us trying to solve the Conundrum, this is a question many in paleontology now say can be answered.

§6.2.1   Testing Fossils For Completeness

Since there is no way of going back in time needless to say there is no definitive way to determine the absolute "completeness" of the fossil record that we do have. Some now extinct organism might never have fossilized and we would never know for instance. Do we have to throw our hands up, declare the fossil record "imperfect" and therefore uninterpretable in the form we find it though? Are there data on which to assess just how incomplete the fossil record is? After more than a century, a literature is at last emerging. Several strategies have been proposed to tackle the problem and debate has even been hosted in the pages of the journal *Nature*.[1-3,4] The strategies for quantitating fossil record incompleteness are: 1) phylogenetic tests, 2) taphonomic tests and 3) historical tests.[3] The utility of different approaches is that a result is more plausibly true if independent methods obtain the same result. What do they show?

1)   *Phylogenetic tests*
These compare phylogenies and cladograms constructed from morphological or molecular characters (amino acid or nucleic acid sequence similarities) with those constructed from the fossil record (§8.2). The problem is how to assess the accuracy of the cladogram that to be used as measuring rod. There is no independent way to be certain the topography revealed by cladistic analysis has historical reality, and even the most basic of its assumptions, that the most parsimonious tree is the true tree, can be challenged (§5.3.3). The point is cladograms are theoretical constructs that assume the fact of evolution already.

What happens if cladograms are held up against the fossil record to assess completeness nonetheless? In a study published in *Nature*, one thousand morphologic and molecular cladograms were compared with the stratigraphic database of *The Fossil Record 2*. The fossil record is adequate.[5] The quality of the fossil record did not decline with increasing age either.[6] The authors concluded: "Assessments of congruence between stratigraphy [the study of rock strata with fossils] and phylogeny for a sample of 1,000 published phylogenies show no evidence of diminution in quality backwards in time. Ancient rocks clearly preserve less information, on average, than more recent rocks. However if scaled to the stratigraphic level of the stage and the taxonomic level of the family, the past 540 million years of the fossil record provide uniformly good documentation of the life of the past... Early parts of the fossil record are clearly incomplete, but they can be regarded as adequate to illustrate the broad patterns of the history of life."[7]

2) *Taphonomic tests*

These directly assess the reliability of the fossilization process. Early estimates seemed to confirm Darwin's pronouncements of incompleteness with approximations of 1% or less of all species considered recoverable from the record.[8,9] And yet the dinosaur fossil record has been estimated to have a completeness of around 25%.[10] But before getting bogged down in numbers controversies that consume even the experts, stand back a moment. What are the relevant and important questions we need answers to, because perhaps those ones are not elusive or difficult.

A first question in testing for reliability, especially as pertains evidence of macroevolution, is whether the chronological order of any two fossils in the record can be trusted. Does the fossil record preserve their true sequence? To ask this another way, if two species A and B coexisted, what's the possibility that the order of their first appearance and extinction could be spuriously reversed by the fossil record? It turns out that when even only one specimen of each is known, the probability of the sequence being correct is greater than it being incorrect, and this probability rises rapidly with increasing number of specimens.[11,12] It also turns out the concern is an infrequent eventuality.

Making the assumption the mean lifetime of a species during the Phanerozoic eon is 6 Myr, and taking 600 Myr as the duration of the Phanerozoic, Christopher Paul at the University of Liverpool has calculated that only about 3% of the species that lived in the Phanerozoic coexisted at any one time. Therefore he says: "only in 3% of random comparisons is there any possibility whatsoever that the first or last occurrences may be in the wrong order with respect to the true original sequence of evolution or extinction. Put another way, the usual 95% confidence limits would occur if the average lifetime of a species had been about 10 Myr throughout the Phanerozoic, and this is assuming that all species which coexisted give unreliable data which is as a manifestly false assumption...This theoretical argument confirms that the fossil record preserves the sequence of evolutionary events reliably...Note, too, that this result is not affected by the completeness or otherwise of the record. Indeed if the fossil record consisted of just two fossils, the chances are overwhelming that they would be preserved in the right order!"[12]

A second question is what proportion of animals that have lived has left a fossil record. A simple if crude way of estimating is to calculate the proportion of creatures with skeletons, knowing skeletized animals are more likely to leave a fossil record than soft-bodied (§5.3.2). Calculations demonstrate that just over 8 percent (99,800 out of the total of 1,239,129 known living species in 1977) are skeletized.[13] Since then biodiversity estimates have been revised upwards by more than an order of magnitude (§1.2), but until data indicates the actual ratio is altered, I take it we can assume it remains. This calculation was for living organisms however. If we are to apply it to the relative proportions of extinct ones, it requires assuming that hard and soft-bodied creatures have remained in relatively similar proportions over geologic time. There is support for this from the Burgess Shale Lagerstätten which shows 14% of the genera and 5% of the total individual specimens are skeletized.[1,14] On the basis of these assumptions and this methodology, the entire fossil record is estimated to be between 5 and 10% complete.[9,15,16] A more comprehensive record than these calculations indicate is provided by comparison of Lagerstätten quality over time backward through the Phanerozoic (in which both soft and hard bodied creatures are fossilized). Thus Lagerstätten can be used as a reference standard to measure the completeness of other less extensive contemporaneous fossil accumulations. Interestingly no decrease in fossil record quality over time is found with that analysis either.[17]

In summary, the answers to both our questions indicate the fossil record is more likely than not to be reliable. Another comment about Lagerstatten like the Burgess Shale is so obvious it would not be worth mentioning, if were not that we find ourselves exploring the fossil record for another pattern than what we see. It's that the pattern Lagerstatten declare is the same pattern we find elsewhere, only better. Simon Conway Morris, among the foremost experts on the Burgess, estimates "[t]he existing [Burgess] collections represent approximately 70,000 specimens. Of these, about 95 percent are either soft bodied or have thin skeletons". At least here the record is as complete as we might hope for. What's the pattern you ask? Sudden appearance, stasis and rare intermediate forms.

Another method of taphonomic testing is to calculate what proportion of living creatures are represented in the fossil record, remembering biotic heterogeneity is at or very close to its peak at the present time (Fig.5.4). Assays of the living intertidal fauna in Friday Harbor, Washington found that between 16%-67% of organisms at the genus level are represented in the fossil record.[18] In another study over 79% of the extant families and 97% of the extant orders of terrestrial vertebrates are found as fossils.[19] In still another study, James Valentine at Berkeley has shown that 89%, 83% and 77% of extant Californian Province bivalve and gastropod families,

270

genera and species respectively, are represented in the fossil record. He concludes: "At least 85% of durably skeletonized living species may have been captured in the record."[20] These studies suggest the fossil record, at least as regards current biotas, is for our interpretation purposes complete enough to interpret.

A still third taphonomic method is based on the distribution of stratigraphic ranges. Mike Foote and David Raup at the University of Chicago used this technique to assess the fossil record completeness for Jurassic bivalves, Upper Cambrian-Lower Ordovician trilobites, Ordovician-Devonian crinoid species and Cenozoic mammals. They found fossil completeness varied from 60 to 90%.[21] The 90% completeness for bivalve species is similar to results obtained independently by other methods also.[22]

3) *Historical tests*

These assess the impact of new fossil discoveries on the existing knowledge base. The logic goes like this. A newly discovered fossil belongs either to a new unknown taxon or one already described. As more and more fossils are found so the proportion of new taxa needing to be erected for a new fossil is an indication of the proportion of fossils still undiscovered in the ground. If the rate of description of new life forms flattens off, it is an indication the bulk of the available record has already been found. The notion is called the "collector curve" and is the familiar observation that when collecting flora or fauna most specimens are novel at first but with increasing collecting, so more effort and more time is required to add new varieties to the list.

The discovery rate of new dinosaur species is representative both of the thinking and the method. Collector curves of new dinosaur species in Europe since the first was found in 1824 have leveled off since the 1920s consistent with a logistic curve, while in China over a similar geographic area but where investigation began a century later, there is evidence for a collector curve also, with a decline over the 1990's from an exponential rise in the 1970s.[3] This data indicates an even higher completeness than taphonomic testing suggests.[10] The approach requires that the confounding effect of synonyms and taxonomic renaming over the years be excluded of course. The method has been applied to a wide variety of other life forms, with similar results, indicating the fossil record is robust as regards completeness. Consider the following three studies as representative.

Christopher Paul evaluated mollusk and echinoderm fossil databases published between 1925 and 1997 looking for descriptions of new taxa. He found 1.8% of 249 families, 11.5% of 631 genera and 41.8% of 1158 species were new descriptions. This data suggests the records of these groups were already as high as 98%, 88% and 68% complete at the familial, generic and species levels by 1925.[1,23]

John Sepkoski at the University of Chicago did a similar analysis on marine animals, comparing the known fossil database in 1982 and 1992.[24] He too found that the pattern remained essentially the same. He said: "As a result of ten years of library research, half of the information in the compendia has changed: families have been added and deleted, low resolution stratigraphic data have been improved, and intervals of origination and extinction have been altered. Despite these changes, apparent macroevolutionary patterns for the entire marine fauna have remained constant. Diversity curves complied from the two databases are very similar, with a goodness-of-fit of 99%...Both numbers and percentages of origination and extinction also match well, with fits ranging from 83% to 95%. All major events of radiation an extinction are identical."[24]

Moving from sea to land, the completeness of the fossil record for tetrapods was assessed in a third study by comparing the published fossil databases in 1900, 1933, 1945, 1966 and 1987 with each other.[25] While knowledge increased substantially, (for instance a doubling of known diversities in all groups and the origins of several being pushed backward in time), the authors found this new data was distributed randomly with respect to time and that there was no change in the original pattern of the fossil record. The authors quite curiously did not emphasize this in their paper. Instead they concluded in muted understatement shrouded in jargon the following remarkable observation: "The six databases of fossil tetrapod diversity considered here, taken as snapshots of paleontological knowledge over the past 100 years, all show the same general pattern of numbers from Late Devonian to Late Cretaceous, followed by a rapid increase in the Tertiary."[25].

Why, you ask is the notion fossils are unreliable indicators of the true history of life still mainstream thinking in science then? In reviews of the quality of the vertebrate fossil record subsequent to these studies, paleontologist Michael Benton at the University of Bristol is explicit:

"The fossil record is good enough to document aspects of the evolution of life, and it can be assumed that errors and gaps are randomly distributed with respect to time."[26] "The fossil record is, then, good enough to read empirically, as a valid indicator of the true history of life."[3]

He gave the bottom line in a review of this matter writing in *Nature*: "Experience shows that major changes in the dating of fossils [with advancing discoveries] do not occur at the level of geological systems or stages, but at the finer divisions of substages and zones. Likewise, orders and families are often relatively stable, while new discoveries constantly alter the definitions of genera and species of fossils. The stability of longer time intervals and larger taxonomic categories perhaps reflects an adequate (if incomplete) fossil record...It is important to distinguish between "completeness" and "adequacy". Early parts of the fossil record are clearly incomplete, but they can be regarded as adequate to illustrate the broad patterns of the history of life."[7] Christopher Paul agrees: "To my mind there is no doubt that the fossil record is quite adequate to assess evolutionary hypotheses, such as ancestor-descendant relationships, and certainly it is adequate to reveal the history of life on earth...Independent estimates of the completeness of our knowledge of the fossil record of echinoderms, trilobites, mammals and mollusks suggest that >60% of fossil species, >80% of fossil genera and >90% of fossil families have already been discovered."[1] James Valentine agrees as well: "The fossil record is adequate to determine the general patterns of diversity of genera and higher taxa across geological time, for most groups of organisms."[27] What can we say in summary?

§6.2.2 Mirror, Mirror On The Wall...

We find formal analysis of the fossil record integrity shows it more than adequate to interpret. The pattern it has is real. While the number of fossils has markedly increased since Darwin's day, there have been few surprises in the overall pattern already established back then. New finds have either been closely related to previously known forms or unique, and not the intermediate forms he predicted.

Quantitative analysis of fossil record completeness by independent methodologies concordantly indicates a level of fossil record completeness demanding assumptions of adequacy before inadequacy. The significance of this can hardly be underestimated. It upends the established view of an "incomplete" fossil record held for over a century. Why was it held in the first place, you ask? No formal analysis was done. This acceptance of adequacy remains controversial dare I say, and it carries weighty implications, which I submit are the only reason why. At least we know it can't be because of fossil evidence.

It bears mentioning that the paleontologists cited above are not creationists or ID advocates. They hold positions at the top echelons of their science. Michael Benton is the editor of *Fossil Record 2*, Christopher Paul is President of the Palaeontographical Society and the Palaeontological Association, and James Valentine is Active Professor

Emeritus at Berkeley University. There's another strange irony. It's that if fossil record reliability is true we've made a full circle back to Darwin and his skepticism. Specifically to discover that on the most basic question of paleontology - what is the true pattern of the fossil record, this science has not advanced in almost one and a half centuries. This was the paleontological thinking in Darwin's day as he wrote defending it from the fossil evidence:

> "That the geological record is imperfect all will admit; but that it is imperfect to the degree required by our theory [of evolution], few will be inclined to admit."[28]

Most paleontologists say the opposite now.

It's also worth reflecting on the logic of the argument that incompleteness prevents interpretation of the fossil record. No science is ever based on complete knowledge, not neontology which is the study of living things, nor surely paleontology which is the study of dead things. "Incompleteness" of knowledge has not prevented hypotheses and conclusions be tested and drawn in neontology, why is paleontology somehow exempted? It is true that as long as one fossil remains in the ground it could add to our knowledge and the record to that extent be incomplete, but what we already know about fossil organisms far outweighs what we don't.

The comparison between paleontology and neontology is illuminating because it speaks to a much bigger issue. How and why an appeal for immunity from the known fossil evidence for "incompleteness" can be reasonably made, and so how evolution science is conducted.[1,23] Estimates of total species biodiversity for extant creatures in the 1960's and 1970's were between one and two million for instance, but today are at least 10 times higher (§1.2). Yet we don't hear biologists complaining their knowledge is incomplete and therefore existing patterns unreliable. Only an estimated 10% of all living species have even been described (§1.2.1). This is only a tiny fraction, at very most 3 to 5% of all species in the Phanerozoic.[23] Thus the databases of neontology and paleontology are both incomplete. The difference is paleontologists claim incompleteness as reason to reject existing evidence as an illusion.

If all this seems provocative, recall Darwin formulated his theory in spite of the fossil evidence, and the science of paleontology followed suit ignoring the innumerable examples of stasis and sudden appearance in the record for those of gradualism which can easily be counted (*LOE#2, LOE#3*). It was not until 1971 that this thinking was even challenged in the paleontologic community, and it still remains controversial (§6.3).[29]

274

On a related point there is something very illogical about a view that the fossil record pattern is untrustworthy because it is imperfect and inadequate, and yet also be certain about evolution from fossil evidence! If the fossils are "imperfect" how can they declare evolution a "fact"?

There are two even more obvious arguments to make against the imperfect record excuse. The first seems superfluous to even say. It's that gaps in the record are the rule and systematic throughout all forms of life, and have been from before Darwin's day (§5.5.2-6). Claiming the fossil record is "imperfect" as regards confirming evolution for particular lineages is plausible, but not for most all the lineages of life. Gaps are not just systematic either. The bigger problem is they are largest between the largest taxa divisions (Fig.5.5a vs. b). This is the opposite to what Darwinism predicts. It's an observation made before.[30] For example while there are multiple postulated transitional forms in the horse lineage, this is not the case over much longer evolutionary distances, say between fishes and amphibians or between flying creatures no matter whether insect or bird or bat, and non-flying ones. An imperfect record caused by a fossilization sampling error should show the reverse. The reason is the larger the gap, the more the evolutionary intermediates there should be.

The second argument is just as obvious. Logic demands by Occam's razor we accept the simplest hypothesis as consistent with the facts of biology. Which is that the fossil record is true and the gaps are real. Instead we are told the opposite is more reasonable. In other words evolution is true and the fossil data follows. What's the evidence for that idea? At least we know it can't be the fossil evidence.

The biggest question of course is why paleontologists are satisfied fossils are evidence of evolution today when Darwin was not. Since it cannot be because of the fossils (§5.5), it must be because of a changed opinion about fossils. What evidence does that rest on? Claiming incompleteness as answer is just an excuse to reject all the fossil evidence we do have that refutes macroevolution for evidence we don't have that doesn't. We discover in attempting to reconcile the fossil record with evolution theory Charles Darwin, to use Niles Eldredge's words, "essentially invented a new field of scientific inquiry – what is now called "taphonomy" – to explain why the fossil record is so deficient, so full of gaps, that the predicted patterns of gradual change simply do not emerge."[31] Evolution theory drives the data and the science is the point not the other way round.

Imperfection of the record is demanded of course, because the fossil evidence rejects the idea (§1.3;§5.5.2-5). As Darwin declared: "He who rejects this view of the imperfection of the geological record, will rightly reject the whole theory"(p.478). Read *The Origin* with me as he continues: "For he may ask in vain where are the numberless transitional links which must formerly have connected the closely allied or representative species...?...[H]e may urge the apparent, sudden coming in of whole groups of species. He may ask where are the remains of those infinitely numerous organisms which must have existed long before the Cambrian system was deposited?"(p.478)

Since he declared the fossil incomplete all this evidence becomes irrelevant because it's illusory. Darwin could dismiss it and continue as if it never existed, as he did: "Passing from these difficulties, the other great leading facts in paleontology agree admirably with the theory of descent with modification through variation and natural selection."(p.479) This would seem a selective interpretation of evidence.

It continues. If you cannot believe that, listen to some experts. Here are three. George Gaylord Simpson says: "As to the basic question, whether evolution is a fact, even in Darwin's day the fossil record already gave an answer...The much greater knowledge of the fossil record since Darwin has conclusively supported that view. Despite evident gaps in the record, it has also produced conclusive evidence of evolution within species and not only within faunas and floras as a whole."[32] Another is Douglas Futuyma who makes the case for evolution in his book *Science on Trial*. He says objections to evolution are a challenge to science itself, a claim we consider much later (§22). Consider for now what he says about fossils though. "The gaps we see today among living species exist because of the extinction of intermediate forms."[33] Don't laugh. He offers no evidence for the selective obliteration. He continues: "Many gaps so far haven't been filled in by fossils, but these are chiefly in groups that do not fossilize well, or groups such as the major phyla of animals that diverged in the very remote past, for which the fossil record is poor." Others find more even fossil intermediates than Dr. Futuyma can imagine and Michigan State University philosopher Robert Pennock is one. He says: "Given the difficulties involved in fossilization in the first place the record of intermediate forms is remarkably good and continually getting better."[34] The disconnect with the evidence is deliberate. These men are elite scientists (§25).

It's also worthwhile debunking some of the myths purporting to explain away the most obvious of all of the "gaps" in the record, the Cambrian explosion. Robert Pennock does not see a gap. "The Cambrian may reckon as a sudden explosion from the point of view of geologic time, but we are still talking about millions of years and this could have been quite sufficient for evolution."[35] Perhaps that's all you need to hear to believe, but what if you want evidence? Douglas Futuyma offers a mechanism. "At first glance, it seems as if all the major groups of animals arose in a very short time, but this is clearly an illusion; 700-million-year-old Precambrian rocks have a rather diverse fauna, and the very fine grained Cambrian shales of British Columbia show that there was an enormous diversity of animals that lacked skeletons. Very possibly this "rapid" diversification of animals in the Cambrian was due to the rapid evolution of hard parts by groups that had evolved long before." [36]

So the Cambrian explosion is an illusion too. This is an old argument. It's Darwin's. Namely that in the Precambrian period "it is indisputable ...the world swarmed with living creatures". In Richard Dawkins' view it is "a gap that is simply due to the fact that, for some reason, very few fossils have lasted from periods from 600 million years ago."[37] There is no fossil evidence, so it requires believing in "some reason". Why should we believe in an unknown reason? It also requires believing in swarms of PreCambrian creatures despite every fossil digging effort before and since 1859 that says no. The question, and the interesting thing, is where and on what evidence the idea is grounded? The "diverse fauna" Dr. Futuyma refers to above is the Ediacaran fauna of course. We know his opinion is false (§5.5.2-3). Stephen Jay Gould pointed this out years ago, but Dr. Futuyma chooses to ignore him too. Gould said, "But this [Ediacaran] fauna can offer no comfort to Darwin's expectation for two reasons. First, the Ediacara is barely PreCambrian in age...Second, the Ediacara animals may represent a failed, independent experiment in multicellular life, not a set of simpler ancestors for later creatures with hard parts."[38]

Futuyma's argument the Cambrian explosion is an illusion is because of "the rapid evolution of hard parts," he says, a view shared by Richard Dawkins[37] and Robert Carroll (e.g. "Because of the absence of bone, there is no fossil record of the earliest vertebrates or their immediate ancestors").[39] The notion is belied by Ediacaran fossils having no hard parts and yet being preserved in deposits world-wide, not to mention the exquisite soft part detail of the Cambrian deposits in the Burgess Shale and Chengjiang. The fossils of many insects are well represented in the record and dated by the science at 60 Myr ago when the mammalian record vanishes. If it is claimed that insects fossilize better, then why are there no fossil intermediates between the insect orders? There are so many insect species they swamp every other (Fig.1.2). What can we conclude?

The claim the fossil record is "incomplete" was nothing but an excuse for all the fossil evidence refuting evolution. It began with Darwin and continues in just as much unreason today. We find ourselves rejecting the inadequate record argument and accepting the fossil pattern as real, as saltational in character. In rejecting anagenesis we reject Darwinism and so stare design in the face as the appearance of nature suggests. But there is a fully naturalistic explanation for the pattern. Macroevolution remains very much alive. We just needed to look harder.

§6.3 THE SECOND WAY: PUNCTUATED EQUILIBRIUM ACTUALLY PREDICTS THIS!

The alternative evolutionary explanation for the fossil record's deviation from the pattern predicted in *The Origin* says phyletic evolution (anagenesis) is not the main pattern of evolutionary change (§2.2). Instead it is by a process called punctuated equilibrium (PE) affectionately called "punk eek". This theory about evolution theory was first proposed in 1971 by Niles Eldredge and further developed by Stephen Jay Gould and Steven Stanley soon after.[29,40-44] PE holds most evolutionary change has a rapid tempo and is by means of divergent branching of lineages in speciation. It occurs in geographically isolated, and therefore reproductively isolated, small groups on the periphery of a population. There they are released from the inhibitory effects of gene flow and genetic homeostasis on large populations and can undergo rapid evolutionary change (§2.4.4). The reason is being small populations this allows rapid fixation of a trait. This evolution is not left in the fossil record because it happens in peripheral isolates, small in number, compared to the mainstream species population. When a new form appears in the fossil record suddenly, it represents a migrational phenomenon as a newly emerged species from a small geographic area disperses, successfully competing and conquering ecological niches over a wider territory to be captured by the fossil record.

The rapid species splitting or "speciation" of PE is to be contrasted with the gradual tempo and in toto transformation of a single lineage that is phyletic gradualism conceived of by Darwin and shown by his figure in *The Origin* (Fig.2.2 vs Fig.2.1B). PE occurs in a geologic instant that Gould operationally defined as 1 percent or less of the organism's later stasis existence.[43] Specifically it is the geologic evidence of allopatric speciation as first proposed by Ernst Mayr, the notion new species are formed as a result of a founder effect by geographic isolation. As a result it is fully consistent with the Modern Synthesis (§2.1).[43]

Allopatric speciation, a word from "other father" meaning a new species from a separate parentage, is to be contrasted with sympatric speciation in which speciation occurs within a species population without geographic isolation (§2.2.1). Steven Stanley encapsulated the effect on the fossil record as: "Macroevolution is decoupled from microevolution."[45] Gould called it "a speciational theory of macroevolution"[46] We need to see the reasons and the reasoning for this seeming contradiction to accepted thinking for more than a century after Darwin and *The Origin*.

The conceptual difference between the two types of evolutionary trend, anagenesis and the rapid species splitting of PE, can be represented graphically (Fig.2.1-2). Since the groups in which PE occurs are small and isolated and ephemeral, there is no fossil record left of the evolutionary intermediates. The sudden appearance of new forms, the gaps and the stasis are explained. PE predicts the fossil record evidence we observe! Dr. Gould explains: "What should the fossil record include if most evolution occurs by speciation in peripheral isolates? Species should be static through their range because our fossils are the remains of large central populations. In any local area inhabited by ancestors, a descendant species should appear suddenly by migration from the peripheral region in which it evolved. In the peripheral region itself, we might find direct evidence of speciation, but such good fortune would be rare indeed because the event occurs so rapidly in such a small population. Thus, the fossil record is a faithful rendering of what evolutionary theory predicts."[47] There are two matters to consider.

The first has to do with the reaction of the scientific community to this announcement. While the reaction and the debate that ensued was an internal one among biologists, it is of interest to us because it speaks to a more important issue. Whether or not this new theory had substance. In other words whether Darwinism really was deficient before PE as PE advocates claimed. This is our first question. If the answer is "Yes" then the second question is for you. How could evolution be a truth for more than a century and fossils be hailed as the evidence? Recall statements like "Fossils demonstrate that evolution is a fact".[48]

We already know the answer to the first question. The difference between what Darwin said the fossil record should be like "if this theory be true" and what it was like, was rarely even admitted much less formally addressed. The reason is the fossil record pattern was regarded, scientifically, to be an illusion (§6.2). There's no question evolution theory was deficient before PE. Gould's words are worth hearing because he thought so too: "The modern synthesis, as an exclusive proposition, has broken down on both of its fundamental claims...I well remember how the synthetic theory beguiled me with its unifying power when I was a graduate student in the mid-1960s. Since then I have been watching it slowly unravel as a universal description of evolution...that theory, as a general proposition, is effectively dead, despite its persistence as textbook orthodoxy."[41] The extraordinary thing is the observation PE claims to explain took over a century to admit as even present (1859 to 1971). This has never been taken as reason for embarrassment much less impugning evolution as true. How can that happen in a science? (§19)

PE would seem to have provided an explanation to all the difficulties posed by the fossil record for evolution we found (§5.5). This is good news then. So how did the evolutionist community react to its arrival?

Not with welcoming arms, but scorn and derision. Like the characterization of Gould as: "a man whose ideas are so confused as to be hardly worth bothering with, but as one who should not be publicly criticized because he is at least on our side against the creationists",[49] and PE as "evolution by jerks". To which Gould, never at a loss for words, retorted: gradualism was "evolution by creeps"!

The rejection of PE came mostly from those who shared the views of Richard Dawkins at Oxford, a group called the "ultra-Darwinists". The very public Dawkins vs. Gould hostility has one of its roots here. Actually Dawkins argued the controversy over PE and anagenesis was merely a misunderstanding from "overblown rhetoric". The entire firestorm (if evolution scientists like him did not think there was one, it certainly was how both the technical and popular media saw it), was really all confined to a teacup. Dawkins wrote dismissively about PE in *The Blind Watchmaker* in a chapter entitled "Puncturing Puctuationalism." He gave two reasons why he had punctured PE away.

The first was that Darwinian theory if not Darwin himself, had fully accounted for the observation Gould and his colleagues were making so they were wrong to sensationalize their "discovery". Dawkins said: "The theory of punctuated equilibrium is a minor gloss on Darwinism, one which Darwin himself might well have approved if the issue had been discussed in his time...What needs to be said now, loud and clear, is the truth: that the theory of punctuated equilibrium lies firmly within the neo-Darwinian synthesis. It always did...The theory of punctuated equilibrium will come to be seen in proportion, as an interesting but minor wrinkle on the surface of neo-Darwinian theory."[50] He wrote these words decades ago. His prediction remains unfulfilled.

What both Dawkins and Gould completely ignore of course is Darwin considered the gaps in the fossil record supposedly now explained by PE, as a contradiction to his theory. He said they were "fatal to my views" (p. 448) and its "most obvious and serious objection" (p. 406) (§2.2). Thus PE cannot remotely be "a minor gloss on Darwinism". And as for Dawkins' claim this issue was not "discussed in his time", gaps and sudden appearances were explicitly addressed by Darwin from the first edition of *The Origin*, they were laid out in an entire chapter, explicitly identified as a contradiction to his theory, and he said he could not solve it any way than by declaring the current fossil evidence false. The problem is not Darwin that failed to "discuss" the problem, it's that the science community accepted macroevolution in spite of his words! This has not stopped selected quotation from Darwin to imply the opposite however and the following example fished out by Dr. Dawkins is probably the best: "Many species once formed never undergo any further change...and the periods, during which species have undergone modification, though long as measured by years, have probably been short in comparison with the periods during which they retain the same form." Selective quotation is all it is, as Gould was swift to say. Even George Gaylord Simpson's "quantum evolution" did not address the problems which PE now claimed to have solved. Since Simpson was Gould's doctoral mentor, I take it Dr. Gould was best placed to say what Simpson believed, and it was not in PE.

What is going on here? It's that gaps are not compatible with Darwinism but gaps are what the fossils declare, yet this science is certain evolution is true. Richard Dawkins made the alternatives plain: "Both schools of thought [gradualists and punctuationalists] agree that the only alternative explanation of the sudden appearance of so many complex animal types in the Cambrian era is divine creation, and both would reject this alternative."[51] The question is why they would reject "divine creation" if it is the only alternative answer, and if it is consistent with the evidence? What is the reason for their certainty? At least we know it can't be fossils.

Having claimed Darwinism accounted for the observations of the punctuated equilibrists all along, Dawkin's second point to puncture PE was to declare that Gould and his colleagues were really gradualists at heart! So in the end he made his argument by Anschluss. Specifically by (1) completely ignoring the distinctions in evolutionary tempo Gould and his colleagues were at pains to point out; (2) mischaracterizing their observations; and (3) allowing them only two speeds of evolutionary tempo one of which was a miraculous saltation. Here is an example from *The Blind Watchmaker* using this numbering:

"The theory of punctuated equilibrium, by Eldredge and Gould's own account, is not saltationary [1]. The jumps it postulates are not real, single generation jumps...The theory of punctuated equilibrium is a gradualist theory [2], albeit it emphasizes long periods of stasis intervening between relatively short bursts of gradualistic evolution...Everyone that is not a saltationist is a gradualist [3], and this includes Eldredge and Gould, however they may choose to describe themselves [2]."[52]

Since Gould and his followers were obviously not claiming miracles Dawkins's conclusion was they were really gradualist brothers. There was no controversy after all. Punctuated equilibrium was punctured. If this has you scratching your head, you need to look at the argument from another angle, from Dawkin's angle. Perhaps we can understand his reasoning better by considering what it is about PE that motivates him. It's an insight into how evolutionary biology thinks and regulates itself and the power of the evolutionary paradigm too. Why was there such publicity over the publication of the theory of PE? Dawkins gave three reasons and none had anything to do with the pursuit of scientific truth! We are beginning to understand his argument. He said:

"The reason in fact it [PE] has received such publicity…is simply that the theory has been sold – oversold by some journalists – as if it were radically opposed to the views of Darwin and his successors. Why has this happened?

There are people in the world who desperately want not to believe in Darwinism [just as there are people who desperately do want to, like him]. They seem to fall into three main classes. First, there are those who, for religious reasons, want evolution to be untrue [No reason why this might apply to Gould and followers though, and they were the only ones advocating PE]. Second, there are those who have no reason to deny that evolution has happened but who, often for political or ideological reasons, find Darwin's theory of its *mechanism* distasteful. Of these, some find the idea of natural selection unacceptably harsh and ruthless; others confuse natural selection with randomness, and hence 'meaninglessness', which offends their dignity; yet others confuse Darwinism with Social Darwinism, which has racist and other disagreeable overtones. [No reason this applied to PE advocates either. The PE vs. gradualism debate was conducted at least as much in the technical literature of science as in the popular media, and it was done in the name of scientific progress, not scandal]. Third, there are people, including many working in what they call (often as a singular noun) 'the media', who just like seeing applecarts upset, perhaps because it makes good journalistic copy; and Darwinism has become sufficiently established and respectable to be a tempting applecart."[53]

So by default the last reason is the real reason. The controversy at least in the minds of ultra-Darwinists had no scientific merit. It was empty sensationalism attracting creationist attention to the appearance "something is wrong with Darwinism". If that's how he felt, he had every incentive to whitewash as he did even if PE had solved a problem as serious as Gould said. His actions and his reasoning at least appear congruent with such motives (§25).

The decades since then have demonstrated the cold reception PE's messengers received was as myopic as it was misguided. Rather than a "minor gloss on Darwinism" PE shares center stage in any discussion of fossil record pattern or evolution tempo today. Writing in *Science* in the same year as Dr. Dawkins above, anthropologist Roger Lewin offered his opinion: "There is an old saying about the reception of a new idea, which goes as follows. At first it is dismissed as being wrong; then it is characterized as being against religion; and finally it is said to be something everyone knew all along. So it is with the notion of punctuated equilibrium, which has been the subject of much lively debate among evolutionary biologists for a decade and a half. The hypothesis, once rejected as being wrong or at least anti-Darwin, now appears to have entered the last of these stages."[54] There have been two major effects of this internal debate within paleontology.

The first has been to draw the attention to the inconsistencies between Darwinism and the evidence of the fossil record. To the observation stasis was the conspicuous feature of the record and yet was ignored, Stephen Stanley said: "Omission of this message from most discussions of the gradualistic and punctuational views would seem to represent the forest not being seen for the trees."[55] Gould said: "When Niles Eldredge and I proposed the theory of punctuated equilibrium in evolution, we did so to grant stasis in phylogenetic lineages the status of 'worth reporting' – for stasis had previously been ignored as nonevidence of evolution, though all paleontologists knew its high relative frequency"[56] "Of the two claims of punctuated equilibrium – geologically rapid origins and subsequent stasis – the first has received most attention, but Eldredge and I have repeatedly emphasized that we regard the second as more important. We have, and not facetiously, taken as our motto: stasis is data."[43]

The second has been to spotlight the relationship between macroevolution and microevolution and a denial of the former from the fossil evidence (§5.5.2-4). Needless to say this is a stunning admission if you are an evolutionist. No wonder there was a firestorm.

Before inspecting PE more closely to see whether it really can be punctured, we should clarify what it is not. It's not "quantum evolution", a term coined by George Gaylord Simpson to describe rapid anagenetic change rather than speciation which is what PE does. It's not "stabilizing selection", namely natural selection that removes the extreme forms of individuals from a population and so keeps it relatively homogeneous, because stasis can be demonstrated over millions of years and through different environmental changes. Under the anagenetic model this should convert one species to another. Furthermore stasis is the rule of the record, not gradualism as predicted by the anagenetic model and by Neo-Darwinism (§2.2). Nor is PE some saltationist phenomenon within the theory of Darwinism. PE proponents have vigorously defended themselves from these claims and those of Dr. Dawkins above, and with good reason:

> "No belief was more central to Darwin's thinking than gradualism; this may be the one point on which all Darwin scholars agree. Darwin's faith in gradualism was greater by far than his confidence in natural selection, though he often conflated the two – as in this clearly invalid statement: 'If it could be demonstrated that any complex organ existed, which could not possibly have been formed by numerous, successive, slight modifications, my theory would absolutely break down'...Did his contemporaries or descendants ever read Darwin as a saltationist?"[43]

But what is PE really? Have we got any closer to a tangible, measurable or testable explanation for the contrary character of the fossil record and for the cause of all the diversity of life? Can PE be punctured? Sure it can.

The first problem is whether PE can be reconciled with the fossil record. For the answer, we need look no further than the Cambrian. It can't. The top-down problem of the record and the lack of precursors is just as much a contradiction to PE as to Darwin's anagenesis. PE operates at the level of species and novelty accumulates up the taxonomic level from species to phylum level. Except we see sudden appearance and all the phyla appear at once. PE does not have nearly that power.

The second problem is PE lacks explanatory power. As James Valentine and Douglas Erwin pointed out decades ago: "neither of the contending theories of evolutionary change at the species level, phyletic gradualism or punctuated equilibrium, seem applicable to the origin of new body plans."[57] Gould has conceded their point: "I recognize that we know no mechanism for the origin of such organismal features other than conventional natural selection as the organismic level."[46]

The third problem with PE is as bad as the other two. How are we to tell it's true? The problem arises because as a solution, it's not far removed from the first response to the evidence of the fossil record which was Darwin's (§6.2). The reason is it too holds the fossil record is so imperfect it cannot ever show us the direct evidence of evolution, namely all those intermediate evolutionary steps. The difference from Darwin's response, is that one declares the gaps cannot be investigated and tested for reality by definition. They are explained as rapid speciation by small numbers of individuals in geographically isolated groups which cannot be found. The gaps in the PE theory are an absence of information after all. So this theory requires believing in an invisible process. Evolution causes wonderful changes but it's always going on somewhere else when you want to see, and we can't see that with fossils. PE must be believed with more than faith when the saltations are huge, say between orders, classes and phyla, which is the real issue evolutionary theory needs to explain from the fossil record. How does one prove PE has occurred, say that two morphologically different species on either side of a gap are related as ancestor and descendant in single phyletic lineage, even assuming migration or immigration of a new species into an area is somehow excluded? Not by investigating the gaps.

If the gaps cannot be investigated what about the other feature of PE, stasis? Can stasis be tested for biological reality in the fossil record? Gould concedes the validation of the theory "will rely primarily upon the documentation of stasis. Gaps can always be attributed to the traditional argument of imperfection and are rarely decisive...but stasis can be studied."[43] The trouble is that stasis is compatible with design, and the fossils supposedly confirm evolution. Anyway stasis taken by itself as he suggests is just that, stasis. Not evolution, which is change. Another problem is to explain is how such remarkable stasis occurs. If the average species is found in the fossil record for about 5 million years, how come natural forces which surely changed in one or more ways (e.g. climate, prey, predator and competitor species) caused no evolution when forces unknown could produce wonderfully remarkable punctuations at other times? Was the environment really "stable" over the times of stasis to explain why no evolution occurred? We are told stasis is the result of stabilizing selection of the established form. But is that true? The observation of the stasis is the evidence for the claim. More worrisome, how can it be refuted because we are without objective data. The observation *is* the evidence. James Valentine sums up:

"It has turned out that many patterns have been different from the predictions [of Darwin], sometimes strikingly so, and that these patterns cannot be argued away as artifacts of an incomplete fossil record. The gradualistic pattern of evolution predicted by Darwin is simply not found – simply did not occur – in many cases. The circumstances acting to produce the patterns we do find – the punctuational pattern of species lineages and the sudden introductions of major new body plans – are still not understood."[58]

And so we discover PE stands unique in science, unique as a scientific theory that explains why there is no scientific evidence for the theory. It is a scientific theory embodying everything in the fable of the Emperors' New

Clothes. PE does not remove the need for fossil evidence. All it does is "explain" why there is no evidence. Alan Templeton at Washington University in St. Louis has pointed to where the science is in all this. "It is critical to realize that punctuated equilibrium is merely a description, not a mechanism or a process."[59] All in all, PE seems an elegant post hoc description of the fossil record pattern to explain away contradictions to evolution but it is not a testable model, it's no more powerful than natural selection at the species level, and it's contradicted by the pattern of the fossil record at least in the Cambrian. Believe PE theory by all means but fossils can't also be the evidence that confirms evolution by PE "a fact".

§6.4 IN CONCLUSION: HOW SOFT CAN HARD EVIDENCE GET?

In summary the fossil record in Darwin's day showed stasis, sudden appearance and not even one intermediate form though he declared gradualism the rule of the record and "the number of intermediate and transitional links, between all living and extinct species, must have been inconceivably great" (p.408). He said "whole groups of species sometimes falsely appear to have been abruptly developed...which if true would be fatal to my views," (p.448) and "[i]f it could be demonstrated that any complex organ existed, which could not possibly have been formed by numerous, successive, slight modifications, my theory would absolutely break down."(p.261)

Darwinism was formulated in spite of the fossil evidence and Neo-Darwinism was accepted by the Modern Synthesis in spite of it also. Stasis was explained as stabilizing selection but this is an explanation that is also all the evidence. Darwin coined the term "living fossils" for those organisms most demonstrative, but this is to invoke an oxymoron. Sudden appearance and the paucity of fossil intermediates was rejected as an illusion caused by an incomplete and unreliable fossil record.

The difference between Darwin's day and today is today we're told fossils confirm evolution as a fact yet the pattern of the record is the same (*LOE#2*, *LOE#3* Table 3.1). Macroevolution is denied by the fossil record evidence however, and Darwin's own words can be used to show it. Fossil evidence was found after evolution was formulated, not before, and was claimed conclusive before the theory of punctuated equilibrium was devised to explain why evolution was true although the fossil evidence contradicted it.

So if you want to believe in macroevolution from fossils you either have to agree with Charles Darwin that the fossil record can't be trusted because the intermediates have still not been found, or you have to agree with Stephen Jay Gould that it's because PE hides the evolution from sight.

The problem with Darwin's idea is that it's falsified by formal analysis (§6.2.1). Neither he nor the Modern Synthesis architects, nor the generations of paleontologists who parroted this excuse for the evidence even bothered to test if his claim was true. Even more seriously, believing that the fossil evidence is "incomplete" as reason for rejecting the evidence is like trusting in a selective memory. Why would the fossil record be incomplete only at the nodes - the intermediates – when they are the fossils supposedly in the most abundance?

As for the problem with the alternative option, Gould's idea, this requires believing in a scientific theory that explains why there can be no scientific evidence for the theory. PE can't explain the top-down pattern of the fossil record, or explain the absence of precursors for the Cambrian explosion, and it has no additional power to produce the rapid biologic novelty of the Cambrian.

The two ways of explaining the fossil record evidence away are more similar than different nonetheless, because both declare the appearance of the fossil evidence illusory. There is a greater truth than the plain appearance of the objective data of the fossil record they both say. How then can fossils declare evolution "a fact"? At least we know the evidence for evolution cannot be because of fossils. It must lie elsewhere then and we find it in the next chapter.

REFERENCES

1. Paul, C. R. C. in *The adequacy of the fossil record* (eds S.K. Donovan & C.R.C. Paul) 1-22 (John Wiley & Sons, 1998).
2. Meehl, P. E. Consistency tests in estimating the completeness of the fossil record: a neo-Popperian approach to statistical paleontology. *Minnesota Studies in the Philosophy of Science* 10, 413-473 (1983).
3. Benton, M. J. in *The adequacy of the fossil record* (eds S.K. Donovan & C.R.C. Paul) 269-303 (John Wiley & Sons, 1998).
4. Smith, A. http://www.nature.com/nature/debates/fossil/fossil_frameset.html, 1998). 1998).
5. Benton, M. J. *The Fossil Record 2*. (Chapman & Hall, 1993).
6. Raup, D. M. Taxonomic diversity during the Phanerozoic. *Science* 177, 1065-1071 (1972).
7. Benton, M. J., Wills, M. A. & Hitchin, R. Quality of the fossil record through time. *Nature* 403, 534-537 (2000).
8. Newell, N. D. The nature of the fossil record. *Journal of Paleontology* 33, 264-285 (1959).
9. Durham, J. W. The incompleteness of our knowledge of the fossil record. *Journal of Paleontology* 41, 559-565 (1967).
10. Dodson, P. Counting dinosaurs: How many kinds were there? *Proceedings of the National Academy of Sciences USA* 87, 7608-7612 (1990).
11. Paul, C. R. C. in *Problems in phylogenetic reconstruction. Systematics Association Special Volume* Vol. 21 (eds K.A. Joysey & A.E. Friday) 75-117 (Academic Press, 1982).
12. Paul, C. R. C. in *Evolution and the fossil record* (eds K.C. Allen & D.E.G. Briggs) 99-121 (Belhaven Press, 1989).
13. Nicol, D. The number of living animal species likely to be fossilized. *Florida Scientist* 40, 135-139 (1977).
14. Conway Morris, S. The community structure of Middle Cambrian Phyllopod bed (Burgess Shale). *Paleontology* 29, 423-467 (1986).
15. Valentine, J. W. How many invertebrate fossil species? A new approximation. *Journal of Paleontology* 44, 410-415 (1970).
16. Newell, N. D. Adequacy of the fossil record. *Journal of Paleontology* 33, 488-499 (1959).
17. Allison, P. A. & Briggs, D. E. G. Exceptional fossil record: distribution of soft tissue preservation throughout the Phanerozoic. *Geology* 21, 527-530 (1993).
18. Schopf, T. J. M. Fossilization potential of an intertidal fauna: Friday Harbor, Washington. *Paleobiology* 4, 261-271 (1978).
19. Denton, M. *Evolution: a theory in crisis*. 189-190 (Adler & Adler, 1986).
20. Valentine, J. W. How good was the fossil record? Clues from the Californian Pleistocene. *Paleobiology* 15, 83-94 (1989).
21. Foote, M. & Raup, D. M. Fossil preservation and the stratigraphic ranges of taxa. *Paleobiology* 22, 121-140 (1996).

22 Harper, E. M. in *The adequacy of the fossil record* (eds S.K. Donovan & C.R.C. Paul) 243-267 (John Wiley & Sons, Chichester, England, 1998).
23 Paul, C. R. C. (http://www.nature.com/nature/debates/fossil/fossil_frameset.html, 1998).
24 Seposki, J. J. Ten years in the library: new data confirm paleontological patterns. *Paleobiology* 19, 43-51 (1993).
25 Maxwell, W. D. & Benton, M. J. Historical tests of the absolute completeness of the fossil record of tetrapods. *Paleobiology* 16, 322-335 (1990).
26 Benton, M. J. Models for the diversification of life. *Trends in Ecology and Evolution* 12, 490-495 (1997).
27 Valentine, J. W. The fossil record: a sampler of life's diversity. *Philosophical Transactions of the Royal Society of London* 330, 261-268 (1990).
28 Darwin, F. *The Life and Letters of Charles Darwin*. Vol. 3, 464 (John Murray, 1888).
29 Eldredge, N. The allopatric model and phylogeny in Paleozoic invertebrates. *Evolution* 25, 156-167 (1971).
30 Denton, M. *Evolution: a theory in crisis*. 191-2 (Adler & Adler, 1986).
31 Eldredge, N. *Reinventing Darwin: The great debate at the high table of evolutionary theory*. 95-6 (John Wiley & Sons, 1995).
32 Simpson, G. *Fossils and the History of Life*. (Scientific American Books, Inc, 1983).
33 Futuyma, D. J. *Science on trial*. 58-9 (Sinauer Associates Inc., 1995).
34 Pennock, R. T. *Tower of Babel*. 153 (The MIT Press, 1999).
35 Pennock, R. T. *Tower of Babel. The evidence against the new creationism*. 154 (The MIT Press, 1999).
36 Futuyma, D. J. 73-74 (Sinauer Associates Inc., 1995).
37 Dawkins, R. *The Blind Watchmaker*. 229-230 (W.W. Norton & Company, 1996).
38 Gould, S. J. *Wonderful Life*. 58-9 (W.W. Norton & Company, Inc., 1989).
39 Carroll, R. L. *Vertebrate Paleontology and Evolution*. 24 (W.H. Freeman and Company, 1988).
40 Eldredge, N. & Gould, S. J. in *Models in Paleobiology* (ed T.J.M. Schopf) 82-115 (Freeman, Cooper, 1972).
41 Gould, S. J. Is a new and general theory of evolution emerging? *Paleobiology* 6, 119-130 (1980).
42 Stanley, S. M. *Macroevolution Pattern and Process*. 2 (Johns Hopkins University Press, 1998).
43 Gould, S. J. in *Perspectives On Evolution* (ed R. Milkman) 83-104 (Sinauer Associates, Inc., 1982).
44 Eldredge, N. *Macroevolutionary dynamics*. 82-115 (McGraw-Hill Publishing Company, 1989).
45 Stanley, S. M. *Macroevolution, Pattern and Progress*. 187 (Johns Hopkins University Press, 1998).
46 Gould, S. J. *The Structure of Evolutionary Theory*. 55 (Harvard University Press, 2002).
47 Gould, S. J. *The Panda's thumb*. 184 (W.W. Norton and Co, 1980).

48 Simpson, G. *Fossils and the History of Life.* 125 (Scientific American Books, Inc, 1983).
49 Smith, J. M. Genes, Memes and Minds. *The New York Review of Books* 42, http://www.nybooks.com/articles/1703 (1995).
50 Dawkins, R. *The Blind Watchmaker.* 244,250-1 (W.W. Norton & Company, 1996).
51 Dawkins, R. *The Blind Watchmaker.* 230 (W.W. Norton & Company, 1996).
52 Dawkins, R. *The Blind Watchmaker.* 244 (W.W. Norton & Company, 1996).
53 Dawkins, R. *The Blind Watchmaker.* 250-1 (W.W. Norton & Company, 1996).
54 Lewin, R. Punctuated Equilibrium is now old hat. *Science* 231, 672-673 (1986).
55 Stanley, S. M. *Macroevolution, pattern and progress.* 88-9 (Johns Hopkins University Press, 1998).
56 Gould, S. J. *An Urchin in the Storm.* 37-8 (W.W. Norton & Company, 1987).
57 Valentine, J. W. & D.H., E. in *Development as an evolutionary process* (ed R.A. Raff and Raff, E.C.) 96 (Alan R. Liss, 1985).
58 Valentine, J. W. in *Phanerozoic diversity patterns* (ed J.W. Valentine) 3-8 (Princeton University Press, 1985).
59 Templeton, A. R. in *Perspectives On Evolution* (ed R. Milkman) 15-31 (Sinauer Associates, Inc., 1982).

7. Looks are important

§7.1 HOMOLOGY AS SIMILARITY IN NATURE

We consider here the last of the five attributes that gives life the appearance of design. Like diversity, unity, complexity and adaptiveness, it's among the most basic observations you can make in biology (§1.2.2-5). It is the observation of shared similarities among different living things. All four-limbed animals have the same skeletal construction pattern that is variations on a theme of five fingers for example, not two or six or some other number (Fig. 7.1). All living things use the L-isomer of amino acids to make proteins, not the equally available and chemically similar D- isomer (Fig.8.1). The genetic code is universal through all of biology, no matter whether you're a bacterium or a bacteriologist. Why should these similarities be? Whatever the cause, it must also be the solution to the Conundrum we seek. So let's go down to the marketplace of ideas to hear what answers are on offer.

Despite the din and the dust we know our task is easy because there are only two alternatives (§1.3). Gathered to the right in the square are all the design advocates. They say biological similarity across species represents "design parsimony", meaning components of a designed object are commonly reiterated in other designed products for reasons of economy or utility or artistic style. In regard to the latter, shared patterns can at times even be recognized as the trademark "signature" of a particular intelligent designer, something surely familiar from everyday life in clothing lines and other commercial product branding. It is from shared patterns of composition that we can recognize a painting as say a van Gogh, or a song as written or performed by a certain band or singer, or a building the work of a certain architect or architectural period. Applying the same reasoning in biology, shared patterns among different species of living things are consistent with ID as their cause. There is no formal disproof of this explanation as it stands. This has not been a problem in scientific practice however, because in being a supernatural explanation it has already been rejected. Why exactly? On grounds of implausibility for being unnecessarily mystical. How? Because scientists ever since Darwin say an entirely naturalistic and tangible alternative with an explanatory mechanism called common ancestry accounts for the observation too. That's when, to our left in the square, evolution advocates step forward to thunderous applause.

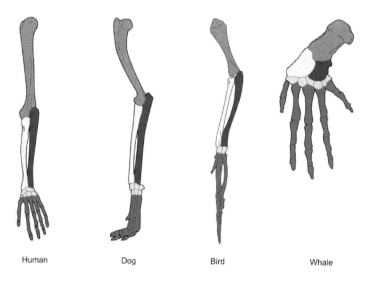

Fig.7.1. **Shared pentadactyl construction of the tetrapod limb.** The similarities are striking and yet the creatures are so different (human, dog, bird and whale) and they use the limb so differently (reaching, running, flying or swimming). How did they get there? (Image credit: Волков Владислав Петрович).

A word of caution if you want to rush off and join them. It's that if it emerged there instead was no evidence for a natural explanation of similarity in biology, and if common ancestry was not tangible or mechanistic or explanatory from the evidence, the same logic demands this would be evidence demonstrating design. Creationists would then not be so crazy, but this is just hypothetical thinking of course. How could anything supernatural be objectively demonstrable by the natural sciences? Speaking of which, the science and education establishment say similarities across taxa such as these show they were inherited from a common ancestor by evolution. There's no doubt about it either. This is the fourth line of evidence for macroevolution in our catalogue (*LOE#4*). Thus we discover the scientific solution to the cause of the differences or diversity among living things (*LOE#2*), as well as the cause of the similarities among living things (*LOE#4*), is evolution.

Consider *LOE#4* again: *The striking morphologic similarities demonstrated by comparative anatomy in very different species (called "homologies") are evidence for evolution because a common ancestry explains their presence as inherited traits.* The task here is to inspect that claim. It's a fascinating story that will show you why there really is in-your-face evidence for supernatural design in biology, how evolutionary biology does business exactly, and how evolution remains a scientific fact despite the evidence. You don't believe it? Then come and see why at least in biology, looks really are important.

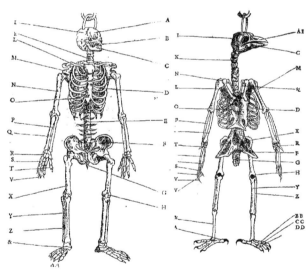

Fig.7.2. Skeletal pattern of human and bird. Why should they be so similar when they are also so different? (Drawing by Pierre Belon, 1555.[1] By kind permission The Natural History Society of Northumbria).

What we wrestle with here is the meaning and the cause of biological "sameness". Consider the body plans of humans and birds (Fig.7.2). Despite their differences they share a highly similar skeletal pattern. This drawing was made over 450 years ago. The observation in science is much more ancient than even that. Aristotle (384-322 BC) reported it 1,900 years earlier still. In *Historia Animalium* he said: "There are living beings such that all the parts of one recall the corresponding parts of others."[3] The observation of biological similarity, like the observation of biological diversity, is surely as old as mankind itself (§1.3). The question for you is what does it mean?

Consider another example of biological similarity. Explain why all four-legged creatures have a similar "pentadactyl" (five-fingered) pattern of limb construction (Fig.7.1). Knowing how different frogs, dogs, lizards, birds, bats, whales and humans are, and knowing how differently they use their limbs, how did these similarities get there?

Charles Darwin asked the same question in *The Origin*. He said: "We have seen that the members of the same class, independently of their habits of life, resemble each other in the general plan of their organization...This is one of the most interesting departments of natural history, and may almost be said to be its very soul." (p.579) He asked what it could mean: "What can be more curious than that the hand of a man, formed for grasping, that of a mole for digging, the leg of the horse, the paddle of a porpoise, and the wing of the bat should all be constructed on the same pattern, and should include similar bones, in the same relative positions?"

In 1843 this ancient observation of structural correspondence by parts or organs in otherwise different organisms was finally given a name by Sir Richard Owen (1804-1892), the first Director of the British Natural History Museum in London. He called it called "homology" from a word derived from the Greek *homologia* meaning "agreement". Owen defined a homologue as "the same organ in different animals under every variety of form and function."[4] Hang on to that definition. Today the term homology applies not only to similarities across taxa in morphology, but also similarities in physiology, biochemistry and even behavior.

What did "homology" mean to Richard Owen and to biology before Charles Darwin? The common features were attributed to the existence of a hypothetical "archetype" or blueprint from which they were derived. A series of drawings of the vertebrate skeleton by Richard Owen in 1848 shows this thinking (Fig.7.3). Biological similarities were explained as fidelity to a shared pattern of construction. This typological concept of homology was creationist of course and it had a history in science at least dating back as far as Aristotle. Then came Charles Darwin.

§7.2   HOMOLOGY REDEFINED BY EVOLUTION

Darwin took this supernatural explanation of homology head on and gave it an entirely naturalistic one. He even changed its definition. Writing in *The*

Origin he quoted one of his supporters Sir William Flower, Richard Owen's successor as the Director of the Natural History Museum, who said of homology:

> "We may call this conformity to type, without getting much nearer to an explanation of the phenomenon, but is it not powerfully suggestive of true relationship, of inheritance from a common ancestor?" (p.580).

It was a fundamental inference. Then Darwin answered Flower's question for him:

> "What can be more curious than that the hand of a man, formed for grasping, that of a mole for digging, the leg of the horse, the paddle of a porpoise, and the wing of the bat should all be constructed on the same pattern, and should include similar bones, in the same relative positions?...The similar framework of bones in the hand of a man, wing of a bat, fin of the porpoise, and leg of the horse, - the same number of vertebrae forming the neck of the giraffe and of the elephant, - and innumerable other such facts, at once explain themselves on the theory of descent with slow and slight modifications." (p.579, 635)

Rather than the metaphysical notion of a shared pattern of design, Darwin proposed there was an ancient ancestor from which the similar characteristics were inherited. The solution seemed brilliantly simple. "On my theory," Darwin said, "unity of type is explained by unity of descent" (p.261). Homology was the consequence of common ancestry. Biological similarity, just like biological diversity, was the consequence of evolution! This was how his naturalistic mechanism displaced the prevailing supernatural explanation in science. Twenty-two centuries of design in biology died that day, because what Darwin said seemed so much more plausible. Some evolution believers like Sir William Flower continued to believe in Christianity, but it was a different religion to that of their fathers or to what the Bible said (§1.3). Today this belief is called theistic evolution and much later we will look at it ourselves (§26.3.3). As for Darwin, he scorned a creation and all like Owen who thought they could reconcile supernatural thinking with scientific conduct, saying:

> "Nothing can be more hopeless than to attempt to explain this similarity of pattern in members of the same class, by utility or by the doctrine of final causes. The hopelessness of the attempt has been expressly admitted by Owen in his most interesting work on the 'Nature of Limbs'. On the ordinary view of the independent creation of each being, we can only say that so it is; - that it has pleased the Creator to construct in each great class on a uniform plan; but this is not a scientific explanation. The explanation is to a large extent simple on the theory of the selection of successive slight modifications..." (p.580-1)

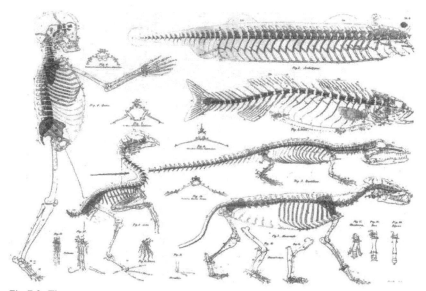

Fig.7.3. **The typographic concept of homology** demonstrated by Owen's "archetype" pattern of the vertebrate skeleton (top right) published in 1848 and representing skeletal iterations shared by fish, reptile, mammal, bird and human.[5] Darwin replaced "archetype" with "ancestor".

Setting aside also for later the logic of his argument and what scientific evidence if any he had for his claim, much less dismiss offhand "the Creator" as "not a scientific explanation" without offering evidence why, so compelling was Darwin's explanation that his interpretation of homology now became its definition! Darwin redefined homology as: "[t]hat relation between parts which results from their development from corresponding embryonic parts" (p.657). In other words "homology" became biological similarity caused by common ancestry, which had caused similar development from the same embryonic tissue.

Recalling Owen's original definition of homology that was a statement of directly observable fact, we see Darwin had made two leaps. First he postulated similar structures arise because they are inherited from a common ancestor, and second he provided a causal mechanism of how that happened. Similar morphology was caused by similar embryonic development. This would seem to follow intuitively. If structures are similar in different adult species and they get there by development, they would be expected to have developed similarly. It seems an elegant theory, except for two limitations.

The first is his naturalistic explanation was no more than the injection of a temporal dimension into the ancient observation of similarity. The new "explanation" was merely Darwin's now naturalistic hypothesis of what had happened in biological history past. Where was the evidence? In giving his postulated mechanism, Darwin had also paradoxically moved homology from something that was an unambiguous observable fact, measurable sameness, to something that would forever be a hypothesis, common ancestry.

The second limitation is that for a completely naturalistic explanation he never addressed the question at the heart of the observation. What exactly is biological "sameness"? What does it mean to regard two structures in different organisms as "the same"? Darwin left this as unanswered as Owen.[7] Even setting these two limitations aside, all was not well, even then. There were two other problems and they concerned his embryologic explanation of homology.

The first is that Darwin ignored the existing evidence from embryology refuting his claim. More than a decade before Richard Owen had instead showed homology "may exist independently of...similarity of development".[8] The evidence included his observations "[t]he femur of the cow is not the less homologous [similar] with the femur of the crocodile, because in the one it is developed from four separate ossific centers, and the other from only one such center."[5] Darwin never countered the contradiction. It also went unmentioned in *The Origin* in every one of its six editions. It's unlikely Darwin was not aware either since he debated Owen. How could this have happened in science?

Knowing Darwin's opinion of Owen is perhaps to know the reason. If Darwin really thought "[n]othing could be more hopeless" (p.580) than Owen's "so it is" science for being void of a naturalistic mechanism, then the power of his "simple" naturalistic alternative had to be more compelling than all the evidence denying it. Owen's data supported supernaturalism and so was "not a scientific explanation", and so must be false. All that was required was more research to overturn the contradiction, like the contradiction of the fossil record that Darwin also chose to dismiss (§5.6). If Darwin did not hold this view in 1859, it certainly is the mindset today (§7.4).

The second problem was greater. It's that his theoretical explanation for the phenomenon of homology had become its definition. The problem is before you use a definition you need to be sure that it's true. The reason is a definition is a filter of observations, by definition. It's to put on rose-colored spectacles and only see rose thereafter, or put on common ancestry spectacles and forever after see only common ancestry and evolution. What was Darwin's evidence for the evolutionary interpretation of biological similarity? Not evidence, but another of his unproved postulates. Namely that embryologic development is inherited.

Darwin developed the arguments for evolutionary homology and evolutionary embryology together in *The Origin* (Chapter XIV), and the concepts have remained linked ever since, in accord with intuitive evolutionary reasoning that they have a mechanistic relationship.[9,10] Darwin's view was that embryos have similar morphologic stages of early development and a "community of embryonic structure reveals community of descent" (p.599). In other words, he said evolutionary history can be seen in shared pathways of embryonic development. We explore the validity of this idea later (§9.3-7). For now it's enough to see that using the postulate from embryology as the evidence for the postulate from homology, as the evidence for evolution, is empty because it's circular. We saw this kind of circular reasoning for fossils earlier, another supposedly independent line of evidence for macroevolution that was not (§5.3,6).

Darwin was a Lamarckian so transmission of similar morphologic structures to offspring during development was easily explained. Now we know similar morphologic features are not inherited, genes are inherited. What's interesting, is this discovery which only came later in the Modern Synthesis, made no difference to the supposed truth in science of either of Darwin's original two theses. Namely that homology was due to inheritance from a common ancestor, and that inherited paths of embryologic development were the proximate cause (§9.1,2). The new knowledge from genetics was simply plugged into Darwin's ideas and the Lamarckism removed. Darwin did provide a "scientific" explanation for homology though and to use his words, because shared embryologic development as its cause is testable. Writing at the end of Chapter XIV, he boldly presented his explanation of homology as solid evidence for evolution:

> "On this same view of descent with modification, most of the great facts in Morphology become intelligible...On the principle of successive slight variations...being inherited...we can understand the leading facts in Embryology; namely, the close resemblance in the individual embryo of the parts which are homologous...and the resemblance of the homologous parts or organs in allied though different species...the several classes of facts which have been considered in this chapter, seem to me to proclaim so plainly, that the innumerable species, genera and families, with which this world is peopled, are all descended...from common parents, and have been modified in the course of descent, that I should without hesitation adopt this view, even if it were unsupported by other facts or arguments" (p.610-1)

It sounded good but it was still only a postulate. The "great facts" were facts all right but the only difference was they were now given his "intelligible...scientific explanation," read naturalistic interpretation, of what significance they had historically. Creationists begged to differ over these same facts (§9.3-7), the existing evidence that was Owen's data was at least on its face a disproof, and Darwin's naturalistic explanation still needed naturalistic validation, namely some objective evidence!

Darwin's claim of a common embryonic source tissue as the cause of similar structures across taxa could be tested. It was and found false (§7.3;§9.3). This too made no difference to the supposed truth of his claim that homology was caused by common ancestry. Darwin's postulate that a similar derivative embryonic tissue was the proximate cause of homology was simply excised out of evolution theory like Lamarckism earlier, and the phylogenetic mechanism of the observation became the new definition of the observation. Evolution theory was refined as new information became available and earlier hypotheses became untenable. Today the word "homology" does not mean biological similarity as Owen used the term, but biological similarity actually caused by common ancestry. "Homology is the central concept for *all* of biology,"[11] and as David Wake at Berkeley says, "*the* most important concept in contemporary biology" now.[12] This is powerful stuff but how far have we really come since Richard Owen? We need to flesh out the details and the history to tell.

The history of homology in science can be summarized very simply. An ancient observation of biological sameness got an explanatory mechanism by Charles Darwin, but his explanation was merely a postulate awaiting evidence. Next this postulated mechanism became the definition of the similarity, and today the postulated mechanism is the observation. The appearance of design and where that came from, with Darwin's core evolutionary conception of homology, namely that it represents similarity caused by common ancestry, is now everywhere a self-evident scientific truth.[13] Take *Encyclopedia Britannica* for example, hardly the place we expect to find myth, which defines homology as: "similarity of structure, physiology, or development of different species of organisms based on their descent from a common evolutionary ancestor". But how do we know that's true? How can we know common ancestry is the cause of homology? Yet it is the definition. While you are considering this, how can we know if other observations of biological similarity taken by science as not from common ancestry but which developed independently by coincidence through natural selection and called "homoplasy", are not? (§5.3.3) A final observation to consider is that if the "central concept for all of biology"[11] is defined as evolutionary to begin with, we should not be surprised when we are informed evolution in biology is "a scientific fact" (§3.1).

So what's the evidence for homology as the evidence for common ancestry? Darwin did not produce any and explanations that are alternative interpretations of observations don't count. We will have to make an inspection ourselves. Tackling this question will be easier if the inquiry is done in two parts.

The first task is to consider Darwin's postulated proximate mechanism of homology. Do homologous structures arise from "corresponding embryonic parts"? They do not and as we have already seen this was known before *The Origin* was written. We need to see this data ourselves because it gets to the heart of the problem we are investigating. We need to explain biological similarity with a naturalistic mechanism and we need to be able to reconcile all of the data from biology.

Second, after a better appreciation of what genetics and development have to say, we will inspect the claim that Darwin's postulate about homology as evidence of common ancestry is true and what this means for the scientific status of evolution. From now on "homology" will have the meaning it has in biology now - similarity actually caused by common ancestry, and not just biological similarity as Owen first defined it. So, by what mechanism does biological similarity come about? At least we can tell by what mechanism it does not.

§7.3. BIOLOGICAL SIMILARITY DOES NOT RESULT FROM SIMILAR DEVELOPMENT OR SIMILAR GENES

The evidence refuting Darwin's claim that similar structures or shared patterns in biology are from shared pathways of embryologic development was marshaled in 1958, after the Modern Synthesis had already been formulated.[14,15] In another irony, the evidence was collated by the Director of the British Natural History Museum, Sir Gavin de Beer. De Beer held the same post as Owen and was refuting Darwin's claim like Owen, only a century later here was an evolutionist doing it. How a delay that long could happen in a science despite a disproof back in 1848 is another question worth asking. The answer should become apparent with more probing.

In a nutshell de Beer showed that structures taken to be homologous yet arise from 1) different regions or segments of the embryo; 2) different tissues of the embryo; 3) different embryonic inductive mechanisms; and 4) different genes of development. This is extraordinarily puzzling. Think about it. Similar structures are generated by non-similar mechanisms. How can that happen? A few examples are given below to show this. The really interesting observation is not this extraordinary discovery however. It's how the science has dealt with it. How it was reconciled within the original paradigm of Darwinism which it refuted. Watch how hard in spite of his great academic stature even Sir Gavin wrestled. It is to see how progress in this science could go backwards from Owen to de Beer to today. It also provides a wide-open window for interested observers like us to see how science is practiced when evolution is already a fact before the facts are in. There are four conclusions to face from the facts of biology here, and this is no matter whether you think evolution is a fact or design is a fact. Here they are:

1. *Homologous Features Arise From Different Parts Of The Embryo*
It turns out the pentadactyl limb develops from different regions of the embryo in different creatures (Fig.7.1). In the newt for instance, the limb develops from trunk segments 2-5, in the lizard segments 6-9 and in man from segments 13-18.[16] De Beer explained the inconsistency on the grounds there was a "displacement of homologous structures" but that is to employ a nomenclature that is a contradiction in terms, not to mention his explanation was now without a naturalistic mechanism. The explanation was the evidence and although this discovery was to him "astonishing", he still could not shake himself to interpret his findings in any other way. The interpretation of biological similarity as evolutionary was unaltered by his results, even though he had just disproved Darwin's mechanism.

Actually de Beer said he had "no doubt" about evolution, but not having evidence he was forced to write his science in diktat and unreason:

> "There is no doubt whatever that the forelimb in the newt and the lizard and the arm of man are strictly homologous, inherited with modification from the pectoral fin of fishes 500 million years ago [Diktat]. They have identical elbow and wrist joints and their hands end in five fingers. The bones and muscles they contain also correspond. [Here comes the unreason...] But a minute examination of their comparative anatomy reveals the astonishing fact that they do not occupy the same positions in the body."[2,16].

Another example of supposed homology divorced from a similar developmental mechanism is the structure that develops into the posterior part of the skull called the embryonic occipital arch. It arises between segments 5 and 6 in frogs, 6 and 7 in newts, 7 and 8 in shark, and 9 and 10 in reptiles and mammals.[16] The neural tube is another example. The neural tube is one of the defining characteristics of the phylum Chordata, the phylum to which humans belong. It usually arises by an invagination of the neural plate to form a hollow canal, but in bony fish (teleosts) and lampreys, the neural tube is formed by a solid rod of cells that first sink beneath the ectoderm then hollows out secondarily.[17] The vertebrate alimentary canal is a supposed homologous structure too. Yet it's variably formed from the roof of the embryonic gut cavity (in sharks and teleost fish), the floor (in lampreys and newts), both the roof and floor (in frogs), or from the lower layer of the blastoderm (in birds and reptiles). The alimentary tract has different embryonic origins even among different species of insect.

Consider the homologous urogenital system. In fish and amphibians it arises from the mesonephros, but in reptiles from a different, independent structure called the metanephros. In summary, similar adult structures in different taxa arise from dissimilar regions of the embryo. How does that happen by common ancestry?

2. Homologous Features Arise From Different Embryonic Tissues

The fertilized eggs of all animals with tissue complexity beyond that of the jellyfish develop three tissue layers as embryos and are therefore called "triploblastic". The innermost layer is called the endoderm, the outermost the endoderm and the middle layer is called the mesoderm. From these three all of the tissues and structures of the adult develop.

In general the ectoderm develops into the epidermis ("outer layer of skin"), nervous system, sensory organs and the nephridia ("kidneys"); the endoderm into the gut, liver and lungs; and the mesoderm into dermis ("inner layer of skin"), cartilage, bone, teeth and muscle, coelom ("body cavity") and oviducts or genital ducts.

But only in general, because in newts the jaw cartilage, visceral arch, skull and teeth are derived from ectoderm, tooth enamel from ectoderm or endoderm and none derive from the mesoderm.[18] The allantoic and amniotic membranes have a different embryonic origin in mammals compared to birds and reptiles. The columella (middle ear bone) of the chick may variably arise from ecto-mesenchyme or mesodermal mesenchyme,[19] and the orbitosphenoid bone develops as a membrane bone in amphibians, but in other vertebrates it develops as a replacement bone.[20] In sharks (cartilaginous fish) the notochord is formed from a cell sheet that involutes into the interior of the embryo, in teleosts (bony fish) from aggregating mesenchymal cells, in amphibians by a cell sheet, and in the chick it's formed from aggregates of mesodermal cells which migrate individually into the interior of the embryo.

On an even more fundamental level, cleavage and blastulation (the two next steps in embryonic development after fertilization) are different across organisms and although there are some similarities of gastrulation (the step when the gut develops), it too is fundamentally different in other ways (§9.3). For example although salamanders and frogs gastrulate similarly at a macroscopic level, the cell movements are different. For salamanders the precursors of the mesodermal cells are on the surface of the gastrula but beneath the surface for some frogs.[21].

Other seemingly equivalent structures like gametes arise variably from endoderm or mesoderm, and cartilage and bone can variably arise from ectoderm/neural crest or mesoderm.

Insects differ even among themselves in their development. They develop either by short germ band (by which abdominal segments are generated progressively after the blastoderm stage), long germ band (in which all segments are formed and specified together by the end of the blastoderm stage), or intermediate germ band (in which some segments are specified before and some after the blastoderm stage).

In summary similar structures in different creatures can arise from dissimilar embryonic tissues, or as de Beer said: "correspondence between homologous structures cannot be pressed back to similarity of position of the cells of the embryo or the parts of the egg out of which these structures are ultimately differentiated." How is that caused by inheritance and common ancestry?

3. Homologous Features Arise From Different Embryonic Induction Mechanisms

Consider some examples. Meckel's cartilage, which forms the lower jaw, is induced by the pharyngeal endoderm in amphibians, the cranial ectoderm in birds and the mandibular epithelium in mammals.[9,22] The development of the eye lens from the ectoderm differs among anuran (frogs and toads) and urodele (newt) amphibians,[23] and even among closely related species. For example the lens demands induction by the underlying optic cup in the frog *Rana fusca*, but not in the closely related hybrid *Rana esculenta*.[24]

Laboratory regeneration experiments demonstrate developmental differences which yet result in structural similarities of the end product. If the lens of an axolotl eye (a urodele amphibian) is removed, it fully regenerates from the iris. The original lens had a different origin however, arising during embryonic development from ectoderm (specifically, dedifferentiated embryonic iris cells).[25]

During development lizard and axolotl tails arise from undifferentiated cells and do not require a nerve supply, but if they are subsequently removed they require a nerve supply to regenerate because it happens as a result of de- and re-differentiation of cells caused by chemical mediators released from nerve endings.[25] Anurans which develop directly from eggs so bypassing the tadpole stage are called the *Eleutherodactylus*, (and are the largest vertebrate genus on earth), have a different morphogenetic sequence of development from anurans that develop indirectly, even though their adult morphologies are the same.[26] Similarly directly developing sea urchins (*Heliocidaris*) share the same adult morphology but have different morphogenetic sequences of development to sea urchins which develop indirectly.[7,27] Chondrogenic condensations differ in the limbs of *Ambystoma mexicanum* and *Titurus marmoratus*.[7,28]

Somites are one of the distinguishing features of the chordate phylotypic stage, yet the pattern of somite formation from the mesoderm differs between salamanders and frogs and even among frogs.[29-31] In reviewing amphibian development, University of Virginia biologist Ray Keller concluded: "the somitic developmental pathway as a whole is not conservative and has been capable of accommodating the use of a number of quite different morphogenic processes, all leading to similar ends".[29]

In all tetrapods except urodeles (newts and salamanders), the digits develop with tissue between them initially which dies off later by apoptosis to leave the formed digit behind. In the urodeles the digits develop separately however, without apoptosis.[32]

The neural crest is the embryonic precursor of several adult structures, among them the sensory and autonomic peripheral nerves, the skull, skin pigment cells and the adrenal medulla. Neural crest cells must therefore migrate widely through the embryo during development. How they do it differs widely across taxa.[33] In chickens and mice they migrate through the anterior but not the posterior half of the somite, in the frog *(Xenopus laevus)* mostly through the posterior half and in the zebrafish along the entire length of the somite before converging on its center. In the salamander and the chicken migration is after neural tube closure, while in mice it is before. In one species of frog (*Xenopus*) neural crest migration begins after neural tube closure while in another (*Bombina orientalis*) it's before, and patterns of cranial neural crest migration are also different among frog species *B. orientalis*, *Rana* and *Eleutherdactylus coqui*.

In summary the number of different ways by which supposedly homologous structures are formed increases further. Similar, and yet formed from different locations, different tissues and different induction mechanisms. De Beer concluded:

> "All this shows that homologous structures can owe their origin and stimulus to differentiate to different organizer-induction processes without forfeiting their homology."

Only he had no evidence how on earth this could happen. How then could he be so certain that homology could not be forfeited? (§19;§20)

4. Homologous Features Need Not Be Produced By The Same Genes
One example is the butterfly wing pattern of *Papilio dardanus*, in which females mimic the wing patterns of two subfamilies of butterflies toxic to their predators.[2,34] Several different female mimetic forms occur, each remarkably similar to a corresponding toxic species in one of two subfamilies (the Danainae and Acraeinae). By crossing the three different *P.dardanus* mimetic morphs of three Danainae subfamily species, the elegant simplicity of a dominant inheritance pattern of single alleles was seen, rather than the offspring having an intermediate pattern that would no longer mimic. So far so good in explaining this as due to mutation, natural selection, common ancestry, homology and evolution.

The problem appeared when one of the female morphs (*P.d.trophonius*) was crossed with a fourth morph that mimics the other subfamily species, Acraeinae (*P.d.planemoides*). Their offspring have a wing pattern unlike either parent and the same as that of *P.d.niobe*, another mimic of Acraeinae. This occurs despite there being a *niobe* determining gene, different from that found in *trophonius* and *planemoides*.

De Beer used a different example to make the point homologous features are formed by different genes though. The same gene, *even-skipped*, is important in the embryogenesis of the fruit fly *Drosophila* but not in the grasshopper *Schistocera Americana* or the wasp *Aphiudus ervi*, even though the segments concerned are homologous.[35,36] Homologous homeobox genes *distal-less*, *engrailed* and *orthodenticle* also have different expression domains and developmental roles among echinoderms.[37] (§9.3.5)

Not only can the same gene give rise to different homologous structures, different genes can result in the formation of the same homologous character. For example mutations to the *eyeless* gene in the fruit fly *Drosophila* cause loss of the eye in the offspring. However the eye develops again several generations later as a result of the action of other so-called "modifier" genes, even though *eyeless* remains just as mutated. We review the genes of development later (§ 9.3.5). This is how de Beer concluded back in 1938: "it is clear the *characters controlled by identical genes are not necessarily homologous* [and that] *homologous characters need not be controlled by identical genes*".[32,38] He wrote again in 1971 concerning the discovery that different genetic mechanisms can account for similar and supposedly homological features: "It is now clear that the pride with which it was assumed that the inheritance of homologous structures from a common ancestor explained homology was misplaced; for such inheritance cannot be ascribed to identity of genes...But if it is true that through the genetic code, genes code for enzymes that synthesize proteins which are responsible...for the differentiation of the various parts in the normal manner, what mechanism can it be that results in the production of homologous organs, the same 'patterns', in spite of their *not* being controlled by the same genes? I asked this question in 1938, and it has still not been answered."[39]

The silence continues. De Beer concluded: "characters controlled by identical genes are not necessarily homologous...Therefore, homologous structures need not be controlled by identical genes, and homology of phenotypes does not imply similarity of genotypes."[15] What does that mean? I leave the problem with you for now. We take it up again later.

In summary what De Beer discovered all those decades ago, and what Richard Owen had demonstrated a century still before that, is Darwin's definition of homology as "That relation between parts which results from their development from corresponding embryonic parts", is "just what homology is not."[18] This is extraordinary.

What is more extraordinary is de Beer still believed Darwin had it right! Dalhousie University embryologist Brian Hall is a doyen of this field today and he thinks so also.[40-42] Among his accolades is The Kawalevsky Medal, awarded to the eight most distinguished scientists of the twentieth century in comparative zoology and embryology. This is an elite group that includes Louis Pasteur and Charles Darwin. Dr. Hall's opinion reviewing the field was to inform us Sir Gavin had it right and nothing had happened to change his conclusion. (But look what he has to concede about homology! Does it make sense? We look later §7.4.3). He said: "De Beer's analysis and summary of the situation are as compelling now as they were a quarter of a century ago. Variability of developmental processes on the one hand, and the constancy of homologous structures on the other, render any single concept of homology that attempts to unite the two as an uncomfortable alliance between the constancy of the final pattern and the variability of the developmental processes"[43]

Fact is Darwin's definition, which was really no more than a postulate, disintegrates at every level and de Beer's data declares it plainly no matter what Sir Gavin or Dr. Hall say. Homologous structures have different 1) embryologic topologic origins, 2) tissue origins, 3) modes of induction and 4) genetic origins. Before *The Origin* Richard Owen demonstrated biological similarity "may exist independently of ...similarity of position of development."[8] In other words, far from "hopeless" as Darwin said, Owen was right all along. De Beer concluded his thesis in an echo of Owen - homology is dissociable from the pathways of embryonic development. Yet there was a century of science between these two men, a century that had believed otherwise simply because of the convicting power of what Darwin said. The discovery made no difference to the acceptance that homology was caused by common ancestry however. How could that happen? We need to unravel this some more.

De Beer did not relinquish Darwinism although after his analysis homology was without a mechanism. He wrote: "The fact is that correspondence between homologous structures cannot be pressed back to similarity of position of the cells in the embryo, or the parts of the egg out of which the structures are ultimately composed, or of developmental mechanisms by which they are formed...homologous structures can owe their origin and stimulus to differentiate to different organizer-induction processes without forfeiting their homology."[15]

In other words homology is true already and evolution is a fact already, and neither can be changed by the evidence, not even by the evidence of his own experiments disproving them. This is not empirical science, it's naturalistic philosophy with all engines firing (§19). The purpose of de Beer's experiments in his mind, like the purpose of evolutionary biology since Darwin, is to explain how evolution happened. That's why incompatibilities of the data as blatant as these are not taken as a rejection of homology or even of macroevolution, but as even a further illumination. If scientific theories are meant to be tentative, this one is not (§21.1). Evolution remains true because the alternative is supernatural design, and that answer is always scientifically false (§3.1;§18;§19;§20).

What can we conclude so far? Aristotle's simple observation twenty-three centuries ago turns out to have been deceptively simple. And yet what we inspect here in sameness across biology is an utterly basic, in-your-face observation about the real world. Look out of the window, down at your pets, up in the sky or take walk in a park. You cannot help but see patterns of similarity across otherwise very different forms of life. The similarities arise in non-similar ways and are caused by non-similar genes and non-similar developmental mechanisms. We are confronted with an unexpected and seemingly intractable and irreducible problem. How do we explain it with an empirical naturalistic mechanism? Come on! We see what the science has done to solve it next.

§7.4   PUTTING HUMPTY HOMOLOGY TOGETHER AGAIN

Do we understand what the problem is? Dissimilar developmental mechanisms and genes, generate similar biological structures. From idiosyncrasy comes consistency. Although there are conserved genes and conserved developmental pathways across life, these are unable to account for the conserved structures we speak of because they are not causal (§9.3.5).[44] The attempt to uncover the cause, to reconcile genetics and embryology with biological similarity in evolution, is the mission of a whole new discipline in biology called evolutionary developmental biology (§9.1). It's a mission so far unmet. The problem is shared genes and shared developmental pathways are the essential evidence needed to invoke shared ancestry as reason for the shared structures, but an empirical basis for the notion has been absent since a day before Darwin (§7.3;§9.3.5). The problem is insuperable because if structural similarity has not arisen by similar developmental pathways or genes, evolution has no mechanism to offer because evolution operates obligately through genes, and structures arise by development.

How is homology explained in the science today then? The dissociation between genes and homology is considered to be due to polygeny and pleiotropy. The reason common ancestry is taken to explain the shared patterns is because structural similarities across taxa are striking, and since evolution must have been their cause, it's much more likely that they evolved only once and were inherited, than each one independently multiple times over to arrive at such a similar end result.

This explanation might seem more satisfying but looking closer even that is not a causal explanation. Why? Because a structure in an ancestor cannot "cause" a structure in a descendant species even if it is identical. The reason is homologous characters are not inherited, genes are inherited. We need to get from gene to structure by development and back to gene again, for it to be ancestrally caused. We need evidence of a naturalistic mechanism, an empirically testable mechanism, for the observation of biological similarity. But instead of evidence comes silence.

In short and to make a circle to Owen and even further back to Aristotle, let me tell you if it's not clear already. Biological similarity is not evolutionary. This is not a philosophical view and it's not a religious view. It's the empirical view. It's bald declaration of the evidence of molecular and developmental biology. The implications are immense. But first, what happened to Darwin's evolutionary definition of biological similarity as the evidence of common ancestry?

It remains mainstream thinking in biology. Homology has been studied either from the historical phylogenetic perspective or from the embryologic developmental perspective. We must consider them both. In light of what we've discovered from embryology and genetics rejecting homology as inherited, if evolution remains sovereign can they be put together with homology again?

§7.4.1 Phylogenetic Approaches to Understanding Homology

Although the evidence rejects Darwin's embryologic explanation for homology, it turned out his notion of biological similarity as evolutionary was not in the least bit threatened by de Beer's discoveries. Instead the problem was taken to lie with homology! In the end De Beer concluded homology was "an unsolved problem" because embryology had been dissociated from common ancestry and even entitled his grand treatise with these words. But if his title and his evidence affirmed an empirical and logical deadlock, at least he was convinced otherwise. For although he had no idea of how biological similarity could happen, he was certain it came from evolution somehow. He said: "it is homologous organs that provide evidence of affinity between organisms that have undergone descent with modification from a common ancestor, i.e. evolution." [45]

In other words homology was the evidence of the evolution that had caused it. But without any mechanism, how could he be sure? There is no way of course unless this tale of Darwinism is wagging the dog of homology in circular reasoning. The question then, is how that could happen scientifically? (§19;§20) If the plain history of homology as a concept in science will not convince you of how it happened, here it is from the lips our expert Brian Hall: "homology is such a central concept, second only to evolution".[40] In other words evolution is what determines the truth of homology which determines how the evidence of biology is to be interpreted as evolution. Now you're seeing it.

Alec Panchen at the University of Newcastle upon Tyne is another expert. He feels the same way. No need to pick on him either because as he says, "Most biologists associate homology with the fact of evolution."[6] But what does that mean, if the evidence does not demonstrate a mechanism by which common ancestry could cause it? Axel Meyer at the University of Konstanz in Germany gives the answer:

> "Homology describes the inevitable evolutionary phenomenon that the similarity of structures among different organisms is due to the commonality of their descent...Why should one be interested in homology? Because it is one principle, maybe *the* unifying principle, of evolution...The fact that homologous structures exist provides one of the strongest lines of evidence for evolution..."[46]

In other words since evolution is already true, common ancestry must be true and so biological similarity by common ancestry must be true. It must. As a result all observations of biology become interpretable from this perspective and all the evidence supports evolution. Evolution is a "scientific fact" (§3.1). Without a mechanism it's empty wishing of course, and knowing the evidence of biology roundly rejects the idea, it's worse. It's willful blindness (§7.3;§9.4). If the consequences of this absurdity do not already suggest themselves, just answer this question: What's the only way homology can end up on this view?

Here is Dr. Hall to give us the answer: "Despite the importance of homology as the 'hierarchical basis of comparative biology' (the subtitle of a book on homology edited by him) and its central place with evolution in biology, definitions of homology abound."[50] You see how it ends up? As editor of that multi-authored book written by a host of the leading lights in the field he concluded: "In many respects, however, homology remains in 1994 where it was 45 years ago when Szarski concluded: "After examining the present status of the concept of homology, one arrives at disquieting results. A basic term of one of the most important zoological sciences, that of comparative anatomy, cannot be exactly defined."[47,48] In other words homology is a central concept of biology, yet it is defies any definition compatible with all the data of biology. That sounds ridiculous? It is ridiculous. That's how homology ends up. Take a closer look at how this science works.

Sir Gavin considered his data demonstrating dissimilar embryonic development causing homologous structures as "unsolved" examples of "anomaly" (to the preeminent evolutionary homology paradigm of common ancestry), but never a disproof.[15] What about the contradiction before his eyes? What about his "question"? Namely:

"what mechanism can it be that results in the production of homologous organs, the same 'patterns', in spite of their *not* being controlled by the same genes? I asked this question in 1938, and it has still not been answered."?[39]

De Beer saw his problem not as resolving the contradiction, but as refining the criteria that could establish evolution and homology as present in the first place, and so biology has continued. Common ancestry remains the explanation of homology, without evidence of a mechanism.[13,49-54] Oh there are theories of mechanism, but these are just theories about evolutionary theory when what we need is empirical evidence and an empirical mechanism consistent with all the claims evolution is "a fact". For instance Ernst Mayr defines homology as: "A feature in two or more taxa is homologous when it is derived from the same (or corresponding) feature of their common ancestor,"[50] and Douglas Futuyma defines it as "the possession by two or more species of a trait derived, with or without modification, from their common ancestor."[51] But how do you demonstrate the common ancestry to demonstrate the evolution to demonstrate the homology? Round and round you go...But if you assume one of them is "a fact" already, then there is no problem. Then round and round you go finding more and more evidence, never realizing this is a world without reality. This is the world, and the worldview, of twenty-first century science (§3.1;§11.5;§18;§20;§26.3).

Common ancestry as cause of biological similarity is riddled with bedeviling contradictions, even assuming macroevolution is true. This may not be public knowledge, but it is to the experts. Here is Brian Hall: "'Homology' as a concept has become increasingly elusive during the course of the 20th century." When said in the same breath as it being the central concept of biology, we had better watch our heads.

To understand why homology could have become "increasingly elusive" with increasing evidence, and even more curiously, the reason for this staying completely unrecognized, is to appreciate the path homology has taken in its march of scientific progress backwards since Darwin. We know that path already.: Biological similarity that was Owen's homology in 1848 and a directly observable fact, was given a phylogenetic and embryological mechanism by Darwin in 1859. Although that mechanism was only a postulate, namely common ancestry though a shared anlage, it became the new definition. Later this was amended to a purely phylogenetic definition, and today that definition is the observation. Today only what is by (postulated) common ancestry is homologous, or to say this another way, believing is seeing. Like Emperor's New Clothes (§20).

Remember too biological similarity is not necessarily homology now. Only similarity caused by common ancestry is homology. Homology is to be distinguished from the similarity considered "analogy", which is similarity by virtue of structures in two organisms performing the same function that did not arise from a common ancestor. But how do you tell? The wing of a bird and the wing of a butterfly are taken to be analogous, but the wing of a bird and the hand of a man are homologous. The wing of a bird and the wing of a bat are analogous as wings (since evolution theory declares they did not develop from an ancestral winged ancestor as bird wings evolved later and independently from an ancestral forelimb), but they are homologous structures when considered as forelimbs. Biological similarity that is homology must be distinguished from other instances of biological similarity that are not due to common ancestry but are "homoplasy", and considered caused by convergent evolution, parallelism and conversion.

This is the reverse of Darwin's deductive method for homology. He inferred common ancestry from similar structures. Nowadays common ancestry is established first, then evolutionary homology inferred after. I told you. Believing is seeing. Evolutionary homology lives on without a mechanism like a banshee without a body. The reason this is still science is evolution is a scientific fact and therefore homology must be a fact also. The mechanism is the definition. Evolution can still be inferred and still be held true, because the divergence from the ancestor over time has also resulted in the erasure of the evidence. Evolution explains why there is evidence and it also explains why there isn't evidence. We saw this for the fossil evidence but don't take my word for it. Here's Rudolf Raff at the University of Indiana who will. He is editor of *Evolution and Development* and author of the seminal text *The Shape of Life* which explores the links between evolution and developmental biology. Don't worry be happy.

"Should all this be cause for despair? No. Biologists have historically used homologies to trace evolutionary histories and phylogenetic relationships. A deeper understanding reveals that this cannot be a tidy programme. That is on the face of it unfortunate, and has generated much handwringing on the usefulness of the homology concept. However, where developmental homologies are difficult to identify because of process shifts in ontogeny, we are actually being told interesting things about evolution. Ambiguities in development of homologues in embryos reveal where and how evolutionary changes occur and thus, although confounding, are difficult Rosetta Stones needed to understand how evolutionary novelties arise."[55]

What the "sameness" of homology is exactly, is not addressed either. As in Darwin's day it's merely a postulated biological history, an injection of time that shifts sameness back to an ancestral creature, without ever clarifying exactly what it is. With this strategy we should not be surprised if illusions get erected, in science, and then take on lives of their own. The question we asked and still waits, is what the evidence is that the working definition of homology as common ancestry is true? Ever since Darwin it's just been assumed. And homology is the basis for the detection of evolution. This is a profound observation. Homology is the evidence of evolution, and evolution is the evidence of homology. The concept of homology not only applies to extant life but fossils too of course (§5.3.3;§5.5.5). Without notions of homology, fossils are inconceivable as evidence for evolution. You merely assumed homology in our inquiry of *LOE*#2 and *LOE*#3 earlier (§5;§6). Ridiculous is what it is without homology however (§7.3).

This is how ancient biological artifacts displaying similar structures become evidence of common ancestry. Fossils show evolution by demonstrating homology. They have to, and the reasoning goes all the way back to Darwin. Arguments for evolution from his evolutionary interpretations of the separate fields of paleontology, embryology and comparative anatomy seemed to unify them all into an elegantly simple,

mechanistic scheme - common ancestry. This became the key to reconstructing phylogeny and reconstructing the course of evolutionary history and still is. In his words:
"all the organic beings, extinct and recent, which have ever lived, can be arranged within a few great classes...according to our theory...the best, and, if our collections were nearly perfect, the only possible arrangement, would be genealogical (p.599)...the arrangement of all organic beings throughout all time in groups under groups... the rules followed and the difficulties encountered by naturalists in their classifications, - the value set upon characters, if constant and prevalent, whether of high or of the most trifling importance...the wide opposition in value between analogical or adaptive characters, and characters of true affinity; and other such rules; - all naturally follow if we admit the common parentage of allied forms, together with their modification through variation and natural selection, with the contingencies of extinction and divergence of character." (p.609)

This how the ancient science of comparative morphology with a legacy at least as old as Aristotle became the science of phylogenesis after Darwin, and common ancestors replaced Owen's archetypes. Yet as Colin Patterson at the British Natural History Museum has pointed out, and he is an evolution advocate, these were "changes in doctrine rather than in practice: the conclusion about forelimb structure in the common ancestor of birds and bats was reached not by observing that ancestor, but by reconstructing it as an abstraction in the same way that Owen reconstructed his archetypes, from the homologous features characterizing systematic groups."[56] In other words this was a philosophical, not an empirical move by Darwinism (§19). Not that there's a problem with subjectivity. Convention is the basis of the Linnaean system (§1.1). The problem is Darwinism claims not to be subjective nor a classification of convenience. It claims to be explanatory and true and the pathway by which all life has come to be (§3.1).

Ever since Darwin comparative anatomists have concerned themselves with using homologous features, the supposed evidence of common ancestry, to infer phylogenetic relationships and thereby reconstruct the path that evolution took. This is how evolution becomes "a scientific fact". The question is how to identify these shared characters (homologies) to a reasonable probability? In 1952 Adolf Remane proposed three criteria: 1) topographic similarity (shared relative positions on a shared overall plan), 2) similarity in structural details between the shared characters, and 3) intermediate forms between them.[57] These criteria were later echoed by Leigh Van Valen.[58] Later in the 1950s the German entomologist Willi

Hennig further developed the methods into what is now called phylogenetic systematics, or cladistic analysis (§5.3.3). With the translation of his work into English in 1966, it became internationally regarded as the key to unlocking evolution, a status it retains still.[59]

Hennig argued that since evolution was the process by which an organism's form and behavior were transformed over time, the way to reconstruct evolution's path was to track evolving lineages by tracking the evolving particular characters on an organism. These are their homologies of course, the shared character traits between different organisms. Those homologies shared by different lineages he called symplesiomorphies, and those that were shared and derived from a common ancestor (and therefore representing a new lineage) he called synapomorphies (Fig.5.2). Which character is a synapomorphy and which is a symplesiomorphy thus depends on which taxon rank the comparisons are being made at, but since all ultimately inherited, all are ultimately synapomorphies at some taxonomic level. The point is that since homologies are ultimately evidence of common ancestry, they can be ultimately be equated to synapomorphies.

In 1982 Colin Patterson at the Natural History Museum in London reinforced this by idea providing three criteria that needed to be satisfied for the detection of homology: 1) similarity (i.e. topographic correspondence); 2) the absence of conjunction (two homologous structures cannot be found together in a third creature); and the most important criterion which was 3) congruence with other homologies. Congruence was the only way to discriminate homology (conservation - common ancestry) from homoplasy (convergence). How is that done exactly? He explains for us: "This test depends on the equivalence of homology and synapomorphy, with the corollary that homologies specify groups that are rendered monophyletic by them. In testing a proposal of homology by congruence, one checks the distribution of the feature (what species does it occur in and group together?) against the distribution of other supposed homologies...A true homology will circumscribe a group that is congruent with those specified by other homologies."[52]

It is to reinforce the importance of Hennigian cladistic analysis then. Cladistics can be applied to any trait, living or fossil, morphologic or molecular. We saw its importance for fossil analysis earlier (§5.3.3), and will again (§10.2.3). Cladistic analysis is directed toward providing the branching evolutionary relationships of all extant and fossil life, ultimately to the common ancestor. Features are organized into monophyletic groups on trees to obtain the best fit using the concept of parsimony (i.e. the hypothesis requiring the fewest assumptions). The nodes of the analysis are empty because the creatures there are hypothetical. And as the number of branches and taxa increases, so the number of possible solutions multiplies rapidly. For 10 taxa this is 34 million possibilities and for 20 it is an overwhelming 10^{21}, only one of which can be true! We should take a closer look at this methodology.

While the statistics of cladistics are rigorous and the rules of analysis of characters are rigorous, cladistic analysis is a black box. The premier method of demonstrating evolution is not empirical. Why is it a black box? The methodology depends on a valid independent recognition of homologous characters for analysis. How do we know their recognition is valid? How do we know the evolutionary interpretation accorded them is valid? If these are false the whole analysis is false no matter how statistically "correct" the "answer". Answers are worse than wrong because they are biologically misleading. Manfred Laubichler at Princeton University describes this particular limitation of Hennig's theories: "But while his methods are operational in the sense that they allow for a logically consistent reconstruction of phylogeny, they still depend on a prior assessment of the sameness of characters, i.e., of homology. Hennig defined homology to include all the transformed states of a character. This definition, however, still requires independent criteria for the assessment of homology. Hennig employed a variety of methods to identify homologues, such as paleontological evidence, but he also heavily relied on Remane, whose homology criteria then became the auxiliary criteria in the context of Hennig's definition."[57] Is there anything apart from evolution that we can we be certain about, is the point and the problem.

The limitations and uncertainties arise because cladistics operates under three separate assumptions, all of which must be true if its analysis is to be true biologically. First, homologous characters are assumed to have been correctly identified for analysis. Second macroevolution is assumed to be true to permit the analysis in the first place. Third the most parsimonious hypothesis is assumed to be the path that evolution took. Concentrating on just the first of these assumptions, the problem of subjective determinations of ancestry is readily declared as homologies and synapomorphies that at first might be convincing, can on analysis later be seen to be homoplasies merely by using additional or alternative traits in the character matrix that is analyzed. Sometimes adding just one additional character to a matrix is all it takes to convert a homology to a homoplasy or vice versa. This is a shocking discovery. It's worth hearing it from a few more of the experts to be sure.

Rudolph Raff says "interpretation of shared derived features is not always an unequivocal matter. Two anatomists may well interpret the same feature in diametrically different ways. The consequence is that features may be falsely homologized and incorrect inferences made about directions of change...Finally, character conflicts are inevitable: one character is consistent with one phylogeny, but a second character is consistent with a different phylogeny. When this occurs, there is no a priori way to test which character is correct. One has either to depend on parsimony, that is, to consider the tree supported by the greater number of characters to be correct, or have some independent means of testing characters. There is often no way to weigh features objectively."[60]

David Mindell at the University of Michigan and Axel Meyer at the University of Konstanz in Germany describe the use of homology like this: "Initial hypotheses of homology are used in constructing character matrices for phylogenetic analyses, and are generally based on similarity in features, such as structure, position and function for morphological and molecular traits, or assessment of similarity in preliminary alignments for molecular sequence traits...These initial hypotheses can be revised following phylogenetic analysis, with homologs denoted as shared derived characters for various monophyletic groups when character changes are plotted on trees. Subsequent hypotheses of homology are used, implicitly or explicitly, in nearly all comparative analyses of evolutionary patterns (phylogeny) and processes (mechanisms of change). Conventionally, natural selection has been inferred where nonhomologous similarity for traits in different taxa is associated with shared environmental and ecological constraints. Major limitations of the comparative approach include taxon-sampling problems, uncertainty associated with phylogeny and lack of knowledge about historical change in environmental influences."[13]

Lastlyhere is Mark Siddall at the American Museum of Natural History: "Ancestors can never be distinguished from sister taxa without putting a premium on negative evidence...The reason why ancestors cannot be distinguished from sister taxa is that a phylogenetic analysis which happens to include an ancestor can only be thought to do so by virtue of lack of any difference between it and the branch to which it connects; it lacks automorphies. If putative ancestor-X, has an autopomorphy, it is seen as distinct from the ancestral internode from which it derives. It cannot be an ancestor, since it does not reside on the internode. If it lacks autopomorphies, it could be postulated as an ancestor, but this would be due to the lack of an autopomorphy (that is *lack* of evidence). Thus there can be no *positive* evidence of ancestry. Lack of evidence can only allow it as a possibility."[61]

How is common ancestry demonstrated then? It's always a hypothesis and it's done using some other criterion, generally fossils, then inferring the relevant structures as homologous afterward. But this is still to use homology to find homology using evolution, then turning around and calling it evidence. Absurd is what it is. If this is not clear, retrace the steps and the sequence of the reasoning required. An inference (that common ancestry explains the similarity) rests on a second inference (that a particular creature is the common ancestor) and a third inference (that a series of aligned fossil creatures is really related by ancestry) which rests on fourth inference (that macroevolution is a reality by extrapolating from microevolution). This is called science but it's silliness. This is not even to mention that common ancestors, the creatures at the nodes of cladistic trees, are hypotheses. Common ancestry is always a hypothesis scientifically. How then can homology be a fact? And if homology is the central concept of biology and homology is what demonstrates evolution, how can evolution be a "fact"?

We have still to address the other two assumptions of cladistics. If the discussion so far has not already demonstrated the hocus pocus here, is there perhaps a more direct way to assess the reality or unreality of Hennigian cladistics and systematics? Of course. We've seen biological similarities across taxa defy a naturalistic explanation. Cladistics proceeds by applying statistics to supposed homologous character traits, but if no naturalistic mechanism can explain biological similarity, any analysis of inferred homologies by inferred evolution has no biological reality. Why does it yet thrive in science? First because evolution is a fact by a law in science called The Law Emperor's New Clothes (§20), and second because whatever solution cladistics finds, it can be explained in the evolution paradigm. It's either homology or its homoplasy and evolution does both.

However appealing the Neo-Darwinian view homology is from shared ancestry, it rests entirely on the assumption macroevolution is true. It cannot therefore be taken as "evidence" for evolution without employing circular reasoning. Those making the case for evolution with arguments from homology, including fossils, ignore this inconvenient truth like a frisky hound its leash. Absent a mechanism "evolution" is merely a metaphysical interpretation of an observation of biological similarity, like Owen's typological notion of archetype, but with an important difference. This interpretation of homology is claimed to be mechanistic and true.

We saw earlier how Brian Hall summed up the situation, that homology "cannot be exactly defined" but we did not hear how he ended. This is how: "It does not therefore appear possible to have a single concept or definition of homology that embraces both pattern and process." Knowing the evidence we knew this was coming. All the same, it's the admission of a complete failure by evolution science to explain the "central concept of all biology". We are worse off than we were with Owen. This science has gone a very long way backwards.

Ingo Brigant is a philosopher of science at the University of Alberta. He weighs in on the phylogenetic approach to homology. "Is the phylogenetic approach, which makes reference to common ancestry of characters, able to give a better definition than ideal morphology with its notion of the archetype? A characterization of homology as resemblance or similarity to be explained by common ancestry might use a comparable notion of similarity as phenetic approaches. Homology now becomes a subset of similarity - that similarity with a specific phylogenetic background. But this does not give an account of how the relevant kind of resemblance is to be understood. Instead of giving a definition of homology such accounts are better viewed as expressing the tenet that homology, which somehow relates to a specific type of perceived similarity, is rooted in common ancestry. Thus homology becomes a part of evolutionary theorizing...As homology is a relation between parts of individuals, we ultimately need an account of homology that defines when a morphological structure of an individual and a part of its progeny are derived from one another. True enough, everyone perceives these characters as perfectly similar and in fact homologous. But this does not yield an account of what constitutes homology...Phylogenetic approaches to homology do not explicate what the same character in ancestor and descendant is. For this reason, they do not offer a definition of homology that is clearer or more explicit than that of idealistic approaches."[62] Hocus pokus, mumbo jumbo is what it is, but even he can't say it. The reason is he believes in evolution as much those who disbelieve him about the coherency of phylogenetic approaches to homology.

In probably the best indication of the bankruptcy homology is now taken to be different on different levels of analysis. Homology is thus scientifically speaking, now you see it now you don't. W. Joseph Dickinson wrote this assessment of homology for *Trends in Genetics*: "there is not necessarily a simple relationship between homology of molecules (or even pathways) and homology of the anatomical features in whose development those components participate. In other words...molecular similarities in the developmental mechanisms that produce specific organs are not, by themselves, strong evidence for homology of those organs. The central point...is that questions of homology can be examined at multiple levels and that homology between a pair of structures can simultaneously be present at some levels but absent at others."[44]

Along with other esteemed experts in the field Louise Roth at Duke, Axel Meyer and Ehab Abouheif at Stony Brook, Peter Holland at the

University of Reading and Rudolf Raff at Indiana University, he provided the current consensus statement for the field: "homology at one level does not necessitate homology at another."[53] Homology now refers independently to molecules (genes and proteins), developmental origins (embryology) and structures (anatomy). This is not a surprise knowing the data because the data shows them irreconcilable. And all this was once a simple elegant hypothesis in *The Origin*. Since homology and evolution are both true already, homology can independently now refer to the same structure (1), the same phylogenetic origin (2), the same molecules (3), the same developmental origins (4), the same developmental constraints (5), the same genetic information (6) and the same epigenetic landscape attractors (7). If this seems convoluted and contrived you're not being scientific. You're the kind of person Richard Dawkins calls...Theodosius Dobzhansky calls...and Stephen Jay Gould calls... Well, you know. Is that a naked buttock cheek I see, or an Emperor's robe? Perhaps you have to be a child to tell. Can we perhaps understand homology better another way?

§7.4.2   Developmental Approaches To Understanding Homology

While the canon of homology is that it's caused by common ancestry, how exactly that is effected is unclear. Since structures have to arise by development, embryology would seem a good place to look for evidence of evolution and we will (§9). For now however, can homology be reconciled in development with synapomorphy somehow?

Evolutionary homology requires shared genes and shared development to invoke a naturalistic mechanism operating by inheritance. In 1984 Duke University zoologist Louise Roth acknowledged this imperative in her criteria by which homology was to be diagnosed: "A necessary component of homology is the sharing of a common developmental pathway. Homologues must, to some extent, follow similar processes of differentiation which, one infers, depend on the same batteries of genes."[63]

We know this wish is a pipe dream (§7.3). She abandoned the demand in an about face four years later, affirming "homologues...do not always develop in a similar fashion."[10] The imperative for a plausible heritable mechanism remained nonetheless. Enter Günter Wagner at Yale and his idea of "biological homology".

He points out that developmental sameness as cause of homology is rejected by the evidence of embryology and genetics and the goal of his biological homology is to retrieve a logical explanation for this (seeming) paradox. Namely "to explain why certain parts of the body [homologues] are passed on from generation to generation for millions of years as coherent units of evolutionary change." In other words the task of research is to retrieve evolution from the evidence of biology, after assuming evolution and common ancestry a fact first. Biological homology assumes causal common ancestry as a starting point, and therefore rests entirely on phylogenetic concepts of homology. If that notion is false his ideas are even more at sea. Biological homology is not a different point of view then, just a different focus. It focuses on what keeps homologous structures the same in evolution rather than what causes them to change by evolution. We know what causes structures to change by evolution. It's natural selection and random mutation. What keeps them the same?

Two mechanisms have been postulated. The first is stabilizing selection effected by the environment. In other words deviation is penalized by natural selection. The second is the action of heritable so-called "developmental constraints". Wagner regards homologous structures as "quasi-autonomous parts" and postulates that sameness is due to "morphostatic" mechanisms that are different to those generating the structures, and which instead are the factors inherited by common ancestry. He postulates these mechanisms, the developmental constraints, are currently hidden from biological science and need to be uncovered. This is his hypothesis: "If the maintaining mechanisms are affected by genes other than the generative ones, then developmental pathways can vary while the adult pattern is kept the same by the compensatory action of the morphostatic mechanisms."[7]

This is back to front reasoning to invent forces to explain the paradox of the evidence. It would seem more logical to question why the evidence is a paradox but this never happens. Now that is a paradox! Dr. Wagner explains the research program of biological homology: "It is proposed that the structural identity of at least some morphological characters is determined by the stabilizing (morphostatic) mechanisms which maintain the characters, rather than by generative processes. This possibility implies a new research program in experimental morphology: to determine the contribution of generative and maintaining mechanisms to the structural organization of a character."[7]

What is homology from the perspective of biological homology then? "Structures from two individuals or from the same individual are homologous if they share a set of developmental constraints, caused by locally acting self-regulatory mechanisms of organ differentiation."[9] He reconciles the paradigm of evolutionary homology with the contradictory evidence from embryology by loosening the definition then. This is scientific progress going backward even faster than it already was, because his definition is inscrutable. Rather than clarifying the situation, his leads to even more questions. What are "developmental constraints"? When are they the same in different individuals? What are "self-regulatory mechanisms"? The answers are left unexplained and his definition is not even open to analysis, much less disproof.[64] By the way, the notion of "developmental constraint" already has a meaning, at least in the scientific literature if not in nature. Here it is defined by evolutionary biologist and Dean of Sussex University, John Maynard Smith: "A developmental constraint is a bias on the production of variant phenotypes or a limitation on phenotypic variability caused by the structure, character, composition, or dynamics of the developmental system"[65]

This is not helpful because what is needed from Wagner and his theory about evolution theory is a specification of what exactly this constraint is, to render a character homologous by his definition. For instance – a "bias on the production of variant phenotypes" – what phenotype? For Wagner's definition to have meaning it presupposes the two structures being compared are homologous already.[62] Why should constraints be shared at all and why can they be compared? The evolutionary notion of homology of structures and development is smuggled in is the point. In sum, the definition is an "empirically empty concept,"[66] to quote Olivier Rieppel at the Field Museum of Natural History in Chicago. This is because structural invariance is explained by developmental constraints, which is explained by structural invariance. The definition is not only intangible and empty, it's circular and for two separate reasons. The important purpose of the definition is seen to be in its corollary and of course that is what he was trying to wring out all along. If development is shared it makes homology evolutionary and then homology is no longer "an unsolved problem". But Wagner's claim is no advance. Oxford University marine biologist Sir Alister Hardy has the last word: "The concept of homology is absolutely fundamental to what we are talking about when we speak of evolution – yet in truth we cannot explain it at all in terms of present day biological theory."[67] Why is design so hard to see?

§7.4.3   Homology Today - Now You See It Now You Don't

Not that any of this has stopped attempts to explain homology within the evolution paradigm. If anything efforts have been redoubled.[42,68] This has had fascinating and amusing results because of the inevitable absurdity. For a start because of the impasse by which development cannot explain homologous features, biologists can't even agree about what homology is!

We saw this earlier. Homology is said to be seen from some perspectives, but not from others. It's now you see it now you don't. In his 1989 *Annual Review* paper Gunter Wagner complained "homology has a firm reputation of being an elusive concept".[9] Again, we speak here not of the arcane but an obvious feature of life, even to non-biologists and which even biologists say is "the central concept for *all* of biology."[11] Brian Hall summarized the hopeless intractability with his question: "Do we possess a single concept of homology transcending (uniting) all of biology?" and answers "No."[69] The answers to other questions he provided "[b]y way of summary and to indicate the compass of homology as viewed today,"[69] are given below. Look at the state of the science:

> 1. "Do we possess a single concept of homology transcending (uniting) all of biology? *No.*
> 2. Can a single definition of homology apply to all elements and all levels in the hierarchy of biological complexity? *No.*
> 3. If homology is defined differently at different levels in the biological hierarchy, then how do (or should?) such definitions relate to one another? *Opinions will vary on the answer to this one.*
> 4. Should the definition(s) incorporate criteria to recognize homology, and/or provide explanations of mechanisms through which homology can, or should be established? *No*
> 5. Should homology encompass explanations of proximate causation (i.e., ontogeny or physiology) and explanations of ultimate causation (i.e., phylogenetic change through evolutionary time)? *Again, opinions will vary.*
> 6. Must homology always be tied to knowledge of evolutionary origins (descent with modification) of the cells, organs, organisms, species, populations, communities under consideration? *Yes, but some would say no.*
> 7. Must homology always be tied to structure, even when behavioral or physiological characters are under investigation? *No, but some would say yes.*
> 8. Must homologous features always share common embryologic development, or can homologous features arise from nonhomologous (nonequivalent) developmental processes? *No and yes, but some would say yes and no*".[48]

The scientific explanation of biological similarity is so muddled it has become absurd. The science says it's because of the complexities of evolution. We know the real reason is much more simple than that.

§7.5   HOMOLOGY IS THE EVIDENCE FOR MOST OF THE
       EVIDENCE OF MACROEVOLUTION

Homology is fundamental to evolutionary biology. This is not because patterns of structural and other similarity across species are so an obvious feature of nature they demand explanation by evolution if evolution is true, but because the observation is itself the evidence of evolution. Homology is the evidence for most of the evidence of macroevolution. Botanist P.F. Stevens puts it this way: "without some similarity, we should not even dream of homology."[70] Here it is from Rudolf Raff: "Homology is the basis for all evolutionary comparisons. Phylogeny, molecular evolution, the evolution of developmental features, all can be understood only through comparisons in which homologies are correctly assessed."[71]

Biological similarity we all can see, is equated with synapomorphy and synapomorphies become the measurable evidence of evolution as input data for cladistic analysis, which in turn is the basic method for the demonstration of evolution as phylogeny and systematics. Round and round we go looking for evolution, and finding it every time. Biological similarity as evidence of evolution (i.e. homology) becomes ultimately how biology is understood. Dobzhansky never did exaggerate when he said: "Nothing in biology makes sense except in the light of evolution".[72] Evolution is an inference from biological similarity that becomes a scientific fact. Niles Eldredge spells out the reasoning in detail:

"Now, how do these simple suppositions establish the very fact of evolution? Simple. Looking around at the more than ten million species currently on earth, we see a nested pattern of resemblance linking them all. And that is exactly what we would expect to see given Darwin's "descent with modification". If all life has descended from a single common ancestor, there should be some features that have been inherited by absolutely all living things. And that's what we see. Fundamental biochemical pathways and, perhaps more graphically, the macromolecule RNA is common to all forms of life (DNA is nearly as ubiquitous, but is not found in some bacteria.)

Take any branch of life and you will see the same thing. We share hair, mammary glands, placentas, and three middle ear bones with other mammals; in our more remote past, we shared with birds and reptiles a special system of liquid-holding membranes in our eggs – the amniote egg. Reptiles, birds, and mammals share a four-leggedness with amphibians, and all these groups share a backbone with the various groups of fishes, and so and on, back into the dim recesses of time, all the way back to all of life sharing its RNA.

This detailed pattern of resemblance linking every living thing is the very best evidence that life indeed has had a long evolutionary history. We see in this pattern the outcome of a simple, natural process. Evolution is a thoroughly scientific proposition: if evolution is "true", we would predict that such a pattern must be there. It is."[73]

And so although homology is listed as only one of the lines of evidence for macroevolution in our catalogue, it's really the evidence for most all of the evidence. Take a look (Table 3.1). Rather than independent lines of evidence as we might have assumed, homology is demanded to interpret fossils (*LOE#2-3*), structural homology (*LOE#4*), molecular homology (*LOE#5*), molecular evolution and systematics (*LOE#6*), comparative embryology (*LOE#7-10*) as well as human evolution *LOE#11*, and which is really a special case of *LOE#2*, *LOE#4* and *LOE#5*. So much for there being consilience. It all comes down to deciding where homology comes from then. And we discover there's no naturalistic explanation. Well, well.

Homology as representing biological similarity by common ancestry is an unsolved problem in science. Today there are multiple "levels" of homology and multiple concepts of homology. An obvious question follows. Why is homology taken as evidence of common ancestry when there is no empirical support for a naturalistic mechanism? Because similarity in biology is obvious to everyone and evolution is already a fact. The power of the paradigm is even taken as the evidence it's true. David Mindell and Axel Meyer in reviewed homology for *Trends in Ecology and Evolution* to summarize the state of the art: "Although first used to describe 'the same organ under every variety of form and function', most biologists now use homology to denote hypotheses of evolutionary relationship among traits of organisms. This is justified by the fact that Darwinian evolution has replaced fidelity to archetype as the explanation for similarity among traits in different species."[13]

But where is the evidence? Homology *is* the evidence! Homology is equated with synapomorphy, by definition. We make an important observation. If homology has no evolutionary mechanism neither do synapomorphies. If this is so the great power of cladistics that underpins much of evolutionary biology is without a biological reality either.

Alec Panchen at the University of Newcastle has a very different take on the claim made by Mindell and Meyer above. He says:

"Most modern biologists associate homology with the fact of evolution. Homologous structures in two different organisms are so recognized because it is proposed that the nearest common ancestor of both had a corresponding structure: indeed definitions of homology frequently involve community of descent. Historically this lacks any backing."[6]

Making opposite claims as they do needless to say, at least one of these points of view must be wrong. You can choose.

§7.6 EVOLUTION EVIDENCE LYING IN THE EYE OF THE BEHOLDER

We are left with having to explain sameness in biology ourselves. If it's not evolutionary then it must be by design, just as it appears (§1.3). If there's a single line of evidence that demonstrates evolution is bankrupt, it's the inability to explain biological similarity by a naturalistic mechanism. If there is a single line of evidence that demonstrates the science is deluded to perversity, it's the way this is denied.

This suppression of the truth about homology leads to all kinds of webs and tangles when the evidence must be engaged. Absurdity always follows when truth is held hostage. Look at two textbook examples yet everywhere taken as classic instances of homology to see this. The question for you as we look is: What's the evidence really?

§7.6.1 A Hands-On Look At The Evolutionary Hand

We return to the pentadactyl limb with which we began this chapter. Mark Ridley's textbook *Evolution* uses it as evidence for evolution just as Darwin did, only he does it as if de Beer's data didn't exist (§7.3). No need to single him out as others do too (§25).[74-77] There are still more problems though, but first listen to the experts make the arguments for evolution from homology (*LOE#4*). Dr. Ridley is here to take us by the hand. He says:

"The evolutionary explanation of the pentadactyl limb is simply that all tetrapods have descended from a common ancestor that had a pentadactyl limb and, during evolution, it has turned out to be easier to evolve variations on a five-digit theme, than to recompose the limb structure...If species have descended from common ancestors, homologies make sense; but if all species originated separately, it is difficult to understand why they should share homologous similarities. Without evolution, there is nothing forcing the tetrapods all to have pentadactyl limbs."[78]

Do you believe him and *LOE#4*? Here are four reasons not to:

(1) The "it has turned out to be easier" hearsay reasoning is no mechanism. It's as metaphysical as Owen's and even recalls Darwin's mocking "just-so" arguments, "we can only say that so it is", that also were without evidence (§7.2).

(2) As regards the postulate of an ancestor with 5 digits from which the pentadactyl pattern was "simply" inherited, Dr. Ridley forgets to tell us most ancient tetrapods had seven (*Icthyostega*) or eight (*Acanthostega*) digits,[79] subsequent tetrapods almost never have five and there are no intermediates to link the various limbed creatures claimed to be related to a common ancestor. Why was this not worthy of mention? (§25)

(3) His argument it is "difficult to understand" homology any other way is a rhetorical device irrelevant to a search for scientific truth and his last sentence ignores counter-arguments to evolution made for over one and a half centuries. Whether he disagrees with them or not is his privilege and his right, but they still need to be acknowledged and addressed, not ignored especially when they explain the evidence. The reason he and his colleagues do of course is because evolution is a fact and they know it, and such arguments about design are taken to be religion which is always false because it invokes a supernatural, and a supernatural is always objectively speaking false (§1.1;§19;§21;§23). But design is exactly what might be "forcing the tetrapods all to have pentadactyl limbs" and the pattern of ID is all around. Is there evidence to support the notion of ID apart from the absence of evidence of any natural cause of biological similarity? We will need to see (*Part III*).

(4) No attempt is made to provide a mechanism for the assertions about evolutionary homology even though they conflict with the embryological evidence. This is what is demanded, a naturalistic explanation with evidence for evolution, to defeat the alternative explanation that is ID.

What is the fossil evidence regarding the supposed homology of the vertebrate limb? Consider the link between fish and amphibians. While the fish fin and the amphibian leg are considered homologous, the problem for evolution is both appear abruptly in the fossil record without linking transitional forms (§5.5.5). As Harold Booher has quipped, "The closest fossils linking a fin and a leg come from a fish with a fin and an amphibian with a leg".[80]

Even more importantly, although the pentadactyl structure is shared there is no mistaking a fin from leg. Crossopterygians and early amphibians of the genus *Ichthyostega* are considered the closest links between fishes and amphibians because of similarities in teeth and skull bones but a crucial difference between them is ignored. Fin fossils are not connected to the bones of the vertebral column and do not have elbow/knee joints to permit gait. Legs on the other hand do. Amphibian fossils have elbow/knee joints and the limb is connected to the vertebral column by articulating pectoral and pelvic bones.

The notion of an evolutionary ancestor is a problem for another reason. Most ancient tetrapods had seven and eight digits[79] and thus "[t]he significance of polydactyly lies elsewhere than in evolutionary origins".[81]

There is still another problem. While there is intuitive appeal in an evolutionary explanation of the similarity of fore or hind limbs across species, what of explaining the evolutionary cause of the correspondence between fore and hind limbs within the same animal? They are as similar structurally to each other as one might make with another species, and as Owen pointed out in an elegant set of drawings in 1848 similar to Fig.7.3. How did they arise from an ancestor, when as de Beer pointed out, "forelimb and hindlimb cannot be traced back to any ancestor with a single pair of limbs"?[82]

It comes as rather a shock to find Rudolf Raff's assessment back in 1996 remains true. "The basis for that commonality is still not resolved."[83] The similarity of repeated structures like limbs along the body axis is what Owen called serial (or iterative) homology. How is this kind of homology resolved for evolution? Rudolf Raff again: "Unless we choose to believe that the first terrestrial vertebrates leapt up on the land like fishy kangaroos, we do not suppose that the forelimbs are homologous to the hindlimbs because they evolved from them."[83]

De Beer called serial homology a "misnomer, because it is not concerned with tracing organs in different organisms to their representatives in a common ancestor."[82] So the absurd conclusion limb homology in the evolution paradigm leads to, gets post hoc resolution simply by changing the definition. Fore and hind limbs are not homologous!

The same logic applies to other iterative structures that look similar.[9] Two hairs on an animal have a different cause for their similarity than two hairs compared across two species. This seems contrary to common sense. Fore and hind limbs on the same creature are instead considered "homoplastic", analogous in that they may perform similar functions, but not related to a single common ancestor.[2] The evolutionary consensus is they evolved independently from pectoral and pelvic fins in gnathostomes.[81] The scientific explanation of the supposed homology of the pentadactyl limb is a fish story therefore, and in more ways than one.

327

§7.6.2   An Earful of Jaw

In the middle ear of all mammals are three tiny bones arranged as a chain of delicately poised levers. They amplify vibrations of the eardrum from sound waves and transmit them to the fluid-filled cochlear of the inner ear as a pressure wave. Without this impedance matching system only 0.1 % of sound in air would become sound in cochlear fluid. Named for their characteristic shapes, at the one end is the malleus or hammer which is attached to the eardrum, at the other end is the stirrup or stapes which in the human is the smallest bone in the body with a foot plate about 2 mm in size that moves in and out of the cochlear like a plunger, and between them both is the incus or anvil. The malleus and incus supposedly evolved from two jaw bones of a reptile, respectively the quadrate bone of the upper jaw and the articular bone of the lower jaw, and the stapes evolved from the gill arches of fish as did the jawbones of all the vertebrates.

These claims have undisputed standing in science. "One of the most celebrated cases of embryonic homology," is what Swarthmore College biologist Scott F. Gilbert says;[84] "one of the most demonstrative examples of how comparative anatomy can determine homology of structures inherited from common ancestors in evolution,"[85] says Sir Gavin de Beer and Stephen Jay Gould gushed: "we have no finer story in vertebrate evolution."[86] This should be a treat then if we look more closely.

So gather around because Dr. Gilbert is here to tell how a fish bone became an ear bone. He lectures directly from his beautifully illustrated and currently the pre-eminent text of the field, *Developmental Biology* (available for free through PubMed). Follow along with his drawings:

"First, the gill arches of jawless (agnathan) fishes became modified to form the jaw of the jawed fishes. In the jawless fishes, a series of gills opened behind the jawless mouth. When the gill slits became supported by cartilaginous elements, the first set of these gill supports surrounded the mouth to form the jaw. There is ample evidence that jaws are modified gill supports. First, both these sets of bones are made from neural crest cells. (Most other bones come from mesodermal tissue). Second, both structures form from upper and lower bars that bend forward and are hinged in the middle. Third, the jaw musculature seems to be homologous to the gill arches of jawless fishes.

But the story does not end here. The upper portion of the second embryonic arch supporting the gill became the hyomandibular bone of jawed fishes. This element supports the skull and links the jaw to the cranium. As vertebrates came up onto land, they had a new problem: how to hear in a medium as thin as air. The hyomandibular bone happens to be near the otic (ear) capsule, and bone material is excellent for transmitting sound. Thus, while still functioning as a cranial brace, the hyomandibular bone of the first amphibians also began functioning as a sound transducer. As the terrestrial vertebrates altered their locomotion, jaw structure, and posture, the cranium became firmly attached to the rest of the skull and did not need the hyomandibular brace. The hyomandibular bone then seems to have become specialized into the stapes bone of the middle ear. What had been this bone's primary function became its secondary function.

The original jaw bones changed also. The first embryonic arch generates the jaw apparatus. In amphibians, reptiles, and birds, the posterior portion of this cartilage forms the quadrate bone of the upper jaw and the articular bone of the lower jaw. These bones connect to each other and are responsible for articulating the upper and lower jaws. However, in mammals, this articulation occurs at another region (the dentary [mandible] and squamosal [temporal] bones), thereby "freeing" these bony elements to acquire new functions. The quadrate bone of the reptilian upper jaw evolved into the mammalian incus bone of the middle ear, and the articular bone of the reptile's lower jaw has become our malleus. This latter process was first described by Reichert in 1837, when he observed in the pig embryo that the mandible (jaw bone) ossifies on the side of Meckel's cartilage, while the posterior region of Meckel's cartilage ossifies, detaches from the rest of the cartilage, and enters the region of the middle ear to become the malleus. Thus, the middle ear bones of the mammal are homologous to the posterior lower jaw of the reptile and to the gill arches of agnathan fishes."[84]

It sounds so great and evolution sounds so amazing, but how can we tell this story is not like the one about how the leopard got its spots?[87] By looking at the evidence. So let's look.

The problems start with his first sentence. Vertebrates are distinguished from other creatures by possession of a bony skeleton including a skull. The oldest fossil vertebrates are conodonts dated to the Cambrian period. As members of a group of organisms called agnathans, they did not have jawbones. Contemporary agnathans are the lampreys and hagfish. Vertebrates with jaws are found later in the fossil record than agnathans, and are therefore considered to have evolved from them in a derived clade called the Gnathostomata. They consist of placoderms, chondrichthyes (or cartilaginous fish – the sharks and rays), osteichthytes (bony fish) and the Tetrapoda (which includes us). We've hardly begun and we have the first problem.

What's the evidence the vertebrate jaw evolved from a gill arch? For that matter, what's the evidence it evolved from an agnathan ancestor? There are no fossil intermediates between agnathans and gnathostomes. We reviewed the fossil arguments for macroevolution already (§4;§5). We know that to interpret fossils as evolutionary demands common ancestry be a fact already in circular reasoning. How can we tell this fish bone evolution story happened? Here is Colin Tudge writing in his review of the grand sweep of nature called *The Variety Of Life*:

> "Tradition has it that jaws arose from the first three sets of gill arches, each of which, in fish, are compounded from several different bony elements...There are problems with this. For example, it seems that the bones of the gill arches of gnathostomes are not homologous with the bones of the gill arches of agnathans: gnathostomes evidently replaced the original set with a new set. Neither are there appropriate fossils to show us the transition. [Why then should we believe they are ancestrally related but for the "commonsense" of naturalism?] But at least at a commonsense level, the idea is obviously plausible. [OK] There certainly seems *some* connection between gill arches and jaws."[88]

So if we are to believe that jawbones became ear bones, we must first believe that gill bones evolved into jawbones. If you can do that with this much evidence, the rest is going to be easy.

The second problem concerns that "commonsense" assumption, read *a priori* evolutionary interpretation of nature that Dr. Tudge just mentioned. Gill arches and jaws are homologous so the story goes, and the former evolved into the latter. Dr. Gilbert told us the three lines of evidence for this, evidence Stephen Jay Gould has called "multifarious and overwhelming".[86]

The first is that since jaws and gills develop from the same embryonic tissue (the neural crest), they have common ancestry. But we know biological similarity, let alone common ancestry as its cause, can be dissociated from a similarity in development (§7.3). How can shared embryology now be used to infer common ancestry?

The second line of evidence is that gills and jaws are structurally similar, specifically "both structures form from upper and lower bars that bend forward and are hinged in the middle". This would seem an interpretation, not evidence.

The third line given by Dr. Gilbert is circular as it stands, but what he means is muscles attaching to particular branchial arches (the embryonic structures that develop into the pharyngeal region in all vertebrates), can be correlated in fish, reptiles and mammals. I might add they also have similar nerve supplies. We review this later (§9.2.) The parsimonious naturalistic explanation of this complicated and shared pattern of development by such different creatures is that it evolved once and was inherited from a common ancestor, rather than evolving separately and repeatedly to arrive at such a striking similarity across different taxa. These two options exhaust the field of naturalistic explanations. The solution of science is it was the former, but again it is to infer common ancestry because of common embryonic development. So the third line of evidence is really another case of the first line of evidence, and just as empty.

In summary we discover if we are to believe jaw bones became ear bones we must first believe gill bones became jawbones. We discover the evidence for that idea is circular reasoning and just a structural similarity. Namely the bones themselves, their muscle attachments and their nerve supplies. We have spent the bulk of this chapter discovering that not only is there no evidence to explain biological similarity as caused by common ancestry, there's no evidence for any natural cause. If you think a jaw bone becoming an ear bone by evolution is already not an earful of jaw, listen a little more because we're still not done.

The third problem has to do with the evidence invoked for the claim that the fish hyomandibular bone evolved into the reptile and mammalian stapes bone. Both bones develop from the second branchial arch of the embryo. The fish hyomandibular bone serves to brace the jaws against the skull, permitting a strong bite while allowing the jaws greater flexibility than if the upper jaw was fused to the skull, as in reptiles and mammals. What's the purpose of that flexibility, you ask? It permits the fish oral cavity to expand and contract as it pumps water into the mouth and out over the gills for respiration. The reptile stapes is a single middle ear bone, and the mammalian stapes is one of three bones in the middle ear used to transmit sound waves to the inner ear. The hyomandibular bones and the stapes are indisputably correlated by an embryonic similarity in coming from the same part of the embryo and they have correlating muscle and nerve supplies too (§9.2). How structurally alike they really are to infer morphologic similarity, I leave for you to decide. Regardless, to conclude that a similar pattern of development means they have a common ancestry as cause is an evolutionary inference divorced from evidence as we have already seen, and will see again (§7.3;§9.2). The hyomandibular bone is fused just anterior to the otic (ear) region of the fish skull and has nothing to do with fish hearing.

As an aside how a tiny fish can hear at all, being so close in density to the water it swims in, (mean density of fish flesh 1.075g/ml and sea water 1.026g/ml), much less hear directionally (the speed of sound is five times faster in water so foiling the tetrapod solution of determining this by comparing the sound arrival times at each ear), are fascinating physics problems that evolution solved all on its own.

As another aside, fish also have hearing bones called Weberian ossicles, one on each side of the body, that transmit vibration from the gas filled swim bladder to the inner ear as it resonates to vibrations in the surrounding water. This pattern of ear construction involving ossicles even in fishes, and so more biologic similarity across vertebrates, is lost in the forest and fanfaronade of the ear evolution story. The evolution scenario contends that over time the brace function of the hyomandibular bone was replaced by a hearing function and changed into a stapes. What is the fossil evidence?

The earliest tetrapod skull known in detail is that of *Acanthostega* in work done by Jennifer Clack at Cambridge University. Writing in *Nature* concerning this fossil creature she said: "The stapes is likely to have had some auditory function because of the close association between the footplate [a part of the stapes] and the otic [ear] capsule"[89] Considering this to be convincing evidence of ear evolution, Stephen Jay Gould cited her statement and explained its significance to his readers:

> "Such a multifarious bone nearly bursts with evolutionary potential. The stapes may have braced for a hundred million years, but it worked for respiration and hearing only in an incipient or supplementary way. When the cranium later lost its earlier mobility [in fishes] and the braincase became firmly sutured to the skull as occurred independently in several lineages of terrestrial vertebrates – the stapes, no longer needed for support, used its leverage and amplified a previously minor role in hearing to a full-time occupation."[86]

Again we ask what is the fossil evidence? The stapes in *Acanthostega* articulates not against the otic wall of the brain case like the hyomandibular of fishes, but against the fenestra vestibuli (the inner ear) as in all other tetrapods. Where's the evolution? Dr. Clack wrote more recently that: "In its otic region, *Acanthostega* is fairly typical for an early tetrapod, and for this comparison, almost any early tetrapod would serve."[90] In other words this is not a transitional ear, this is a typical but anciently dated fossil tetrapod. As she explained in the same paper:

> "There still remains a morphologic chasm between the construction of the ear region as seen in tetrapod like fishes (tetrapodomorphs) and even the most primitive tetrapods with limbs and digits...A similarly large disparity lies between the otic morphology seen in the earliest tetrapods and that of the earliest amniotes. Although it is still difficult to envisage intervening stages between these, some functional

correlates may be invoked to explain how the ear region subsequently changed as it did [because evolution is already true of course]."[90]

In summary if we are to believe that jawbones became ear bones, we also need to believe that the hyomandibular of fishes evolved into the stapes bone of reptiles and mammals without any direct evidence of that either.

We now know how the stapes supposedly evolved, but what of the two other inner ear bones of mammals? All reptiles whether fossil or living have just one middle ear bone - the stapes, and all mammals have three. The malleus attaches to the eardrum and the incus articulates between the malleus and the stapes. The three bones form chain across the middle ear to transduce and amplify the mechanical energy of the vibrating eardrum caused by sound waves (and causing the malleus to move) to fluid waves in the cochlear of the inner ear. It is worth pointing out all reptiles have four or five jawbones (the articular, quadrate, angular, squamosal and dentary with articulation between the first two bones), while mammals have two different jawbones. The evolutionary view is the mammalian malleus evolved from one of the reptilian upper jawbones (the quadrate) and the incus from one of the lower jawbones (articular) and that they migrated into the middle ear. There are three lines of evidence for this contention, (1) embryonic, (2) paleontological and (3) functional.[86]

(1) The embryonic evidence is the malleus and incus develop from the first branchial arch from which the jaws of all vertebrates develop. This is an old observation. In 1837, C.B. Reichert showed these two bones develop from the posterior part of the precursor embryonic tissue of the first branchial arch (Meckel's cartilage) by detaching and migrating to the ear. In reptiles and mammals, Meckel's cartilage also forms the jawbones. Since the malleus and incus have a common embryonic origin with the jaw, this is taken as evidence of common ancestry from a jawbone. Dr Gilbert concludes: "Thus, the middle ear bones of the mammal are homologous to the posterior lower jaw of the reptile ".

But this is the argument similar structures arising from a common embryonic tissue infers common ancestry. It was an argument first made by Darwin in spite of evidence (§7.2) and comprehensively discredited since (§7.3). Dr. Gould goes even further though, arguing the migration of the malleus and incus during development is evidence of their evolution from the reptile quadrate and articular jawbones, because they are reenacting or "recapitulating" evolutionary history during embryogenesis. As he says it: "Thus every animal records in its own embryonic growth the developmental pathway that led from jawbones to ear bones in its evolutionary history"[86]

This is an intriguing contention we take up later, but for now notice the only evidence for his claim is the evidence of his interpretation (§9.3). The tensor tympani, the muscle which pulls on the malleus to dampen movement of the ossicular chain by excessive noise, is thus an "ear" muscle but it is supplied by a "jaw" nerve, a branch of the trigeminal cranial nerve. This is used as evidence that a jaw structure evolved into the ear. It's not evidence so long ancestry cannot be correlated with shared embryological pathways however (7.3). The reason for the anatomy is easily explained by understanding embryonic development. The mandibular division of the trigeminal nerve supplies muscles that develop from the first branchial arch and both the tensor tympani (and the malleus to which it attaches) are derived from it.

What is forgotten in making this argument is in an identical way the facial nerve not only supplies the face muscles but an "ear" muscle, the stapedius. It attaches to the stapes (stapedius) and dampens movement of the stapes to protect the cochlear from loud noise in a way similar the tensor tympani does at the malleus. The reason for this apparent quirk of anatomy is simply that the facial nerve supplies the muscles derived from the second branchial arch, wherever they end up. We do not need to invoke the evolution of stapedius from a fish face, just as we do not need to invoke the evolution of the malleus and incus from a reptile jaw. By the way, another "ear" muscle is also found in the mouth. The tensor veli palatini is in the soft palate. It opens the Eustachian tube to equalize the pressure in the middle ear and is also supplied by the trigeminal nerve (and thus we know it developed from the first branchial arch).

(2) The paleontological evidence is to demonstrate a reptile could disarticulate its jaw and permit those bones to move into the ear yet be able to chew all the while. The solution is to show a double jaw articulation. The evidence Stephen Jay Gould used is that some cynodont therapsid fossils have an articulation between the squamosal bone and a part of the lower jaw called the surangular, and others with the squamosal bone in addition to the quadrate-articular jaw joint. Additional evidence is that two of the earliest fossil mammals, *Morganucodon* and *Kuehneotherium* have a malleus and incus still articulating with the jaw. However there are no reptile fossils with more than one inner ear bone. There is also no fossil evidence for a structure that might be considered reasonably an intermediate to the mammalian Organ of Corti, which, after all, is essential to interpret any of the movement of these supposedly evolving ear bones.

(3) The functional evidence represents observations that modern day reptiles like snakes "listen" with their jaws. They rest the lower jaw on the ground to transmit vibrations from jaw to quadrate to stapes. While these creatures are not taken as ancestral to mammals, they support the notion

that listening with the jaw is not as crazy as it sounds. What's crazy is not the idea but its belief despite the evidence to the contrary. In this we also need to give a naturalistic explanation for why creatures would evolve in this way. It's not enough to wave the arms and claim vagaries of mutation and natural selection. There seems no selective advantage developing three ear ossicles for instance as birds hear just as well with only one. We close by returning to Aristotle's ancient and profound observation. Tell me what similarity in biology is exactly?

§7.7 WHAT EXACTLY IS BIOLOGICAL SIMILARITY?

The crux of the homology enigma is that it's a pattern of similarity or stability (shared features across diverse species), paradoxically arrived at by variability (different developmental processes). This is a demonstration of a discontinuous, digital construction of biology across kinds of creatures. What are the various evolutionary explanations for the phenomenon nonetheless?

The first is common ancestry, a notion which began with Charles Darwin and which remains axiomatic in the mainstream of science.[13] It equates homology with synapomorphy and holds to a conjecture that the stability of patterns across species is explained by an inheritance from a common ancestor. There is no mechanism for this opinion and the evidence of embryology and genetics rejects it (§5.3;§7.2-3;§9.4-5). Evolutionary homology is an illusion.

The second is to regard biological similarity to be merely a pattern not a process yet retain the evolutionary interpretation of that pattern, namely common ancestry (homology). This is the view held by Dalhousie University embryologist Brian Hall. It's similar to how the fossil record is explained away by punctuated equilibrium. Hall says homology is a pattern which should not be conflated with a process. This is not what happens when an evolutionary interpretation is retained however. He explains his belief all the same:

"My position is that homology is a statement about pattern which is separate from statements about processes that produce those patterns. Often homologous structures arise through the same developmental program. But as developmental programs evolve (genes are substituted, parts of causal sequence are lost or altered, and so forth), many homologous structures no longer share the same set of developmental sequences. Similarly (or conversely?), there are many cases in which the same genes direct the development of different (analogous) structures...homologous structures can be produced by equivalent and non-equivalent developmental processes."[91]

The problem with his view is not his understatement of the problem that supposedly homologous structures have "non-homologous" development, nor the unsubstantiated speculation that it's by evolution. It's that his interpretation of the pattern of homology as evolutionary, while also declaring it is not a process, is an oxymoron. Evolution *is* a process. What we have here is another empty claim believed in spite of the evidence.

The third is the concept of "biological homology", a view advanced by Louise Roth at Duke and Günter Wagner at Yale (§5.5).[9,10] It goes like this: "Structures from two individuals or from the same individual are homologous if they share a set of developmental constraints, caused by locally acting self-regulatory mechanisms of organ differentiation." It confines homology to structures with homologous development (so acknowledging that evolution demands this), but solves the problem of the contradictory data by ignoring it. What precisely these "constraints" might be, is left unanswered. The notion can only apply to those circumstances in which the mechanisms of development are already known, and to avoid the contradictory data the "definition must be qualified further because of the variability of development...only those aspects of development are relevant for a biological homology concept that cause developmental constraints on the further adaptive modification of the structure."[9]

As Brian Hall says this definition is "non operational" because "the central portion of the definition [genetic constraints] is not amenable to analysis". The "constraints" are mystically responsible for causing similarities by inheritance somehow, and not surprisingly have spawned fanciful theories including 'genetic piracy' (genes previously responsible for one homologous structure can be deputized by another set to cause the feature to appear in another context).[10] This view is another philosophical view then, not an empiric one.

The fourth view, held by Leigh Van Valen, is the view which holds homology is "correspondence caused by a continuity of information". There are two problems with it. The first is the vague notion of "information" to accommodate the anomalies of homology, when what we seek is clarity about a basic observation in biology we all can see. The second is that "correspondence" is the notion that needs clarification by a definition of homology but his solution just restates it. To hear this from an expert: "[r]ather than providing a concrete definition of homology that does not involve a similarly unexplicated term, van Valen's concept of homology is an attempt to encompass the various aspects of homology in one term."[62]

The fifth is the typological view, namely that homology is a pattern of similarity in adult structures divorced from a natural cause. We've heard this one before. We return to Owen and to Aristotle and to design. Only the view in science this time around is different, holding that evolution is true and design is false. It's an extraordinary idea. Alec Panchen believes it. He says: "Evolution, or rather phylogeny, is the explanation of homology, as it is of the phenomenon of natural classification, itself reconstructed from the hierarchy of homologies. It should not be part of the definition."[2]

Although he is an evolutionist, homology for him is similarity alone. Holding to that definition and that belief is an abdication of the evolutionary paradigm because common ancestry, not similarity is evolution. Moreover the obvious patterns of biological similarity demand a mechanism from him as a biologist, and similar structures supposedly variably arising from convergent evolution, parallelism or reversal are not common ancestry and not homologous. His definition is therefore worse than empty "because the multiple causes for similarity would make homology with this defining criterion uninformative for studies of evolution."[13]

§7.7   IN CONCLUSION: EVOLUTIONARY LOOKS THAT LIE IN THE EYE OF THE BEHOLDER

Patterns of similarity across nature are an objective phenomenon and one of the five reasons for its appearance of design (§1.2). Engaged as we are in dissecting a debate about an illusion, we should at least begin by establishing what is indisputably real (§1.3). The consensus of science is biological similarity is the inheritance of shared traits from a common ancestor and is, as Scott Gilbert says: "*the* most important concept in contemporary biology"[12] or as David Wake says: "the central concept for *all* biology. Whenever we say that a mammalian hormone is the "same" as a fish hormone, that a human sequence is the "same" as a sequence in a chimp or a mouse, that a HOX gene is the "same" in a mouse, a fruit fly, a frog, and a human – even when we argue that discoveries about a roundworm, a fruit fly, a frog, a mouse, or a chimp have relevance to the human condition – we have made a bold and direct statement about homology."[11]

When the word "homology" was used above it meant biological similarity caused by evolutionary ancestry. The observation is itself the evidence of evolution.

Actually the observation of homology is also the evidence for most of the other evidence of macroevolution. Mark Ridley spells this out: "Homologous similarities between species provide the most widespread class of evidence that living and fossil species have evolved from a common ancestor. The anatomy, biochemistry, and embryonic development of each species contain innumerable characters like the pentadactyl limb and the genetic code: characters that are similar between species, but would likely be different if the species had independent origins [*LOE#2-8*].

Yet we know biological similarity has no natural cause (§7.3). As we found for fossils this is at once a demonstration that nature is discontinuous in its construction, as it is a disproof of common ancestry.

The interesting thing is biological similarity as representing evidence of common ancestry and evolution is a truism in science.[13] This is despite the demonstration the mechanism Darwin first proposed (Lamarckian inheritance and inherited developmental pathways) is false, or that homology may arise by from different parts of the embryo, different embryonic tissues, different developmental pathways or by different genes, and despite an inversion of Darwin's original reasoning, to now infer homology of structures by first inferring their common ancestry on other grounds. Darwin's contention was to infer common ancestry from the similar structures.

Whereas Darwin used similarity of morphologic traits as evidence of common ancestry without further discrimination, current evolution theory assumes common ancestry for other reasons, then picks similarities it determines are ancestral and which are synapomorphies, and rejects other similarities it considers homoplasy. There is no agreement as to what homology is among biologists under the weight of all this, and yet most all are certain evolution causes it. Homology is an evolution look that lies in the eye of the beholder.

Why is evolution a fact? At least we know it cannot be because of structural similarity across nature.

If biological similarity is not evolutionary, then it represents similarity by design. In Owen's words: "HOMOLOGUE. The same organ in different animals under every variety of form and function." Design accounts for the curious finding that the causative mechanisms of similar features across taxa are more diverse than the features themselves. The idea is not refutable if formulated this way and it also explains the evidence.

Tim Berra at Charles Darwin University in Australia demonstrated this argument for ID in his book *Evolution and the Myth of Creation*. I should also say he did it unintentionally. He was making the case that ID is irrational. The fallacy he invoked however, has since been called dubbed

"Berra's blunder" by Berkeley law professor Phillip Johnson and should be preserved for future generations of students of the philosophy of science. Dr. Berra was using the fossil species in the human evolutionary line as his example. We have yet to meet them but their specifics are not necessary for now (§10.6 if you are desperate):

"The accelerating pace of hominid fossil differences is truly dazzling. In Darwin's time, only a few Neanderthal remains were known and they were misunderstood. Today we have a whole cast of characters in the drama of human evolution. These fossils are the hard evidence of human evolution. They are not figments of the scientific imagination. If the austrophithecines, *Homo habilis* and *H. erectus* were alive today and if we could parade them before the world, there could be no doubt of our relatedness to them. It would be like attending an auto show. If you look at a 1953 Corvette and compare it to the latest model, only the most general resemblances are evident; but if you compare a 1953 and a 1954 Corvette, side by side, then a 1954 and a 1955 and so on, the descent with modification is overwhelmingly obvious. This is what paleoanthropologists do with fossils and the evidence is so solid and comprehensive that it cannot be denied by reasonable people."[92]

Exactly. The patterns of similarity among the cars are a show of ID by car makers, not naturalistic evolution of cars. At least we know his unbelief of ID is not because of ignorance about the patterns of similarity in biology.

Two recent multi-authored books, one the proceedings of a symposium by some of the leading lights in the field[42,68] have attempted to bring order to what Berkeley's David Wake above has called "the problem of sameness in biology".[54] They are a revealing insight of how an evolutionary explanation of similarity is retained in spite of all the evidence in ways that are perverse because of the absurdities that necessarily follow when the data is eventually engaged. Sameness is a fundamental feature of nature, and he calls it "the central concept all of biology".[11] He could not be more qualified to speak to us about it either. A past board member of the National Academy of Sciences, editorial board member for the journal *Proceedings of the National Academy of Sciences* and editor of the journal *Evolution*, his opinions were sought in recent multidisciplinary reviews of homology[11,54] and he has written widely including invited commentary in journals like *Science* [11] and *Nature*,[93,94] His analysis of homology is worth hearing in full. You're also just in time for tea. So take a seat and we will listen in:

"Doubters point out that if homology is so important, we all should agree as to what it is. One main criticism is that, like species and adaptation, there is no 'naturalistic' mechanism to determine what

339

species, adaptations and homologues really are. Where is the empirically demonstrated naturalistic mechanism of homology? This is exactly the kind of question that we cannot answer, but more to the point, it is an irrelevant question. The clear implication is that if there is no answer evolutionary theory is in trouble. But the central issue is overlooked – homology is the anticipated and expected consequence of evolution."[54]

There's the evidence. The observation *is* the evidence. He continues by telling us this really is rational after all.

"Homology is not evidence of evolution nor is it necessary to understand homology in order to accept or understand evolution. That there is a genealogy of life on this planet necessitates that there be homology...It is a genealogical necessity, and no naturalistic mechanism is necessary to account for the phenomenon other than inheritance."

In other words similarity is a fact and evolution is a fact, and that is the end of the matter.

"Darwin knew this, and Gegenbaur accepted it, but we are still stuck somewhere between Darwin and Owen, knowing that homology is about evolution, but still wanting a naturalistic explanation. The only way out of this dilemma is to stop talking about homology and instead deal with the real questions that interest us. Systematicists should adopt the terminology of Hennig (1966), who recognized that science requires precise, technical terminology. So-called homology assessment, however, is essentially post hoc process in systematics, and there is no need to use the word homology. Besides, who wants homologues that can become homoplasies simply with the addition of another character to a matrix, or even in an alternative, equally parsimonious tree?...Is homology absolute? That is, can homology be seen as independent of a theory of phylogenetic relationships? No!...Can homology ever be definitively demonstrated? I think not. I view homology as something that follows from the fact of evolution."

He explains what he means:

"There is a continuity, as Darwin perceived and branching and extinction have produced genealogical entities whose phylogenetic relationships are inferred largely on perceived degree of overall homology, especially that portion of overall homology that is uniquely shared. Depending on our outlook we may posit that organs are homologous on some biological grounds and be happy when the proposition passes a phylogenetic test (as with the transition from bones forming a lower jaw articulation in some amniotes to forming inner ear ossicles in others), but all that we have done really is identified a pathway of phylogenetic continuity. All proximal attempts to explain homology in terms of structure, connectivity, topography or morphogenesis, for example, can fail. Is homology any more than phylogenetic continuity? Maybe we need to turn Van Valen's (1992) definition ('correspondence caused by continuity of information') around: the continuity of information necessitates correspondence."

This prompted questions by other symposium speakers. He answered:

"I will clarify my view. I certainly do not want to abandon the concept of homology; rather, I want to stop the fruitless discussion of what it is!...I want to get on with it and to leave behind debates that started when biologists really did not have sufficient biological knowledge to appreciate the causes of biological similarity and when they did not yet understand that Darwin was right in his view that there is one genealogy for all of life. Common ancestry is all there is to homology."

There was another exchange to show there was more than one scientist drinking Mad Hatter's tea and debating Mad Hatter riddles. Actually the conversants were a veritable "Who's Who" in evolutionary biology, among them Gunter Wagner, Alan Panchen, Rudolf Raff, Brian Hall and John Maynard Smith. Clean cup, clean cup…Move down…so listen some more:

Wagner: The reason Reidl refused to define this [homology] is because he recognizes that a premature definition will constrain our ability to learn about reality…It will not be advantageous for us to come up with a definition of homology now because it is too early and we do not understand the phenomenon itself."

Panchen: What we have now is a succession of unsatisfactory definitions and it seems as though we shall stay in this state forever.

Wagner: If the field turns out to be as productive as the field of genetics, then we won't need a definition.

Rudolf Raff: We do have a historical definition of homology, but a more powerful event happened, i.e. natural selection, so that the governing principle of the last century became the explanation of what creates homology…

The "governing principle" of course is that evolution is a fact and that macroevolution makes homology.

Maynard Smith: I don't want to suggest to you what the definition of homology ought to be, but I would say that you are spitting in the wind if you imagine that you can define it as you like. I didn't know what homology was, so yesterday I looked it up in the Collins Dictionary of Biology just to see what we were going to discuss. It doesn't mention Owen's concepts at all, only 'common descent'. I'm not saying this is the correct definition, but it is the meaning that people use…

Wagner:...One of the reasons we are finding it difficult to reach a conclusion is because there is an inherent tension between the phylogenetic definition of homology and the meaning of homology as character identity. It is rarely recognized that the phylogenetic definition of homology presupposes that we know what 'sameness' means, and we don't."[6]

We should leave them.

There is more to this. Consider that even if there were a similarity of development as Darwin claimed, this would still not necessarily exclude ID as cause of biological similarity. "They look the same because they develop the same because they were created to develop the same," would be the claim. But what we have in biology is different. We find similarity across taxa without connecting links. Biological similarity is separately, digitally, constructed. Like we found for fossils. It need not have been this way. Biology has patterns across it demonstrably beyond the power of natural forces. It has been made and obviously so. This pattern is a communication then. Biological similarity is a communication about biology in biology to all who would come investigate her.

Yet as we found for fossils the science blocks its ears and shuts its eyes to impose an artificial paradigm in unreason. In unreason because all can see the appearance of design and all can see the appearance confirmed. Natural forces are demonstrably exhausted, the paradigm is disintegrated but the pattern of similarity is still there (§1.3). This is not ignorance. It's willful blindness. It's also a clenched fist shaken in the face of God. This is a terrifying discovery.

Why is evolution a scientific fact? At least we know it's not because of structural similarity (§7), or the evidence of fossils (§5-§6), or the evidence of natural selection (§4). We need to look elsewhere.

Homology as a concept is not confined to morphologic structures and organs but includes molecular structures too. This includes the chemical that is the gene, deoxyribonucleic acid (DNA). Since inheritance is through genes, and evolution operates through gene mutations and their inheritance is after natural selection, a homology of DNA may be where the truth of evolution lies. In this regard chimpanzee DNA is 98.8% identical to human (§10.1). This is homology that Darwin did not even dream of! Surely at last, this is evidence of evolution. Not to mention getting to the molecular level of biology to get as close to biological processes and to nature as is possible. If evolution is true then it must be true at the molecular level. An inspection of this so-called molecular homology and molecular evolution is where we go next.

REFERENCES

1. Belon, P. *L'histoire de la nature des oyseaux*. (Guillaume Cavellet, 1555).
2. Panchen, A. L. in *Homology. The Hierarchical basis of comparative biology* (ed B.K. Hall) 22-63 (Academic Press, Inc., 1994).
3. Aristotle. Historia Animalium.
4. Owen, R. *Lectures on the comparative anatomy and physiology of the invertebrate animals* (Longman, Brown, Green and Longmans, 1843).
5. Owen, R. *On the archetype and homologies of the vertebrate skeleton*. 5 (John van Voorst, 1848).
6. Panchen, A. L. in *Homology. Novartis Foundation Symposium 222* (eds G.R. Bock & G. Cardew) 5-23 (John Wiley & Sons, Inc., 1999).
7. Wagner, G. P. in *Homology. The Hierarchical basis of comparative biology*. (ed B.K. Hall) 273-299 (Academic Press, Inc., 1994).
8. Owen, R. *Report on the Archetype and homologies of the vertebrate skeleton*. 174 (Murray, 1846).
9. Wagner, G. P. The biological homology concept. *Ann. Rev. Ecol. Syst* 20, 51-69 (1989).
10. Roth, V. L. in *Ontogeny and Systematics* (ed C.J. Humphries) 1-26 (Columbia University Press, 1988).
11. Wake, D. B. Comparative terminology. *Science* 265, 268-269 (1994).
12. Gilbert, S. F., Opitz, J. M. & Raff, R. A. Resynthesizing Evolutionary and Developmental Biology. *Developmental Biology* 173, 357-372 (1996).
13. Mindell, D. P. & Meyer, A. Homology evolving. *Trends in Ecology and Evolution* 16, 434-440 (2001).
14. de Beer, G. *Embryos and ancestors*. (Oxford University Press, 1958).
15. de Beer, G. *Homology: An unsolved problem. Oxford Biology Readers No.11*. 15 (Oxford University Press, 1971).
16. de Beer, G. *Homology: An unsolved problem. Oxford Biology Readers No.11*. 8 8 (Oxford University Press, 1971).
17. Hall, B. K. *Evolutionary Developmental Biology*. 342-3 (Chapman & Hall, 1998).
18. de Beer, G. *Homology: An unsolved problem. Oxford Biology Readers No.11*. 14 (Oxford University Press, 1971).
19. Noden, D. M. The role of the neural crest in patterningof anvian cranial skeletal, connective and muscle tisuee. *Developmental Biology* 96, 144-165 (1983).
20. Bellairs, A. d. A. & Gans, C. A reinterpretation of the amphisbaenian orbitosphenoid. *Nature* 302, 243-244 (1983).
21. Keller, R. E. in *Developmental biology: a comprehensive synthesis Vol 2. The cellular basis of morphogenesis* (ed L. Browder) 241-327 (Plenum Press, 1986).
22. Hall, B. K. Developmental processes unerlying heterochrony as an evolutionary mechanism. *Canadian Journal of Zoology* 62, 1-7 (1984).
23. Hall, B. K. *Evolutionary Developmental Biology*. 182 (Chapman & Hall, 1998).

24 de Beer, G. *Homology: An unsolved problem. Oxford Biology Readers No.11.* (Oxford University Press, 1971).
25 Hall, B. K. *Evolutionary Developmental Biology.* 340 (Chapman & Hall, 1998).
26 Raff, R. A. & Kaufman, T. C. *Embryos, genes and evolution.* 13 (Macmillan Publishing Co., 1983).
27 Wray, G. A. & Raff, R. A. Rapid evolution of gastrulation mechanisms in a sea urchin with lecithotrophic larvae. *Evolution* 45, 1741-1750 (1991).
28 Blanco, M. J. & Alberch, P. Caenogenesis, developmental variability, and evolution in the carpus and tarsus of the marbled newt Triturus marmoratus. *Evolution* 46, 677-687 (1992).
29 Keller, R. The origin and morphogenesis of amphibian somites. *Current topics in developmental biology* 47, 183-246 (2000).
30 Radice, G. P., Neff, A. W., Shim, Y. H., Brustis, J.-J. & Malacinski, G. M. Developmental histories in amphibian myogenesis. *International journal of developmental biology* 33, 325-343 (1989).
31 Malacinski, G. M. Amphibian somite development: contrasts of morphogenetic and molecular differentiation patterns between laboratory archetype species Xenopus (anuran) and axolotl (urodele). *Zoological Science* 6, 1-14 (1989).
32 Hall, B. K. Descent with modification:the unity underlying homology and homoplasy as seen through an analysis of development and evolution. *Biological Reviews* 78, 409-433 (2003).
33 Collazo, A. Developmental variation, homology and the pharyngula stage. *Systematic Biology* 49, 3-18 (2000).
34 Clarke, C. A. & Sheppard, P. M. R. R. R. R. R. R. R. The evolution of dominance under disruptive selection. *Heredity* 14, 73-87 (1960).
35 Patel, N. H., Ball, E. E. & Goodman, C. S. Changing role of even-skipped during the evolution of insect pattern formation. *Nature* 357, 339-342 (1992).
36 Grbic, M. & Strand, M. R. Shifts in the life history of parasitic wasps correlate with pronounced alterations in early development. *Proceedings of the National Academy of Sciences USA* 95, 1097-1101 (1998).
37 Lowe, C. J. & Wray, G. A. Radical alterations in the roles of homeobox genes during echinoderm evolution. *Nature* 389, 718-721 (1997).
38 de Beer, G. R. in *Evolution:Essays on Asepcts of Evolutionary Biology Presented to Professor E.S. Goodrich on his Sevenieth Birthday* (ed G.R. De Beer) 59-78 (The Clarendon Press, 1938).
39 de Beer, G. *Homology: An unsolved problem. Oxford Biology Readers No.11.* 16 (Oxford University Press, 1971).
40 Hall, B. K. in *Homology. Novartis Foundation Symposium 222* 1-4 (John Wiley & Sons Ltd., 1999).
41 Hall, B. K. *Evolutionary Developmental Biology.* (Chapman & Hall, 1998).
42 Hall, B. K. *Homology. The Hierarchical basis of comparative biology.* (Academic Press, Inc., 1994).
43 Hall, B. K. *Evolutionary Developmental Biology.* 2nd edn, 347 (Chapman & Hall, 1998).
44 Dickinson, W. J. Molecules and morphology: where's the homology? *Trends in Genetics* 11, 119-121 (1995).

45 de Beer, G. (eds J.J Head & O.E. Lowenstein) 4 (Oxford University Press, 1971).
46 Meyer, A. in *Homology. Novartis Foundation Symposium 222* (eds G.R. Bock & G. Cardew) 141-153 (John Wiley & Sons, Inc., 1999).
47 Sjarski, H. The concept of homology in the light of the comparative anatomy of vertebrates. *Quarterly Review of Biology* 24, 124-131 (1949).
48 Hall, B. K. in *Homology. The Hierarchical basis of comparative biology* (ed B.K. Hall) 1-21 (Academic Press, Inc., 1994).
49 Simpson, G. G. *Principles of Animal Taxonomy.* (Columbia University Press, 1967).
50 Mayr, E. *The Growth of Biological Thought.* 45 (Harvard University Press, 1982).
51 Futuyma, D. J. *Evolutionary Biology.* 552 (Sinauer Associates, 1986).
52 Patterson, C. Homology in classical and molecular biology. *Molecular Biology and Evolution* 5, 603-625 (1988).
53 Abouheif, E. *et al.* Homology and developmental genes. *Trends in genetics* 13, 432-433 (1997).
54 Wake, D. B. in *Homology. Novartis Foundation Symposium 222* (eds G.R. Bock & G. Cardew) 24-33 (John Wiley & Sons, 1999).
55 Raff, R. A. in *Homology. Novartis Symposium, London 21-23 July 1998* 110-120 (John Wiley & Sons Ltd., 1999).
56 Patterson, C. in *Molecules and morphology in evolution: conflict or compromise* (ed C. Patterson) 1-22 (Cambridge University Press, 1987).
57 Laubichler, M. D. Homology in development and the development of the homology concept. *American Zoologist* 40, 777-788 (2000).
58 Van Valen, L. Homology and its causes. *Journal of Morphology* 173, 305-312 (1982).
59 Hennig, W. *Phylogenetic Systematics.* (University of Illinois Press, 1966).
60 Raff, R. A. *The shape of life. Genes, development, and the evolution of animal form.*, (The University of Chicago Press., 1996).
61 Sidall, M. (www.nature.com/nature/debates/fossil/_3.html, 1998).
62 Brigandt, I. Homology and the origin of correspondence. *Biology and Philosophy* 17, 389-407 (2002).
63 Roth, V. L. On homology. *Biological Journal of the Linnean Society* 22, 13-29 (1984).
64 Hall, B. K. *Evolutionary Developmental Biology.* 348 (Chapman & Hall, 1998).
65 Maynard Smith, J. *et al.* Developmental Constraints and Evolution. *Quarterly Review of Biology* 60, 265-287 (1985).
66 Rieppel, O. in *Homology. The Hierarchical basis of comparative biology.* (ed B.K. Hall) 63-100 (Academic Press, Inc., 1994).
67 Hardy, A. *The Living Stream.* 213 (Collins, 1965).
68 Bock, G. R. & Cardew, G. *Homology. Novartis Foundation Symposium 222.* (John Wiley & Sons, Inc., 1999).
69 Hall, B. K. *Evolutionary Developmental Biology.* 338 (Chapman & Hall, 1998).

70 Stevens, P. F. Homology and phylogeny:morphology and systematics. *Systematic Botany* 9, 395-409 (1984).
71 Raff, R. A. *The shape of life. Genes, development, and the evolution of animal form.*, 37 (The University of Chicago Press, 1996).
72 Dobzhansky, T. Nothing in biology makes sense except in the light of evolution. *American Biology Teacher* 35, 125-129 (1973).
73 Eldredge, N. *Reinventing Darwin: The great debate at the high table of evolutionary theory.* 49-51 (John Wiley & Sons, 1995).
74 Ruse, M. *Darwinism Defended.* 309 (Addison-Wesley Publishing Company, 1982).
75 Grant, V. *The Evolutionary process. A critical review of evolutionary theory.* (Columbia Unversity Press, 1985).
76 Futuyma, D. J. *Science on trial.* 47 (Sinauer Associates Inc., 1995).
77 Mayr, E. *What Evolution Is.* 27 (Basic Books, 2001).
78 Ridley, M. *Evolution.* 46 (Blackwell Science, Inc., 1993).
79 Coates, M. I. & Clack, J. A. Polydactyly in the earliest known tetrapod limbs. *Nature* 347, 66-69 (1990).
80 Booher, H. R. *Origins, icons and illusions.* 89 (Warren H. Green, Inc., 1998).
81 Goodwin, B. in *Homology. The hierarchical basis of comparative biology* (ed B.K. Hall) 229-247 (Academic Press, Inc., 1994).
82 de Beer, G. (eds J.J Head & O.E. Lowenstein) 9 (Oxford University Press, 1971).
83 Raff, R. A. *The shape of life. Genes, development, and the evolution of animal form.*, 36-7 (The University of Chicago Press, 1996).
84 Gilbert, S. F. *Developmental Biology.* 6th edn, (Sinauer Associates, Inc., 2000).
85 de Beer, G. *Homology: An unsolved problem. Oxford Biology Readers No.11.* 7 (Oxford University Press, 1971).
86 Gould, S. J. An Earful of Jaw. *Natural History* 99, 12-23 (1990).
87 Kipling, R. *How the Leopard Got His spots.*
88 Tudge, C. *The Variety of Life.* (Oxford University Press, 2000).
89 Clack, J. A. Discovery of the earliest known tetrapod stapes. *Nature* 342, 425-427 (1989).
90 Clack, J. A. Patterns and processes in the early evolution of the tetrapod ear. *Journal of Neurobiology* 53, 251-264 (2002).
91 Hall, B. K. *Evolutionary Developmental Biology.* 349 (Chapman & Hall, 1998).
92 Berra, T. M. *Evolution and the myth of creationism.* 117 (Stanford University Press, 1990).
93 Wake, D. B. A few words about evolution:building a hierarchical framework on the foundations of Darwinism [Review of] The Structure of Evolutionary Biology, by S.J. Gould. *Nature* 416, 787-788 (2001).
94 Wake, D. B. Speciation in the round. *Nature* 409, 299-300 (2001).
95 Carroll, L. *Alice's Adventures In Wonderland. In Lewis Carroll The Complete Illustrated Works.* (Gramercy Books, 1995).

8. Of molecules, mice & men

§8.1   GETTING DOWN TO DETAILS: THE MOLECULAR
       EVIDENCE OF EVOLUTION

Among the most basic observations you can make in biology is to see the similar morphologic traits reiterated in otherwise dissimilar organisms (§1.2.5;§7.1). To dismiss this appearance of design as instead an illusion of design caused by evolution as science says, there needs to be evidence of a hereditary mechanism. Shared traits should demonstrate shared and therefore conceivably inherited, causal developmental pathways and genes because traits arise by development from genes. But we did not find evidence for either (§7.3). This was a discovery about extant life consistent with our earlier inspection of extinct life. To wit, fossils rejected the notion of macroevolution and common ancestry too - by their stasis, their sudden appearance and the extreme rarity of transitional forms (§5.5). Only the erection of a post hoc theory called punctuated equilibrium to explain why there is no fossil evidence kept Darwin's claim of evolution in the fossil record alive (§6.3). But believing in punctuated equilibrium requires believing in an unseen process for which there can be no independent, confirmatory evidence. In, our search for objective evidence of macroevolution has been fruitless so far.

But perhaps that's because we've not looked closely enough. We need to get down to the molecular level of biology to be sure, because it's there that anatomy, embryology, physiology and behavior ultimately reside. If evolution is true, then it must be true at the molecular level. The converse is true too. This brings us to *LOE*#5 and #6. Refresh your memory on Table

Evidence of molecular evolution

3.1. The discovery of molecular similarities among living things, like shared genes, shared proteins and shared biochemical processes, is everywhere in science now irrefutable evidence of the evolutionary relationship of life. It's even called "molecular homology". Perhaps this is the evidence we've been looking for!

For a start, molecular similarities among living things are still more impressive than morphologic similarities. Living things don't just have similar biochemistry, it's practically identical. The same chemical pathways, the same chemical laws, the same carbohydrate, lipid, protein and nucleic acid monomers and even more remarkably, the same genetic code. Even the stereoisomers are the same. What are "stereoisomers"? Three of the four major organic chemicals of life, namely proteins, nucleic acids and carbohydrates, are constructed in chains like beaded necklaces. The individual monomer "bead" amino acid or nucleotide or monosaccharide that are the building blocks of proteins, nucleic acids and carbohydrates respectively, come in two forms. There is a left handed (levo or L-) and a right handed (dextro or D-) form. They are mirror images of each other in their shape, just like your left and your right hand are, but otherwise are structurally and chemically identical (Fig.8.1). Test-tube chemical reactions produce an equal mix of the D- and L- stereoisomers.

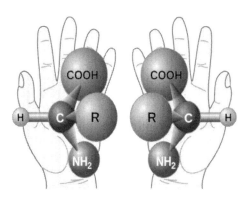

Fig.8.1. D- and L- amino acid stereoisomers. The amino acid monomers of proteins, like the nucleotide and monosaccharaide monomers of nucleic acids and carbohydrates, occur in two different non-superimposable, chiral stereoisomer forms. They are as similar and as different as your left and right hands. In test-tube chemical reactions they occur in equal proportions but in biochemistry only D-nucleotides and only L-amino acids and L-monosaccharides are employed. The laws of physics offer no reason for this phenomenon. (Image credit: NASA).

Now here is the curious thing. Chemical reactions inside cells (or "biochemistry") only involve the D-stereoisomer form for nucleic acids and only the L-stereoisomer form for amino acids and carbohydrates. There's no chemical reason for this selection and the laws of physics don't explain it either (§12.3.1). Yet from bacteria to plants to reptiles to man the patterns and processes of biochemistry are shared in ways like this with

348

only extreme exception. The uniformity of biochemistry is among the most remarkable characteristics of living things. Why should this be?

The answer from science is it's because life is descended from a single common ancestor from which these idiosyncratic molecular specificities were inherited. It's the molecular evidence of macroevolution - *LOE#5* in our catalogue. Mark Ridley at Oxford University explains: "The universality of the genetic code is important evidence that all life shares a single origin...Universal molecular homologies, such as the genetic code, now provide the best evidence that all life has a single common ancestor."[1] Ernst Mayr agrees and says this is the final proof that Charles Darwin was right: "Up to 1859 people had been impressed primarily by the enormous diversity of life, from the lowest plants to the highest vertebrates, but this diversity took on an entirely different complexion when it was realized that it all could be traced back to a common origin. The final proof of this, of course, was not supplied until our time, when it was demonstrated that even bacteria have the same genetic code as animals and plants."[2] Douglas Futuyma says *LOE#5* the best: "The only possible reason for these chemical universalities is that living things got stuck with the first system that worked for them."[3]

There is an alternative possibility they do not consider of course, that they were designed that way, but let's leave that idea for later. At least as far as the science establishment is concerned, the universality of biochemistry is "a frozen accident" to use a metaphor coined by DNA co-discoverer Sir Francis Crick (1916-2004) because it represents the inherited biochemical traits of the original, common Archean ancestral cell.

This explanation might have an intuitive appeal, but how does one test it? The problem is the observation can become the evidence it claims to explain. At least it can't be proved directly since there's no way of going back to recover that ancient ancestral cell of the postulate. We have to look at other aspects of molecular homology to get an answer then. Fortunately there is much more to the concept to permit just such an inspection.

Molecular similarities across nature are so impressive that not only are the reactions and reagents of biochemistry shared, there are similarities among genes and hence the proteins they encode. Cambridge University double Nobel Prize laureate Fred Sanger was the first to discover this in 1955, when the amino acid sequences of insulin from sheep, pig and cow were found to be highly similar.

The "sequence" of a protein is the order of its constituent amino acid monomers along its length, and the sequence of a gene is the order of its constituent nucleic acid bases. There are twenty possible amino acids from which to make a protein and four possible nucleotide bases from which to make a gene (§2.3). Thus amino acid sequences are 5% similar and nucleic acid sequences are 25% similar just by chance. Similarities above this represent similarities in sequence identity.

One of the most impressive is the chromosomal protein histone-4. This molecule is found in creatures as disparate as bacteria, birds, bison and bishops, and I mean the human kind. In humans histone-4 is 103 amino acids long and identical to that found in yeast and chicken, and the pea and the cow forms differ by only two amino acids (i.e. they are ~98% identical). As regards the nucleotide bases of the gene encoding for histone-4, remembering that three nucleotide bases specify one amino acid (§2.3.3), over 200 of the 306 DNA nucleotides are identical across all of life. This is remarkable. How remarkable? The odds of this happening are easy to calculate. Four equally possible DNA bases for 200 positions; that's one chance in 4^{200} or one in 10^{120}. It could not have happened by fluke! This is taken as indisputable evidence of evolutionary common ancestry and is *LOE #5* in our catalogue (Table 3.1).

§8.2   THE RISE OF MOLECULAR PHYLOGENETICS

The ability to compare nucleic and amino acid sequences across life has become a powerful tool in evolutionary biology for at least three reasons. The first is it's mathematical. Similarities can be expressed precisely and even as percentages, in welcome contrast to the subjectivities of assessing structural and morphologic similarities across taxa. How do you express the similarity of a bird skeleton to a human skeleton mathematically, or a frog leg to a dog leg for instance? (Figs.7.1&2)

The second reason is because this analysis is molecular and molecules are what organisms are made of. Molecules in the end are what account for the diversity and the complexity and the wonder of biology. The analysis goes in at biology's ground level, is the point. This is right where we need to be too.

The third reason is because comparisons are taken as permitting an inference of the "evolutionary distance" between taxa, and so is a tool to decipher how it was that evolution generated the living things in question. Actually it's "the most important principle in molecular evolution",[4] namely "the degree of similarity between genes [and proteins] reflects the strength of the evolutionary relationship between them".[4] In other words the greater the similarity between two gene or protein sequences, the closer is the evolutionary distance between them. This is *LOE#6* in our catalogue.

A whole field in science has developed from this thinking called molecular phylogenetics, dedicated to the construction of evolutionary phylogenies. A molecular foundation for systematics, resting in this way on quantitative evolutionary origins and the evolutionary relationships of species, has great appeal for being incomparably better than Linnaeus's arbitrary morphologic classification (§1.1). The reason is because it's explanatory, assuming macroevolution is true that is. There's also the expectation that as comparisons across taxa become less morphologic and more molecular and quantitative, so more of the precise route by which evolution generated all the variety of life will be revealed. If a valid phylogenetic tree of organisms can be established, so the thinking goes, so we will arrive at how mutations generated all the species on the tree. Evolution operates through molecules after all, so the evidence for evolution should be emphatic there (§2.3). Well let's look.

The various traits of adult living things originate from a single cell, typically the zygote, and are produced by proteins encoded by genes (§2.3). Morphologic similarities inferred as being caused by evolution are therefore ultimately due to the inheritance of similar genes and similar developmental pathways. A particular biological function of a whole organism say a whale or an elephant, can ultimately be reduced to a cell which in turn is reducible to the actions of its constituent proteins, which in turn is reducible to its genes. The progressive change of morphology and biological function caused by macroevolution in the end is the evolution of proteins encoded by genes undergoing random changes in the sequence of their nucleotide bases (§2.3-5).[5]

There's a nasty little problem with this smooth reasoning and we know it already. Anatomical homology is not explained by molecular homology (§7.3).[6,7] The interesting thing about this, and why you might find this inquiry interesting and perhaps even amusing, is that it has not stood in the way of any of the assumptions of molecular phylogenetics. For example already back in 1985 Stephen Jay Gould proclaimed: "We finally have a method [in gene and protein sequencing] that can sort homology from analogy...Perhaps we should mount the parapets and shout: 'The problem of phylogeny has been solved.'"[8] Before rushing off up there with him, wait to hear a little more.

The reasoning behind molecular phylogenetics was pioneered by Georgetown University biochemist Margaret Dayhoff (1925-1983), author of the seminal text comparing protein sequences across different species.[9] She said biomolecules are like fossils, and are at least as much evidence of evolution: "Many proteins and nucleic acids are "living fossils" in the sense that their structures have been dynamically conserved by the evolutionary process over billions of years. Their amino acid and nucleotide sequences occur today as recognizably related forms in eukaryotes and prokaryotes, having evolved from common ancestral sequences by a large number of small changes." There is a little technicality you should know before we look at the evidence for this claim.

Similarity in a gene or a protein sequence is not always caused by an ancestral relationship. Homology refers only to similarity caused by common ancestry (§7.2). Sequence similarity that may be just as impressive but which is inferred as not caused by direct inheritance, is called homoplasy. The latter is taken to have happened either by convergent evolution (when the ancestral condition is different), parallel evolution (when it is the same) or reversion (which is a return to an ancestral condition not present in the recent ancestors) (Fig.5.2b). Similarity of genes adjudged from homology (common ancestry) arises either by "orthology" by which genes in two species are considered descended from a single gene of a common ancestor, or by "paralogy" which is caused by the inheritance of an ancestral gene that has undergone one or more duplications. Bluntly, gene and protein sequences can be highly similar yet not be considered to reflect a common ancestry. Despite this the similarity of nucleic acid or protein sequences is commonly referred to as "sequence homology" without reference to ancestry, even in the molecular biology literature. In this use of the term "the protein is 70% homologous" is taken to mean 70% of the aligned amino acids are identical to that of another protein. Ross MacIntyre at Cornell University explains: "most molecular evolutionists are content to evaluate sequence comparisons in terms of "similarity" and tacitly to assume they are homologous".[10] But they are not, and repeated attempts have been made to preserve clarity in the literature. Among them are the efforts of the University of California Irvine's Walter Fitch, one of the doyens of this field of molecular evolution. He says:

> "It is worth repeating here that homology, like pregnancy, is indivisible. You either are homologous (pregnant) or you are not. Thus, if what one means to assert is that 80% of the character states are identical one should speak of 80% identity, and not 80% homology."[11]

And so we discover evolution theory insists the degree of sequence identity across nature is not the real issue. The real issue is the ancestry that gave rise to this spectacular and unmistakable finding, some of the time. Pairs of equally similar protein sequences are adjudged totally different evolutionary causes, either homology or homoplasy, by inferential determinations made on other grounds than sequence similarity. If this is to be used as evidence of evolution, it's circular. The concern is compounded when we consider the statistics of getting the sequence similarity we see in nature. For a DNA gene strand n bases long with 4 equally probable bases at each position there are 4^n possible sequences or 2^{2n}. In the same way a protein sequence of n amino acids for which there are 20 possible, there are 20^n possible alternative sequences assuming all the amino acids are equally available. A typical gene contains a sequence of 1000 bases. The number of possible such sequences is 4^{1000} or around 10^{600}. In case that number doesn't appear big, and I mean absurdly big, know there are only 10^{80} electrons in the Universe.[12] The chance of blindly arriving at a particular specific sequence by chance, even for a very small molecule, is effectively impossible. This is a stunning observation.

How did that molecule get to be discovered by evolution in the first place? The odds of getting a specific sequence of 200 base pairs that generates a protein domain in cytochrome c that functions so well it's conserved across all nature is 1 in 4^{200} or 1 in 10^{120}. What can we say so far? Three things. Ponder them before we move on.

(1) There are extraordinary molecular similarities across taxa. It's an unmistakable feature of life, one that gives it the appearance of design and which can be measured mathematically, and taken as evidence of common ancestry and macroevolution (*LOE#5*;Table 3.1).

(2) Differences between molecules across taxa are taken as evidence of evolution, specifically evidence of their evolutionary divergence (*LOE#6*). In other words the similarities among molecules are the evidence of evolution (*LOE#5*) and the differences among molecules are evidence of evolution (*LOE#6*)(§20).

(3) In the name of evolution we move from an objective mathematical analysis of molecular similarity across taxa to an inference of what that similarity means – homology or homoplasy. This is an inference made independent of the similarity, although the similarity was the basis for the molecular evolution inference in the first place.

It's time now to look at the data for *LOE#5* and *LOE#6* ourselves. What does molecular evolutionary biology declare? Is this macroevolution at last? Is it even internally coherent?

§8.3   MOLECULAR EQUIDISTANCE AND SEPARATION, NOT MOLECULAR EVOLUTIONARY LINKS

The techniques to decipher the sequences of proteins were discovered a century after *The Origin* was published, in the 1960s. The reasoning went like this. Since biological structures like fingers and arms and eyes and ears evolved, and since they can be reduced to the actions of molecules, so the molecules that made them must have evolved and the evidence for evolution should be in these molecules. If evolution is a fact, then molecular evolution must be a fact. The relative difference in amino acid or nucleic acid sequences across species can be quantitated and presented on a matrix as percent difference, which should allow us to see this.[13,14]

Cytochrome c is an electron transport protein for energy metabolism. At around 100 amino acids long, it is ideal for this analysis because it's also small and ubiquitous in nature. The amino acid sequence of the human form compared with twenty-three other species across the spectrum of life is shown in Fig.8.2.[15] Four features are evident:

1. The percentage difference in amino acid sequence from the human form ("0%" top left corner of the table) becomes greater with greater taxonomic distance. Move down the extreme left column to see this. Human cytochrome c differs by 1% from rhesus monkey, 10% from pig and sheep, 14% from turtle, 20% from tuna, 27% from fruit fly, 38% from wheat, 41% from yeast and 65% from the bacterium *Rhodospirillum*. This is what one might have predicted on morphologic grounds, and it also appears consistent with evolutionary divergence. In fact in most places nowadays it's not even called sequence difference but "sequence divergence", at once describing and explaining the observation. Emanuel Margioliash (1920-2008) at Northwestern University spent much of his life investigating this protein. He explains the concept for us:

"It appears that the number of residue differences between the cytochrome c of any two species is mostly conditioned by the time elapsed since the lines of evolution leading to these species originally diverged."[16]

This is just a restatement of *LOE#6* of course.

Physicist and science author Paul Davies at Arizona State University discusses this further in his book *The Fifth Miracle*. Dr. Davies is a recipient of the prestigious Templeton Prize, named after mutual fund philanthropist Sir John Templeton (1912-2008). The Prize is awarded annually for "progress toward research or discoveries about spiritual realities". What is interesting is not that Dr. Davies won a religion prize though he is a physicist. Nor that so much religious significance is attached to this science. We know that already and we know why. What's interesting is the size of the prize for a discovery in religion. Set at over 1.5 million USD and always more than the Nobel Prize, here's an incentive to think up new

	Human	Monkey	Pig	Horse	Whale	Chicken	Penguin	Turtle	Snake	Bullfrog	Tuna	Carp	Lamprey	Fly	Silkworm	Bean	Sunflower	Wheat	Candida	Yeast	N. crassa	R. rubrum
Human	0	1	10	12	10	13	13	14	13	17	20	17	19	27	29	40	38	38	46	41	43	65
Monkey	1	0	9	11	9	12	12	13	14	16	20	17	19	26	28	41	38	38	45	43	43	64
Pig	10	9	0	3	2	9	10	9	14	11	16	11	13	22	25	41	40	40	45	41	43	64
Horse	12	11	3	0	5	11	12	11	19	13	18	13	15	22	27	42	43	41	46	42	43	64
Whale	10	9	2	5	0	9	9	9	21	11	18	13	14	22	25	41	39	39	45	41	43	65
Chicken	13	12	9	11	9	0	2	8	18	11	17	14	17	23	26	43	41	41	47	44	44	64
Penguin	13	12	10	12	9	2	0	8	19	12	18	14	18	24	25	43	41	41	47	43	44	64
Turtle	14	13	9	11	9	8	8	0	21	10	16	13	17	22	26	44	41	44	45	43	43	64
Snake	13	14	19	21	21	18	19	21	0	23	26	25	26	30	29	45	44	45	48	45	44	65
Bullfrog	17	16	11	13	11	11	12	10	23	0	14	13	20	23	27	45	44	44	47	44	44	65
Tuna	20	20	16	18	18	17	18	16	26	14	0	8	18	22	30	46	43	45	47	45	45	66
Carp	17	17	11	13	13	14	14	13	25	13	8	0	12	21	27	46	43	43	46	45	45	64
Lamprey	19	19	13	15	14	17	18	17	26	20	18	12	0	14	30	47	45	46	50	47	47	65
Fruit fly	27	26	22	22	22	23	24	22	29	23	22	21	14	0	14	44	41	41	44	45	45	65
Silkworm	29	28	25	27	25	26	25	26	29	27	30	25	30	14	0	41	40	40	41	42	42	65
Bean	40	41	42	43	41	43	43	41	45	44	46	45	47	44	41	0	10	13	45	43	43	66
Sunflower	38	38	40	43	39	41	41	39	44	41	43	41	44	41	40	10	0	13	47	45	43	67
Wheat	38	38	40	41	39	41	41	42	43	44	46	42	46	42	40	13	13	0	47	45	44	66
Candida	46	45	45	46	45	47	46	45	47	47	48	47	50	45	43	47	47	45	0	25	39	72
Yeast	41	43	41	42	41	44	43	43	45	43	45	45	47	45	44	43	45	42	25	0	38	69
N. crassa	43	43	43	43	43	44	44	43	44	44	45	45	47	45	44	43	48	44	39	38	0	69
R. rubrum	65	64	64	64	65	64	64	64	65	65	66	64	65	65	65	66	67	66	72	69	69	0

Fig. 8.2. Matrix of cytochrome c amino acid sequence differences across 22 species representing all biology. Percent difference from the human (top left) normalized to how many amino acids differ per hundred. Observations 1-3 as per text. (Data from Dayhoff, 1972).[9]

ideas about God if ever another one were needed. But don't leave. Dr. Davies is here to explain the evolution evidence in cytochrome c. He says:
"By comparing the sequences of amino acids in cytochrome c taken from different species, we can make an estimate of the evolutionary distance they have traveled from each other. To give a concrete example, human cytochrome c is identical to that of rhesus monkeys save for a single amino acid, but there are forty-five differences between human and wheat cytochrome c. Everyone knows that humans are more closely related to monkeys than to wheat; this study shows by how much...Generally speaking, the farther apart two species are genetically, the longer ago they diverged on the tree of life."[17]
This looks like macroevolution all right! Is this really the evidence at last?

2. We also see from the matrix that organisms can be grouped into subclasses or "kinds" by a similar percent sequence identity. This is shown by the boxes on the grid. Read horizontally now and not just vertically. Box divisions are seen to fall across taxa roughly consistent with the degree of difference one might have estimated on morphologic grounds. Is this evidence of macroevolution too? Look again, like Australian molecular biologist Michael Denton did back in 1985.[18]

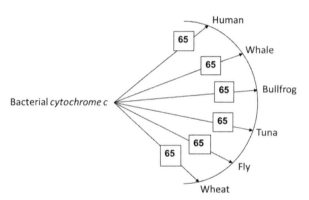

Fig.8.3. **Molecular equidistance, not molecular evolutionary trends.** Percent difference in cytochrome c amino acid sequence compared to the bacterial form. The bacterial molecule is as different from wheat (65%) as it is from human or fruit fly or tuna or frog or whale. There are no links, and the wheat sequence is not "more primitive" than the whale. Instead there is a pattern of orderliness across nature of mathematic precision. (Data from Dayhoff, 1972[9] and figure modified from Denton, 1986[18]).

It becomes apparent there are no intermediate or transitional sequences linking the subclass boxes. They are discontinuous and discrete, like islands. This is the opposite of what evolution predicts! For example read left to right across the bottom row now. Bacterial cytochrome c is not only discontinuous across taxa, the sequence differs from every eukaryotic species by a remarkably similar amount (64-72%) no matter whether it's yeast or insect or fish or bird or human. This is despite cytochrome c sequences varying among the eukaryotes themselves by as much as 50%. What does this mean?

It means bacterial cytochrome c is equally isolated from all other taxa, without intermediates linking to any of the eukaryotes (Fig.8.3). In other words where we should see molecular evidence of evolution the most, looking up from the base of the evolutionary tree, we find the opposite. Separation, even to equidistance. No evolutionary links. No evolution.

The same observation is seen at taxonomic levels above the prokaryotes. Take yeasts for example. Reading horizontally across the fourth row from the bottom, *Candida* cytochrome c is as different from the human as it is from bullfrog, sunflower and wheat. The same for plant and insect cytochrome c.

Read across the fruit fly row now (9th from bottom). Its protein sequence differs from that of cyclostome (lamprey), fish (tuna, carp), amphibian (bullfrog), reptile (rattlesnake, turtle), bird (chicken) and mammal (whale, dog, horse, pig, monkey, human) by a conspicuously similar amount (21-29%). The human sequence is as similar to fruit fly as is that of fish.

Consider the teleost (bony) fishes. Tuna and carp cytochrome c are as different from the human as they are from horse, chicken, turtle, bullfrog and fruit fly. The point is the evolutionary sequence we all know: cyclostome → fish → amphibian → reptile → mammal (*LOE#3*) is not seen at the molecular level where we should see it the best.[18] Molecules are the last level of biology we can appeal to.

In light of this discovery the first observation we made above reading down the left hand column is seen to be one of typology, not evolution because there are no links joining the taxa. *LOE#5* is rejected. We found the same thing for fossils (*LOE#2* & *LOE#3* §5.5) and for structural similarity (*LOE#4* §7.3).

3. Observe now the within-box (subclass) percentage similarities. The differences between a particular species and the others in its subclass box are just a few percentage points different. This is extraordinary. Why? It's a pattern of precision. This is the opposite of a stochastic or random process that is evolution. We make the same observation of molecular similarity as we did of anatomic similarity: Idiosyncratic patterns of ordering across

nature. Patterns without evidence of a natural mechanism. Patterns of design.

4. In corollary, the matrix also contradicts the evolutionary notion of an ancestral sequence because molecules are separated by a similar degree. The lamprey cytochrome is no more different from tuna than it is from horse. Lamprey and fish are no more "primitive" at a molecular level than amphibians are from mammals. *Rhodospirillum* cytochrome c is equally different from human as it is from bullfrog or tuna fish or fruit fly (Fig.8.3). Thus the evolutionary notion of "primitive" or "advanced" at the molecular level, *LOE#6*, is rejected too. In other examples of this we see silkworm is no more primitive from human than is rattlesnake, fruit fly from human than lamprey, or *Candida* than whale or pig or chicken. This is not a phenomenon peculiar to cytochrome c either. Similar results are found for other proteins, as perusing any protein atlas will show. In summary, matrices of protein sequence similarity across nature show separations into distinct, disjoint subgroupings to mathematical precision. There are "kinds" not links. We can say in conclusion that macroevolution is not apparent at the molecular level, right?

Wrong. Evolution is much more powerful than the evidence to the contrary. You should know that by now! (§1.5;§4.10;§5.6;§6.4;§7.8). Evolution was true before molecular biology was even known! This is when the inquiry gets interesting, and it is to take up the second task of our investigation - why evolution is a scientific fact in spite of the evidence. As you know, if evolution is a fact there HAS to be molecular evolution.

In *Biology*, the acclaimed college textbook by Peter Raven and George Johnson published by McGraw-Hill for instance, we read "molecules reflect evolutionary divergence. You can see that the greater the evolutionary distance from humans...the greater the number of amino acid differences"[19]. This is *LOE#5* of course and a simplified matrix similar to Fig.8.2 for β-globin is shown as the evidence. I say simplified because observation 1 above is made for the students but no y axis is plotted and observations 2-4 are not mentioned (Fig.8.4). The students are deceived.

More interestingly, evolution is true even in the minds of those in science who do make observations 2 and 4! This is when studying biologists is even more interesting than studying biology. It turns out the seeming flat contradiction to molecular evolution we just found is actually one of the major research fields dedicated to finding more evidence of evolution. It's all based on the so-called "molecular clock hypothesis," and it finds plenty of evolution in the data above. We just weren't looking hard enough.

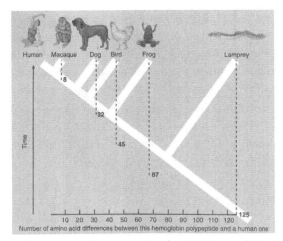

Fig.8.4. The molecular evolution of β-hemoglobin (*LOE#5*) according to *Biology* by Raven and Johnson.[19] The number of amino acids differing from the human 146 long amino acid sequence is shown. Students are told: "Comparison of the DNA of different species strong evidence for evolution. Species deduced from the fossil record to be closely related are more similar in their DNA than are species thought to be more distantly related". The caption to this figure reads: "Molecules reflect evolutionary divergence. You can see that the greater the evolutionary distance from humans (white cladogram), the greater the number of amino acid differences in the vertebrate hemoglobin polypeptide". How true this is can be seen by reading horizontally compared to vertically in Fig.8.2. (Reproduced with permission Raven and Johnson, 2002 ©McGraw-Hill).

§8.4   THE MOLECULAR CLOCK HYPOTHESIS

How can we account for the almost identical molecular divergence of the subgroups from each other as shown by observations 2 and 3 above by evolution? (Fig.8.2-3) Only by every cytochrome c molecule in every species within a particular subclass on the grid undergoing mutational change at the same rate after their divergence from the common ancestor. After each divergence at a solid grid line, the different species in the box it circumscribes must separately undergo mutation and result in the same net amount of change relative to those species in the box and to others in other boxes. This process must apply not only to cytochrome c of course but every other protein in the cell that arose by evolution too. It turns out different proteins differ to different degrees across taxa. For example frog β-hemoglobin is 46% different from human but its cytochrome c 17% different (Fig.8.2; Fig.8.4). Thus each protein in the cell evolves at its own unique rate, but at the same rate across different taxa. It's the basis of evolution theory's so-called "molecular clock hypothesis". Take a look.

Reading the left column in Fig.8.2 reading top to bottom and so working up from the base of the standard evolutionary tree: prokaryote → eukaryote → plant → animal → vertebrate, since cytochrome c in eukaryotes is equally different from prokaryotes (~65%) it must have

Evidence of molecular evolution

evolved at exactly the same rate in all eukaryotes after their divergence off the tree from prokaryotes. Consider this the first node of the tree. Similarly since the sunflower cytochrome c is equally different from all higher taxa (38%), the same net amount of evolution of this protein happened independently in all animals after their divergence from the ancestral plant at the second node. The same applies to the evolution of the animals after the third node of the tree, as insects are equidistant from lamprey as they are from humans. Since cytochrome c in all vertebrates at the evolutionarily lowly lamprey level is as distant from all the non-vertebrates (19%) as from the lamprey cytochrome, the cytochromes of all other vertebrates also all evolved at the same rate after their divergence at the fourth node. We see the evolution of amphibians (e.g. bullfrog cytochrome is equally different from reptiles, birds and mammals) and birds (chicken and penguin equidistant from mammals, reptiles and amphibians) had to happen in the same way too.

In summary, the equidistance of molecules across kinds can only be explained by evolutionary terms if the rate of evolutionary change of a protein is exactly the same in all organisms after their divergence from the common ancestor. The problem is how mutations could accumulate so uniformly in this way to produce such a wonderfully ordered pattern. Recall mutations do not accumulate by absolute time but by generation time, and generation times differ by orders of magnitude across nature. It's a few minutes to many decades and thus they must vary across taxa by tens to hundreds of thousand fold (§2.4). And yet their constituent proteins can differ by only a few percentage points, a result already not plausibly attained even with generation time differences of several fold difference (Fig.8.2). Although nobody has yet come up with an objective explanation how such a fantastic process might occur exactly, it's believed all the same. Maybe it's true nonetheless. We can at least test if this belief is internally coherent with what it claims, and this might be all we need to tell.

§8.4.1 How It Ticks and How It's Set

The molecular clock was originally proposed in 1962 by two-time Nobel laureate Linus Pauling (1901-1994) and his colleague Emile Zuckerkandl. They postulated that the number of amino acid differences in any given protein increased with an increasing time of divergence from the common ancestor.[20] It might seem intuitive but it's a highly simplistic idea by any measure knowing the complexity of protein interactions. It also assumes macroevolution is already true. If macroevolution is false then it's as misleading as cladistics (§5.3.3;§7.4.1;§10.2.3). The purpose of the molecular clock today is taken to reconstruct the evolutionary history of molecules, first by determining the branching order from the common ancestor, and then the dates of the branches. The latter is done by making an assumption the molecular evolutionary rate has been approximately constant over all of geologic time.

The molecular clock hypothesis can be formally stated in the words another pioneer of this field, Japanese National Institute of Genetics biologist Motoo Kimura (1924-1994): "For each protein, the rate of evolution in terms of amino acid substitutions is approximately constant per year per site for various lines, as long as the function and tertiary structure of the molecule remain essentially unaltered."[21] The hypothesis holds that although mutation is a stochastic process (§5.3.1), and although it is related to generation time, when rates of change are measured over long periods proteins and nucleic acids evolve at a constant rate relative to absolute time in a particular lineage. Can we believe it?

The early work uncovering evidence for molecular evolution was done with protein sequences not gene sequences because protein-sequencing technology was available first (1953 vs. 1977). The results seemed to give a resounding confirmation of the molecular clock hypothesis, at least at first. For example the analysis of the cytochrome c protein sequence is shown in Fig.8.5. Elaborate phylogenetic trees have been derived from it and championed still today.[22] Another prototypical example of this evolution evidence is the α-globin and β-globin proteins. They join with heme to form α-hemoglobin, the oxygen transporter of adult red blood cells. As seen in Fig.8.5 there is a linear correlation of sequence divergence across time, just as predicted by the molecular clock hypothesis.

This is how the molecular clock "ticks": each tick is another accumulated mutation at some rate that is constant for that particular protein. Thus each protein is itself a ticking clock over geologic time in the cell. But how is the clock set to enable evolutionary comparisons to be made retrospectively like this by the science? There are two ways.

The first is by using the fossil record, and specifically by "knowing" the time of divergence of two species from the common ancestor. This is an inference and it also assumes macroevolution of course, but with that date the rate of sequence change over time, called the molecular evolution rate of the protein or gene, can be derived. To see how this works, consider the amino acid sequences of a protein shared by two different species - lysine-proline-leucine in species A and lysine-glycine-leucine for that protein in species B. If the two species diverged from a common ancestor one million years ago, the substitution of glycine for proline in species B occurred at an evolution rate of one amino acid change per two million years. (Twice the time from the extant species back to the common ancestor since evolution is acting on both diverging sequences simultaneously). Conversely knowing the molecular evolution rate from such calculations, a construction of phylogenetic relationships and branch points for an evolutionary tree can be done, without the need for fossil evidence after the calibration. The assumption is the greater the number of substitutions, the greater is the evolutionary distance between the molecules.

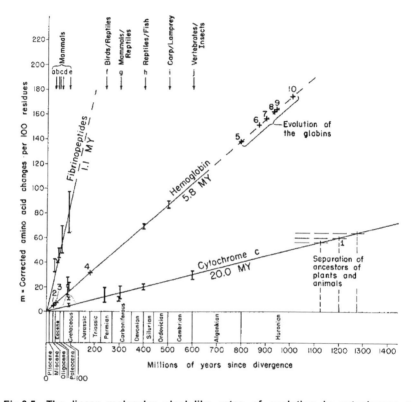

Fig.8.5. **The linear molecular clock-like rates of evolution in cytochrome c, hemoglobin and the fibrinopeptides**. All kinds of inferences were required for this drawing beginning with macroevolution a fact and all life descended from a single common ancestor (*Macroevolution & Common Ancestry Inferences*, §3.2). After that molecular evolution has to be a fact, or as the author of this figure Richard Dickerson at Caltech says: "protein molecules contain a record of their evolutionary history that is fully as informative as that obtained from macroscopic anatomical features".[23] It's a small step thereafter to insert trees to the seeming molecular evolution and fossil record data as seen above, below and to the right of the three plots. Molecular macroevolution is now indisputable and reinforcing. (Reproduced from Dickerson 1971 with permission).[23]

The second method not only requires that macroevolution be true, but the course it took be known also. It's called the "relative rate test" and it uses three sequences in the comparison, one of which is "known" to be an outgroup. The clock hypothesis becomes a very powerful tool this way, even compared to fossil data. Penn State University biologist Blair Hedges says the molecular clock is actually the best way to reveal macroevolution:

"[F]ossilization is rare. Another problem is that the number of characters known for fossil species is often limited. Many extinct species are named from teeth, jaw fragments, or other small remnants. This poses problems in distinguishing one fossil species from another, and in trying to determine relationships using a limited number of characters...Fortunately molecules can give us an additional perspective...Each living species has thousands of genes and millions of nucleotide sites that can be sampled for evolutionary studies...Molecular clocks...also have an advantage over fossil clocks because they start counting mutations at the actual time of divergence. Fossil-based estimates of divergence time, on the other hand, are minimum estimates only."[24]

No wonder the molecular clock hypothesis is applauded the way it is, and accolades include: "[t]he discovery of the evolutionary clock stands out as the most significant result of research in molecular evolution",[25] and "[m]olecular clocks have revolutionized evolutionary biology".[26]

The legitimacy of the praise rests on an expectation there is a clock that can keep time of course. It also rests on the assumption the macroevolution that powers the clock is true too. You may have recognized already that if the clock is set to the fossil record, which we know is interpreted on the basis of inferences and assumptions, then molecular homology is not independent of the fossil record (§5.3). Neither of them is independently objective and both rest on inferences macroevolution is true. Circular reasoning is not the only problem with the molecular clock though.

Another is its assumption that mutations at branch points in phylogenetic trees remain unchanged after they have occurred. Multiple superimposed mutations at a particular locus could occur over time however, and a new mutation be erased by a subsequent mutation at the same locus either in the same species or a descendant. This would cause an underestimation of the true evolutionary rate. If we refer back to cytochrome c for instance, for which a full 65% of the amino acids of the protein are changed between bacteria and all other species, this is more than just a wrinkle in the reasoning (Fig.8.2).

The next observation to become apparent as more proteins were sequenced, is that the rates of molecular evolution were different among different proteins in the same cell. Every protein has its own molecular clock, ticking at its own unique rate. This is was apparent to us earlier from the different slopes of the lines for hemoglobin and cytochrome c in Fig.8.5. Since hemoglobin sequences differ more across taxa than does cytochrome c, it follows that hemoglobin evolves faster. Speeds differ enormously. Other examples are listed in Table 8.6. Histone protein H4 is famously stable, and changes at a rate of 0.01 per amino acid site along its sequence per billion years. Insulin on the other hand has molecular evolution 40 times greater, and for the fibrinopeptides, the cleavage products of fibrinogen generated by the blood clotting cascade, it's over 800 times still faster again.[21] These disparities are explained by functional constraint and dispensability. How can this happen?

Protein Name	No. of amino acid changes per 10^9 yr
Fibrinopeptides	8.3
Pancreatic ribonuclease	2.1
Lysozyme	2.0
α-Globin	1.2
Myoglobin	0.89
Insulin	0.44
Cytochrome c	0.3
Histone H4	0.01

Fig.8.6. Variable rates of molecular evolution for different proteins. The rates expressed as number of changes per amino acid site per 10^9 years.
(Data from Kimura 1983 with permission).[21]

Two mechanisms are proposed. The first is by natural selection of advantageous mutations which improve fitness. This is classic Neo-Darwinism with which we are very familiar (§2.2-3). The other is by so called "genetic drift", meaning mutations with no effect on fitness, also called "neutral" mutations. Which of these is the predominant mechanism of molecular evolution continues to be disputed by the two camps in the field, the "selectionists" and "neutralists", but neither has been able to provide a convincing solution to the problem.

The problem selectionists need to explain is how the rate can be so uniform for a particular protein across all phylogenetic lines, and why the rates differ so much. For instance how can fibrinopeptides have the greatest evolutionary rate in light of their known biological function? What selective constraints are operating? Another problem is that so-called living fossils have sequences that do not significantly differ from supposedly much more recently diverged organisms. For instance snapping turtles are held to have diverged 57 Myr ago and alligators 35 Myr ago, but they have cytochrome c sequences that differ by 14%[9] and 13%[27] respectively from the human sequence and which are not significantly different from supposedly much more recently diverged organisms, like the chicken and horse (Fig.8.2). Do the neutralist explanations fare any better? We need to inspect the idea of genetic drift in more detail to know.

§8.4.2 The Neutral Theory of Molecular Evolution

It was Motoo Kimura who provided the mathematical formulation of the molecular clock based on genetic drift and that is now called the neutral theory of molecular evolution. To Dr. Kimura's credit and consistent with a scientific hypothesis, the model is both predictive and testable. It rests on two assumptions.

The first is that mutations (i.e. amino acid or nucleotide substitutions) are relatively constant over time because they are literally, the ticking of a clock.

The second is that substitutions are neutral, meaning they have little or no functional consequence on the fitness of an organism. Changes in amino acid or gene sequences by evolution are therefore dependent on mutation rate and time, not natural selection. Needless to say, this is a massive conceptual break with Darwinism! If the alternative alleles of a gene are neutral to natural selection, their relative frequency in the population occurs by random sampling called "genetic drift". Kimura demonstrated that the rate of substitution of neutral alleles is the same as the rate of mutation of those alleles. For the mathematically inclined the proof is as follows:

> The rate of nucleotide substitution k in a diploid population of size $2N$ is the number of new mutations μ, multiplied by their probability of fixation in the population u.
> Therefore $k = 2Nu\mu$.
> The probability of fixation is the reciprocal of population size, so $u = 1/2N$. Substituting for u in the equation, $k = (2N)(1/2N)\mu = \mu$.)

The neutral theory does not deny natural selection. It just holds favourable mutations are so rare that they have no significant effect in the overall rate of evolution of nucleotide or amino acid substitutions. Similarly, the bulk of mutants are deleterious and removed by natural selection without affecting the rate of the molecular evolution. In Kimura's words:
the "theory does not deny the role of natural selection in determining

the course of adaptive evolution, but it assumes that only a minute fraction of DNA (or RNA) changes are adaptive."[28]
It holds that most evolutionary change occurs by random drift between forms that are equivalent, or "neutral" to the action of natural selection. No wonder selectionists and neutralists fight! How can theories so opposite yet be both true in evolution science? (§20)

The differences between them boil down to the differences in relative frequencies of neutral and adaptive mutations, but how does one determine their absolute frequencies to resolve the dilemma? Actually a direct test is "impossible," says Mark Ridley. As a result the verification of the neutral theory's modification of Neo-Darwinism has degenerated to narrative debate by neutralist and selectionist viewpoints, not proof or disproof. It comes down to personal scientific opinion or words, in other words. We have seen plenty evolution faith already, but this is the first time we find interdenominational differences. Dr. Ridley explains:

"The case for the neutral theory mainly consists of the reasoning that the conditions needed for natural selection to produce the observed patterns of molecular evolution are false, or contradictory, whereas neutral drift runs into no such difficulties; and the case for the selective interpretation mainly consists of showing that the difficulties are illusory and that the neutral theory does not fit the facts."[29]

There should be more to the differences than this. Let's take a closer look to see if light can be shed on the matter nonetheless.

§8.5   A CLOCK THAT CANNOT KEEP TIME!

Initial protein sequence comparisons seemed to confirm the idea of a constant clock-like rate of change in amino acid substitution over all of geologic time. Dr. Kimura even cited rate constancy as the best evidence for the neutrality of mutations.[21] The huge variation in evolution rates, over 800 fold (Table 8.6), was explained by functional constraint and dispensability. As more genes and proteins were sequenced however, it became clear that rates were not just different in different proteins, they were different across different species *for the same gene or protein*. In other words the molecular clock can't keep time. Even a stopped clock is right twice a day, but this clock is can't even do that. Take a look at three representative examples of supposed clear-cut molecular evolution to see what I mean - cytochrome c, relaxin and hemoglobin, and we will inspect the machinery of the clock at the same time.

§8.5.1   Not Exactly Swiss

Since cytochrome c is around 100 amino acids long, the percentage difference across species can for the sake of argument be considered the number of amino acids that are different. Thus 39 amino acids are the same across all of nature, one third of the amino acids are invariant. The neutral theory predicts nucleotide substitutions are random and stochastic, like radioactive decay, and that the number of substitutions has a Poisson distribution with a mean of kt, where k is the rate of substitution and t is the age of the lineage. For a Poisson distribution the mean and the variance are identical i.e. the ratio or dispersion index R = mean/variance = 1. If the data fit a Poisson process, the variance in the number of substitutions between different lineages should not be greater than the mean. Thus the dispersion index (R) becomes is a simple test of the machinery and the internal coherence of the molecular clock. What does it show?

Analysis of 20 different proteins has found R to vary from 0.32 to 43.82. In more than half R was greater than 1, and the average was just under 7.[30,31] What's even more interesting is that the response by neutralists has not been to reject the notion nor even to suggest an overhaul. It has been to claim 1) faster rates of evolution, 2) lateral gene migration or 3) that phylogenetic relationships and divergence times used to make the calculations are incorrect, and therefore the derived molecular evolution rates invalid. Since ancestors are hypothetical and phylogenetic relationships are inferred, there is no direct evidence to resolve this in an objective manner. Like the fossil evidence for evolution the evidence for these post hoc explanations of the data is the explanation so maybe it's true, maybe it's not. Is there another way to test the accuracy of the molecular clock?

Consider the "relative rate" test alluded to earlier. Here the number of different amino acid or nucleic acid substitutions between two closely related taxa is compared with that of a third and supposed more distantly related taxon. This analysis requires no knowledge of the divergence times of the taxa in question. The molecular clock breaks down under this analysis just as easily however. To see this take a look at two examples - relaxin and hemoglobin.

§8.5.2   Telling The Time By Relaxin

Relaxin is a hormone that widens the mammalian birth canal to allow the fetus to exit the pelvis. In 1986 biochemist Christian Schwabe at the

Medical University of South Carolina used relaxin homology to draw attention to the ad hoc status of the arguments above to maintain the status of molecular evolution against the inconsistencies of its predictions with the evidence of biology.[32] Amino acid sequence homologies (i.e. the inference the sequence similarity is explained by common ancestry) for relaxin across species reproduced from his paper is shown in Table 8.7.

	Human	Pig	Rat	Shark	Skate
Human		46	46	48	35
Pig	46		54	50	31
Rat	46	54		37	25
Shark	48	50	37		42
Skate	35	31	25	42	

Table 8.7. Matrix of relaxin sequence differences across taxa. Numbers shown as percent amino acid sequence similarity. (Reproduced from Schwabe, 1986 with permission).[32]

The percent similarity of relaxin homologies between human and pig and human and rat is identical at 46% (i.e. 54% different). It recalls the observation made for cytochrome c among "subclasses" on Fig.8.2 earlier, a pattern of orderliness to mathematic degree. Considering that pig, rat and human are held in evolution theory to have diverged from one another only recently, a 54% difference between them is large. The explanation is that it's from rapid evolutionary change in the hormone after divergence from the common ancestor. But this is no explanation. It's just a restatement of the observation with an untestable mechanism. In other words, a semantic illusion made by a simplistic explanation. If we are to believe it we must accept that shark relaxin, which according to the fossil record diverged about 500 Myr ago, is almost as different from pig as pig is from human although mammals supposedly diverged only 70 Myr ago.

This requires not only different speeds of molecular evolution by the same protein molecular clock, but the requirement of pig relaxin to continue to mutate after its function was acquired, and also maintain that function despite a 54% change to evolve into the human sequence. But this is not the only implausibility.

The minke whale (*Balaenoptera acutorostrata*) differs from Bryde's whale (*Balaenoptera edeni*) by only three residues, but the latter differs from pig by only one. To cap it all in 1999 it was shown that the lowly tunicate *Ciona intestinalis*, with a fossil record dating back to the Cambrian, has relaxin of identical sequence to that of pig.[33] Based on evolutionary theory, the phylogenetic distance between tunicates and pigs is essentially all of eukaryote evolutionary time. Yet their sequences are identical while those of supposedly closely related mammals differ at more than half of their amino acid positions and are almost identical in whales that are more distantly related, or so evolution says (§5.5.4.1). How can an evolutionary tree be constructed here? To add to these problems, shark relaxin has been shown to have biological activity in mice and in guinea pigs! It widens their pelvic bones although these structures, by evolutionary theory, developed hundreds of millions of years later. The notion of evolutionary molecular homology, at least for relaxin, is a morass of contradictions and incoherence.

§8.5.3   Hemoglobin Horology

In light of α-hemoglobin being supposed such a clear-cut example of molecular evolution (Fig.8.5), we should look more closely at its evidence too. Hemoglobin transports oxygen. In mammals the molecule is a tetramer (i.e. four protein chains bound together) of two alpha-like and 2 beta-like globin chains each attached to a heme molecule containing iron to which the oxygen binds. Humans have three alpha-like genes and five different beta-like globins to choose from. As regards the alpha globins, there are two adult α types called $α_1$ and $α_2$ and a third form found only in the embryo called ζ. As regards the beta-like globins, the five different forms are called ε, ^{G}γ, ^{A}γ, δ and β. Like the alpha-like globin genes, the beta-like globin genes are expressed at different times of human development and specifically during early embryogenesis, fetal stage, infancy and adulthood. The adult form of hemoglobin is $α_2β_2$. The human α genes and the β genes are clustered together on different chromosomes (16 and 11 respectively) and aligned along the chromosome in exactly the same order in which they are expressed during development, in quirky orderliness.

Since the α and β globin gene sequences are so similar, it is concluded they originated as a duplication from a single ancestral gene around 500 million years ago and somehow ending up on different chromosomes.[34]

Another question is what the steps were along the way to $α_2β_2$ from 500 million years ago, the supposed time of evolutionary vertebrate origin. We should see an accumulation of globin mutations at a constant rate in the vertebrate evolutionary tree and the accumulation of alpha and beta mutations should be related to each other across species. No such pattern is evident (Fig.8.8). "These numbers are a direct and obvious disproof of the whole concept of protein phylogenetics," says Cambridge mathematician, astronomer and panspermia advocate Sir Fred Hoyle. The key assumption of neutral theory, in Kimura's words, that "the rate of evolution in terms of amino acid substitutions is approximately constant per year per site for various lines", is not what we find. This contradiction is not taken as reason for doubt about molecular evolution either. Instead it's explained away, unassailably, by holding that the rate of evolution is faster in some species than in others for a particular protein or gene. So maybe it's true, maybe it's not. At least two things have become clearer.

	Human	Mouse	Rabbit	Dog	Horse	Cow
Human α		18	25	23	18	17
β		27	14	15	25	25
Mouse α			27	25	24	19
β			28	30	36	39
Rabbit α				28	25	25
β				31	25	30
Dog α					27	28
β					30	28
Horse α						18
β						30

Fig.8.8. Matrix of α-globin and β-globin sequence homology across taxa. (Reproduced from Hoyle, 1999 with permission.)[35]

The first is that macroevolution remains unshakable despite molecular evolution predictions being refuted. We saw this already when the fossil record refuted Darwin's assumptions (§5.4-5), and when the notion of morphologic homology was found riddled with internal contradictions from genetics and from embryology (§7.3). The interpretation of the data is changed to fit the assumption that evolution is already a fact (§3.1;§19;§20).

The second is that the focus for research has been to explain why the evolutionary molecular clock is not manifest in the observations, not whether the evolution driving the clock is even there. In fact the variability in the 'tick' of the molecular clock, an oxymoron if ever there was another like "living fossils", anywhere, but especially in science, has prompted several hypotheses to account for it. The problem to explain is why mutation rates are not constant for a given protein across different species. This has been variously attributed to between-species differences in:

1) Generation time (shorter life cycles accelerate the clock because it takes less time to fix the mutation in a population),
2) Population size (smaller populations require less time to fix a new mutation in a population and accelerate the clock),
3) DNA repair efficiency (less fidelity of DNA Polymerase function or other intracellular mechanisms which result in poorer repair of mutations accelerates the clock), and
4) Metabolic rates between species (the claim being organisms with high metabolic rates have higher rates of DNA synthesis and consequently higher rates of mutation so accelerating the clock).
5) Changes in function of a protein over evolutionary time, particularly after gene duplication. The increased rate of evolutionary change of the globin genes after duplication is claimed as example of this (in circular reasoning).
6) Incorporating slightly deleterious mutations into the theory. Such mutations behave as neutral in small populations, but are selected against by natural selection in large populations. This has been called the "nearly neutral theory" of molecular evolution and was first proposed by Tomoko Ohta, a colleague of Motoo Kimura at the National Institute of Genetics.[36,37]

We should inspect each of these hypotheses for explaining why the molecular clock does not behave like a clock, because if they are all false the obvious explanation that the clock does not behave like a clock is because there is no clock. The reason there's a clock of course, is there must be a clock if macroevolution is a fact, and it is a fact (§3.1).

§8.6   WHY CAN'T IT CLOCK? TESTING THE MOLECULAR CLOCK HYPOTHESIS

§8.6.1   Flies In The Face Of Evolution

How plausible are these six explanations for why the molecular "clock" can't keep time? Francisco Ayala, past President of the American Association for the Advancement of Science and another Templeton Prize winner, tested them all when he analyzed the molecular evolution of the enzymes glycerol-3-phosphate dehydrogenase (GPDH) and copper-zinc superoxide dismutase (SOD).[31,38,39] Amino acid sequences were compared in 27 different dipteran (two-winged insect) species from 2 families, 3 genera and 5 subgenera and then compared with 9 other species from 3 mammalian orders (man, mouse and rabbit), 3 metazoan phyla (nematode, arthropod and chordate) and 3 multicellular kingdoms (fungus, plant, and animal). The results are shown in Table 8.9.

GPDH is a crucial enzyme in the glycerophosphate cycle that provides energy for dipteran flight muscles. As can be seen reading down the extreme right column of Fig.8.10 and moving up the taxonomic levels, the rate of amino acid replacement in GPDH by evolution (for convenience expressed in units of 10^{-10} amino acid replacements/site/year) is 0.9-1.1 between Drosophila species and subgenera, rising to 2.7 between dipteran genera that diverged only a little earlier, to 4.7 between dipteran families, 5.3 between mammals, 4.2 between animal phyla and 4.0 between kingdoms. There is a 15 fold different rate of evolution across different Drosophilids. Rather than a constant rate of amino acid replacement, the clock both speeds up and slows down with increased evolutionary time. This is a watch for the March hare.

Taxa Compared	Divergence Date (Myr)	Evolution Rate*	
		SOD	GPDH
Between Drosophila groups	45 ± 10	16.6	0.9
Between Drosophila subgenera	55 ± 10	16.2	1.1
Between fruit fly genera	60 ± 10	17.8	2.7
Between mammals	70 ± 15	17.2	5.3
Between fruit fly families	100 ± 20	15.9	4.7
Between animal phyla	650 ± 100	5.3	4.2
Between multicellular kingdoms	1,100 ± 200	3.3	4.0

Table 8.9. Rates of protein evolution for Superoxide Dismutase (SOD) and Glycerol-3-phosphate Dehydrogenase (GPDH). *The evolution rate is measured as amino acid replacements per site/year x10^{-10} after correction for multiple replacements. The same species are compared for both genes. Standard errors are typically 1-5% of the mean. Myr is million years since divergence. (Reproduced from Ayala, 1999 with permission.)[38]

These results should be compared with those of another cellular protein, superoxide dismutase (SOD) which is an enzyme that scavenges toxic free oxygen radicals. Now the pattern is reversed! Referring to Table 8.10 third column from left, the rate of amino acid evolution slows down with increased evolutionary time from around 16 for Drosophila species and subgenera, to 3.3 between the multicellular kingdoms. There is also a dramatic deceleration in rate for those species diverging at 60 to 1,100 Myr (from 17.8 to 3.3), a time during which the rate of GPDH evolution by contrast is relatively steady. The widely fluctuating rates of the molecular clock for GPDH and SOD, and their going in opposite directions, are not explained by hypotheses 1-4 above because they occur in the same species. Such influences should produce changes to GPDH and SOD mutation rate in the same direction.

Other rebuttals to hypotheses 1-4 above could be advanced. For instance generation times (hypothesis 1) cannot account for the discrepancies as mammals are seen to diverge at the same rate as fruit flies in the case of SOD, and even fivefold faster for GPDH. There is also no evidence to suggest that the function of these key enzymes has changed over time to support hypothesis 5.[38] This leaves hypothesis 6, which unlike the original neutral theory, has no predictive power because it claims the vagaries of natural selection as post hoc cause of any deviation in molecular evolution rate. And so it is also unassailable, a far cry from the original neutral theory advanced by Kimura that rested on mathematical foundations and was testable and on testing shown false. Fact is, rather than the promise of the molecular clock to reliably reconstruct phylogeny, we see the opposite. Based on the rate of protein evolution of GPDH, the divergence of animal phyla occurred at 2,500 Myr and that of the multicellular kingdoms at 3,990 Myr. Based on SOD however, they were at 211 Myr and 224 Myr. The molecular clock is completely unpredictable which is to say there is no clock, by any definition of the word.

Related results that also do not fit the predictions of the molecular clock come from other quarters. Three species of Drosophila (*D. melanogaster, D. simulans and D. yakuba*) are closely related fruit flies that supposedly descended from a common ancestor 5 Myr ago. Toshiyuki Takano at the National Institute of Genetics in Mishima, Japan has performed a comparative analysis of DNA sequence evolution in these species.[40] The dispersion index (R) was greater than 1 for 9 of 13 genes for synonymous substitutions, 8 of 13 for nonsynonymous substitutions, and in only 1 gene was it <1 for both types. Interestingly, no correlation between dispersion indices for the two kinds of substitution was seen. High rates of substitution for one kind of substitution were found with either high or low rates for the other. In a further surprise, there was also no correlation between the rates of substitution and the location of genes to chromosomal regions with known low or high rates of recombination and polymorphism.

Dr. Takano's interpretation of the unpredictable sequence changes was to blame natural selection, postulating "change(s) in action of natural selection" and "[f]luctuating selection intensity" as cause. Francisco Ayala's assessment of the data is much simpler: "It seems, therefore, that once again only fluctuating and unpredictable natural selection can account for these erratic patterns of molecular evolution."[38] Resorting to a "fluctuating and unpredictable" process to explain why the evidence of biology denies the predictions of the molecular clock hypothesis is no advance in understanding. It's also hardly a mechanism and might just as well be mystical, or non-existent. Like The Emperor's New Clothes. Why should we believe there's a molecular clock again? (§19)

§8.6.2　　　　　　　Losing Time With Albumin

Another test of the assumptions of the molecular clock hypothesis is done by measuring the molecular evolution of the albumin gene.[41] Albumin is part of a family of genes that consist of four evolutionarily related members (considered to be so because they are so similar), specifically albumin (ALB), vitamin D binding protein (DBP), α-feto protein (AFP) and α-albumin (ALF). They are held to have arisen by a series of duplications from a common ancestor, first to form DBP and the precursor of a second lineage, which reduplicated to give rise to ALB and a common precursor to AFP and ALF. Rather than a constant rate of molecular evolution, the homologous proteins have different rates of evolution, varying from 36 amino acid replacements/100 Mya for DBP to 54 for ALF. This is also not what the molecular clock predicts! The countering post hoc explanation is that this is due to hypothesis 5 above. Namely that the molecules have changed in function over time, notwithstanding the fact that those of AFP and ALF still are unclear, but even that explanation is rejected after a further analysis.

　　The proteins each consist of 3 separate domains of similar length. The rates of evolution of the component internal domains differ even more than they do for the protein taken as a whole! There is also no apparent trend in any of this tangle. In domain I they vary from 25%-65% and rank AFP>ALF>ALB>>DBP. For domain II they vary from 37-52% and rank AFP>DBP>ALF>ALB, and for domain III they vary 37-69% with a ranking ALF>>DBP>ALB≈AFP. Instead of the prediction of the molecular clock of constant rate of change following divergence from a common ancestor, on 3 levels different variation in unpredictable directions is seen. Between the four members of a closely related gene family across the same species, for the same protein across species, and internally between component homologous domains. This was the conclusion of the authors of that study: "Even if the molecular clock runs at a constant rate but the constancy is in reference to any altering function of a protein, the clock is intractable...Models are only models; they are only as good as the underlying assumptions. And if the underlying assumptions (unknowns) is great than the number of equations, a rigorous solution is but an illusion. This seems to be the case with the molecular clock."[41]

A similar incongruous variation in evolutionary rates among component domains of a protein is seen in hemoglobin molecules of the brine shrimp *Artemia*.[42] *Artemia* hemoglobin is a dimer of two covalently bound, very similar polypeptides encoded by two genes that are held to have duplicated. Each polypeptide has 9 domains that can be easily identified to permit homologous sequence identity comparisons between the polypeptides. The two polypeptides differ at 11.7% of the amino acid sites. However the differences across the 2 polypeptides for each of the nine domains are different, varying by factors of up to 2.7 fold. At the nucleotide level it is just as baffling. As with the *Drosophila* genes above there is discordance between synonymous and non-synonymous substitutions which vary in different directions, even for the same domain. Hypotheses 1-4 are rejected as they occur in the same lineage, and since each polypeptide is used in the same protein, hypothesis 5 is rejected too.

§8.6.3   The Paradox of the C Paradox

One last piece of evidence to reconcile is the so-called C paradox. It's called a paradox because it's contrary to the phylogenic tree posited by evolutionary theory. Evolution science does not consider that reasoning to be contrived or inverted. The observation is the following. While most of the DNA in the genomes of Bacteria and Archaea code for proteins (e.g. 88% in *E.coli*) it appears that as much as 97% of the DNA in vertebrates does not. To what extent and how this DNA has function has still to be clarified. The term used to refer to the total amount of DNA per haploid genome of an organism is called the "C value". The C value varies widely, both across and within taxa, and so much so it's called the C-value paradox because it varies without relationship to morphological complexity or gene number or evolutionary phylogenetic hypotheses.

For instance there's about 200 times more DNA in amoeba, *Amoeba dubia*, (670 million kb) than in man (3.3 million kb). The pea (*Pisum sativum*) has about as much DNA as man, and about tenfold less than the lily *Lilium longiflorum*. There is a twofold variation in C value among mammals with the aardvark and the Muntjak deer representing the opposite extremes, a tenfold variation among frogs and toads, one hundred fold among insects and three hundred and fifty fold among the teleost (bony) fish.[43] Chromosome number varies widely too: 48 in potato, 24 in tomato, 20 in corn, 8 in Drosophila, 40 in mouse, and 46 in man. These differences are not accounted for by differences in the number of genes because these also vary widely: about 25,000 in man (*Homo sapiens*), 80,000 in mouse (*Mus musculus*), 50-100,000 in fish (*Fugu rubripes*), 14,000 in fruit fly (*D. melanogaster*), about 34,000 in the potato (*Solanum tuberosum*) and 4,288 in *E.coli*).[43,44]

§8.7   IN CONCLUSION: GARBAGE IN GARBAGE OUT

In sum evolution as demonstrated by molecular homology (*LOE#*5, *LOE#*6) is swamped in contradictions and incoherence, and even the most basic of the observations is an unexplained puzzle. Specifically "[t]he puzzle for evolutionary biologists is that the amount of DNA is not associated with genetic or morphologic complexity – some 'simple' organisms, such as amoebae, can have enormous genomes."[43] The most interesting observation of all though, is that the notion of molecular evolution is not rejected nor the truth of macroevolution questioned.

Molecular clocks and their implications and their reasoning are published in the best scientific journals and even whole journals are dedicated to the idea as if it were empirical and true, like the *Journal of Molecular Evolution, Molecular Biology and Evolution,* and *Molecular Phylogenetics and Evolution.* Such is the power of molecular evolution that even Dr. Ayala, having debunked the clock by his analysis of SOD and GPDH, still will not reject it in logic scientifically so unfathomable it is surely only perverse. Listen to his lament:

> "Whither the molecular clock? Should it be abandoned altogether? The theoretical foundation originally proposed for the clock, namely the neutrality theory of molecular evolution, is untenable…Without theoretic underpinnings to buttress the clock, we will remain unable to anticipate when the clock may be erratic or how erratic it may be…I am not however, willing to propose that we give up altogether the molecular clock. There are many evolutionary issues concerning both timing and topological relationships between species for which molecular sequence data provide the best, if not the only dependable evidence."[38]

This is not grounded in evidence and his own data shows why. Which leaves only philosophy. Evolution is already true so the clock must be true. Exactly how just needs to be discovered (§19).

The research continues today, even toward developing multiple protein clocks so they may be used concurrently to improve accuracy in an assumption there is greater reliability when there is consensus about an unpredictable result. To say this scientifically in the words of experts:

"To infer origination times for species from molecular data requires the assumption that the rates of nucleotide substitution in a given gene remain constant or nearly constant with time. This is an astonishingly simplistic view given the complexity of biological processes. In spite of evidence that this hypothesis is incorrect, very little has been done to develop more accurate methods of molecular clock calibration."[45]

It should come as no surprise to learn phylogenies constructed on morphologic similarities and those constructed on molecular similarities conflict.

What then are we to make of the molecular clock? The evidence of biology does not support the idea, even if we grant the macroevolution supposedly powering it as true. If macroevolution is false, the entire enterprise is rubbish. For all the complexities of the calculations invoked, and all the elaborate phylogenies derived from supposed examples of molecular homology, they have no grounding in biologic reality. The internal incoherence is itself a disproof.

And so we find that just as there are morphologic similarities across nature, both in extant life and fossil life without evidence of a linking mechanism, for the former no hereditary mechanism and for the latter a fossil record of stasis and sudden appearance, so now we find it at the molecular level. Molecular similarity but no intermediate molecular forms. So no links, and no more basic level of biology to appeal to.

This time we can also quantitate the observation of the patterns of molecular similarity across biology as probabilities. For a small protein, say of 100 amino acids for which we will assume 20 equally possible amino acids could be found at each position, the chance of natural construction producing that sequence is 20^{100} or 10^{130}. To get technical and account for synonymous mutations and different amino acid frequencies, it's less, around ~10^{125} (§12.4.2), and assuming that somehow only L-amino acids get incorporated into the growing chain without cross inhibition, the odds are ~1 in 10^{65}. This is still no chance and which is my point, and this is for just one small protein.[46] The odds are multiplied with each additional protein required for living. The multiplied probability for only two small proteins is already hopelessly beyond even the total number of atoms in the Universe multiplied by the number of nanoseconds since the supposed time of the Big Bang.

There is a related point that also gets to the impossibility of stochastic random processes to generate these highly ordered conformations of matter that are the molecules of life. Not just the absence of links, but how were they discovered so well by chance in the first place? Complex purposeful arrangements of amino acids into functional molecules, but no stepping-stones. No molecular intermediates, no evolutionary links. The same pattern as we found for fossils (§5.5) and for structural similarity across nature (§7.8). *LOE#5* and *LOE #6* are rejected. Notions of a molecular clock and molecular homology persist in science, and with sophisticated mathematical modeling, that finds more and more evidence of evolution. It's just garbage in garbage out. As for us, we will have to look elsewhere for evidence of evolution. It's time to look at embryology.

REFERENCES

1. Ridley, M. *Evolution.* 48 (Blackwell Science, Inc., 1993).
2. Mayr, E. *One Long Argument.* 23 (Harvard University Press, 1991).
3. Futuyma, D. J. *Science on trial.* 205 (Sinauer Associates Inc., 1995).
4. Page, R. D. M. & Holmes, E. C. *Molecular Evolution.* (Blackwell Science Ltd., 1998).
5. Ridley, M. *Evolution.* (Blackwell Science, Inc., 1996).
6. Dickinson, W. J. Molecules and morphology: where's the homology? *Trends in Genetics* 11, 119-121 (1995).
7. Abouheif, E. *et al.* Homology and developmental genes. *Trends in genetics* 13, 432-433 (1997).
8. Gould, S. J. A clock of evolution: we finally have a method for sorting out homologies from "subtle as subtle can be" analogies. *Natural History* 94, 12-25. (1985).
9. Dayhoff, M. O. *Atlas of protein sequence and structure.* Vol. 5 D8 Fig 8.2 (National Biomedical Research Foundation, 1972).
10. MacIntyre, R. J. Molecular evolution: codes, clocks, genes and genomes. *BioEssays* 16, 699-703 (1994).
11. Fitch, W. M. Homology. A personal view on some of the problems. *Trends in genetics* 16, 227-231 (2000).
12. Eddington, A. S. *New pathways in science.* 221 (Cambridge University Press, 1935).
13. Dayhoff, M. O. *Atlas of Protein Sequence and Structure.* Vol. 5 Supp. 3 (National Biomedical Research Foundation, 1972).
14. Dayhoff, M. O. *Atlas of Protein Sequence and Structure.* Vol. 5 Supp. 3 (National Biomedical Research Foundation, 1978).
15. Dayhoff, M. O. *Atlas of Protein Sequence and Structure.* Vol. 5 Supp. 3 (National Biomedical Research Foundation, 1972).
16. Margoliash, E. Primary structure and evolution of cytochrome c. *Proceedings of the American Philosophical Society* 50, 672-679 (1963).
17. Davies, P. *The Fifth Miracle.* 76 (Simon & Schuster, 1999).
18. Denton, M. *Evolution: a theory in crisis.* (Adler & Adler, 1986).
19. Raven, P. H. & Johnson, G. B. *Biology.* 452 (McGraw-Hill, 2002).
20. Zuckerkandl, E. On the molecular evolutionary clock. *Journal of Molecular Evolution* 26, 34-46 (1987).
21. Kimura, M. *The Neutral Theory of Molecular Evolution.* (Cambridge University Press, 1983).
22. Baba, M. L., Darga, L. L., Goodman, M. & Czelusniak, J. Evolution of cytochrome c investigated by the maximum parsimony method. *Journal of Molecular Evolution* (1981).
23. Dickerson, R. E. The structures of cytochrome c and the rates of molecular evolution. *Journal of Molecular Evolution* 1, 26-45 (1971).
24. Hedges, S. B.   (www.nature.com/nature/debates/fossil/_3.html, 1998).
25. Wilson, A. C., Carlson, S. S. & White, T. D. Biochemical evolution. *Annual Review of Biochemistry* 46, 573-639 (1977).
26. Kumar, S. Timeline: Molecular clocks: four decades of evolution. *Nature Reviews Genetics* 6, 654-662 (2005).
27. Barber, M. J., Trimboli, A. J., Clark, M., Young, C. & Neame, P. J. Purification and properties of Alligator mississipiensis cytochrome c. *Archives of Biochemistry and Biophysics* 301, 294-298 (1993).

28 Kimura, M. The neutral theory of molecular evolution and the world view of the neutralists. *Genome* 31, 24-31 (1989).
29 Ridley, M.   147 (Blackwell Science, Inc., 1993).
30 Gillespie, J. H. Lineage effects and the index of dispersion of molecular evolution. *Molecular Biology and Evolution* 6 (1989).
31 Ayala, F. Neutralism and selectionism: the molecular clock. *Gene* 261, 27-33 (2000).
32 Schwabe, C. On the validity of molecular evolution. *Trends in the Biochemical Sciences* 11, 280-283 (1986).
33 Georges, D. & Schwabe, C. Porcine relaxin, a 500 million-year-old hormone? The tunicate Ciona intestinalis has porcine relaxin. *FASEB J.* 13, 1269-1275 (1999).
34 Lewin, R. Evolutionary history written in globin genes. *Science* 214, 426-429 (1981).
35 Hoyle, F. *Mathematics of evolution*. 133 (Acorn Enterprises LLC, 1999).
36 Ohta, T. Slightly deleterious substitutions in evolution. *Nature* 246, 96-98 (1973).
37 Ohta, T. Mutational pressure as the main cause of molecular evolution and polymorphism. *Nature* 252, 351-354 (1974).
38 Ayala, F. Molecular clock mirages. *BioEssays* 21, 71-75 (1999).
39 Ayala, F. Vagaries of the molecular clock. *Proceedings of the National Academy of Sciences USA* 94, 7776-7783 (1997).
40 Takano, T. S. Rate variation of DNA sequence evolution in the Drosophila lineages. *Genetics* 149, 959-970 (1998).
41 Gibbs, P. E. M., Witke, W. F. & Dugaiczyk, A. The molecular clock runs at different rates among closely related members of a gene family. *Journal of Molecular Evolution* 46, 552-561 (1998).
42 Matthews, C. M., Vandenberg, D. J. & Trotman, C. N. A. Variable substitution rates of the 18 domain sequences in Artemia hemoglobin. *Journal of Molecular Evolution* 46, 729-733 (1998).
43 Page, R. D. M. & Holmes, E. C. *Molecular Evolution*. 71-2 (Blackwell Science Ltd., 1998).
44 The Potato Genome Sequencing Consortium. Genome sequence and analysis of the tuber crop potato. *Nature* 475, 189-195 (2011).
45 Pawlowski, J. & De Vargas, C.   (www.nature.com/debates/fossil, 1998).
46 Yockey, H. P. A calculation of the probability of spontaneous biogenesis by information theory. *Journal of Theoretical Biology* 67, 377-398 (1977).

9. Long live "Ontogeny recapitulates phylogeny!"

§9.1 THE SIGNIFICANCE OF EMBRYOLOGY TO EVOLUTION

We look here at *LOE#7-10*, the evidence of evolution from embryology. See what the experts say on Table 3.1. In brief they go like this. Animal species that look very different as adults - say turtles, fish, dogs and humans, have very close morphologic and molecular similarities when they are embryos. Assuming this is true (we will see for ourselves if it really is later §9.3.3), the next question is why. The question is at once answered, and the observation interpreted, as evidence of their common ancestry by evolution. This is *LOE#7* and *LOE#8*.

Embryo similarities across species are taken to show that ancestral structures may be repeated or "recapitulated" during development, revealing the phylogenetic history of the creature in a confirmation of evolution. This is *LOE#9*.

LOE#10 holds that so-called "vestigial organs" like wings on flightless birds, male nipples and the human appendix are evidence of evolution because they represent the degenerate remains of once fully functional structures in their evolutionary ancestors. Ernst Mayr sums up the evidence for evolution like this: "These three phenomena – embryonic similarities, recapitulation, and vestigial structures – raise insurmountable difficulties for a creationist explanation, but are fully compatible with an evolutionary explanation..."[1] The chapter should at least be entertaining then, because I submit the evidence shows the opposite. Take a look.

Embryology is the name for animal development. It begins at fertilization when the embryo is a single cell or "zygote", and it continues to birth or hatching. It also includes the steps of metamorphosis in animals with larval stages like beetles and butterflies. The terms embryology and development are equivalent and can be used even more broadly to refer to all of development from zygote to mature adult. For simplicity's sake the latter meaning will be used, also called the animal's "ontogeny".[2]

Embryology is controlled by genes. As cloning experiments like Dolly the sheep have shown, it's the genes that determine the shape, size and ultrastructure of an organism.[3] Morphological differences between living things are ultimately due to differences in their genes that cause differences in their development. All the features that distinguish a fish from a falcon, an earthworm from an eagle or a man from a mouse. All the diversity of life in fact. But these are the features supposedly caused by evolution! (§1.2.1) Three observations can be made.

The first is that a study of development should at least demonstrate evidence of evolution, but perhaps a lot more. It might reveal the path that evolution took. This is a view not only shared by the mainstream of science, it's taken as self-evident by the mainstream of science. For instance Stephen Jay Gould says:

"Evolutionary changes must be expressed in ontogeny, and phyletic information must therefore reside in the development of individuals. This, in itself, is obvious and unenlightening."[4]

But it is a monumental claim! What, dare we ask, is its empirical basis? Even a preliminary inspection is revealing.

The only way embryology can be both an imperative that "must" demonstrate evolution, and also be "obvious and unenlightening", is if evolution is already a fact. To cut a long story short, we will find evolution is indeed a fact first and paradigm for interpreting all observations in embryology second (§3.1;§9.2). What's the effect? Whatever the observations they will be more evidence of evolution in embryology. Functional structures are evidence of evolution (*LOE#7;LOE#8*), and non-functional structures are evidence of evolution (*LOE#10*). Embryo similarities are evidence of descent, and embryo differences are evidence of divergence, but both are evidence of evolution. In fact even if a veritable maze of contradictions were found it would still be evidence of evolution. It would be evidence of its complexity and so the need for constructs to accommodate apparent incongruities, constructs like hierarchical assessment for instance (§7.4.3;§9.3.5). This is how evolution in embryology is an experimental scientific fact. If this seems ridiculous, that's because it is ridiculous. If you can't believe it could be happening in twenty-first century science, just let me show you.

The second observation is if development is so important to evolution and so "obvious" to the science, why have we not heard about it before? The reason is it was not part of the Modern Synthesis.[5] (§2.1) That's a surprise! It's one some senior in embryology have even taken as reason to impugn the sufficiency of Neo-Darwinism. Which is another (§9.8).[5,6] Still one more, is that at least as far as Darwin was concerned embryology was by far the best of the evidence for evolution! In 1860 he wrote his friend the Harvard botanist Asa Gray saying: "Embryology is to me by far the strongest single class of facts in favor of change of form".[7] Darwin had already said so. In *The Origin:* "Thus, as it seems to me, the leading facts in embryology, which are second to none in importance, are explained on the principle of variations in the many descendants from some one ancient progenitor" (p.600).

The third observation is evolution manifesting in development is essential to evolution theory. The reason is easy to see. Evolution causes new phenotypes, but where do new phenotypes come from? From development. Therefore development has to be evolutionary. But how could evolution do that exactly? Evolution is the interplay between mutations that produce genetic variation and natural selection which acts as a survival sieve of that variation (§2.3-5). We know mutation produces new genes but the problem is natural selection acts on phenotypes, not genes. In other words there's a gap between genotype and phenotype, right between the two forces of evolution mutation and natural selection. So how do you go from a new evolving gene to a new evolving phenotype?

By the mutation causing evolving changes to embryonic development. It's called "developmental reprogramming" in evolution theory (§9.5).[6,8] Embryology is essential to evolution theory because it enacts macroevolution, as University of Wisconsin Howard Hughes developmental biologist Sean Carroll says: "The evolution of form occurs through changes in development".[9] What's the significance of that to us? If we are to solve the Conundrum by macroevolution, right here looking at embryology is where we need to be.

There is a corollary to all this, and perhaps you've realized it already. It's that the stakes for evolution could not be higher here. Development *must* be evolutionary if evolution is true, and the reason is development is what links mutation and natural selection. No link, no evolution. So if, just for argument's sake, it turned out there were no empirical evidence for evolution in development, it would be the discovery of a disarticulation of the evolution mechanism. To discover that would be to discover an emptiness at the very heart of evolution theory, right between natural selection and mutation. That can't be possible can it? Oh yes it can.

You should also know the science community's satisfaction at having demonstrated evolution in embryology is without reservation. This sentiment was even re-vitalized by discoveries beginning in the 1980's, that similar genes are expressed during the development of very different creatures and even as different as fruit-flies and humans.[10,11] Why should that be?

It was the molecular evidence of evolution in embryology! This is *LOE#8* and it has resulted in a whole new discipline in science, one that combined the previously separate fields of evolution and developmental biology. The field is called evolutionary developmental biology now, or more affectionately "evo-devo".[8,12] The multiple books and journals dedicated to evo-devo are one demonstration of a veritable explosion of information from this new discipline, one that has emerged in our time. Scott Gilbert at Swarthmore, Brian Hall at Dalhousie University, Rudolf Raff at Indiana University, Sean Carroll at the University of Wisconsin and Wallace Arthur at the National University of Ireland are among its most notable pioneers. We will meet them all.

The field of evo-devo has already made remarkable discoveries toward understanding developmental pathways across the animal kingdom, discoveries that have been indisputably scintillating (§9.3.5). The progress made at uncovering the evidence of evolution however, or at least objective evidence of it - and the reason for its existence after all - hasn't been even so much as a faint flash in the pan (§9.7.1). That's not what you'll hear from the science, and that's also when this subject gets a lot more interesting.

We will have to look for ourselves of course, but if that charge of a cognitive disconnect is true we have reasons already not to be surprised. We looked at the biological-similarity-equals-evidence-of-ancestry argument in adults earlier and found it empty (§7;§8). Supposed homologous structures were even found to arise by different ("non-homologous" if you will), embryologic and genetic routes and structures considered analogous, from similar (§7.3-4). In other words we already know there is a dissociation between ancestry, genetics and embryology. Why should supposed similar morphologies and shared patterns of gene expression in embryos now be assumed homologous? The science disagrees. You must hear the evidence and the arguments.

§9.2 THE EVIDENCE OF EVOLUTION IN EMBRYOLOGY:
      DARWIN'S 5 CLAIMS

The first question to answer however is how Lines of Evidence #7-10 became evidence of macroevolution in the first place. How they did is as interesting as their evidence and the reason is each can be traced back to Charles Darwin. With the exception of the molecular evidence, all can be found on the pages of *The Origin*. Although more than one and a half centuries have elapseds since then, Darwin's words remain the most eloquent at making the case for evolution from embryology. Listen as he looks. The question for you is, has he seen?

"How, then, can we explain these several facts in embryology, - namely, the very general, though not universal, difference in structure between the embryo and the adult; - the various parts in the same individual embryo which ultimately become very unlike and serve for diverse purposes, being at an early period of growth alike; - the common, but not invariable resemblance between the embryos or larvae of the most distinct species in the same class [1]; - the embryo often retaining, whilst within the egg or womb, structures which are of no service to it, either at that or at a later period of life [4]...the embryonic or larval stages show us, more or less completely, the condition of the progenitor of the whole group in its adult state [3]...I believe all these facts can be explained, as follows... In two or more groups of animals, however much they may differ from each other in structure and habits in their adult condition, if they pass through closely similar embryonic stages, we may feel assured that they all are descended from one parent-form, and are therefore closely related. Thus community of embryonic structure reveals community of descent [2]; but dissimilarity in embryonic development does not prove discommunity of descent, for in one of two groups the developmental stages may have been suppressed, or may have been so greatly modified through adaptation to new habits of life, as to be no longer recognizable [5]." (p.592,598,599)

To summarize, he makes the observation there are differences between the adult forms of animals, but similarities among their embryonic forms (but see *LOE#4-6* earlier where similarities were the evidence!). He then makes five separate claims for inferring evolution from embryology, the same claims biologists make today:

Claim [1]:   Embryonic stages are "closely similar" in very different adult animal species.

Claim [2]:   It's inferred from [1] they must be related by common

Evidence of embryology

descent. This is *LOE#7* & *#8*.

Claim [3]: Embryos reveal what their ancestors looked like. This is *LOE#9*. Darwin spelt this reasoning out more clearly in *Descent of Man*, published in 1871: "the [human] embryo itself at a very early period can hardly be distinguished from that of other members of the vertebrate kingdom...man and all other vertebrate animals have been constructed on the same general model...they pass through the same early stages of development...Consequently we ought frankly to admit their community of descent."[13]

Claim [4]: It is inferred from [2] that embryos often "retain" redundant structures. The existence of these so-called "vestigial" organs, "of no service to it", is explained by an inheritance from evolutionary ancestors for whom these were once of service (*LOE#10*).

Claim [5]: Animals can still be related by descent even if their embryonic stages are dissimilar. This is because they have evolved (diverged) so much so as to no longer demonstrate the evidence.

Are these 5 claims evidence of evolution? If you said yes with science, there are four immediate problems.

The first is what underpins the reasoning is the notion that from observations of biological similarity [Claim 1] we can infer common descent [Claims 2 & 3]. In other words, homology. We looked at that idea as explaining biologic similarity earlier in adults without success (§7.3). The only difference now is similarities are being claimed for immature organisms, not mature ones. Why should this be different?

The second problem is the premise of embryo similarity [Claim 1] on which Darwin made his inferences [Claims 2 & 3] was not his own evidence but evidence from two contemporaries, each named Ernst. The first Erst was Karl Ernst Von Baer (1792-1896) and the second was Ernst Haeckel (1834-1919). The problem is the evidence from the first Ernst was false and the evidence from the second Ernst was forged. The even more interesting thing, is the work of both is found in biology textbooks as if it were true (§9.3;§25.5). We look at the evidence ourselves in a moment and in doing so, will discover the kind of morphologic similarity Darwin speaks of doesn't even exist. You reserve judgment for now however.

The third problem is his argument is not by direct evidence but by inference from evidence. The inference is not the problem. The problem is the inference rests entirely on the truth of two other inferences, the Macroevolution Inference and the Common Ancestry Inference, and both of them rest on the truth of a premise, the Naturalistic Premise (§3.2). What evidence do they rest on?

The fourth problem is evolutionary ancestry is claimed demonstrable by both similarity [1] and by dissimilarity [5] lessening, if not scuttling, the association that was the basis for the reasoning in the first place. Darwin's result is still famously effective however. The reason is because whatever the observation of embryonic stages, similarity or dissimilarity, evolution explains it!

Setting these objections aside, are the five claims Darwin made back then and evo-devo makes today even individually valid? Highly similar gene expression during development of flies and humans alone suggests these evolution inferences have been spectacularly confirmed. As far as experts Scott Gilbert and Rudolf Raff are concerned, such "homologies of process within morphogenetic fields provide some of the best evidence for evolution...Thus, the evidence for evolution is better than ever".[5] This chapter is an exploration what they just said, of *LOE#*7-10, and of Darwin's five claims from where they originally came. We will take them one by one, beginning with the first claim above that embryos are "closely similar" across species. (§9.3-9.7). It's the key premise on which claims 2 and 3 rest. Solve this and you solve the Conundrum! What is the evidence for Darwin's supposed "community of embryonic structure"? Does it mean evolution?

§9.3   CLAIM #1: ARE EMBRYOS SIMILAR ACROSS SPECIES?

Are embryonic stages morphologically "closely similar" across different animal species as Darwin said? The answer is yes and no. The reason is there are similarities and there are differences. Framed like this, the question is not helpful therefore. The relevant questions instead are these. How similar is development across different taxa of life? What is the cause? What does it mean? Is it evolution? It will be useful to tackle them from a historical perspective to observe how the answers in the science developed, and how evolutionary biology thinks and works. We must begin back in antiquity, in Ancient Greece.

§9.3.1   The Historical Background

Aristotle's idea that the world could be ordered into a "Great Chain of Being" called the *Scala Naturae* persisted as orthodoxy for more than two thousand years, right up to the early nineteenth century. He said there is a linear hierarchy to nature like a ladder, with inorganic objects on the lowest rungs and ever more complex life on the rungs above finally to humans at the top (§2.2). Aristotle's schema was not evolutionary. Like Plato and Socrates he saw design but it is hard to have that view of his ladder now, so powerful and so accustomed are we to Darwinism. You see Aristotle's rungs were independent and fixed, and they represented if not a divine plan of creation, at least an ordering of it.

The idea that different creatures on different rungs of the ladder had similar morphologies while they were embryos began in 1811 and 1824 when J.F. Meckel (1781-1833) in Germany and Etienne Serres (1786-1868) in France independently proposed the idea of "parallelism". It was an attempt to offer a mechanism for Aristotle's ladder, and it is another example of biology done backwards - from false philosophy to data. No surprise then where this will end up then, but let's look for ourselves.

Parallelism is the notion that more "complex" animals higher on the ladder repeat or "recapitulate" the adult forms of lower, less complex creatures while they are embryos, and that the lower creatures represent the permanent larval forms of the ones above. The idea holds for example that the human fetus progressively recapitulates through a hierarchy of "lower" forms fish, reptile and simpler mammal on rungs beneath during development. This was not proposed as an evolutionary idea. It was even considered by some, including Harvard's Louis Agassiz, as a divine plan of forming life.[14] Parallelism became highly influential evolutionary thinking later, but it would take one more step.

The next landmark was in 1828 when Karl Ernst von Baer published his "Four Laws of Development". It is a peculiar designation for the work, because they are not remotely laws but instead observations or postulates about trends. Still more peculiar is they are violated more often as the rule than the exception, yet they're roundly accepted as true! (§9.3.4) You'll find them in the major embryology textbooks for instance, and Stephen Jay Gould calls them "probably the most important words in the history of embryology".[15] What's going on and why might this be? First, what do von Baer's laws state? Here they are:

1. The more general characters of the group of animals to which the embryo belongs appear before the special characters.
2. The less general structures in a creature form after more general structures until finally the most specialized structures appear.
3. Each embryo of a given species, instead of passing through the stages of other animals, departs more and more from them.

4. Embryos of higher animals resemble the embryos and never the adults of lower animals.[15,16]

What do von Baer's laws mean? They beg the question more than once which is also peculiar for a law of science. For instance what does "general" mean? Or "group"? And what exactly is a "character"? Setting concerns of ambiguity and circularity aside, this is how Stephen Jay Gould summarizes what Von Baer said and what he meant:

> "Development proceeds from the general to the special. The earliest embryonic stages of related organisms are identical; distinguishing features are added later as heterogeneity differentiates from homogeneity. Recapitulation is impossible; young embryos are undifferentiated general forms, not previous adult ancestors."[17]

Von Baer declared embryos of different adult species were highly similar at their earliest stages, almost identical in fact. To make the point he said: "In my possession are two little embryos...and at present I am quite unable to say to what class they belong. They may be lizards or small birds, or very young mammalia, so complete is the similarity..." (p588-9).

What's the significance of Von Baer and his laws, you ask? They were Charles Darwin's best evidence of evolution. He said: "Hardly any point gave me so much satisfaction when I was at work on the *Origin* as the explanation of the wide difference in many classes between embryo and the adult animal, and of the close resemblance of the embryos within the same class".[18] This is what Darwin said about von Baer's work:

> "generally the embryos of the most distinct species belonging to the same class are closely similar, but become, when fully developed, widely dissimilar. A better proof of this latter fact cannot be given than the statement by von Baer that "the embryos of mammalia, of birds, lizards and snakes, probably also of chelonia [tortoises and turtles] are in their earliest states exceedingly like one another, both as a whole and in the mode of development of their parts; so much so, in fact, that we can often distinguish the embryos only by their size."[19] (p.587)

How was this evidence for evolution? Von Baer's "general" characters were ancestral characters by another name. "Special" characters were those that had evolved subsequently, and the "general" to "special" sequence of development that von Baer declared even to be a law of nature, was evolutionary divergence.[20] Darwin replaced the original notion of recapitulation taken as indicative of Aristotle's ladder of life, with the announcement it was the revelation of evolutionary ancestry. He also reasoned the exceedingly similar embryo morphology among animals was in contrast to the diversity of environments in which embryogenesis occurred. The similarities were unlikely to be adaptive therefore. Why were they there then? They must be inherited. It was evolution! He summarized, in seeming brilliant simplicity, "the embryo is the animal in its less modified state; and in so far it reveals the structure of its progenitor."[21]

But what was the objective scientific evidence for this idea? You know. It was to quote von Baer! Considering the momentous consequences, it was also among the more ironic events in the history of biology, not least because Darwin had reinterpreted von Baer's laws that expressly refuted recapitulation and then cited him as evidence for the evolution idea. Von Baer had no evidence for his claim either by the way, but more of that later (§9.3.3-4). This was how the idea that embryos are morphologically similar early in development first became a science: the father of embryology, Karl Ernst von Baer, said so. How it became evidence of evolution, is because Darwin said it was caused by common ancestry, and he said recapitulation was evidence also. Today both are scientific truths, embodied as *LOE#7* and *LOE#9* respectively. How this happened from such speculative beginnings in *The Origin* is fascinating story about how science is done when evolution is already a fact.

§9.3.2   Haeckel's Embryos & The Biogenetic Law

Von Baer's first two laws remain the observational basis for all the interpretations in embryogenesis. That's why they're so important and that's why Stephen Jay Gould calls them, "probably the most important words in the history of embryology" (§9.3.1).[15] In another twist to this story, I should tell you von Baer was not an evolutionist and not even after Darwin published *The Origin*. Which of them had it right? We need to see the evidence. We know our first question, and it's a very simple one. We asked it earlier and are still waiting for an answer. Are animal embryos indeed (1) morphologically "exceedingly like one another" at (2) the "earliest embryonic stages" as von Baer said and Darwin assumed?

No and no. Neither man had it right. In fact at the "earliest embryonic stages" to quote Dr. Gould, or the "earliest states" to quote Charles Darwin paraphrasing von Baer's laws, embryos are not remotely similar let alone "identical" and you don't have to be an embryologist to tell (Fig.9.1). This is why von Baer's laws are false. You would never think so if you listened to the experts though. Jerry Coyne at the University of Chicago, another expert in this field, for instance says: "Embryos of different vertebrates tend to resemble one another in early stages, but diverge as development proceeds, with more closely related species diverging less widely. This conclusion has been supported by 150．years of research".[22] Despite what he says there's no difficulty telling embryos apart at any stage of development (§9.3.3-4). Von Baer's claim that embryos of reptiles, birds and mammals are barely distinguishable is a lie (Figs.9.1-3,5). The fascinating thing is von Baer's notion is a truism. What's going on? More about this disconnect and how and why it could happen in science in a moment, but the story gets even more bizarre. It happened 15 years later when the evidence was finally found for Darwin's ideas of evolution in embryology.

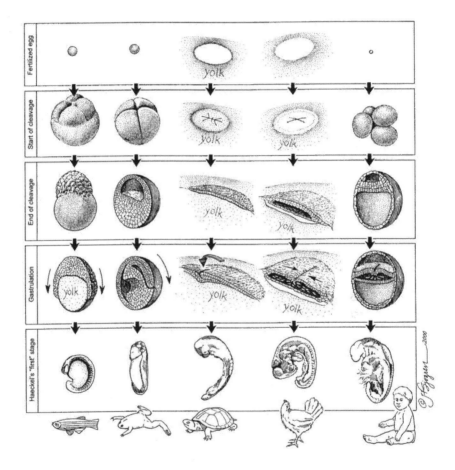

Fig.9.1. **Comparisons of early stage vertebrate embryo morphology across species.** Contrary to von Baer's "laws" and Charles Darwin's claims, dissimilarity in vertebrate development is far more striking than is similarity and obviously so. Compare across the rows representing stages of development in fish, frog, tortoise, chicken and human. The claim for a conserved morphology in development across the vertebrates is a myth and you don't have to be an embryologist to tell. (Drawing © Jody Sjogren, 2000. Reproduced with permission).

The evidence was discovered by University of Jena zoologist Ernst Haeckel (1834-1919). He reformulated Darwinian recapitulation as the "Biogenetic Law" which is another law in the history of the science of embryology that's not a law. Formally stated, it declares that "ontogeny" or development of the individual from fertilization to maturity, results from "phylogeny" which is the evolutionary lineage of that individual. Haeckel encapsulated it as the familiar phrase "ontogeny recapitulates phylogeny". By this he meant animals sequentially repeat or recapitulate ancestral adult morphologies during their development. By this law animals have virtually identical early stages of development because higher more "complex" descendant animals recapitulate ancestral adult stages in sequence during development. Thus embryogenesis is a re-run of an organism's evolutionary history, or as Haeckel said it, "the rapid and brief ontogeny is a summary of the slow and long phylogeny".[23,24] He explained that the stages of an animal's development are like a portrait gallery of its evolutionary ancestors. Like Darwin, Haeckel merely reinterpreted von Baer and parallelism and embryology from an evolutionary perspective. The important difference is Haeckel produced evidence for the idea.

Haeckel said new adult morphologies arise by evolution only as a result of the addition of new terminal stages of development to the end of an abbreviated inherited sequence of development of the ancestor. This abbreviation or "telescoping" as it was called, of the acquired terminal steps by recapitulation Haeckel called "condensation". He said it was due to the acceleration and/or deletion of steps of development. The reason for the abbreviation? To make room. Otherwise more complex higher animals would require an impossibly long embryogenesis to work through their multiple inherited stages of development from all their ancestors. Haeckel summarized these ideas in his book *Evolution of Man*, published two decades after *The Origin*. There he explained:

"Ontogeny is the short and rapid recapitulation of phylogeny; or, somewhat more explicitly; that the series of forms through which the individual organism passes during its progress from the egg cell to its fully developed state, is a brief, compressed reproduction of the long series of forms through which the animal ancestors of that organism...have passed from the earliest periods of so called organic creation down to the present time."[25]

Haeckel was enormously influential in other ways as well. He defined the stages of metazoan development, and coined a nomenclature for the science of embryology. He related each stage of development to an equivalent recapitulated ancestral form and many of his terms are used still today. Some you know. The stages of development began with the "cytula" (his now obsolete term for zygote) whose ancestral form he said was the *Amoeba*. This is followed by the "morula" stage (when the embryo has divided sufficiently to become a solid ball of cells), and its ancestral form he called the *Synamoeba*. Next follows the "blastula" (when the morula has become a hollow sphere), and its ancestral form he called *Planaea* (Blastaea). The blastula is followed by the "gastrula" and its ancestral form he called the *Gastræa*. The *Gastræa* was the hypothetical common metazoan ancestor and his conception of biology has also been called the Gastræa Theory to indicate this.

Haeckel identified additional stages for the embryonic development of vertebrates, namely the coelomic pouch ("coelomia"), followed by the neural tube and notochord ("chordula"), followed by the segmented mesoderm ("spondula") stage.[16] These three latter terms are obsolete but vertebrates indisputably share the major sequential stages of zygote, morula, blastula, gastrula and neurula. Since these stages are seen for all metazoans except, and explain it to me with evidence please, sponges, Haeckel declared this was evidence of their common ancestry and that metazoans are monophyletic. This thinking can be seen in the very first phylogenetic tree of the animal kingdom that he drew in 1866 (Fig.1.3).

The take home point as you may have realized already, is like his phylogenetic tree drawing Haeckel's explanations of embryos and embryology were all conjectural. He constructed an elaborate theory that was no more than his inferences of evolution already a fact, but needing demonstration in embryology. He even said embryos were "more informative than fossils"[26] at demonstrating evolution, and the reason was because by recapitulating they literally reconstructed phylogeny for the observer and showed the path taken by evolution. Exceptions to this wonderful revelation of a creature's phylogeny he called "cenogenesis", whereby ancestral stages were either dropped (which he called "heterochrony") or new stages were added (which he called "heterotopy" §9.5). In other words the data could never contradict his theory. It was recapitulation or it was cenogenesis and evolution did both.

Haeckel said the primitive (or "palingenetic") characters in development were informative of ancestry but not the novel ones because these were cenogenetic. This is another interesting observation, because cladistics now operates from the opposite assumption. Specifically it is the new characters in Hennigian phylogenics which occupy "center-stage position"[8] in evo-devo and are regarded as informative, and the reason is they're taken to be synapomorphies (§5.3.3;§7.4.1;§10.2.3). Haeckel wrote extensively about these ideas and others too, but for our purposes in attempting to solve the Conundrum they can be reduced to these three:

393

(1) There is a highly conserved stage of development across different animal species and it is caused by common ancestry.

(2) Phylogeny is revealed in ontogeny by the recapitulation or repetition of inherited ancestral characters. In other words ancestry is declared in development through stages representing character traits of ancestral creatures.

(3) Recapitulation occurs by the addition and condensation of new stages of embryonic development sequentially added to the end of the inherited ancestral program.

You can see now how the ideas of Ernst von Baer (the first idea above) and Charles Darwin (the second and third) were appropriated by Dr. Haeckel. Next we need to know if any have reality. To assist the reader unfamiliar with biology, the first and the second ideas are everywhere truisms, as *LOE*#7-#9 respectively (Table 3.1). We begin our examination with Darwin's claim 1 from back in §9.2, namely that embryos are similar across taxa because of evolution. We find the premise is false, that it was based on a forgery, and yet it remains a scientific truth.

§9.3.3   The Phylotypic Stage Is Not Highly Conserved

The Biogenetic Law is a fascinating idea. Not only did it give Darwin's ideas of evolution in embryology a mechanism, it provided an explanation for the events of embryology as well. It has one little problem though. It's false.

It turns out that while stages of development are shared, particular embryo morphologies at shared stages are at least as different as they are similar (§9.3.3-4;Fig.9.1). Even more confusingly to the evolution claim, the morphologic and the molecular similarity that does occur is arrived at by different mechanisms (§9.4). Don't act surprised. We found the same thing for adults (§7.3).

Brian Hall at Dalhousie University acknowledges this paradox, at least if you believe in macroevolution it's a paradox, but he does not acknowledge the implication for the coherence of evolution theory. And so renders a quite remarkable opinion in thundering understatement. You heard it two chapters ago. He says: "the stages [of embryonic development] are often more highly conserved than the developmental processes that produce them."[27] In other words stages of development are conserved (zygote, morula and blastula for instance), but are formed differently in different species. If that doesn't sound loud to you then explain it naturalistically. If the similar morphology of embryos is inherited but the mechanisms are different, how can we infer evolution as the cause? We need a heritable mechanism to explain what is at least the appearance of a disarticulation. More theory to account for a contradiction of evolution theory, doesn't count. I mean an evolution mechanism with evidence. Come on!

Two consequences follow and they will trace out the rest of our inquiry as interweaving themes. The first is the reaction of science to the evidence. You guessed it. Evolution in embryology stays a fact and recapitulation also. To say recapitulation scientifically today you say, "there are parallels between phylogeny and ontogeny".[28-30] How that happened exactly we will discover shortly but it was not because of any new evidence (§9.3.3-9.4).

The second theme is what the observation made by Dr. Hall means for finding the true solution to the Conundrum. We take up both themes to see what empiric basis they have momentarily, but the anomaly at least injects some doubt into the seeming obvious inference of ancestry from supposed embryo similarities. How similar development across the animal kingdom really is, in other words the extent to which Darwin's claim 1 that there is "community of embryonic structure" is valid, is the fulcrum on which the homology inference of his Claim 2 and *LOE#7* & *LOE#8* rest. We've still not seen any scientific evidence. We return to von Baer's first and second laws to look at what Haeckel did to provide it.

Haeckel was a Lamarckian so it was easy for him to explain how the condensation and addition of terminal stages to pathways of embryonic development could be inherited (§2.2). With Lamarckism's debunking, Darwin's views of recapitulation returned to the status of descriptive evolutionary interpretation absent empirical explanatory mechanism, the way they were in *The Origin* before Haeckel. Terminal addition was impossible because genes present from conception direct development. Thus the notion of recapitulation as "ontogeny recapitulates phylogeny" in the way Haeckel conceived it was disproved by Mendelian genetics.

It turned out Haeckel was not debunked by geneticists however, but by embryologists. Well, at least initially. In 1922 embryologist Walter Garstang (1868-1949) called the Biogenetic Law "demonstrably unsound" because "ontogenetic stages afford not the slightest evidence of the specifically adult features of the ancestry." As if that were not already explicit enough, Sir Gavin de Beer (1899-1972) tediously repeated experiments demonstrating violations of the Biogenetic Law as if expecting a different result eventually declaring: "Recapitulation, i.e. the pressing back of ancestral adult stages into early stages of development of descendants [i.e. recapitulation in the way Haeckel said it happened] does not take place." The interesting thing is evolution in embryology remained as true as ever (§7.3;§9.1). Recapitulation remained true too, but in some as yet unknown formulation. Even Walter Garstang declared "recapitulation is a fact" despite his strident objection to the Biogenetic Law! It was only Haeckel's mechanism for recapitulation and specifically his insistence that it happened though terminal addition, that was rejected in the end. Recapitulation was resurrected in the 1970's and is now a central concept in evo-devo. It's *LOE#9*. This was principally the work of one man, Stephen Jay Gould (§9.5).

Another central concept in evo-devo is the von-Baer-Darwin-Haeckel idea that embryos across species are very similar early in their development (*LOE#7*). What's the evidence for that? Contemporary expert Michael Ruse is here to show us. He uses the embryos of humans and dogs as evidence. He says: "Why should man and dog have similar embryos? The answer lies obviously in common ancestry..."[31]

Dr. Ruse supports his claim by showing us scientific evidence, the external morphologies of dog and human embryos (Fig.9.2 left panels). Not just merely similar, they look almost identical. His claim of evolution is thus believable just as Darwin speculated in *The* Origin:

"however much they may differ from each other in structure and habits in their adult condition, if they pass through closely similar embryonic stages, we may feel assured that they all are descended from one parent-form, and are therefore closely related". (p.599)

Let's look at that evidence.

The evidence Dr. Ruse shows is not embryo photographs, but embryo drawings and they're not recent drawings, they're drawings made more than a hundred years ago. Charles Darwin used them also and to make the same point in *Descent of Man*, and declared human and dog embryos "can hardly be distinguished". Darwin said the drawings were "carefully copied" from "two works of undoubted accuracy" and that Ernst Haeckel had "analogous drawings".[32] How similar dog and human embryos really are can be seen from the right panels of Fig.9.2. Something fishy is going on, and it brings us at last to an exploration of the evidence for a near identical conserved stage of vertebrate development that is the premise of *LOE#7* and *LOE#9*, and that is Darwin's claim 1 from back in §9.2. What is that evidence you ask?

Nothing more than a set of drawings by Ernst Haeckel, first published in *Generalle Morphologie* in 1866 and *Anthropogenie* in 1874 (Fig.9.3).[34] The problem is not their age. It's that they're forgeries. Until recently these century-old drawings were used in science textbooks as if they were true, and even books authored by some of the biggest names in biology. Like Ernst Mayr (*What Evolution Is* 2001), Douglas Futuyma (*Evolutionary Biology* 1998), Scott Gilbert (*Developmental Biology* 1997), Rudolf Raff

(*The Shape of Life* 1996), *Gray's Anatomy* (1995), Bruce Alberts and James Watson (*Molecular Biology of the Cell* 1994) and Michael Ruse (*Darwinism Defended* 1982) amongst others.[35-39]

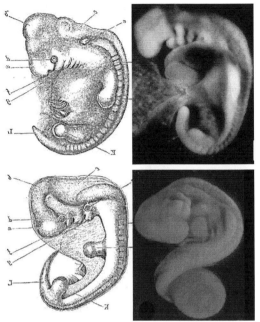

Fig.9.2 Are human and dog embryo almost identical? Drawings from Charles Darwin's *Descent of Man*, 1871[32] showing human (left top) and dog (left bottom) embryos reversed from original also reproduced in *Darwinism Defended* by Michael Ruse with the caption: "Why are they so similar if man and dog have not descended from a joint ancestor".[31] Darwin used them to make the same point claiming "The [human] embryo…can hardly be distinguished from that of other members of the vertebrate kingdom".[32] Right panels show what they really look like. (Photographs from Richardson, 1997 Fig.8e,k with permission).[33]

Serious objections to the drawings were made, even when they were first published. Harvard's Louis Agassiz wrote in the margin of his copy in 1868: "Naturally – because these figures were not drawn from nature, but rather copied one from another!"[40] Ludwig Rutimeyer accused Haeckel that year of using the identical drawing for three different species, to which Haeckel later confessed.[24] Another anatomist, Wilhelm His, accused Haeckel of several separate counts of fabrication in 1874 and of the drawings in Fig.9.2 which were Plates IV and V of *Anthropogenie*[34] he said, "The majority of figures in embryo plates IV and V are invented."[24] Adam Sedgwick was another who protested.[40] In 1894 he said: "There is no stage of development in which the unaided eye would fail to distinguish between them [vertebrate embryos]…a blind man could distinguish between them".[33]

More recently Richard Goldschmidt called Haeckel's work describing new species of Medusae [jellyfish] and Radiolaria [a type of marine plankton] as "almost the only factual contributions Haeckel made to zoology".[41] Objections were made by Walter Garstang and Sir Gavin de Beer, but these also were ignored. How could this blindness by science have happened and still be happening? You know why. The reason for the irrepressible success of his drawings is obvious. They're such good evidence! They're the demonstration of an amazing Darwinian community of embryonic structure, almost identical. The figures are also a ringing confirmation of the Biogenetic Law. The evidence shows common embryonic stages across taxa, a conserved stage at which the defining features of the phylum appear first, and shows that development reveals evolutionary history, in other words phylogeny, through recapitulation.

The supposed most conserved stage of embryonic development is called the "phylotypic" stage and it is represented by the top row of Fig.9.3. For the current definition of the phylotypic stage, here is Bath University embryologist Jonathan Slack. It's "the stage of development at which all major body parts are represented in their final positions as undifferentiated cell condensations, or the stage after the completion of the principal morphogenetic tissue movements, or the stage at which all members of the phylum show the maximum degree of similarity."[42] Thus we find the name of a mere stage of embryo development, already fully frontloaded with descriptive evolutionary significance. Haeckel's drawings are unambiguous though: vertebrates develop from an almost identical phylotypic stage embryo. Take a closer look at Fig.9.3. Three unmistakable features are evident:

1. Reading horizontally along the top row is to see an extra-ordinary resemblance between the embryos of different species. This is evidence of a conserved stage of development or *LOE#7*. To hear this evidence in the words of contemporary experts turn to Table 3.1, but in Ernst Haeckel's words "we are unable even to discover any distinction between the embryos of these higher Vertebrates and those of the lower, such as the Amphibians and Fishes".[24] Looking at his data, you have to agree. It's also a confirmation of what Darwin had claimed: "if they pass through closely similar embryonic stages, we may feel assured that they all are descended from one parent-form, and are therefore closely related" (§9.2). You have to agree with Darwin's original inference it means evolutionary ancestry as well.

2. Reading across the figure is to see recapitulation, thus confirming the Biogenetic law. Specifically we find fish fins evolving to limb buds and

fish gills to human lungs. Haeckel wrote about this remarkable observation:
> "The fact is that an examination of the human embryo in the third or fourth week...shows it to be altogether different from the fully developed Man, and that it exactly corresponds to the underdeveloped embryo-form presented by the Ape, the Dog, the Rabbit, and other Mammals, at the same stage of their ontogeny. At this stage it is a bean-shaped body of very simple structure, with a tail behind, and two pairs of paddles, resembling the fins of a fish, and totally dissimilar to the limbs of Man and other animals, at the sides. Nearly the whole of the front half of the body consists of a shapeless head without a face, on the sides of which are seen gill fissures and gill arches as in Fishes. In this stage...the human embryo differs in no essential way from the embryo of an Ape, Dog, Horse, Ox, etc., at a corresponding age."[24]

You have to agree with this observation too. This is *LOE#9* and his first few sentences are *LOE#7* again. For contemporary voices that echo him, see Table 3.1. No wonder these drawings are so popular in books that teach evolution.

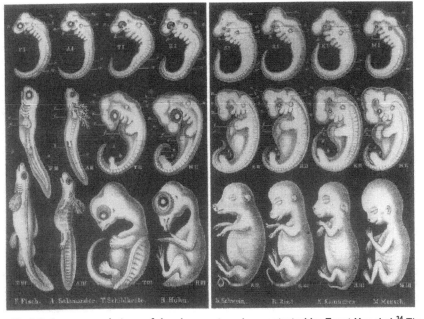

Fig.9.3. **A conserved stage of development as demonstrated by Ernst Haeckel.**[34] The species from left to right are fish, salamander, turtle, chicken, pig, cow, rabbit and human. The successive stages of vertebrate development are read from top to bottom. The top row represents the conserved phylotypic stage.

3. Reading vertically within a particular species column is to see the progressive divergence of creatures during their development, confirming von Baer's first and second laws.

Haeckel did not report the exact timing in development of his specimens, but this omission never stood in the way of a wholesale acceptance of his data by the science community. Still more extraordinary, instead of attempting to replicate his data subsequent workers simply reproduced his drawings, like Michael Ruse did in Fig.9.2, and even added their own definitions of the timing. To add to the confusion, definitions of the supposed shared morphologic stage in vertebrate development have been different among different investigators, even to the extent there is still no agreed upon definition of the "phylotypic stage" that is supposedly so similar. The list can be reduced to one of three basic types: 1) pattern based definitions, 2) process based definitions and 3) character based definitions.[43] Since this is such a critical point in evolution theory, here they are for the detail oriented. Jump ahead if this feels like baggage being loaded on your back.

1) Pattern-based: It is the period when there is there is the least amount of variation among a group of organisms.[44,45] This is today conceptualized by an hourglass model of evolutionary development with the similar phylotypic stage at the mid stages of development represented by the narrow waist, and the very different early and late stages of development by the wide top and bottom of the hourglass (Fig.9.4). It's the first evidence we find for why Baer's first two laws are false for they predict a cone shape, from similarity to greater dissimilarity). More about this later (§9.3.4).

2) Process-based: It is the period of the most interactive signaling between embryo components.[45,46]

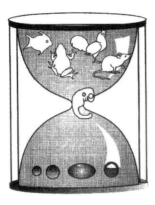

Fig.9.4. The hourglass model of vertebrate development. The "phylotypic egg-timer" is a metaphor for the great differences in early and late development with a relative reduction at the time of the tailbud stage which corresponds to the neck of the hourglass. (Reproduced from Duboule, 1994 with permission).[44]

3) Character-based: Chief of these have been the "pharyngula stage" definition proposed by Dartmouth zoologist William Ballard in 1981 (and taken as the stage when the paired pharyngeal arches and pouches first appear)[47]; the "early somite stage" by Wolpert in 1991 (when somite segregation begins)[48] and the "tailbud stage" by Jonathan Slack in 1993 which we heard him define above.[42] Character based definitions are closely related to pattern based definitions in that they attempt to attach quantifiability and refer to the time in development when the pharyngeal arches, paired appendage buds, the heart and the tail bud appear.

By the way the concept of a phylotypic stage, a stage of embryonic development shared by all members of a phylum, is not restricted to vertebrates although they are by far the popular example, and it is another of the legacies of Haeckel's drawings. But what does the "phylotypic stage" actually mean?

Ubiquitous opinion today, regardless of definition preference above, is not that it is just a conserved stage of development or even that it is archetypal, but in Brian Hall's words "[t]he phylotypic stage is the physical embodiment of the link between ontogeny and phylogeny."[16] It represents evidence of the ancestral condition no less. Recapitulation. Here, in the flesh, is *LOE#9*. We look at the evidence for this claim next. It is among the most spectacular of the displays of the Emperor in his New Clothes. Please put your sunglasses on.

Haeckel's influence has been phenomenal, and it continues simply because he put his finger on a fundamental requisite of evolution. Recapitulation has to happen in some sense if evolution is true. Haeckel's appeal was extraordinary even in his day and not just in the science and academic community either. One of his books, *Weltratsel,* was "among the most spectacular successes in the history of printing"[49] and was translated into twenty-five languages. It sold almost half a million copies in Germany alone. As you might have gathered from this much popularity, Haeckel did not limit his opinions to embryology or even biology. He extrapolated his evolutionism to contemporary society, even proposing Chancellor Bismarck be awarded a doctorate in Phylogeny, in accord with Germany's position at the top of the Phylogenetic tree.[50]

In Haeckel's defence, making extrapolations from the drawings was hardly unreasonable because the evidence of his drawings was so clear. There was no significant difference between human embryos and salamanders or chickens or fish embryos from whose ancestors human beings had come (Fig.9.2). Humans evolved from slime that had learnt to wriggle, then squirm, then be beast, and was continuing to evolve purposelessly in a never-ending, morally meaningless fight for the fittest.

Human evolution views are inherently discriminatory, needless to say. They have to be if some human traits are fitter than others (§11.5). Sadly but not surprisingly Haeckel's influence contributed significantly to the subsequent rise of Nazism in his country.[51] The story gets even more weird.

Rather than getting a delayed debunking from science in the twentieth century, Haeckel's ideas received a boost and it came from a most unlikely quarter, from molecular biology. The year was 1993 and the occasion was a remarkable demonstration by Jonathan Slack and his colleagues, then at Oxford University, of an extraordinarily conserved pattern of gene expression across the animal kingdom now called the homeobox (or *Hox*) genes (§9.3.5).[42] He proposed an organism could even be taxonomically defined as being an animal on this basis. Since then genes with similarities to *Hox* have been found in plants and yeast however. Slack considered this genetic character to be a synapomorphy of the animal kingdom, and he called it the "zootype".[42] Interestingly the zootype is most clearly evident at the phylotypic stage! This is what gave Haeckel's idea of a conserved stage in development that boost.

Slack's observation appears to be a stunning demonstration of a link between genotype and the phylotypic˚ stage and importantly for evolution theory, a confirmation of phylogeny. It was big news in *Nature* where the report was first published too. There the authors concluded:

> "...the phylotypic stage is of critical importance not only for defining individual body plans, but also for relating these body plans across the whole animal kingdom...the persistence of the zootype indicates common ancestry...we believe that there was once a primordial multicellular ancestor of all existing animals and this ancestor was the first organism to possess the zootype."[42]

This notion represents a large chunk of the dataset for *LOE*#8. The fact that vertebrate embryos have strong resemblances by gross morphologic or molecular character traits is not in dispute. Their origins and their meaning is what is in dispute. Biological similarity is not new to us anyway. We investigated it in adults earlier (§7.3). We know vertebrate embryos share the same sequence of stages of development like fertilization, zygote, morula and blastula, and we know that vertebrates at the phylotypic or gastrula stage share structures like pharyngeal pouches, somites, optic anlagen, neural tube and a notochord (Fig.9.1-2). The question is how similar they really are when formally inspected, and whether or not any similarities mean a causal evolutionary relationship. We know what science says (*LOE*#7). Let's look for ourselves now.

First, are vertebrates at the tailbud or phylotypic stage as conserved as Haeckel's drawings indicate? If there is recapitulation to support evolution theory, embryos should look similar and should use similar developmental mechanisms reflecting inheritance as claimed. This is Darwin's "community of embryonic structure" (§9.2). Haeckel's drawings and these evolutionary implications have merely been assumed as true. It took over a century before attempts were even made to replicate his result.[24] How could such a thing happen in a science? This is not a question for you, because you have seen how.

History declares recapitulation an idea never deduced from data, but an inference imposed on embryology by faith in evolution from a time before there was knowledge of genes or DNA or for that matter, Darwinism. If evolution were not true the idea of recapitulation would be ridiculous. How can terminally differentiated tissues that are the adult ancestor, be represented by undifferentiated tissues that are the developing embryo in recapitulation, yet be directed by genes that remain the same from zygote to adult? A series of three papers authored by Michael Richardson at the University of Leiden has finally and conclusively demonstrated the von Baer-Darwin-Haeckelian notion of highly conserved morphology in vertebrate development, and now called the phylotypic stage, false. Look at the evidence of embryology now.

The first paper was published in 1995.[52] It was really just a literature review, compiled by comparing already published normative stages of vertebrate embryonic development across different species. Thus most if not all of the information was widely accessible to anyone interested and had been, for some time. That it was not used to challenge Haeckel's drawings in over a century of comparative embryology, again begs the question how science could have behaved so badly. The answer to this question was a mystery to Dr. Richardson. He said: "One puzzling feature of the debate in this field is that while many authors have written of a conserved embryonic stage, none has cited any comparative data in support of the idea. It is almost as though the phylotypic stage is regarded as a biological concept for which no proof is needed".[33]

You know why (§3.1;§19;§20;§25). What Richardson showed is rather than similarity among embryos of different species, most events in development are highly variable in their timing (Fig.9.5). He concluded his paper like this:

> "The data reveal striking patterns of heterochrony [i.e. changes in developmental timing §9.5] during vertebrate evolution. These shifts in developmental timing have strongly affected the phylotypic stage, which is therefore poorly conserved...This is contrary to the impression created by Haeckel's drawings, which I show to be inaccurate and misleading."

Evidence of embryology

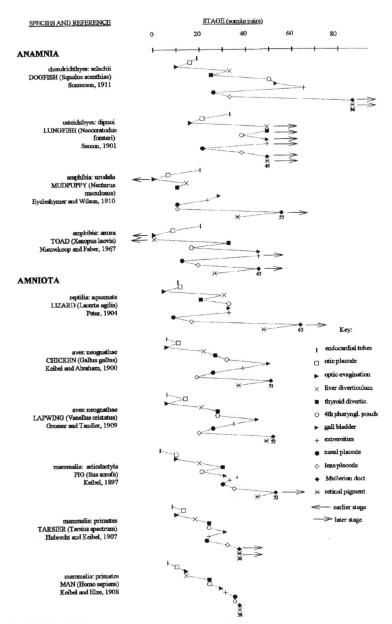

Fig.9.5. Variability across taxa is the rule in development. Developmental sequences in various species plotted against somite count and compared to human. Although some embryo features have relatively reproducible timing (such as the optic evaginations and otic placodes appearing early in development, thyroid and pharyngeal pouches in intermediate stages, and the gall bladder diverticulum and Mullerian duct anlagen appearing later), variability rather than conservation is the rule throughout development as these line plots show. (Reproduced from Richardson, 1995 with permission).[52]

The second paper was published in 1997.[33] It is a meticulous study of the comparative external anatomy of 39 different vertebrate embryo species selected at the time of advanced somite segregation (i.e. before organ differentiation, tailbud elongation and segmentation), and thus intended to be directly comparable to Haeckel's first stage drawings on the top line of Fig.9.3. Tellingly the authors had difficulty being more precise with the timing definition of the phylotypic stage because of the variability in morphology. This was an observation they had made in the first paper using normative tables from other workers, it is implied by the varying definitions of this stage which we heard earlier, and now was independently confirmed. This data, some of which is reproduced in Figure 9.6, should be compared directly with Haeckel's top row of Fig.9.3 (ignoring cow for which data was not provided and for which dog is substituted).

Rather than embryos being almost identical at the tailbud stage, they were morphologically very different. In fact differences were found for every parameter that they measured and specifically in:

1. size (a thirteen fold variation),
2. differential growth (the technical name is "allometry", e.g. the mesencephalic vesicle is the most prominent brain vesicle in the chicken but it is the prosencephalic vesicle in the rat),
3. developmental timing of structures (called "heterochrony" §9.5, e.g. the heart and branchial arches of the zebrafish have not formed yet these are supposedly the conserved feature of the phylotypic stage and phylum. As a converse example marsupial forelimb development is advanced rather than still as limb buds)
4. somite or body segment number (numbering from 11 to 60), and
5. body plan differences (e.g. no paired fin or limb buds ever develop in the lamprey).

Haeckel never gave the species names of the animals he drew (for how much difference that makes see Fig.9.6a and for other examples see Richardson 1997)[33], nor the timing of his specimens, nor his sources for the drawings, but he did claim to have his own specimens which makes the discrepancies only the more disturbing.[24,33] What can we say?

Haeckel's phylotypic stage drawings are not a fair comparison of the stages of vertebrate development for three reasons. First they do not show the earliest stage of embryogenesis (fertilization, cleavage and early blastulation), but kick off when development is already well underway at the so-called phylotypic stage (Fig.9.3). Because these earlier stages are so different morphologically from each other they alone are a disproof of the Biogenetic Law(Fig.9.1).

Second Haeckel omitted to show all the classes of vertebrate, leaving out the agnathans (jawless fishes) and the chondroicthytes (cartilaginous fish), and of those he did show half are mammals and all are from one order (namely placentals, omitting monotremes and marsupials). The selection strengthened his thesis. His work was sloppy that much is clear, but that his drawings were believed nonetheless is more the problem of a gullible science community and why that might be, than the problem of a rogue scientist.

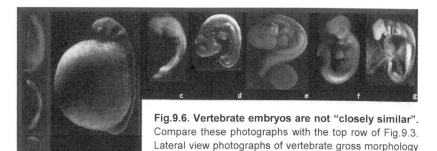

Fig.9.6. Vertebrate embryos are not "closely similar". Compare these photographs with the top row of Fig.9.3. Lateral view photographs of vertebrate gross morphology at the phylotypic stage 10x magnification. From left to right a) salmon *Salmo salar* (top), sea lamprey *Petromyzon marinus* (middle) and flying fish *Exocetus sp.* (bottom), b) hellbender salamander *Cryptobranchus allegheniensis*, c) European pond terrapin *Emys orbicularis*, d) chicken *Gallus gallus*, e) domestic dog *Canis familiaris*, f) rabbit *Orctalagus cuniculus* and g) human *Homo sapiens*. (Photographs from Richardson, 1997 Figs. 3,5,6,7,8 reproduced with permission, Springer).[33]

Third the main problem was not Haeckel's sloppiness but what Richardson had shown. His work was forged. Haeckel drew idealized embryos he believed must exist from his views of evolution-not-yet-a-reality. As Wilhelm His had said back in 1874 Haeckel cannot claim poor technical skill because he used drawing prisms and had his own specimens.[24] Why is this so hard for scientists to admit? Some do, like Brian Hall who says: "Haeckel's famous and oft-reprinted figures of conserved stages were figments of idealism and imagination rather than a reality derived from observation."[16] Such admissions are the exception however. We will deal with the scandal later. First the science.

Haeckel's drawings and ideas continue to be published and used as evidence for evolution, and not by the lunatic fringe of science but even by the elite of science. More interesting, is even to those like Stephen Jay Gould who acknowledge the work to be "fraudulent"[40] and Sir Gavin DeBeer who at least realized it was false, recapitulation, is still a truism, albeit slightly modified from the way Haeckel conceived it (§9.4). Even Michael Richardson believes in it! (§9.4) In fact Dr. Richardson could not conclude his stunning paper in anything but this muted monotone:

"We find that embryos at the tailbud stage – thought to correspond to a conserved stage – show variations in form due to allometry, heterochrony, and changes in body plan and somite number. These variations foreshadow important differences in adult body form...The wide variation in morphology among vertebrate embryos is difficult to reconcile with the idea of a phylogenetically conserved tailbud stage...Heterochronic variation makes it impossible to define a conserved stage at which all vertebrate embryos have the same combination of organ primordial present."[33]

Translating this science-speak into English, a supposed scientific "fact" for well over a century was not just false it was counterfeit.[24,40,53,54] The evidence rejects morphologic embryo conservation to the degree claimed. Equally problematically and a point Dr. Richardson incomprehensibly never made, there was no indication of an evolutionary trend in the variation they did find. Morphologic comparisons of the internal anatomy of the embryos have never been published, but one surmises they would show even more differences than surface anatomy does as Fig. 9.5 suggests.

The third paper was published the following year.[55] It was a formal comparison of somite number across different embryo species. It therefore represented the first mathematical analysis of Darwin's "community of embryonic structure". Admittedly this was basic mathematics but for a field to then hamstrung by a "lack of a quantitative framework",[24] it provided an easy way of comparing embryos with numbers rather than the "by-eye" subjective descriptive assessments. What was found? You guess.

Somite number was extremely variable. Teleost fish had from 26 to 200, sharks up to several hundred, amphibians from 6 to 285, reptiles up to 565, birds 37 to 53 and mammal tails varied from 3 to approaching 100, and in all this variability there was no evolutionary trend.

The fourth paper was published in 2002.[43] It was a comparative analysis of the relative timing of stages of development across taxa. It is therefore a more definitive attempt at comparative quantitative embryology because the authors compared not just somite number, but all of the major developmental events over a standardized time frame across different species. If we accept the claims of Darwin, von Baer or Haeckel, or the more recent hourglass notion of development (Fig 9 3;§9 3 4), morphologic variation should be least at the middle of vertebrate development at the time of the phylotypic stage. The results were unambiguous. The only surprise is that an analysis had not been done before and why that could have happened. No need to guess what was found either.

407

It was not just a rejection of conservation during the phylotypic stage, but the opposite of what the phylotypic stage notion predicted. Morphologic variation to quote the authors "is actually greatest in the middle of the mid-embryonic period. This strongly contradicts the pattern predicted by the hourglass definition of the phylotypic stage...we argue against the existence of a phylotypic stage...in vertebrates."[43] Strange days indeed. What can we say in summary?

1. Artificial character based definitions of the "phylotypic stage" create the illusion of a morphologically conserved vertebrate stage for that very reason. It's the consequence of a subjective cherry-picking of particular characters, namely those that define the supposed phylotypic morphology, while ignoring others both externally and internally. This is driven by an evolutionary mindset going back to Ernst Haeckel and before that, to Charles Darwin.

2. Haeckel's work was a forgery of the worst kind - academic. However it was perpetuated for more than a century not by deceit but by a fawning science community held fast in a philosophical straightjacket that declares evolution already a fact. Haeckel's data fit that naturalistic paradigm perfectly. Theories and data are only rejected when they become inconsistent with the data, and recapitulation must be true if evolution is true. In fact recapitulation is still true now even though Haeckel is known to be a fraud! (§9.5,§25.5)

3. As we found for adult morphologic similarity (e.g. the pentadactyl limb), embryo morphologic similarity is not evolutionary either. While there are indisputable similarities among embryos across species, their differences are at least as great as these similarities and nowhere near similar to what is claimed, or what the premise of *LOE#7* demands. Ironically at the supposed most similar or "phylotypic" stage according to evolution theory, they are the most variable.[43] As Richardson pointed out:

> "it is a big leap from embryonic resemblance to a suggestion that all higher taxa are characterized by highly conserved morphologic stages. As yet, there are no persuasive comparative data, from a wide range of clades, to support this notion."[56]

I should tell you there is a sizable and a serious scientific literature by the leading figures in evo-devo, that actually explains why the phylotypic stage is so conserved by evolution.[45,57,58] If the experts can be so wrong and still stay "scientific", how will we know when they are right?

Similarity between embryos across taxa is critical to Darwin's evolution inference though (his claims 2 and 3). As he said this: "if they pass through closely similar embryonic stages, we may feel assured that they all are descended from one parent-form, and are therefore closely related". The inference is that if two embryos look similar, it is evidence of common ancestry because the chances of random mutation and natural selection having caused such similarity independently are less. We find morphologic similarities among embryos are tenuous and the notion is based on a scientific forgery.

The dubious degree of similarity between embryos across taxa scuttles the intuitive idea of a morphologic "community of embryologic structure" which is the basis for a claim of common ancestry, and so rejects the supposed evidence for evolution by a "community of descent". To cap it all, and let you in to a little secret, morphologic and molecular similarities among embryos arise by different mechanisms (§9.4). How can this be evolutionary? Lines of Evidence #7 and #9 are empty and they are false but we are not done with them yet. We need to go back to Ernst von Baer and to his laws.

§9.3.4 Von Baer's Law Is Not A Law: Mechanisms Of Development Are Not Conserved

It may have occurred to you as an omission already that Haeckel in holding to his Gastræa theory and Biogenetic law which demanded recapitulation from the earliest embryonic stages, did not show drawings of conserved vertebrate development before the tailbud stage in Fig 9.3. In other words he did not provide comparisons of zygote, cleavage and blastula stages, showing them as similar across species as at the tailbud stage. Von Baer's first and second laws demand it also. It turns out, as we know, that contrary to those "laws" embryo morphology across species differs from the very first stages of development (Fig.9.1). Rudolf Raff identified this problem in 1996 in his review *The Shape of Life*. The interesting thing is he not only ignored the history of embryology on this matter, he sidestepped its obvious problem for evolutionary embryology. You have to listen to his reasoning to see where he is coming from:

> "Adult structures derived from further development of the phylotypic stage are indeed homologues in having historical continuity in evolution. However, the phylotypic stage...has the troublesome characteristic of being the most evolutionarily conserved stage of development, but also being attainable through non-conserved developmental processes [why exactly is this "troublesome"?]...Early development in vertebrates is divergent, and so is late development. But the phylotypic stage is conserved. We are faced with a paradox. Body plans are clearly stable over long evolutionary spans. If they were not,

we would not recognize persistent taxonomic categories such as phyla and classes. Yet basic elements of body plans are attained by different developmental pathways"[59]

Paradox indeed. But only if the phylotypic stage really is conserved and only if evolution is a fact. An empirical mind would call this a contradiction, not a paradox. His observation is also an echo of what we heard for biological similarity in adults earlier. Embryological similarity needs a naturalistic mechanism and there is no evidence for even one.

Haeckel's Biogenetic law is falsified these grounds alone of course. It's demise never demanded the debunking of Lamarckism as Stephen Jay Gould and others would have us believe.[14,60] It's falsified because embryos differ from the earliest stages of development. Markedly (Figs.9.1; 9.2;9.5;9.6). To hear this from Rudolf Raff: "eggs, cleavage, gastrulation and germ layer formation are very different in amphibians, birds, and mammals."[61] Von Baer's first and second laws are false.

That embryos differ from the earliest times of development is another problem for the notion of recapitulation and Darwin's claim of a "community of embryonic structure" from which to infer common ancestry and so evolution (§9.2). Logic that ignores the manifestly obvious divergent morphologies at the earliest stages of development (Fig.9.6) to claim recapitulation because of supposed similarities at the later tailbud stage is surely contrived. William Ballard at Dartmouth University, who proposed the "pharyngula" definition for the phylotypic stage (§7.2.3) expressed his misgivings at this decades ago, but the compulsion of the evolutionary imperative was far louder than his words. He said:

"At the core of the trouble here is the reading of too much meaning into homologies...Thus the energies of investigators and particularly students is diverted into the essentially fruitless 19th century activity of bending the facts of nature to support second-rate generalities of no predictive value...To bolster the partial truths in Von Baer's generalities by insisting that the eggs of vertebrates are more like one another than their "blastulas", the blastulas more like one another than their "gastrulas", and to homologize all theoretical "functional blastopores" where "invagination" is taking place would be running the risk of assuming what is not yet demonstrated - that the genetic, physiologic, and cell-behavior processes going on are the same in time and nature... Only by semantic tricks and subjective selection of evidence can we claim that "gastrulas" of shark, salmon, frog and bird are more alike than their adults...It seems wise, in the meantime, to avoid assumptions of uniformity drawn not from precise observation but from antique homological theory."[62]

Now here comes the interesting thing. In spite of all this von Baer's first two laws have still not been rejected. Here is a panel of experts for you to ask. Brian Hall says:
> "Characteristics of the type or phylum appear first, followed by those of the class, order, etc., until finally, species characteristics emerge. Von Baer thus provided embryological criteria and an embryological rationale for taxonomic organization."[16]

Jerry Coyne says:
> "Embryos of different vertebrates tend to resemble one another in early stages, but diverge as development proceeds, with more closely related species diverging less widely. This conclusion has been supported by 150 years of research."[22]

And here is Wallace Arthur:
> "Von Baer's divergence applies only after the 'phylotypic' stage...we should not abandon vonbaerian divergence altogether, especially as the hourglass is a very asymmetric one, with the point of the constriction close to the beginning."[8]

The latter might be from *Nature*, but it's still full of contradiction. Setting aside the difficulties defining when the phylotypic stage is exactly, advocacy for a law that consistently deviates from its tenets but can be excused because deviation is brief, is surely a new measure for a law of science.

So why all the gymnastics by a supposedly skeptical empirical science, just to hold onto Von Baer's laws? You know why (§19). We saw how Stephen Jay Gould feels about them (§9.3.1). This is how Brian Hall feels:
> "von Baer built on this scheme with a four part law that applies across the animal kingdom...Yet despite the repeated assertions of the uniformity of early embryos within members of a phylum, development *before* the phylotypic stage is very varied. Not that von Baer's law is invalid. Rather embryos do not necessarily display the greatest similarity at the outset of development."[16].

Even the change from the von Baerian cone of development that Darwin used to the hourglass pattern did not threaten the truth of evolution (Fig. 9.4). It just required a different post hoc explanation to account for it. Here is Jonathan Slack's explanation in *Nature*:
> "It is generally assumed that this variability of early stages results from adaptation to particular types of reproductive strategy or to the demands of embryonic nutrition."[42]

The explanation is the evidence. Reconciling von Baer's "laws" with the different reality of empirical biology just required a reinterpretation of those laws, and what he really must have meant. Evolution always remains true no matter what happens (§20). Here is Mark Ridley in still another example of the silliness that results when truth is suppressed:

"For phylogenetic inference, von Baer's first law is most important. It states: "The general features of a large group of animals appear earlier in the embryo than the special features." Cartilage, for example, is found in all fish – in cartilaginous fish such as sharks, rays and dogfish as well as in bony fish. Cartilage is a general character; bone is a special character, being found only in bony fish. Von Baer's law predicts that, in bony fish, cartilage will appear earlier in individual development, and will transform into bone. This prediction is correct.

Von Baer's law can be given an evolutionary interpretation and then put to phylogenetic use. [Hold on, here we go] The characters von Baer called "general" are, in evolutionary terms, ancestral, and his "special" characters are evolutionarily derived. Thus, the successive transformations from general to special forms of a character are evolutionary changes between ancestral and derived character states. By the embryological criterion, cartilage is inferred to be an ancestral state, and bone a derived state. The bone in bony fish, therefore, evolved from a cartilaginous ancestry. In general, if we have a group of species and a list of homologous characters, then (if they are the kind of characters that undergo development – i.e., not things like chromosomal bands) the relative ancestral and derived character states can be inferred from their order in development." [*LOE#9*]

The embryological criterion works only when von Baer's law is correct...[and it's a law?] The law's scope has never been systematically studied. [What!] It is widely accepted to have some truth, and for this reason it can be used in phylogenetic inference [So that's why!] It is also known to have exceptions, however. Clearly, von Baer's law should be applied only where we can be reasonably confident it is valid, and inferences made from it should, if possible, be tested against other classes of evidence."[20]

In other words he says this law is only a law when he says so. The fact is von Baer's law is not only not a law, it's not even true. And this for Darwin was the best of the evidence for evolution![7] Darwin's claim 1 that embryos are "closely similar", is rejected. The irrationality lives on however as the basis for inferring Lines of Evidence #7 and #9.

Before leaving we should take a look at the molecular similarities among embryos as evidence of evolution. This is *LOE#8*. Perhaps we have missed something here. It's molecules that make structures and species after all. There is a striking conservation of gene expression during development across the animal kingdom, even in creatures as different as insects and human beings. Is this evidence of evolution? The science community says yes. Michael Purugganan at Carolina State University summarizes the current consensus:

> "The study of the molecular evolution of development revolves around asking two key questions. How do developmental genes evolve? And what are the interconnections between changes at these regulatory genes and the evolution of developmental processes."[63]

It sounds reasonable, except evolution is taken as a given already. The problem with this assumption and where it leads is obvious to us by now. He continues by telling us the current understanding of evolution in molecular developmental biology is poor:

> "Despite the central relevance of developmental evolution to the study of morphological diversification, we know very little about the molecular evolution of developmental genetic pathways and the genes that comprise them. The molecular evolutionary approach is central to an emerging evolutionary developmental biology, as it provides detailed outlines of evolutionary histories and mechanisms that are difficult to obtain by other means, and permits in-depth analyses of the dynamics that characterize the evolution of developmental systems."

In other words evolution is true and evolution revealed by embryology is true, but what is not known is exactly where exactly that evidence is in molecular embryology. This is what will depend on the evidence, when it is uncovered. This is how the molecular biology of development discovers that evolution is a scientific fact. But don't take my word for it. See the evidence for yourself. The story starts more than a century ago, with the description of a rather bizarre set of mutations.

§9.3.5 *Hox* Gene Conservation: Phylogeny Verified Or Hoax?

In 1894 Cambridge University geneticist William Bateson (1861-1926) coined the term "homeosis" for plant and animal malformations where a structure or body region develops the identifying characteristics of a structure normally found elsewhere. Examples in the fruit-fly *Drosophila* are the duplication of thoracic segments and wings, or wings developing where the balancing organs should be, or legs developing where the antennae should be (Fig. 9.7).

These deformities are now known to be caused by single mutations to genes controlling the development or "patterning" of these structures. They are called "homeotic" genes, and in *Drosophila* were first sequenced in the 1960's and 1970's. The surprise came when genes of very similar sequence were found expressed in the development of other creatures, in fact almost universally throughout the animal kingdom, sponges and protozoa excepted. They are called *Hox* genes, a contraction of <u>H</u>omeob<u>ox</u>, the name for a DNA sequence motif they share. In a nutshell what was found is animals are constructed from a zygote by a highly similar set of genes in the way the same toolkit of a few wrenches, screwdrivers and hammers can construct multiple different things. This is an amazing discovery by any measure for why should that be? Science says it's because *Hox* genes are the homologous genes of a common evolutionary ancestor of all animals (*LOE#8*). The notion is now accepted most everywhere.[65] Let's inspect this evidence of evolution then.

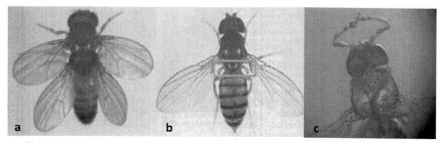

Fig. 9.7. **Homeotic mutations.** Normal *Drosophila* (b) has two wings on the anterior thorax and two halteres (boxed) on the posterior thorax. An *Ultrabithorax* mutation (a) transforms the halteres into wings. The thorax and wings have been duplicated. The *Antennapedia* mutation (c) causes legs to develop where the antennae should be. (Image credit: A and B Slattery et al, 2011[64] and C by Toony, Wiki Commons).

First what do these genes do exactly? They encode transcription factors which are proteins that regulate the expression of whole suites of other genes by variably turning them on or off. *Hox* genes are gene switches. This is how a single gene mutation can cause such dramatic effects. But *Hox* genes are much more interesting than this.

Hox genes have the extraordinary property of what's called spatial and temporal colinearity. Spatial colinearity means these genes are only expressed in discrete domains along the antero-posterior axis of the embryo, actually in the same order as they are aligned along the chromosome. The genes at one end of the DNA molecule pattern the head end of the embryo, those at the other pattern the tail, and the rest coordinately in between. Temporal colinearity refers to the timing of their expression during development - you guessed it. These genes are expressed in the same order as their alignment along the chromosome, beginning at the head region and proceeding sequentially toward the tail as development proceeds (Fig. 9.8). Deviations from this behavior have been found with the sequencing of more animal genomes.[66] These represent important details but they are not important to the arguments made for evolution and therefore not to us here either. How spatial or temporal colinearity is generated is not known. What is known however, because it is so obvious, is that this is another observation of a quite exquisite order in biology that supposedly got there just by chance.

Hox genes all share a sequence or motif called the "homeobox" which codes for a 60 amino acid segment also called the "homeodomain". This is the part that binds to DNA to turn target genes on or off. For example the *Drosophila Hox* gene *Ultrabithorax* controls as many as 170 other genes this way.[68] There is a 60 to 70% sequence similarity of the homeobox/homeodomain across the animal kingdom, but sequence similarity is even higher when more restricted "within-class" comparisons of *Hox* genes are made.

This was another extraordinary discovery: *Hox* genes are found in clusters along the length of the chromosome (Fig 9.8). What is extraordinary in this, and which follows as a consequence of *Hox* colinearity, is that *Hox* genes are more similar to genes in the equivalent position of other creatures than they are to genes within its own cluster within the same species! Such similar genes in similar relative positions across clusters are called "paralogy groups". The number of *Hox* genes differs among species as does the number of *Hox* clusters, but both are low in keeping with the metaphor they are like a toolkit by which the construction of an embryo is effected. The cephalochordate *Amphioxus* has one cluster, *Drosophila* has two (containing a total of nine genes), and man and mouse each have 4 (containing a total of 39 genes). The *Hox* genes can be grouped into 13 such paralogy groups according to DNA sequence similarity and relative position along a particular cluster. What is this taken to mean?

The *Hox* genes, and what is considered the paralogue of *Hox* the *Para Hox* gene cluster, are taken to have arisen by duplication from an ancestral sequence of a postulated primitive common ancestor called *Urbilateria*. This creature is considered to have had a single cluster that remained single in "simple" creatures like *Amphioxus*, but which duplicated twice to produce the vertebrate condition of four. Interestingly those four clusters in vertebrates are somehow now on separate chromosomes by evolution. The fact that not one of the four mammalian clusters contains genes from all 13 of the paralogy groups, and one species of fish (the pufferfish) has four clusters while another species of fish (the zebrafish) has seven, is ascribed to gene loss or gene duplication, also by evolution.

Evidence of embryology

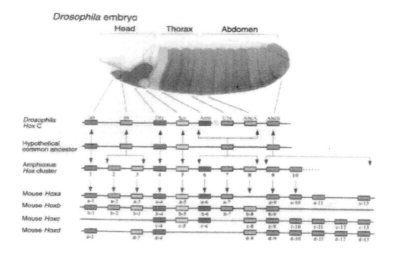

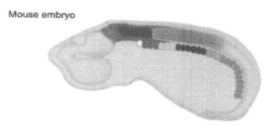

Fig.9.8. *Hox* genes in insect and mouse showing spatial colinearity, gene conservation and paralogy groups. The figure shows the orientation of the genes along the chromosome and their expression pattern in the embryo. (Drawing from Carroll, S.B. 1995 with permission).[67]

Thus the conservation of *Hox* genes across the animal kingdom is even more striking and enigmatic than it first appeared. The reason is it's found at multiple separate levels. To review, *Hox* genes are conserved in 1) sequence identity across the animal kingdom, 2) alignment order along the chromosome, 3) in the timing of their expression in development and 4) the location in which they are expressed in an embryo. To get a better appreciation of what this means, chicken *Hox* genes and chicken regulatory elements will function when inserted into mouse embryos, and the enhancer region of a human *Hox* gene (*deformed*) will function when inserted into a fly. Given this exquisite conservation *Hox* genes are everywhere in science now considered a "shared ancient structural blueprint"[69] of development inherited from a common ancestor, indicative of the monophyletic nature of the animal kingdom or the "zootype" (§9.3.3) and so demonstrative of evolution (*LOE*#8).[42] But are they really?

416

Since only a relatively few genes are responsible for all of the diversity of animal life and since they are also so similar, if evolution is the cause then it can't be *Hox* that's evolved, it has to be the way in which *Hox* is regulated that's evolved. Where is the evidence of evolution exactly? What we really have is a situation where supposed homologous molecules (*Hox*), and therefore homologous molecular processes, generate analogous structures. A contradiction to evolution. And yet the science community says it is not. This is instead explained as the evidence why homology is hierarchical. In other words homology at one level is not necessarily homology at another (§7.4). This is not an objective scientific explanation of course, but a post hoc philosophical construct invoked to preserve rationality. It's also a very good one because it's unassailable. The explanations of *Hox* causing evolution are really the evidence of evolution. This is really what *LOE#8* is. Two specific examples are instructive in demonstrating this.

The first concerns one of the *Drosophila Hox* genes called *distal-less*. Ehab Aboueif and Gregory Wray at Duke University review the evidence for us. Watching them commentate is also an opportunity to see how scientists think about these things and how they see the matter. Our experts begin by explaining what the science of evo-devo does. "The field of evolutionary developmental biology provides a framework with which to elucidate the 'black box' that exists between evolution of the genotype and phenotype by focusing the attention of researchers on the way in which genes and developmental processes evolve to give rise to morphological diversity."[70] In other words evolution in embryology is a fact but there is a "black box", a common metaphor in science for an unknown mechanism, between the genotype and the phenotype to demonstrate that fact. We knew this already but here it is from the mouths of experts who don't see the contradiction (§3.1;§9.1).

Next they tell us evo-devo is directed to revealing how evolution happened by providing a framework by which all the data from molecular biology can be interpreted. Evolution can never be falsified therefore (§20). Contradictions concerning distal-less are instead the evidence of an evolutionary hierarchy (§7.3-4). Here is all that illogic, in their words:

"*Distal-less* protein is a transcription factor that plays an important role in organizing the growth and patterning of the proximodistal axes of limbs in *D. melanogaster*, and is expressed in the distal regions of developing limbs. Thus, it came as something of a surprise when *distal-less* expression was detected in the 'appendages' of animals from five additional phyla: (1) chordate fins and limbs, (2) polychaete annelid parapodia, (3) onychophoran lobopodia, (4) ascidian ampullae, and (5)

enchinoderm tube feet. The question of course is how this *Hox* gene conservation can yet be empirically reconciled with common ancestry."
They continue:

"These results raise an interesting evolutionary question: do these similarities in gene expression indicate that these different appendages are homologous and are therefore derived from an appendage possessed by the most recent common ancestor of these six animal phyla? Although there has been, and continues to be, much debate surrounding this question, most researchers are coming to the conclusion that homology is hierarchical, and that features that may be homologous at one level of biological organization, such as genes and their roles, do not necessarily indicate homology at other levels, such as morphological structures. This point becomes clearer when distal-less expression is examined in a historical framework, and the homology of different biological levels is defined independently."

Independent assessment is indeed a good way to discover if you are dealing with reality or illusion, but the examination they propose is not independent. The reason is because the "historical framework" they mean is the irrefutable presumption that the "history" is evolution. What else could they discover but more evidence of evolution? The inspection they propose will be more than adequate to show us fact from phantom here.
They continue:

"The fossil record of vertebrates clearly indicates that the earliest members of this group lack limbs entirely. [§5.6,§6.4 for how clear this is really] This kind of historical evidence indicates that the appendages of arthropods, annelids, echinoderms and chordates, as morphological structures, are not homologous. In contrast, comparisons of sequence and expression data indicate that the *distal-less* gene, and possibly its role in patterning the proximodistal axes, is homologous in all of these phyla. Thus, we have an evolutionary scenario in which a nonhomologous structure is being patterned by a homologous developmental gene and by implication, a homologous process."

Or to say it more clearly, we have an obvious contradiction even granting the claim fossil evidence "clearly indicates" limb evolution. Homologous processes cause "non-homologous" - they can't bring themselves to say it - analogous, structures. Considering that homology means common ancestry this is not paradoxical, it's self-refuting.

And yet their statement is taken as if it were a truth. We find the following in the *Encyclopedia of Life Science* for example: "It has usually been understood that the *Hox* genes are homologous, that their clusters are homologous, that their developmental mechanisms (regulation of A/P patterning) are homologous, but that the anatomical structures specified in different phyla are not necessarily homologous."[71] The scientific way out? Learn from the experts:

"This scenario can be interpreted in at least three ways: (1) distal-less was part of a proximodistal patterning system that patterned a limb-like outgrowth in the ancestor of six phyla, and was then recruited into appendage development independently in several phyla; or (2) distal-less was part of a genetic network that patterned axes in a completely unrelated structure in the ancestor of these phyla, and was subsequently recruited independently to pattern the proximodistal axis of nonhomologous appendages; or finally (3) the recruitment of distal-less into appendage development in these phyla was completely coincidental, and the similarities observed in the expression domains of these genes are instances of convergent evolution. It is not possible to distinguish between these interpretations until more comparative data become available."

In other words there is no naturalistic explanation but the explanation itself. The question is how do we know that any of their proposals is correct? How would "comparative data" objectively ever tell us? This is not seen as a muddle but as more evidence of evolution because this is how they conclude:

"This and other studies highlight the insights to be gained by incorporating evidence from paleontology, metazoan phylogeny, multiple taxa, and multiple levels of biological organization."

The "insights" are more evidence of evolution and of the hierarchy of homology, not its internal contradictions. Emperor's New Clothes.

The second example concerns the *Hox* gene variably called *eyeless* in *Drosophila*, *Pax-6* in mouse and *Aniridia* in human, and which are taken to be homologous genes across these taxa and thus molecular evidence of evolution from a common ancestor (*LOE#8*). Ernst Mayr is our expert this time:

"It had been shown by morphological-phylogenetic research that photoreceptor organs (eyes) had developed at least 40 times independently during the evolution of animal diversity. A developmental geneticist, however, showed that all animals with eyes have the same regulatory genes, *Pax 6*, which organizes the construction of the eye. It was therefore at first concluded that all eyes were derived from a single ancestral gene with the *Pax 6* genes. But then the geneticist also found *Pax 6* in species without eyes, and proposed that they must have descended from ancestors with eyes. However, this scenario turned out to be quite improbable and the wide distribution of *Pax 6* required a different explanation. It is now believed that *Pax 6*, even before the origin of eyes, had an unknown function in eyeless organisms, and was subsequently recruited for its role as an eye organizer."[72]

Translating, we have no clue how it happens but that it is evolutionary we already know. How are we to make sense of all this?

To look in the right direction for evidence of evolution we need to look not at *Hox* but downstream to the regulatory networks where development is coordinated. This conceptual gap between *Hox* gene conservation on the one hand and phenotype diversity on the other is what evolution evidence must illuminate. What is that evidence exactly?

It's quite simple really. Evolution has caused co-option, the recruitment of a regulatory gene to new functions. Sean Carroll at the University of Wisconsin and John True at Stony Brook explain the concept:

"Co-option occurs when natural selection finds new uses for existing traits, including genes, organs and body structures. Genes can be co-opted to generate developmental and physiological novelties by changing their patterns of regulation, by changing the functions of the proteins they encode, or both."[73]

This is the molecular equivalent of exaptation, a term coined by Stephen Jay Gould by which a particular supposed ancestral structure is co-opted for another function by evolution (§15.4.2). They explain further:

"The story of evolution is about the appearance and disappearance of species and traits of various kinds: organs, structures, pattern elements, cell types, physiological processes and genes. These traits can either evolve de novo or may be derived from ancestral characters. Although de novo invention certainly fueled early evolution, the broad distribution of conserved proteins and motifs across the tree of life indicates that current diversity of phyla of modern organisms has extensively involved new combinations and modifications of pre-existing molecular characters. The acquisition of new roles by ancestral characters or new characters from old ones is known as co-option."

What promised to be an interesting exposition gets cut short however, at least if you were wanting evidence, when they tell us what is story and what is actually known here. "As fundamental as this process has been in evolution, little is known about the mechanisms by which co-option of gene function takes place and whether particular modes of co-option are responsible for important episodes of change in the evolutionary history of complex organisms."[73] In other words the evidence is the explanation. Specifically co-option exists as a process because evolution is true and *Hox* genes must be homologous because they are so similar and so ubiquitous.

Since *Hox* genes are few in number their downstream processes are the site and seat of evolution in development. Little is known about how exactly this happened, but happen it did because evolution is a fact and these few genes in the face of all the diversity of the animal kingdom is also a fact. More research is needed to sort the evolution out. Whether that research is even on the right track is in no doubt either, although all we hear is an appeal to the absence of evidence.

In that regard Dr. Carroll recently wrote a book to show how *Hox* genes confirm macroevolution.[74] It was listed among the top science books of the year. Yet nowhere in his book were we told the details of how his wonderful claim of macroevolution by *Hox* happened exactly, or even could have happened by an offer of specific explanatory details. He simply pointed to the potential combinatorial power of using a few different *Hox* gene switches in the way four DNA bases of varying order can specify so much protein diversity, as if that were the evidence. He did tell us this evidence of evolution through differential control of *Hox* gene switching happened in the non-coding part of the genome he called the "dark matter of the genome", making the comparison this was like ordinary matter compares to dark matter (§13.4.2). The metaphor was apt. Biology in darkness. *LOE#8* is empty and is rejected.

§9.4 CLAIM #2: IS ANCESTRY IS INFERRED BY SIMILARITY IN DEVELOPMENT?

We have evaluated Darwin's Claim 1 that embryos are "closely similar" to discover that while vertebrate development shares stage and morphologic similarities across taxa, the similarities are nowhere near what Darwin or Haeckel or von Baer claimed. Differences are at least as impressive as similarities. This removes the basis for inferring common ancestry as the cause. Darwin's Claims #1 and #2 are rejected too.

Morphologic and molecular similarity among embryos is taken by evo-devo to be most evident not at the beginning of development as von Baer's laws assert, but roughly midway at the tailbud or phylotypic stage (Figs. 9.2;9.3;9.6). In other words the naked-eye gross morphology supposedly converges toward and away from a pattern of similarity in hourglass fashion (Fig.9.4). This is yet taken to be caused by common ancestry. Objective analysis rejects the idea and anyway, the processes by which this supposed similar morphology is attained differ (§9.3.3;Fig.9.5). In regard to *Hox* genes and evidence of supposed molecular similarity caused by evolution, "[t]he study of many animal systems, both vertebrate and invertebrate, suggests that the mechanisms used to establish *Hox* transcription are nearly as diverse as the body plans they specify."[69] It's a puzzle to be sure, but not one that comes as a surprise to us and also one that does not imply evolution (§7.3).

While in *Hox* a highly similar set of genes is seen to be employed during development, they do different things in constructing different organisms like a toolbox is used to construct different things. The science says this is explained by common ancestry. The question is how that could happen plausibly by a naturalistic mechanism? With such differences in process behind supposed embryo morphologic similarity, not to mention different morphology now absent Haeckel's forgeries and von Baer's empty claims of similarity, how can there be a "community of embryonic structure" to permit the inference that an ancestor-descendant relationship is the cause? Darwin's Claims #1 and #2 are rejected on grounds of dissimilarities of morphology and molecular mechanism. Objective analysis does not confirm similarity sufficient to warrant an inference of ancestry. Rarely this will be admitted, as Duke biologist Kathleen Smith does: "The idea that stages are directly comparable across taxa has its origins in Haeckel and also in the pre-Darwinian comparative embryologists, and is demonstrably false...virtually all detailed comparative developmental studies show that a regular progression of stages with detailed equivalence across taxa at higher levels simply does not exist."[75].

You won't find that said in the textbooks however (§25.5). What we see is different mechanisms employed by similar genes, so really representing dissimilarity. The problem for evo-devo is the same as we found earlier for homology in adults. It's the need for a naturalistic mechanism, and any with objective evidence will do. If similar genes and similar gene actions were found, it would make an evolutionary inference easy because genes are the effectors and ultimate mechanism of evolution. Instead we see similar genes used in different ways in different organisms. Developmental mechanisms and structures change rather than the genes, just when we thought evolution was the changing of genes, which caused changing of developmental mechanisms, which caused changing structures. Needless to say shared structures and use of shared genes is fully compatible with the reiterated features of design economy and ID. Design is only refuted when a more parsimonious naturalistic explanation accounts for the same finding. The evidence of evo-devo by all its absurdities and contradictions is scientific evidence of design.

The interesting thing is these problems for evolution theory have proved no barrier to community of descent persisting as a scientific truth in embryology. The reason is since evolution is true there must be evidence of evolution in development somewhere. Even Michael Richardson sees evolution in his embryos. He writes in *Nature*: "Indeed, fish and human embryos do look similar because they share primitive features – 'symplesiomorphies' in modern terms."[53] To arrive at this conclusion in the face of his data takes your breath away (§9.3.3). While we might grant him the notion of similarity (but see the morphological differences even among fishes Fig.9.4a), we do not grant him a leap from this tenuous claim of similarity to a conclusion of symplesiomorphy. There is no evidence for that either. How can there be? If the naked eye evidence of morphologic similarity we grant him is not borne out by similarity when comparisons are made at the quantitative, ultrastructural and process level, and by his own research mind you,[33,43,52,55,56] it's surely misleading similarity if it is to be used to infer ancestry because it's really dissimilarity. LOE#7 and LOE#8 are rejected.

The focus of evolutionary developmental biology is the explanation of how evolution causes novel morphology. It began in 1922 with the observation by Walter Garstang that the evolution of new animal forms is in essence the evolution of embryologic development. In his words: "ontogeny does not recapitulate phylogeny, it creates it". The task of evo-devo is to recreate or reconstruct the path evolution took. Evolution itself is not a matter of inquiry because that is a fact already. This gets interesting. We know that to get from mutation to selection we need development, but how do we get from development to evolution? This is not as simple as it might seem. For while development and evolution are both processes of change, they are fundamentally different. Development is programmed and tightly controlled while evolution is a result of random undirected processes filtered by natural selection. How can they be related in an evolution paradigm? By recapitulation. Thank goodness for Ernst Haeckel. And you thought he was dead! We look next at how Line of Evidence #9 works.

§9.5 EXHUMING ERNST HAECKEL – THE PARALLELS BETWEEN ONTOGENY AND PHYLOGENY

Haeckel's notion of ontogeny recapitulating phylogeny by terminal addition was rejected because it was shown that genes already present from conception direct development. Recapitulation as concept and as a process was not rejected however, only those postulates that had to do with terminal addition. The reason it survived is because evolution remained true. Recapitulation just needed a new mechanism. Stephen Jay Gould resurrected it in 1977 after de Beer had seemingly debunked recapitulation along with the Biogenetic Law for posterity, even calling Haeckel's conception a "mental straitjacket which has had lamentable effects on biological progress"[76]. Gould performed this feat with his enormously influential book *Ontogeny and Phylogeny* that defined the current working paradigm for the field.[5,75]

The idea there is a repetition of ancestral forms in development in some form, in other words that "parallels exist between stages of ontogeny and phylogeny", is a basic truth in evo-devo. It is *LOE#9*.[8,44,77-82] Here it is in the words of Wallace Arthur: "taking a broad view, both von Baer and Haeckel captured elements of the truth: evolution leads both to embryonic divergence and, in some lineages, to a lengthening of the ontogenetic trajectory leading to more complex adult phenotypes with greater numbers of cells, their embryos passing through simpler, quasi-ancestral forms...The ontogeny of the individual mouse recapitulates the broad levels of complexity, but not the precise ancestral forms (either embryonic or adult), that the mouse lineage passed through in the course of its evolution."[8]

Let's follow recapitulation's rise to see how we got to this current state of affairs. We need to wind time back to 1977. We are in Dr. Gould's office on the Harvard campus. He indicates the starting point for his reasoning in the introduction to his book. Actually we read it at the start of our inquiry earlier:

"Evolutionary changes must be expressed in ontogeny, and phyletic information must therefore reside in the development of individuals. This, in itself, is obvious and unenlightening."[4]

The notion rests entirely on the assumption evolution is a fact. Yet Dr. Gould never considered it necessary anywhere in his 500-page thesis to review, marshal or even allude to the evidential basis of that foundation on which he constructs his thesis. The reason it's important is if evolution is false, the edifice he builds is going to be false also.

So debunked was Haeckel by de Beer before Gould's writing that ideas of recapitulation were held with scientific sheepishness. Why were they held at all then? You know (§20). Gould relates his experiences of this faith-in-evolution-over-the-facts-in-biology by his peers while he was writing *Ontogeny and Phylogeny*. "I have had the same, most curious experience more than twenty times," he said. "I tell a colleague that I am writing a book about parallels between ontogeny and phylogeny. He takes me aside, makes sure that no one is looking, checks for bugging devices, and admits in markedly lowered voice: "You know, just between you, me, and that wall, I think that there really is something to it after all."[83] This is infatuation against reason's better judgment, done scientifically. Here is Rudolf Raff at it: "No matter how much we deride the naiveté of Haeckelian recapitulation, and however we redefine the mechanism that connect evolution and development, there is a shadow of truth in the idea."[84] The fact is recapitulation is an imperative if evolution is true. Ernst Haeckel is as much a part of evolutionary theory as Ernst Mayr, George Gaylord Simpson, Theodosius Dobzhansky and Charles Darwin, and always will be.

Gould states the purpose of writing his book in its opening pages. He discovered there was truth in Haeckel's ideas that needed to be retrieved, revived, reformulated and presented as "a central theme in evolutionary biology" with "no apology". He says:

"I hoped at best, to retrieve from its current limbo the ancient subject of parallels between ontogeny and phylogeny. And a rescue it certainly deserves, for no discarded theme more clearly merits the old metaphor about throwing the baby out with the bath water. Haeckel's biogenetic law was so extreme, and its collapse so spectacular, that the entire subject became taboo...But I soon decided that the subject needs no apology. Properly restructured, it stands as a central theme in evolutionary biology...The starting point for a restructuring must be the recognition that Haeckel's theory requires a *change in the timing of developmental events* as the mechanism of recapitulation. For Haeckel, the change was all in one direction – a universal acceleration of development, pushing ancestral adult forms into the juvenile stages of descendants. Our current, enlarged concept does not favor speeding up over slowing down; all directions of change in timing are equally admissible. Paedomorphosis – the appearance of ancestral juvenile traits in adult descendants – should be as common as recapitulation."[4]

The concept he presents has an appealing simplicity. He also has a solution which explains recapitulation without having to invoke Haeckelian, and therefore Lamarckian, terminal addition. Evolution can occur in only two ways. This is how:

"Evolution occurs when ontogeny is altered in one of two ways: when new characters are introduced at any stage of development with varying effects upon subsequent stages, or when characters already present undergo changes in developmental timing. Together, these two processes exhaust the formal content of phyletic change; the second process is heterochrony."[85]

"Heterochrony" refers to a change in the timing of an ancestral event or character, which may therefore appear earlier or later or proceed at a different speed in the development pathway of the descendant. It is a relative term then, relative to the ancestor and the ancestor is always hypothetical. But since evolution is "a fact" it becomes reasonable to employ, as proxy or by default, comparisons between the development of living taxa already assumed to be related by evolution. This is circular reasoning.

Heterochrony might be contrived, circular and legless, but it is an immensely important concept in evo-devo all the same. Dr. Gould continues and the italics are all his own:

> "All parallels between ontogeny and phylogeny fall into these two categories: If a feature appearing at a standardized point of ancestral ontogeny arises earlier and earlier in descendants, we encounter a direct parallel producing [Haeckelian] *recapitulation* (the descendant repeats in its own ontogeny the sequence of stages that characterized ancestors at their standardized point)...If a feature appearing at a standardized point of ancestral ontogeny arises later and later in descendants, we encounter an inverse parallel producing *paedomorphosis* (early features of an ancestral ontogeny are carried forward to appear at the standardized point of a descendant) [or reverse recapitulation]...I must emphasize that classifications based upon addition and displacement completely exhaust the morphological description of how evolution can occur. [Thus the theory of evolution dictates here what observations in biology must discover, rather than conventional science, which is the reverse (§18.7).] Evolutionary changes must appear in ontogeny, and they can arise only by the introduction of new features or by the displacement of features already present. The second process produces parallels between ontogeny and phylogeny; the first does not. Together, they describe the course of morphological evolution. *The continued relevance to modern biology of the great historical theme of parallels between ontogeny and phylogeny rests entirely upon the relative frequency of evolution by displacement rather than by introduction.*"[86]

We should pause to inspect this further.

First, Dr. Gould is unambiguous: "Parallels between ontogeny and phylogeny are produced by heterochrony."[87] The stages of an animal's development and life history run in parallel with the adult forms of their ancestors, a concept called heterochrony. Thus the ancestral condition is reflected in embryogenesis. The process operates in all cases, except those in which totally novel features appear during development. That ancestry and evolution is not only demonstrable in developmental pathways but a "pervasive phenomenon",[88] is assumed to be true and is incorporated into the paradigm by definition. Tests of the paradigm cannot be used to evaluate evolution then, as any apparent anomalies need to be accommodated and explained within the paradigm. Conversely the paradigm is self-fulfilling if it is used to provide evidence for evolution. Evolution remains a fact and the evidence collected confirms it a scientific fact still further. Emperor's New Clothes.

But what does the term "heterochrony" actually mean? It's a critical concept to evolutionary embryology. Rudolf Raff even says heterochrony "has been the single most pervasive idea in evolutionary developmental biology".[80] The word has an interesting history.

It was coined originally by Ernst Haeckel to mean close to the opposite of what it means today, namely exceptions to global recapitulation. Haeckel meant it to refer to a change in the timing of development of one structure or organ relative to others within the same individual in creatures contradicting the Biogenetic Law. De Beer proposed that new morphologies caused by evolution occurred by changes in the timing of sequences in development and he called this heterochrony.[89] Thus it was De Beer who changed Haeckel's meaning of the word to refer to displacement in developmental timing of a character relative not to others in that same individual, but to the time that character appeared in the development of the ancestor. De Beer also argued that evolutionary change need not only occur at terminal stages of development as Haeckel insisted, but at any stage. He also "discarded the concept of recapitulation, as well as phylogeny in and of itself, as the proper focus of comparative developmental biology".[77] To de Beer heterochrony simply meant a change in the timing of ontogeny, without any recapitulatory connotation.

Gould reinvented both recapitulation and parallelism by changing the definition of heterochrony once again. The word is now restricted to shifts in timing of development associated with a parallel between ontogeny and phylogeny. Recapitulation is thus a fact, and heterochrony applies to recapitulatory events during development. Gould defines heterochrony as:

> "phyletic change in the onset or timing of development, so that the appearance or rate of development of a feature in a descendant ontogeny is either accelerated or retarded relative to the appearance or rate of development of the same feature in an ancestor's ontogeny."[90]

Kathleen Smith at Duke explains what it is "Heterochrony involves a shift in the timing of developmental processes so that an event occurs earlier, later, or at a different rate in a taxon compared to its ancestor."[77] The term refers specifically to morphologic differences - structures, organs, shapes and traits between descendants and ancestors, and not to the rates themselves.[78] What is the significance of heterochrony?

First, it demands that evolution be present. Second, heterochrony is the process that blurs the evidence of ancestry in evolution, and so acts to keep the theory of evolution intact when the evidence appears to say otherwise. It's the scientific mechanism by which contradictions to evolution become evidence. Really, but don't take my word for it. Here is Michael Richardson: "Heterochrony alters developmental sequences – and as a consequence, stages – making cross-species comparisons of stages difficult or impossible"[24] "Heterochronic variation makes it impossible to define a conserved stage at which all vertebrate embryos have the same combination of organ primordial present."[33] Thus so long as there is heterochrony, evolution is already present in embryology.

The specific patterns of heterochrony, for the record, are summarized in Fig. 9.8. The first is recapitulation á la Haeckel, when the sequence of events in ontogeny is directly parallel to the sequence of characters in the genealogy of the phylogeny. It is when, as Gould stated it, the "ontogeny of the most advanced descendant repeats the adult stages of phyletic series of ancestors)."[92]. Recapitulation occurs either by an increased speed of somatic development which is called "acceleration" [which was Haeckel's only speed], or by retarded sexual development, called "hypermorphosis". The notion assumes that physical maturation ceases with sexual maturation.

The second pattern is "paedomorphosis" (literally meaning "shaped like a child") which can be considered the reverse of recapitulation. It is retention of the ancestral juvenile shape in a descendant adult, and produced when the "ontogeny of the most remote ancestor goes through the same stages as a phylogeny of adult stages [of the descendant] read in the reverse order ...[as] more juvenile stages of ancestors become the adult stages of successive descendants".[92] It may be helpful to see the concept demonstrated in specific example. Ehab Abouheif and Gregory Wray again:

"Perhaps the most famous example [of contradictions of the Biogenetic Law] occurs in the axolotl, *Ambysoma mexicanum*. This salamander does not metamorphose, and aquatic larval features, such as gills and a tail fin, are retained throughout adult life. Conversely, most of the other salamander species within the same genus undergo complete metamorphosis and lose their gills and tail fins. [Where is the evolution you ask? It's the opposite of Haeckel's acceleration, it's retardation.] Thus, the most parsimonious interpretation of this observation is that the ancestral salamanders of this genus underwent complete metamorphosis and lost their gills and tail fin, whereas the retardation of development observed in Ambystoma mexicanum is an evolutionarily derived feature relative to its ancestors."[70]

Paedomorphosis occurs either when sexual development is accelerated (called "progenesis"), or when somatic development is retarded (called "neoteny").

Timing		Name in de Beer's system	Morphological result
Somatic features	Reproductive organs		
Accelerated	-	Acceleration	Recapitulation (by acceleration)
-	Accelerated	Paedogenesis [=progenesis]	Paedomorphosis (by retardation)
Retarded	-	Neoteny	Paedomorphosis (by recapitulation)
-	Retarded	Hypermorphosis	Recapitulation (by prolongation)

Fig.9.8. The patterns of heterochrony. Reproduced by permission of the publisher from *Ontogeny and Phylogeny* by Stephen Jay Gould, p229, Cambridge, Mass: The Belknap Press of Harvard University Press © 1977 by the President and Fellows of Harvard College.[91]

Gould is quick to call it a process but of course it's nothing even close. A change in timing cannot be a mechanism! Heterochrony is instead a comparative pattern of development, but comparisons of what exactly? Heterochrony is just a placeholder for some other as yet undefined developmental mechanism that works through evolution and begs a naturalistic explanation with details. Gould postulated the changes of heterochrony lay at the level of regulatory genes in some fashion. Absent a defined genetic or biochemical mechanism, his theoretical concept of heterochrony serves no other purpose than to shoehorn observations in embryology, whatever they may be, as evidence of evolution.

The concept of heterochrony has two obvious problems. It stands only if the assumption of parallelism is true and if evolution is true, and it depends on the reliability of circular reasoning regarding its comparisons with ancestrally related and unrelated species. Rudolf Raff reflects on this and says:

"The persistence of the idea that heterochrony is the major mechanism for evolutionary changes is probably due to its appealing simplicity and to the ease with which examples can be fit to the various predicted categories of heterochrony [actually all observations can be accommodated]. Although it is applied widely as an explanation, it is not so certain that heterochrony is a universal mechanism. Heterochronic results are inevitable. We measure all developmental events along a time axis. Therefore, anything that happens to alter the course of development has temporal consequences. Nevertheless, heterochrony has been a very important concept because it has provided a simple unifying mechanism around which data can be ordered." 93

There you have it from the experts. It's just a descriptive construct for ordering data, for interpreting observations in embryology as evolutionary, but the inclination to call it a "mechanism" (Raff) or a "process" (Gould), is irresistible. The reason is because it must be if evolution is true. The fact is it has no defined mechanism. Drs. Abouheif and Wray are back in case you don't believe me:

"Although heterochrony provides an adequate description of the dissociations possible in the timing or onset of development events between ancestral and descendant ontogenies, it fails to provide an adequate mechanistic explanation for these developmental events."

We should stop here and take stock.

In evolutionary theory changes in developmental programming are the link between the effects of mutation and natural selection. The theory holds there are four possible ways by which a mutation could produce an evolutionary change in development. In: a) timing, b) spatial distribution, c) quantity and d) type.[8] We have examined a) or heterochrony already. The term "heterotopy" refers to b) - the development of a feature in a different place, from different cell types or by other different mechanisms relative to that of an ancestor. This is simply a variation in space and complementary to heterochrony which is a variation in time. The word was coined by Haeckel to refer to the different germ layer origins of reproductive organs that arise variably from endoderm or mesoderm in different taxa. Heterotopy was given a scant few lines by Gould in *Ontogeny and Phylogeny* and in Brian Hall's opinion is still "virtually ignored".[94]

As can be surmised from its definition, heterotopy introduces new morphologies different from the ancestor and thus does not lead to parallelism between ontogeny and phylogeny, perhaps the reason Gould did not discuss it further. The concept concerns us however, in that it is hard to conceive of differences that are purely heterochronic or purely heterotopic when comparing patterns of development across taxa.[81] Most changes, even in our imagination, are surely a combination of both temporal and spatial differences and therefore heterotopy as a concept is sure to be revived in the future. We will have to see what the future history of evolutionary biology delivers up. Both heterochrony and heterotopy as currently defined are strictly evolutionary concepts, but of course the terms could be used in a comparative manner between any two taxa without inferring ancestry. Perhaps the most familiar example of recapitulation as evidence of evolution today, is in the notion of gill slits found in all vertebrate embryos reflecting their evolutionary ancestry from fishes. We take a close look at this idea much later (§25.5).

We have now dealt with Darwin's claims 1-3 and *LOE#7*, *LOE#8* and *LOE#9*. What about his claim 4 that is *LOE#10*, namely that evolution can be inferred because the embryo retains "structures which are of no service to it"?

§9.6 CLAIM #4: WHAT IS THE EVIDENCE FOR EVOLUTION FROM VESTIGIAL STRUCTURES?

There is a notion in evolution theory so familiar it has currency even in the popular culture. It's that animals including humans bear the degenerate remains of structures once important to their evolutionary ancestors. This is *LOE#10*.[95-99] Human examples are the appendix as "simply the remnant of an organ that was critically important to our leaf-eating ancestors, but of no real use to us"[95], the coccyx which is " a vestigial tail...It's what remains of the long, useful tail of our ancestors,"[95] erector pili muscles that only give us goose bumps but which raise the hair of wild mammals to make them seem larger when angered, lanugo hair which "can be explained only as a remnant of our primate ancestry: fetal monkeys also develop a coat of hair at about the same stage of development"[95] and the male nipple. Douglas Futuyma explains them for us. "Every organism has vestiges of structures that can only be the useless remnants of past adaptations."[100] Guess where that idea came from?

Charles Darwin. He called such structures "rudimentary" and then used his interpretation as the evidence of their evolution! But we're racing ahead. This is how he originally laid out his argument in *The Origin*:

"Organs...bearing the plain stamp of inutility, are extremely common, or even general throughout nature....In the mammalia, for instance, the males possess rudimentary mammae; in snakes one lobe of the lungs is rudimentary; in birds the 'bastard wing' may be safely considered a rudimentary digit, and in some species the whole wing is so rudimentary it cannot be used for flight. What can be more curious than the presence of teeth in fetal whales, which when grown up have not a tooth in their heads; or the teeth, which never cut through the gums, in the upper jaws of unborn calves?...rudimentary organs...are the record of a former state of things, and have been retained solely through the power of inheritance...Rudimentary organs may be compared with the letters in a word, still retained in the spelling, but become useless in the pronunciation, but which serve as a clue for its derivation." (p.601,608) His view was not from evidence however. He had no ancestor, he could not explain how these organs evolved through disuse by natural selection, and his genetic mechanism was the false intuition of Lamarckism. And yet academia today agrees with his interpretation (Table 3.1). If you do also, consider four objections.

The first is that Darwin's argument rests on a premise these organs are of no adaptive use. It's a premise in logic called a universal negative. The problem is we need to be certain it's true for his reasoning to proceed, but it's hard to be sure. For instance there was a time when it was generally agreed there were 180 human vestigial organs.[101] This was also the opinion of the American Association for the Advancement of Science, the world's largest general scientific society, and which submitted the list as amongst the evidence for evolution at the Scopes trial (§24.1.1). That long list has all but since evaporated however. Arguments for and against some of the most commonly regarded vestigial organs are listed in Fig.9.9. Make of them what you will.

The second objection is that the argument is not only from ignorance, it's circular. It's accepting the idea that the "utility" of a biological structure determines its presence in an organism in the first place. On this premise the appearance of degeneracy that is a vestigial organ is argued in reverse, but what's the evidence for that? To assume the premise is already to assume macroevolution true and natural selection the cause. To apply that premise as the rationale by which to label structures vestigial and then discover evidence for evolution, is to trace a perfect circle.

Perhaps an example is helpful. Jerry Coyne says vestigial organs are "some of the most powerful evidence for evolution."[95] How? Because they "make sense only as remnants of traits that were once useful in an ancestor," he says. But how? It's because of what's "makes sense". Only evolution does. He says: "It is vestigial not because it's functionless, but because *it no longer performs the function for which it evolved*". Which is not only to presume vestigiality but evolution too, and an evolutionary ancestor. Call this "powerful evidence" like him if you must, but it's all just circles.

Third, what's the mechanism by which rudimentary organs arise? To be caused by evolution there has to be a naturalistic mechanism to do it. This

STRUCTURE	EVOLUTIONARY EXPLANATION	COUNTER EXPLANATION
Human coccyx	It is the evolutionary residua of an ancestral tail.[97,102]	The spine must end somewhere. It's needed for perineal muscle attachment and pelvic floor support.
Appendix	It's the vestigial cecum of a plant eating ancestor, or a synapomorphy between apes and humans.	It's a lymphoid organ. Most mammals have a cecum and an appendix. Evidence does not support it as a synapomorphy (§10.4.2).
Male nipple	It is a vestigial remnant of the evolutionary precursor to male and female	Human embryos are initially sexually dimorphic i.e. both Mullerian (female) and Wolffian (male) duct systems develop by design economy and definitive sexual characteristics develop later by hormones to become male or female. The nipple is an indication of embryogenesis therefore, not evolution. It's dubious they are "useless" anyway, being secondary sex characters.
Erector pili muscles	"Humans also have muscles that can make body hair stand on end when an individual is cold or excited. This phenomenon is homologous with the erectile fur of other mammals, which signals alarm, or aggressive intent. In humans, the homologous response results in the vestigial trait called goose bumps."[97,103]	The evolutionary argument requires biological similarity be evolutionary but it's not (§7). Design economy sees no contradiction in shared features.
Wisdom teeth	"The posterior molar or wisdom-teeth were tending to become rudimentary in the more civilized races of man...In the melanin races, on the other hand, the wisdom-teeth are...generally sound"[104] "Why should we have wisdom teeth, unless our jaws have become shorter, so that our ancestor's teeth no longer fit?"[100]	Wisdom teeth have the same chewing function as the other two molars. The problems they cause have to do with diet.

Wings on flightless birds	"the bones of the flightless dodo and penguin are also hollow, as if adapted for flight...Every organism has such vestiges of structures that can only be useless remnants of past adaptations."[100]	Biological similarity is not evolutionary (§7). Wings have other functions like balance, mating rituals, warmth, cooling and swimming. Design economy sees no contradiction in shared features and no contradiction in a non-functionality.

Fig 9.9 **Vestigial organs.** Evolution evidence or not.

was Darwin's mechanism: "It appears probable that disuse has been the main agent in rendering organs rudimentary. It would at first lead by slow steps to the more and more complete reduction of a part, until at last it became rudimentary" (p.606).

At first blush this might seem satisfactory but look again. It cannot explain further change after a structure is no longer used. Consider supposed structures like snake "legs" or whale "hind legs" (§5.5.4). Darwin recognized this problem, and said: "After an organ has ceased being used, and has become in consequence much reduced, how can it be still further reduced in size until the merest vestige is left; and how can it be finally quite obliterated?"(p.607-8). What's the answer? I hope you have a better one than he does, because this was his answer: "Some additional explanation is here requisite which I cannot give" (p608). In summary, the mechanism by which "rudimentary" organs arise is unknown, the answer is not aided by the notion they are unused organs, and yet they are evidence. None of this garners even a passing mention by those who claim vestigial organs are evidence of evolution.[95-99]

Fourth, Darwin takes the idea of organ "inutility" a step further, to "inheritance". It's that a particular "rudimentary" structure in an organism is indeed in the same evolutionary lineage as a purported ancestor. Inferring ancestry is an interpretation of similarities between two creatures and cannot be more than a hypothesis. Thus the underlying assumption of vestigial organs is just an inverse of the argument from homology. We inspected that idea earlier and found it empty (§7;§9.3-4). The demonstration biological similarity is not evolutionary is the best evidence refuting vestigial organs.

The vestigial organ argument is widely taken as demonstrable scientific evidence for evolution by natural selection but we find this science is on a quicksand of subjectivity. Specifically 1) subjective opinion that a structure is useless; 2) subjective opinion that structures exist and are generated in organisms only because of their utility and adaptiveness; 3) subjective opinion that a particular structure in a species is related by ancestry to another selected structure in other species, and 4) subjective faith, because rudimentary organs cannot be explained as arising by being rudimentary (i.e. by disuse), even if we grant the assumption macroevolution by natural selection is true. *LOE#10* is rejected.

The argument for from vestigial organs is peripheral to the case that must be made by evolution to be the solution to the Conundrum anyway. This is because *LOE#10* is really an argument for devolution, for loss of structures and the inverse from what we have been dealing with to now, namely the cause of biological complexity. The problem for evolution to explain is not loss or deterioration of function, but gains. How gains in complexity and diversity occur in biology and how the generation of new structures, organs and life forms happens naturalistically. So-called vestigial organs are a problem for evolution nonetheless because they require a naturalistic explanation and there's not one in view. ID by contrast makes no assumption traits must always be adaptive. Organisms can have structures without nothing more than a creative, artistic or cosmetic purpose. Non-adaptive, quirky and gaudy designs are familiar features of design and not just in the fashion and construction industry, especially when the budget is big.

§9.7   CLAIM #5: IS ANCESTRY INFERRED
          BY DISSIMILARITY IN DEVELOPMENT?

We come to Darwin's fifth and last claim. Here it is again:
"Thus community of embryonic structure reveals community of descent; but dissimilarity in embryonic development does not prove discommunity of descent, for in one of two groups the developmental stages may have been suppressed, or may have been so greatly modified through adaptation to new habits of life, as to be no longer recognizable." (p.599)

It's at once a strength and a weakness. It's a strength because whatever embryos are like they are evidence of evolution. It's also a weakness because this is a subjective descriptive construct and not falsifiable. Embryology can never be independent empirical evidence for evolution in this paradigm because the paradigm presupposes all the evidence to be evolutionary. This one-track evolutionary thinking began with Charles Darwin and never stopped. Haeckel coined the concept of cenogenesis to refer to instances of exception to his Biogenetic law, in his words, to those

435

instances which "falsify and obscure" it,[24] and today "heterochrony" does the same thing. What we are seeing again of course is the conflict between independent observations of objective fact and the greater fact of the evolution paradigm by which the enterprise that is biology is conducted. This scientific faith in an empty philosophy continues even more obviously in the understanding, if it can be called that, of hierarchical homology - anatomic, molecular, and now here developmental (§7.4.3). Vestigial organs are another example and so is recapitulation. Listen to Ernst Haeckel impose evolutionary recapitulation on nature:

> "If [recapitulation] was always complete, it would be a very easy task to construct the whole phylogeny on the basis of ontogeny. If one would like to know the ancestors of each higher organism, including man, and from which forms this species developed as a whole, one would only need to follow the chain of forms of his individual development from the egg onwards; then one could consider each of the existing morphological stages as representative of an extinct ancient ancestral form...But in the great majority of animals, including man, this is not possible because the infinitely varied conditions of existence have led the embryonal forms themselves to be changed and to partly lose their original condition."[23]

Evolution drives the interpretation of data and it comes from the application of a law in science that is faith in the Naturalistic Premise (§19,§20). This is evo-devo. Emperor's New Clothes.

§9.8 IN CONCLUSION: REPEAT A LIE OFTEN ENOUGH AND EVERYBODY WILL EVENTUALLY BELIEVE IT

Wallace Arthur at the National University of Ireland summarized the history and the status of evo-devo for the readership of *Nature*. As a pioneer of this field, the author of several books on evolution and a founding editor of the journal *Evolution and Development*, who better qualified to do it? We have reviewed that history for ourselves. This was his assessment of the ground that we have just covered:

> "Evolutionary developmental biology has its origins in the comparative embryology of the nineteenth century and in particular the work of von Baer and Haeckel...Following a quiescent period of almost a century, present day evo-devo erupted out of the discovery of the homeobox in the 1980's. One principal focus over the past 20 years has been the comparative study of the spatiotemporal expression patterns of

developmental genes.... Although there are many cases of conservation of expression patterns, there are also cases where these patterns differ markedly between closely related species, even when the gene concerned acts at a very early developmental stage. We have thus moved from one extreme to the other: from laws that turn out to be, at best, overgeneralizations, to a situation where it seems that anything goes, that is, any developmental gene, its expression pattern and the resultant ontogenetic trajectory can evolve in any way. If this were true, no generalizations would be possible, let alone universally applicable laws. However, the search for general patterns is fundamental to science and is not easily suppressed, even when the relevant data set looks very complex, as it currently does in evolutionary developmental biology. There are signs, particularly over the last few years that new general concepts are emerging."[8]

Translating, we get an oblique admission from one of the captains of evo-devo that the field is in chaos. What else is a "science" where "anything goes," or wants for "general concepts" after twenty years of work? Of more concern, why should we believe him "over the last few years" things are now improving? You've seen the evidence. You tell me.

Dr. Arthur alludes to our earlier discovery that although embryology produces all the phenotypes of animal life, and although it was the best of the evidence for Charles Darwin, it yet was not a part of the Modern Synthesis (§9.1). We never discussed how such a thing could have happened. How was embryology ejected from evolution theory?

The reason was mostly a person, geneticist and 1933 Nobel laureate Thomas Hunt Morgan (1866-1945).[5] It had to do with his frustration at embryology's lack of objective and quantitative data and specifically its nebulous notion of a "morphogenetic field", the research paradigm at the time. The term referred to a domain of cells that could shape body form. It was a competing alternative to the gene as the basic unit of ontogeny. Genes had not yet been discovered. Morgan declared "genetics...has brought to the subject an exact scientific method of procedure" that will "furnish us today with ideas for an objective study of evolution in striking contrast to the older speculative method as treating evolution as a problem of history".[5]

Science has since reversed itself on both opinions. With the rediscovery of morphogenetic fields[105] and the reincarnation of evolutionary embryology Morgan discarded which is now the "evo" in evo-devo, it will indeed need to find itself "general concepts" and quantitative measures for demonstrating evolution if it is to be more fruitful than its hapless predecessor, but it can't. Which is why we see serious discussion such vacuous philosophies as developmental systems theory can inform the science of embryology.[106]

Going back from history present to history past, Morgan and his then graduate student Theodosius Dobzhansky explained evolution as changes in gene frequency and the inevitable byproduct of population genetics in tangible and mathematic ways. The morphogenetic field by contrast was considered hand waving in a discipline still reeling in a vacuum left by the implosion of the Biogenetic Law. Embryology was ejected from the mechanism of evolution and the focus of embryology shifted from a paradigm of "fields" to comparative developmental gene expression and differentiation.[5] The significance of this, and the point, is genetics cannot deliver up macroevolution without operating through embryology. Neo-Darwinism is deficient as a solution to the Conundrum of Life without embryology, even in theory. You won't hear this from the Ultra-Darwinists, but listen to what the embryologists say. Scott Gilbert, John Opitz and Russell Raff make a rather surprising admission, one we wish we would have heard a lot earlier in our inquiry:

"The Modern Synthesis is a remarkable achievement. However starting in 1970s, many biologists began questioning its adequacy in explaining evolution. Genetics might be adequate for explaining microevolution, but micro evolutionary changes in gene frequency were not able to turn a reptile into a mammal or convert a fish into an amphibian. Microevolution looks at adaptations that concern only the survival of the fittest, not the arrival of the fittest. As [Brian] Goodwin [a biologist at Schumacher College and member of the Board of Directors of the Santa Fe Institute][107] points out, "the origin of species - Darwin's problem - remains unsolved."[5]

Wallace Arthur agrees: "neo-Darwinism does not constitute an overall theory of evolution, as most of its practitioners readily admit. It is not wrong; rather, it is incomplete".[6]

Development is essential to evolution but there is no objective evidence of evolution in development. We discover a missing link in the theory. It is the most important and the most missing of all the missing links of evolution because by it the mechanism of evolution is uncoupled. How do you get from mutation to natural selection without embryology?

We close by returning to Darwin's five claims with which we began. His first claim that embryos are very similar across animal taxa is a lie. Amazingly he never produced data for the idea although the declaration of evolution in embryology is premised on it being true. The "data" was first an empty claim by Ernst von Baer enshrined as a "law" even to this day, and second a series of forged drawings produced by Ernst Haeckel after

Darwin's claim of evolution in embryology had already been made. Embryo similarities are at least as great as their differences and even for *Hox* genes, which are the most impressive of the embryo similarities found so far (§9.3.5). Embryos are as empty of objective evidence for ancestry as we found for adults (§7.3).

Darwin's second claim that morphologic (*LOE#7*) and the later claim by the science that molecular (*LOE#8*) similarities among embryos imply common ancestry breaks down also therefore. There is some morphologic similarity at the supposed "phylotypic" stage but nowhere close to the similarity claimed or to entertain ancestry. The term phylotypic stage is instead another example of how evolution is enforced on biology by definition, and how biologists and their books perpetuate the myth (§25.5). To the contrary of demonstrating heritable traits to support common ancestry and evolution as cause of the similarities among embryos that do exist, we find that dissimilar developmental pathways attain similar external morphology across taxa and that similar molecules and genes across taxa do different things in different taxa. In other words what is really similar, are not the causative mechanisms, but their end result. This is an astonishing paradox. It is the same thing we found in adults. It's also a flat rejection of evolution .

Darwin's third claim that embryos recapitulate features of their evolutionary ancestors (*LOE#9*) was a notion conceived of before Darwinism was even imagined, and supposedly explained the *Scala Naturae*. Darwin reinterpreted that interpretation without evidence, the evidence became Haeckel's drawings for over a century, and today the explanation of recapitulation is the evidence. Though almost as effective as Haeckel's, this evidence is as empty, being constructed with definitions like "heterochrony" and "heterotopy" that accommodate all possible observations of development as being evolution, and thus enforce evolution on embryology.

Darwin's fourth claim, that vestigial organs demonstrate evolution (*LOE#10*) was an idea from a time when Lamarckism was true and his first two claims were true also. Darwin could not explain the existence of vestigial organs by evolution, yet he regarded them as evidence of evolution nonetheless. When used as evidence for evolution today *LOE#10* is as circular an argument in demanding evolution be true, biological similarity be evolutionary, and requiring the assumption be certain these "vestigial" structures have no function, not to mention the assumption that biological structures must have a function by natural selection to exist at all. Where does that idea come from?

Darwin's fifth and last claim is among the best evidence of why and how evolution is a fact from the evidence of embryology. Whatever the observation, embryo similarity (Darwin's claim 2) or embryo dissimilarity (Darwin's claim 5), evolution in embryology explains it. You heard how Wallace Arthur says it nowadays. Embryos "can evolve in any way".[8]

How can we summarize our observations in one sentence? Similarities of embryonic development across nature are discontinuous. It's the same observation we made for anatomical similarity in extant adult organisms, the same observation we made for extinct organisms and the same observation we made for molecular similarity. We see it in mature life (§7), immature life (§9), dead life (§5) and the molecules of life (§8). It exists at every level of life and everywhere in life. It's also an observation that's not a paradox for evolution, it's a rejection of evolution. Why? Because evolution demands uniformitarian, linking evolutionary relationships, a reductionist continuity across taxa, to permit an inference of common ancestry (§1.5.3;§3.6). We have no natural mechanism to explain the observations of order and complexity in development. The evidence instead indicates that the obvious appearance of design in nature is exactly that - obviously real.

A conspicuous final observation about evo-devo has to do with the conduct of the science. And here I don't mean its silliness like an a priori evolutionary framework for its conduct, or post hoc definitions already front-loaded with evolution by which to find more evolution. It's something more serious. It's that Haeckel's fraud is still seldom exposed for what it is to a degree this would be amusing if it were not egregious. This is Wallace Arthur's assessment for example:

> "Studies of the proposed phylotypic stage in vertebrates reveal more variation than was recognized by either von Baer or Haeckel, partly as a result of screening a wider range of species."[8]

He never tells the scandalous remaining reasons, and his contention is obviously untrue as a few species make the point just fine (e.g., Fig. 9.2 or even just Fig. 9.2a). Stephen Jay Gould made no attempt to set the record straight in his grand review *Ontogeny and Phylogeny* either, although Haeckel's work was at the hub of his heterochrony proposal. Given Gould's command of the literature it would seem inconceivable he was unaware. Thus he either believed Haeckel over those accusations to which we say on what rational basis, or he concealed them to which we will say nothing. Gould's definition of heterochrony and his views in *Ontogeny and Phylogeny* remain as true as ever, despite his eventual acknowledgement of Haeckel's embryos as being fakes.[40,78] This is very anomalous behavior by a science (§25). There would seem only two possible ways to explain it.

The first is the view there "is a shadow of truth" in Haeckel's ideas, and that compared to an overwhelmingly greater truth of evolution Haeckel's lies are not worth criticizing. The effect of this scientific expediency is that Haeckel is still praised, even by leaders in the field, to the extent one might wonder whether the truth even mattered. Take Rudolf Raff for instance:

> "One of the few scientific concepts to resonate well in popular culture is the idea that the evolutionary history of ancestors is recapitulated in the development of their descendants...No matter how much we deride the naiveté of Haeckelian recapitulation, and however we redefine the mechanism that connect evolution and development, there is a shadow of truth in the idea."[84]

Here is Brian Hall:

> "Whether Haeckel's Gastræa theory was right or wrong, it represented a brilliant synthesis of recapitulation, Darwin's theory of evolution, comparative morphology, homology and comparative embryology which had revealed common embryological construction from the same germ layers."[26]

Jerry Coyne says Haeckel was guilty "only of sloppiness": "Haeckel was accused, largely unjustly, of fudging some drawings of early embryos to make them more similar than they really are...his "fraud" consisted solely of illustrating three different embryos using the same woodcut."[95] Stephen Jay Gould even declared Haeckel "outstandingly right far more often than random guesswork would allow," setting a new measure for scientific greatness, and then presented a list of examples that supposedly confirm the Biogenetic Law as if it were true! (Listed in [24]). Even whistle-blower Michael Richardson is dazzled. He says: "The Biogenetic Law is supported by several recent studies – if applied to single characters only". Which is like saying the Flat Earth theory is supported by several studies if applied to this side of the horizon only. Dr. Richardson also says: "Haeckel's much-criticized embryo drawings [by him!] are important as phylogenetic hypotheses, teaching aids, and evidence for evolution...Despite his obvious flaws, Haeckel can be seen as the father of sequence-based phylogenetic embryology."[24] Claiming the disgrace of that paternity, the joke is on evo-devo. What does need examination though, is his claim that Haeckel's untruth is somehow useful to teach students about evolution's truth. Why?

Why if not that Haeckel better than anyone bonded ontogeny (the sequential appearance of embryo characters) and evolutionary phylogeny (that these characters are derived from an ancestor). He did it in a way irresistibly satisfying to evolutionary preconceptions and demands. Though evolutionary recapitulation was proposed by Darwin, it was Haeckel who popularized it because he first provided the evidence for it. It remains a truism and the reason is that there must be parallels between ontogeny and phylogeny if evolution is a fact. "Long live Ontogeny Recapitulates

Phylogeny!" This surely is the reason why Haeckel's forgeries persisted as textbook orthodoxy and why they are even now, just "flaws".[24] For though the year was already 1997 and Haeckel's supposedly "obvious" drawing was from way back in 1845, Richardson's revelations produced widespread reaction and surprise in both academic and lay press alike.

The issue it raised had been appreciated a long time before. It is that if you repeat a lie often enough people will eventually believe it. A specific example of this repetition and its effectiveness, is the seemingly unshakable notion of "gill slits" which are supposedly evident during the development of mammals. It is a notion very much alive in minds of even elite scientists today although it is absent a shred of evidence (§25.5).

A second reason why Haeckel's fakes don't get airtime is surely condemnation draws unnecessary attention and from creationists in particular. If so, this is a cover-up, by science. This is why the reaction to Haeckel's fakes is still a concern although they were definitively exposed by Michael Richardson years ago. This is not my conspiracy theory. The reaction to Dr. Richardson's work and his response in the (very public) letter pages of *Science* and *Nature* is more than illustrative of this concern.

Dr. Richardson was the whistle-blower and yet it is his letter, written in response to just that kind of creationist "Told you so" reaction, that is the most revealing of something very rotten here. I do not single him out either. His views are the overwhelming consensus. It's just that given his research of the matter, he would surely be expected to express views most skeptical of the supposed evo- in evo-devo and yet he does not. Far from it, for this is what he said:

"Our work has been used in a nationally televised debate to attack evolutionary theory, and to suggest that evolution cannot explain embryology. We strongly disagree with this viewpoint. Data from embryology are fully consistent with Darwinian evolution...On a fundamental level, Haeckel was correct: All vertebrates develop a similar body plan (consisting of notochord, body segments, pharyngeal pouches, and so forth). This shared developmental program reflects shared evolutionary history. It also fits with overwhelming recent evidence that development in different animals is controlled by common genetic mechanisms. Unfortunately, Haeckel was overzealous. When we compared his drawings with real embryos, we found that he showed many details incorrectly. He did not show significant differences between species, even though his theories allowed for embryonic variation. For example, we found variations in embryonic size, external form, and segment number which he did not show. This does not negate Darwinian evolution. On the contrary, the mixture of similarities and differences among vertebrate embryos reflects evolutionary change in developmental mechanisms inherited from a common ancestor."[108]

The shame in this writing is not the attempt at obfuscation with the "[o]n a fundamental level" deflection that flies in the face of Haeckel's explicit views, or that embryos are plainly different right from the time of fertilization (Fig. 9.1). It's not the characterization of blatant academic fraud as merely being "overzealous" either. It's the refusal of the scientist to consider any alternative paradigm to evolution even when confronted by the evidence of his own experiments. We have seen this same scientific performance by others so don't pick on him (§7.3;§7.4.3;§8.6.1;§9.5). That's how the Emperor always has, and always will scientifically keep his clothes on (§3.1). Dr. Richardson and his peers don't get it because they're always learning but never able to acknowledge the truth. Why this happens exactly is the question for you.

This inspection of embryology concludes our inspection of Lines of Evidence #1-10. The inquiry has only been an overview however. There is no science without detail and it is the details that expose truth from illusion. As a case study to deliver detail the next two chapters will consider the evidence of evolution by an in depth focus looking now at just one species of life to test whether the general observations apply. It is an inspection that applies the observations we have already made concerning microevolution (§4), paleontology (§5,§6), morphologic (§7) and molecular homology (§8) and embryology (§9) to *Homo sapiens*. It is thus a test of the conclusions we have made ,as much as it is another inquiry of the truth of evolution. What is the evidence for human evolution?

REFERENCES

1. Mayr, E. *What Evolution Is.* 31 (Basic Books, 2001).
2. Hall, B. K. *Evolutionary Developmental Biology.* (Chapman & Hall, 1998).
3. Wikipedia contributors. in *Wikipedia, The Free Encyclopedia* cited Jul 25, 2006 (http://en.wikipedia.org/wiki/Dolly_%28sheep%29, 2006 18 July).
4. Gould, S. J. *Ontogeny and phylogeny.* 2 (Harvard University Press, 1977).
5. Gilbert, S. F., Opitz, J. M. & Raff, R. A. Resynthesizing Evolutionary and Developmental Biology. *Developmental Biology* 173, 357-372 (1996).
6. Arthur, W. The concept of developmental reprogramming and the quest for an inclusive theory of evolutionary mechanisms. *Evolution and Development* 2, 49-57 (2000).
7. Darwin, F. *The life and letters of Charles Darwin, including an autobiography.* Vol. 1, 72 (D. Appleton & Co., 1887).
8. Arthur, W. The emerging conceptual framework of evolutionary developmental biology. *Nature* 415, 757-764 (2002).
9. Carroll, S. B. *Endless Forms Most Beautiful* 294 (W.W. Norton, 2005).
10. Orr, A. H. Turned On. *The New Yorker* (24 Oct 2005).
11. Gilbert, S. F. The morphogenesis of evolutionary developmental biology. *International Journal of Developmental Biology* 47, 467-477 (2003).
12. Raff, R. A. Evo-devo: the evolution of a new discipline. *Nature Reviews Genetics* 1, 74-79 (2000).
13. Darwin, C. *Descent of Man.* (Appleton New York ,1872).
14. Bryant, H. N. The threefold parallelism of Agassiz and Haeckel, and polarity determination in phylogenetic systematics. *Biology and Philosophy* 10, 197-217 (1995).
15. Gould, S. J. *Ontogeny and phylogeny.* 56 (Harvard University Press, 1977).
16. Hall, B. K. Phylotypic stage or phantom: is there a highly conserved embryonic stage in vertebrates? *Trends in Ecology and Evolution* 12, 461-463 (1997).
17. Gould, S. J. *Ontogeny and phylogeny.* 486 (Harvard University Press, 1977).
18. Darwin, C. in *The Life and Letters of Charles Darwin* (ed F Darwin) 21 (1902).
19. Darwin, C. *The Origin of Species,* 587 (Random House, Inc., 1998).
20. Ridley, M. *Evolution.* 2nd edn, 478-9 (Blackwell Science, Inc., 1996).
21. Darwin, C. *On the Origin of Species* 1st ed, 449 (John Murray, London, 1859).
22. Coyne, J. A. Creationism by stealth. *Nature* 410, 745-746 (2001).
23. Haeckel, E. *Anthropogenie oder Entwickelungsgeschichte des Menschen. Keimes- und Stammesgeschichte.* 5th ed. edn, (Engelmann., 1903).
24. Richardson, M. K. & Keuck, G. Haeckel's ABC of evolution and development. *Biological Reviews* 77, 495-528 (2002).
25. Raff, R. A. & Kaufman, T. C. *Embryos, genes and evolution.* 13 (Macmillan Publishing Co., 1983).
26. Hall, B. K. *Evolutionary developmental biology.* 2nd edn, 80 (Chapman & Hall, 1998).
27. Hall, B. K. *Evolutionary Developmental Biology.* 129 (Chapman & Hall, 1998).

28 Alberch, P. & Blanco, M. J. Evolutionary patterns in ontogenetic transformation - from laws to regularities. *International Journal of Developmental Biology* 40, 845-858 (1996).
29 Langille, R. M. & Hall, B. K. Developmental processes, developmental sequences and early vertebrate phylogeny. *Biological Reviews* 64, 73-91 (1989).
30 Gould, S. J. *Ontogeny and phylogeny.* 167 (Harvard University Press, 1977).
31 Ruse, M. *Darwinism Defended.* (Addison-Wesley Publishing Company, 1982).
32 Darwin, C. *The Descent of Man and Selection in Relation to Sex.* 1st edn, 15 (John Murray, 1871).
33 Richardson, M. K. There is no highly conserved embryonic stage in the vertebrates: implications for current theories of evolution and development. *Anatomy and Embryology* 196, 91-106 (1997).
34 Haeckel, E. *Anthropogenie: Keimes- und Stammes-Geschichte des Menschen.* (W. Engelmann, 1874).
35 Gerhart, J. C. & Kirschner, M. W. *Cells, Embryos, and Evolution.* (Blackwell, 1997).
36 Gould, J. L., Keeton, W. T. & Gould, C. G. *Biological Science.* (W.W. Norton & Co., 1996).
37 Kardon, K. V. *Vertebrates.* (Wm. C. Brown, 1995).
38 Starr, C. *Biology. Concepts and Applications.* 5 edn, (Wadsworth Group. Brooks/Cole, 2003).
39 Starr, C. & Taggert, R. *Biology. The Unity and Diversity of Life.* 10 edn, (Wadsworth Group. Brooks/Cole, 2004).
40 Gould, S. J. Abscheulich! (Atrocious!). *Natural History* 109, 42-49 (2000).
41 Goldschmidt, R. B. *The Golden Age of Zoology, Portraits from Memory.* (University of Washington Press, 1956).
42 Slack, J. M. W., Holland, P. W. H. & Graham, C. F. The zootype and the phylotypic stage. *Nature* 361, 490-492 (1993).
43 Beninda-Edmonds, O. R. P., Jeffery, J. E. & Richardson, M. K. Inverting the hourglass: quantitative evidence against the phylotypic stage in vertebrate development. *Proceedings of the Royal Society London B* 270, 341-346 (2002).
44 Duboule, D. Temporal colinearity and the phylotypic progression: a basis for the stability of a vertebrate Bauplan and the evolution of morphologies through heterochrony. *Development* Supplement, 132-142 (1994).
45 Raff, R. A. *The shape of life. Genes, development, and the evolution of animal form.*, (The University of Chicago Press, 1996).
46 Galis, F. & Metz, J. A. J. Testing the vulnerability of the phylotypic stage: on modularity and evolutionary conservation. *Journal of Experimental Zoology (Mol.Dev.Evol.)* 291, 195-204 (2001).
47 Ballard, W. W. Morphogenetic movements and fate maps of vertebrates. *American Zoologist* 21, 391-399 (1981).
48 Wolpert, L. *The triumph of the embryo.* (Oxford University Press, 1991).
49 Gould, S. J. *Ontogeny and phylogeny.* 77 (Harvard University Press, 1977).
50 Hall, B. K. *Evolutionary Developmental Biology.* (Chapman & Hall, 1998).

51 Gasman, D. *The scientific origins of national socialism: social Darwinism in Ernst Haeckel and the German Monist League*. (MacDonald, 1971).
52 Richardson, M. K. Heterochrony and the phylotypic period. *Developmental Biology* 172, 412-421 (1995).
53 Richardson, M. K. & Keuck, G. A question of intent: when is a 'schematic' a fraud? *Nature* 410, 144 (2001).
54 Pennisi, E. Haeckel's embryos: fraud rediscovered. *Science* 277, 1435 (1997).
55 Richardson, M. K., Allen, S. P., Wright, G. M., Raynaud, A. & Hanken, J. Somite number and vertebrate evolution. *Development* 125, 151-160 (1998).
56 Richardson, M. K. Vertebrate evolution: the developmental origins of adult variation. *BioEssays* 21, 604-613 (1999).
57 Wagner, G. P. Homologues, natural kinds and the evolution of modularity. *American Zoologist* 36, 36-43 (1996).
58 Wagner, G. P. & Misof, B. Y. How can a character be developmentally constrained despite variation in developmental pathways. *Journal of Evolutionary Biology* 6, 449-455 (1993).
59 Raff, R. A. *The shape of life. Genes, development, and the evolution of animal form.*, 196-7 (The University of Chicago Press., 1996).
60 Gould, S. J. *Ontogeny and phylogeny*. 6 (Harvard University Press, 1977).
61 Raff, R. A., Wray, G. A. & Henry, J. J. in *New Perspectives In Evolution* (eds L. Warren & H. Koprowski) 189-207 (Wiley-Liss, 1991).
62 Ballard, W. W. Problems of gastrulation: real and verbal. *BioScience* 26, 36-39 (1976).
63 Purugganan, M. D. The molecular evolution of development. *BioEssays* 20, 700-711 (1998).
64 Slattery, M., Ma, L., Negre, N., White, K. P. & Mann, R. S. Genome-wide tissue-specific occupance of the Hox protein Ultrabithorax and Hox cofactor Homothorax in Drosophila. *PLoS ONE* 6, e14686. doi:14610.11371/journal.pone.0014686 (2011).
65 Ferrier, D. E. K. & Minguillon, C. Evolution of the Hox/Parahox gene clusters. *International journal of developmental biology* 47, 605-661 (2003).
66 Lemons, D. & McGinnis, W. Genomic evolution of Hox gene clusters. *Science* 313, 1918-1922 (2006).
67 Carroll, S. B. Homeotic genes and the evolution of arthropods and chordates. *Nature* 376, 479-485 (1995).
68 Mastick, G. S., McKay, R., Oligino, T., Donovan, K. & Lopez, A. J. Identification of target genes regulated by homeotic proteins in Drosophila melanogaster through genetic selection of Ultrabithorax protein bonding sites in yeast. *Genetics* 39, 349-363 (1995).
69 Gellon, G. & McGinnis, W. Shaping animal body plans in development and evolution by modulation of Hox expression patterns. *BioEssays* 20, 116-125 (1998).
70 Abouheif, E. & Wray, G. A. in *Encyclopedia of Life Sciences* http://www.els.net/els/subscriber/article/article_main.asp?sessionid=320891927576455115 8&srch=qsearch&aid=A3208919275760001661&term=evolution+of+development (Macmillan Publishers Ltd, Nature Publishing Group, 2002).

71 Gaunt, S. J. in *Encyclopedia of Life Sciences* www.els.net (Nature Publishing Group, 2001).
72 Mayr, E. *What Evolution Is*. 113 (Basic Books, 2001).
73 True, J. R. & Carroll, S. B. Gene co-option in physiological and morphological evolution. *Annual Review of Cell and Developmental Biology* 18, 53-80 (2002).
74 Carroll, S. B. *Endless Forms Most Beautiful*. (W. W. Norton, 2005).
75 Smith, K. K. Heterochrony revisited: the evolution of developmental sequences. *Biological Journal of the Linnean Society* 73, 169-186 (2001).
76 de Beer, G. *Embryos and Ancestors*. 3rd edn, 172 (Clarendon Press, 1958).
77 Smith, K. K. Sequence heterochony and the evolution of development. *Journal of Morphology* 252, 82-97 (2002).
78 Gould, S. J. Of coiled oysters and big brains: how to rescue the terminology of heterochrony, gone astray. *Evolution and Development* 2, 241-248 (2000).
79 Klingenberg, C. P. Heterochrony and allometry: the analysis of evolutionary change in ontogeny. *Biological Reviews* 73, 79-123 (1998).
80 Raff, R. A. *The shape of life. Genes, development, and the evolution of animal form.*, (The University of Chicago Press, 1996).
81 Zelditch, M. L. & Fink, W. L. Heterochrony and heterotopy:stability and innovation in the evolution of form. *Paleobiology* 22, 241-254 (1996).
82 Mayr, E. Recapitulation reinterpreted, the somatic program. *Quarterly Review of Biology* 69, 223-232 (1994).
83 Gould, S. J. *Ontogeny and phylogeny*. (Harvard University Press, 1977).
84 Raff, R. A. *The shape of life. Genes, development, and the evolution of animal form.*, 255 (The University of Chicago Press., 1996).
85 Gould, S. J. *Ontogeny and phylogeny*. 4 (Harvard University Press, 1977).
86 Gould, S. J. *Ontogeny and phylogeny*. 214 (Harvard University Press, 1977).
87 Gould, S. J. *Ontogeny and phylogeny* 228 (Harvard University Press, 1977).
88 Gould, S. J. *Ontogeny and phylogeny*. 221 (Harvard University Press, 1977).
89 de Beer, G. *Embryos and ancestors*. (Oxford University Press, 1958).
90 Gould, S. J. *Ontogeny and phylogeny*. 479-486 (Harvard University Press, 1977).
91 Gould, S. J. *Ontogeny and phylogeny*. (Harvard University Press, 1977).
92 Gould, S. J. *Ontogeny and phylogeny*. 215 (Harvard University Press, 1977).
93 Raff, R. A. *The shape of life. Genes, development, and the evolution of animal form.*, 24 (The University of Chicago Press., 1996).
94 Hall, B. K. *Evolutionary Developmental Biology*. 375 (Chapman & Hall, 1998).
95 Coyne, J. A. *Why evolution is true*. 78,238 (Viking, 2009).
96 Mayr, E. *What Evolution Is*. 12-39 (Basic Books, 2001).
97 Freeman, S. & Herron, J. C. *Evolutionary Analysis*. 2nd edn, 31 (Prentice-Hall, Inc., 2001).
98 Ridley, M. *Evolution*. 57 (Blackwell Science, Inc., 1993).
99 Grant, V. *The Evolutionary process. A critical review of evolutionary theory.* (Columbia Unversity Press, 1985).
100 Futuyma, D. J. *Science on trial*. (Sinauer Associates Inc., 1995).

101 Wiedersham, R. *The Structure of Man: An Index to Past History*. (Macmillan, 1895).
102 Ledley, F. D. Evolution and the human tail: a case report. *New England ournal of Medicine* 306, 1212-1215 (1982).
103 Futuyma, D. J. *Science on trial*. 206-7 (Sinauer Associates Inc., 1995).
104 Darwin, C. *The Descent of Man and Selection in Relation to Sex*. (D. Appleton & Co, 1896).
105 Gilbert, S. F. in *DevBio - A companion to Developmental Biology* (ed S.F. Gilbert) 8th ed 3.3 (Sinauer Associates, 2006).
106 Robert, J. S., Hall, B. K. & Olson, W. M. Bridging the gap between evelopmental systems theory and evolutionary developmental biology. *BioEssays* 23, 954-962 (2001).
107 Goodwin, B. *How the Leopard Changed Its Spots: The Evolution of omplexity*. (Scribner's, 1985).
108 Richardson, M. K. *et al.* Haeckel, embryos and evolution. *Science* 280, 983-985 (1998).

10. Making a monkey out of you

§10.1 FOUR ARGUMENTS FOR THE FACT OF HUMAN EVOLUTION

Most of us have an interest in finding out who our ancestors were, and the further back we can go so the more interesting this inquiry becomes. Tracing one's roots back to the very first ancestor of man is a whole other matter however. The reason is this discovery can never be only a statement of historical fact. It has profound philosophical implications too. If casual observers of the evolution-creation debate do not recognize that, at least evolution scientists do. Douglas Futuyma even thinks this is the cause of the entire controversy. He says: "Evolution wouldn't be such a controversial subject if it didn't touch on our perceptions of ourselves."[1] There is indeed a lot more to what he says than might first meet the eye. The boundary lines of many of the great debates of the day like abortion or same-sex marriage or euthanasia, matters seemingly having nothing to do with paleontology, can be traced back to how the question of human origins is answered (§11.5). And there are only two ways. They are as opposite as oil and water and both have been scientific orthodoxy at different times in the past (§1.2).

Before Charles Darwin scientific consensus was that God had created man, and the reason was the Bible said so. Reading from *Genesis*: *"then the LORD God formed the man of dust from the ground and breathed into his nostrils the breath of life, and the man became a living creature."* (*Gen 2:7*) There was also general agreement that humans were different from every other creature because God had created man "in his own image" but not any other creature (*Gen 1:27*). Today there's still overwhelming consensus in science ("no significant doubt" is how the National Academy of Sciences says it),[2] but now it's to agree man is instead descended from an ape and the first ancestor of man was a microorganism that evolved into a human by chance. In other words the current consensus in science is a bacterium became a brain surgeon.

These two alternatives, a supernatural creation on the one hand and human evolution by natural forces on the other, are not just opposite answers to a question of a historical reality one of which must be true and one of which must be false, they have completely opposite implications for human living. The reason is the meaning and purpose of life lies in the hand of the creator of life. If the creator is a "blind watchmaker" then life has the meaning, purpose and morality of blind chance and natural forces. Creatures are free to live as they see fit. On the other hand if they were created by God then morality and purpose and meaning come from God, not man. To think otherwise is to think the pot makes purpose for itself rather than the potter or that the axe raises itself above him who swings it. Thus the scientific search for man's first ancestor is not only to solve a burning question of history, it's also to discover a whole lot about living. So tell me, where did we come from?

Perhaps it is because of these immense implications that the study of human origins is by far the most prestigious of all branches of evolution research now. This is shown by how discoveries of human fossil ancestors are announced in the most prestigious journals like *Nature* and *Science*, and how the popular media is just as captivated. Everyone, it seems, is interested. Another observation, and one that begs the even more interesting question why that might be, is these announcements frequently fly far beyond the bounds of the actual evidence!

Take *Ardipithecus ramidus kadabba* and *Sahelanthropus tchadensis*, for instance. They are two fossil species recently announced in *Nature* as ape-like ancestors on the evolutionary line to man.[3-5] The reports were accompanied by a media fanfare that included cover page and full length articles in newspapers and popular magazines, as well as detailed artist renderings of what these creatures may have looked like.[6,7] For example in *Time* the reader was treated to a delightful display of big-eared and hairy ape-men on the cover of the July 23, 2001 issue. The evidence on which these drawings was made might surprise you however.

For *Ardipithecus ramidus kadabba* (whose face, head and neck made it to the cover of *Time*) it was a toe bone, a finger bone fragment, a humerus fragment, a clavicle fragment, an ulna, a jaw bone and some teeth. For *Sahelanthropus tchadensis* it was a skull, a jaw and some teeth. There was no fossil record of their skin, hair, nose or ears to provide any clue as to what these looked like either. The creatures were drawn from imagination therefore, and yet they had a hard news reality to make a Hollywood animator proud.

In another example the bottom line appraisal in *Time* of the already giddy speculation in *Nature* about *Ardipithecus*, namely that "[t]he dorsal orientation of this [toe joint] surface may therefore constitute important evidence of a unique bipedal [i.e. two-legged] morphology"[3], I say giddy because this claim was based solely on a particular interpretation of a single fourth toe bone (i.e. without any ankle or leg, pelvis, spine or skull data), the *Time* article flew on further in the thin air to spell out the significance in large emboldened type: "THIS TOE BONE PROVES THE CREATURE WALKED ON TWO LEGS". The reader was left in no doubt human evolution was a scientific truth and here was more evidence to prove it. The interesting thing is all the paleontologists agreed because no objection was voiced. What's going on?

To start our inquiry at the beginning however, what is the evidence for human evolution in the first place? To hear the answer from Maeve Leakey who is among the most prominent paleontologists of our time, "[t]he evidence, in general, is staggering"[8]. It's Line of Evidence #11 for evolution in our catalogue. Have another look (Table 3.1). Why human evolution should be so scientifically certain when it can be never be more than inferential, is an observation begging our inspection already! In the end there are four major lines of evidence for human evolution, arguments in effect.[9-14] Here they are:

Argument #1: Evolution is a fact.
Macroevolution is the only scientific, objective and empiric explanation for the origin and diversity of animals. The evidence is conclusive and thus a scientific "fact" (§3.1). Humans are animals so human evolution is just as certain.

Argument #2: Apes are very similar to humans.
The similarities between living apes and humans are considered evidence of their common ancestry. These similarities are morphologic and molecular and are impressive. For instance as regards morphology, "[a]pes and humans can be matched bone for bone and muscle for muscle"[15]. As regards molecular similarity, an analysis of twelve different proteins from

chimpanzee and man finds their amino acid sequences 99.3% identical.[16] And as regards genetic similarity, chimpanzee DNA is 98.8% identical with the human sequence.[17,18] Michael Ruse explains what all this means:

> "Why should there be these fantastic similarities, given our different life-styles, if we are not descended from common ancestors? Without evolution, the homologies do not make sense...No other explanation makes sense."[19]

The only "other explanation" is ID of course. There is no "other explanation". But ID does not "make sense" so is rejected. His rejection is merely a plea from ignorance. The reason this dogmatism is yet taken as reasonable is because human evolution is already a fact from the evidence of Argument #1. But what does that evidence rest on? (§3.1;§20)

Argument #3: Extinct fossil species with transitional characteristics between extant apes and human beings are found in the fossil record.

These fossils are considered evolutionary intermediates in a pathway from an ape-like ancestor to present day humans confirming evolution theory. For instance *Australopithecus afarensis* and *A. africanus* are fossil species dated, the science says, at between 3 and 4, and 2 and 3 million years old respectively, and *A. anamensis* is dated at 4.2 million years old. While all of them have ape-sized brains, the evidence indicates they were capable of bipedal (two-legged) walking. Bipedal gait is how humans walk. No living ape can do that. Fossil species with such mosaic features of ape as well as human, are inferred as evolutionary intermediates after the ape ancestor of humans. There is no doubt about it in the science community either. The evidence is taken to satisfy scientific skepticism to an even stirring degree. Michael Ruse again: "One could not desire a more beautiful example of a "missing link" [in *A. afarensis*] – something that is truly a being evolving up from the apes to the humans."[20]

Fossils are also claimed to show evolutionary trends in other traits like dentition and brain size and tool artifacts. We are told that in the Miocene there are fossils with clear ape characteristics, in the Pliocene mosaic combinations of ape-like and human-like features and in the Pleistocene, there are fossils of modern man. Douglas Futuyma details the argument and lists the key evidence in his book *Science on Trial*. He gives you the evidence for human evolution in a nutshell here:

> "One "species" blends into another as you go from younger to older fossils...The first australopithecines...unfortunately cannot be dated by radiometric methods, but other geological evidence indicates that they are at most 2 to 3 million years old...*Africanus*, like *afarensis* was fully bipedal but it also had some other more human characteristics such as a

short canine tooth and a parabolic dental arch. It is found in association with "pebble tools" that it apparently made by knocking chips off larger stones. *Australopithecus africanus* had a brain size of about 440cm^3; thus it combined hominid features with an ape-sized brain. As in evolution generally, different human characteristics evolved at different rates. Our erect posture preceded the enlargement of our brain...*Homo habilis*...can be dated...from 2 to 1.6 million years ago...The chief reason for dignifying *habilis* with the name *Homo* is that it had a larger brain: about 600 cm^3 rather than 440 as in *africanus*. Habilis is associated with an extensive industry of stone tools, which become increasingly sophisticated...The next stage in human evolution is represented by some of the first human fossils that were ever found. The "Java man", discovered by Dubois in the 1890's ...was essentially a modern human in every aspect except its brain size, which ranged from 750 to 900 cm^3. The same species, now known as *Homo erectus*...is essentially the same as *habilis*, except for its larger brain, and is larger in more recent than in older specimens. It also is associated with more sophisticated tools than *habilis*, including stone hand axes. *Homo erectus* was essentially modern in size and posture, and almost fully modern in dental characteristics. The face still projected in a somewhat apelike form, but was flatter than *habilis*, and the cranium, although larger, had a low, sloping forehead...By the late Pleistocene, 200,000 to 100,000 years ago, humans of almost fully modern form had expanded into Europe. Their brain size was on average 1200 cm^3, as large as that of many living humans, although not quite up to the modern average (1,400 cm^3). Whether fossils from this period should be considered *Homo erectus* or *Homo sapiens* is entirely arbitrary, especially as from 200,000 years onwards there was a rapid change to the fully modern condition: the skull became rounder, the face, teeth, and brows were reduced."[21]

Argument #4: No evidence of modern man exists in the oldest parts of the hominid fossil record, therefore man must have evolved from some other creature.

The human fossil record extends back just a few hundred thousand years the experts say, and the molecular evidence agrees (§8.4). Yet radioisotope dating, we are told, declares the fossil record began hundreds of millions of years ago (§5.3.1). While Argument #4 does not refute design, it supports the claim that humans evolved from an ancient ancestor of a different morphology, so confirming evolution theory.

§10.2 THE SUPPOSITION OF PALEOANTHROPOLOGY: EVOLUTION MUST BE PRESENT

The demonstration of human evolution would seem certain then. In the words of George Gaylord Simpson: "Certainly, the recognition that we are related to all other organisms is one of the greatest advances ever made in self-knowledge."[22] Setting aside what reasons he could have for saying this, the first question to answer is whether the recognition he makes is even true (§11.5). To tell, we will need to know the hominid fossil record and the appendix at the end of this chapter is a resource to that end (§10.6). You should know the human evolutionary tree continues to be in great flux, for example the number of genera has more than doubled in the past 8 years compared to the previous 150. This is not a problem for us because while fossil nomenclature may become dated, the principles of analysis will not, and they are what concern us. Take a look at §10.6 and I will wait for you here.

There you have it, the human family tree. How are these various fossil species related to each other and to us? How exactly are they evidence of evolution? More particularly, how does the science of human paleontology do its business to get this result? We need to understand the last question before we can inspect the evidence and so answer the other questions. To answer the last question however, and also to cut to the chase, there is a supposition by which paleoanthropology is conducted that I submit is fatal.

Paleoanthropology is a science with evolutionary prejudice. To wit, history declares the notion of human evolution was not formulated from fossil evidence, but fossils and fossil artifacts were found after human evolution was already presumed true (§5.1). This is a colossal difference. Why? Because fossils require interpretation to have meaning (§5.3). What's the interpretation that they get in paleoanthropology? The mindset of Argument #1, namely that the bones must be evolutionary in some way. This removes any alternative to the idea. It's also the mindset of Argument # 2 - namely that whatever the similarities and whatever the differences between fossils, it's evidence of human evolution in some way. This makes the idea irrefutable. That's the interpretation they get. No wonder human evolution is a scientific fact.

We already know two things about fossil analysis from chapter 5. The first is that morphologic similarities between fossil bones, which everywhere are taken to reflect evolutionary relationships, can never be more than inferences because one can never be certain two fossil species aligned chronologically are indeed ancestrally related. This is an inference resting on other evidence. It might still be good evidence but it depends on the grounding evidence. The second thing we know is that the most robust of the analytical methods employed by this science, cladistic analysis, demands evolution be a fact already (§5.3.3). As a result cladistics can never do more than support a hypothesis that two fossil species are related by evolution already.

Thus we discover the evidence of human evolution rests on the notion biological similarity is evidence of common ancestry or homology (i.e. Argument #2), and on grounds that evolution already is true (i.e. Argument #1). We explored these lines of evidence earlier and found them wanting (§3;§7;§8). Yet human evolution is a construct resting on both and into which fossils, considered the best evidence, have been plugged according to their particular shape and inferred age. This enterprise brings us no nearer establishing whether such constructs have biologic reality. Actually, and this is the interesting thing given all that effort, internal contradictions rather suggest the opposite. Take a look.

§10.2.1   Inference Without Hard Evidence

Given the interest in human evolution, it might come as a surprise to discover that Darwin did not even address the subject in *The Origin*. His only allusion was an enigmatic sentence on the third-to-last page where he wrote: "Much light will be thrown on the origin of man and his history" (p.647). It would take twelve more years before he would explain what he meant. Finally in 1871 it came, in *The Descent of Man*. Darwin still had no fossil evidence. Paleoanthropology like paleontology was evolutionary before fossils were ever studied (§1.5.4;§5.3.3). It continues still, because the task of fossil discovery and fossil interpretation is everywhere taken to be discovering the course that evolution took from an ancient ape-like ancestor to a human being. Whether such a thing even happened is not doubted at all.

If not fossils, what was the original evidence used by Darwin to indicate human evolution? Comparative anatomy. Specifically it was the similar skeletal structure of man to the extant apes and the inference it was caused by a common ancestry. Thus the evidence was at once an observation and an inference, and it is Human Evolution Argument #2 (similarity-equals-homology) resting on Argument #1 (evolution-is-already-a-fact). We saw the logical inconsistencies of the homology argument earlier, but let's test the validity of those conclusions arrived at for biology generally, by using *Homo sapiens* as a detailed inspection of one particular species (§7;§8).

The extant great apes consist of two species of chimpanzee (*Pan troglodytes* and *Pan paniscus*), one species of gorilla (*Gorilla gorilla*) and one species of orangutan (*Pongo pygmaeus*) (Fig 10.5). Chimpanzees are the closest living relatives to man. So close, that UCLA physiologist Jared Diamond ignores all their differences to call man "the third chimpanzee"! The idea barely raises an eyebrow now at least in intellectual circles, though it's a pole away from what scientists and teachers said in Darwin's day. Namely that man was "a little lower than the angels" (*Ps 8:5*). Now man is a little higher than the apes!

In another twist, the structural similarities between man and ape that were the original evidence for human evolution were recognized long before evolution was even conceived of. That evidence goes back at least as to 1698, when English anatomist Edward Tyson reported that apes resembled humans more than they did monkeys (by his criteria in 47 and 34 ways respectively)[23]. Linnaeus acknowledged the similarities between apes and humans too, classifying them in the same genus (*Homo*). The break from this typological thinking came with Darwin's claim that morphologic resemblances were not typological or created, but shared traits from a common ancestor of both by evolution.

In 1863 an enthusiastic disciple of Darwin named Thomas Huxley, also called "Darwin's bulldog" because of his enthusiasm(!) and grandfather of novelist Aldous (author of *Brave New World*) and evolution scientist Julian (who coined the term "Modern Synthesis"), formalized this thinking in a collection of essays entitled *Evidence as to Man's Place In Nature*. The title of his book captures the thrust of paleoanthropology, the field his ideas would spawn. In a nutshell Huxley said the Bible had it all wrong. Man is not a God-created creature nor even unique; not even by virtue of a transcendent intelligence or moral and spiritual awareness or even thoughts or expectations of infinity. Instead man was just another animal that arose purposelessly by natural forces, through evolution. Huxley showed the anatomic similarities between man and ape in a drawing of their skeletons (Fig.10.1). This drawing can be considered the archetype for every ape-morphing-into-man evolution figure since, and which are so iconic of this science.

Huxley outlined his interpretation and its philosophical implications. From now on the differences between man and beast were going to be far less important than their similarities. This is the consensus today, but Huxley said it first and he said it like this: "But if Man be separated by no greater structural barrier from the brutes than they are from one another then, there would be no rational ground for doubting that man might have originated...by the gradual modification of man-like ape. Man is, in substance and in structure, one with the brutes."[24]

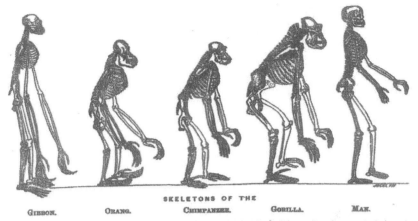

Fig.10.1. **Comparative skeletal anatomy of man with the great apes.** (Published by Thomas Huxley in 1863.[24] Figures to scale except gibbon (far left) which is 2x).

What evidence did Huxley's and Darwin's claims of human evolution rest on? There were three lines of evidence:
(1) Comparative anatomy of man with ape (Argument #2), in other words man as a specific example of *LOE#4* (§3.2);
(2) Human embryonic development whereby the embryo was considered to recapitulate through phases of early development similar to and thus inherited from lower ancestral vertebrates. It was an older version of *LOE#9* (§9.2-4);
(3) Vestigial structures with man as a specific example of *LOE#10* (§9.5).

The evidence was all circumstantial. Nevertheless to Darwin and his believers, and in another example of the naturalistic conviction disconnected from data taking hold of science, it was to "declare in the plainest manner, that man is descended from some lower form, notwithstanding that connecting-links have not hitherto been discovered."[25] With the exception of the very first hominin fossil finds, namely Neanderthal bones in 1856 and others unrecognized as Neanderthal in 1829 and 1824, every hominid fossil has been discovered and interpreted in the light of this preconceived idea that man is descended from an ape. Ironically not even Neanderthal fossils contributed to the evolutionary thinking because they were considered fully human, by Darwin and Huxley included.

What is the significance of such a mindset studying fossils? The risk of bias. In paleontology interpretation is everything, even after meticulous measurements of bone shape and all estimates of bone age (§5.3). Interpreting fossils from a preconceived evolutionary perspective is to see evolution in circular self-fulfilling prophecy because a fossil cannot of itself indicate a genetic relationship with any other to prove its ancestry, much less its evolution. With these expectations by paleoanthropology, the interpretation of a bone will always be evolutionary and so more evidence of evolution will be found. Research has been done with the expectation of finding bones that show a transition from ape to ape-man to man-ape to modern human, and this is just what has been found. The amazing thing is why the evidence of human evolution is yet laden with contradictions and different interpretations by different experts (though the conviction human evolution happened stays unswerving), but more about both later.

In summary opinions believed about what the ancestors of modern man were like before fossil research began are very similar to those held now. Brilliant intuition or self-fulfilling prophesy? We will need to inspect the fossil data to decide, but for now, and speaking of the sagacity of visionaries without fossils, Darwin wrote in *The Descent of Man* that the common ancestor of ape and man was a "hairy quadruped, furnished with a tail and pointed ears, probably arboreal in its habits, and an inhabitant of the Old World", and Ernst Haeckel inserted "Ape-Men" below man into his tree of life (Fig. 1.2). He also wrote in *The History of Creation* (1868) the common ancestor was the "hypothetical primaeval man...who developed out of the anthropoid apes...with slanting teeth...the hair covering the whole body was probably thicker [than that of modern man]...their arms comparatively longer and stronger; their legs, on the other hand, knock-kneed, shorter and thinner...their walk but half erect." These visions of human ancestors of course, were all conjectural back then.

§10.2.2   Hard Evidence By Inference

When did inferences from comparative anatomy and homology progress to hard evidence, namely the fossil evidence that is now Argument #3? And when did fossils assume the importance in human evolution they do now? Our guide to getting these answers is Ian Tattersall, Head of the Anthropology Department at the American Museum of Natural History. He asked them in his book *The Fossil Trail - How We Know What We Think We Know About Human Evolution*. To let you in to a little secret and to the answer to the second question it was in 1912, over half a century after *The Origin* was published. It was not because of any new fossil evidence either. It was because of a fossil forgery now called the Piltdown man. He explains:

"Paradoxically, however, it was Piltdown, in its role as "missing link", that for the first time brought the human fossil record squarely into the public eye and established it as a major source of media interest. And it was Piltdown that established the central place of the fossil record in understanding the mysterious process by which modern humans had emerged from an "ape" ancestry."

Did this change the need for inferences in paleoanthropology? He continues:

"This much is unquestionably on the plus side; but it's also undeniable that paleoanthropology is unique among the subdisciplines of paleontology in ascribing an almost iconic significance to each new fossil that appears. Rarely do paleontologists working on other groups of organisms feel it necessary to undertake wholesale overhauls of their beliefs every time new fossil members of those groups come to light. Paleoanthropologists often do, however, and it's fair to claim that this has been detrimental to their science. But unusual as this paleoanthropological tradition might be, it equally adds excitement and at least the appearance of progress to the study of human evolution – which is maybe what we want most out of the study of our own ancestry. And for better or for worse, this way of doing business can be traced back to the embarrassing Piltdown "specimen".[26]

He shows us the power of explanation and the continued instability of a branch of science notwithstanding one and a half centuries of its work. That "wholesale overhauls" of "beliefs" can continue to result from data that has "iconic" status is hardly something to extol as "exciting", at least if it is to really be "science". Dr. Tattersall is far too modest. This kind of thing is not "unusual" in science, it's unique.

The reason it happens and the reason it's tolerated and even extolled, is because of the paucity of absolute certainties in a discipline that seeks to recreate histories of relationships from bones that cannot do so themselves, when their evolution is already certain. Paleoanthropology must rely on "soft" data instead, which is plausible inference and interpretation. The hard evidence comes by packing the soft evidence. Six pages on in his explanation, Dr. Tattersall shows us how easily bias can come when the data is soft:

"Dart [the discoverer of the Taung skull §10.6.2] was the first student of the human fossil record who seems to have felt viscerally the drama of the biological history of mankind. He was certainly among the first to perceive that the story of our species is one of individuals, populations, and species living, striving, and dying as part of an enormously complex web of organic life, and to see that this story is not complete if we do not let ourselves go well beyond the teeth and bones that are our primary evidence of it." [27]

The primary evidence of teeth and bones he considers to be the "story of our species". In other words, fossils indicate human evolution before research even begins. Perhaps it's no surprise then that there is no fossil specimen for any of the ancestors to the four extant great apes, but an ever-increasing collection claimed as the ancestors of *Homo sapiens* (Fig.10.5).

Tattersall demands the science "go well beyond the teeth and bones that are our primary evidence" (i.e. actively go to inference and interpretation), otherwise "this story [of human evolution] is not complete". This is how to pack the soft evidence into hard evidence. This is the considered view of one the foremost authorities in the world, and co-editor of the current definitive text for the field (*The Human Fossil Record*).

Paleontologist and science writer Roger Lewin even justifies emotionalism over the evidence. In discussing the exhibits displayed at the 1984 *Ancestors* exhibition, an unprecedented and still unsurpassed display of over forty original hominin fossils brought from all around the world to the museum at which Dr. Tattersall is Curator, Lewin described the scene:

> "It was like discussing theology in a cathedral," said Michael Day, a British anatomist and longtime associate of the Leakey family. "To be in the same room with all these relics was for many workers an emotional event," agreed Christopher Stringer, an anthropologist at the British Museum (Natural History). "Sounds like ancestor worship to me", was the comment of a sociologist of science who was observing the events. It is hard to imagine a group of, say biochemists becoming quite so emotional in the presence of their favorite experimental organism, *Escherichia coli*. There *is* a difference. There *is* something inexpressibly moving about cradling in one's hands a cranium drawn from one's own ancestry." [28]

Berkeley University law professor Philip Johnson has responded to this unscience. He says:

> "Lewin is absolutely correct, and I can't think of anything more likely to detract from the objectivity of one's judgment. Descriptions of fossils from people who yearn to cradle their ancestors in their hands ought to be scrutinized as carefully as a letter of recommendation from a job applicant's mother." [29]

In summary paleoanthropology is a historical science that in its final analysis relies not on measurements of fossils (which would offer an objective, refutable "scientific" interpretation), but on their historical interpretation which is inferential and always evolutionary. Those brilliantly exact are hard to discriminate from those banally false. How hard is it to discriminate between them, you ask? Back at the American Museum of Natural History in New York, Dr. Tattersall gives us further insight into the problems that must be overcome for us to tell:

> "When I was a graduate student, I occupied a desk in a basement storeroom of the Peabody Museum of Natural History [at Yale]: one of the nation's great repositories of fossil vertebrates. As an aspiring paleontologist I was impatient to learn the arcana of the paleontologist's science. I would watch, mystified, as visiting scholars bent intently over specimens they had retrieved from collection drawers. Measurements they took of skulls and bones seemed pretty straightforward...Intricate anatomical descriptions I understood too...But what of the other notes that these professionals so industriously made, as they pored over the fossils for hour after hour? What were they doing that allowed them to understand these documents of ancient life, beyond simply describing them?...Eventually I plucked up the courage to ask a distinguished scholar the crucial question: "How *does* one study fossils? How *does* one understand what they tell us about the history of life?" The answer? "You look at them long enough and they speak to you."[30]

So what is the way out of the nebulousness and emotionalism that even he sees clearly? Tattersall's solution is the cladogram because it is an interpretation that by contrast he, like the rest of science, contend can be tested. *Nature* editor Henry Gee agrees:

> "Without cladistics, paleontology is no more of a science that the one that proclaimed that Earth was 6000 years old and flat – and then had the effrontery to claim divine sanction for this view".[31]

Which is to tell us a lot about paleontology already. He's a paleontologist! How well cladistics can be tested and how reliable it really is, is the all-important question we inspect in a moment. Dr. Tattersall describes how paleoanthropological analysis should properly proceed:

> "If a hypothesis is to be scientific, it has to be proposed in such a way that it can at least potentially be proved wrong: it has to be objectively testable...the only kind of testable paleontological hypothesis is the cladogram, which simply tells you which taxa are most closely related to which others. It says nothing of the relationships involved...If you add ancestry and descent (which usually involves time) to your cladogram, you get what's called an "evolutionary tree"...But since it's not actually

possible to prove ancestry (or, in some cases, to disprove it), trees are not only more complex statements than the cladogram you started with; they're also not testable."[32]

And there you have the problem of modern paleoanthropology.

To see this another way, a cladogram demonstrates by a branching diagram how closely related particular fossil species are. This is done by inference, by stipulating which character traits are derived and which are to be primitive, which characters will be assessed and with what weighting compared to what others, and also stipulating beforehand that they are related by evolution in some way (§1.4;§5.3.3;§7.4).

Cladistics assumes the fact of evolution so it's circular to use it as evidence for the fact of evolution. Cladograms can do no more than show one species more similar to another based on the character states selected in advance for comparison. Characters can be defined, given a topology and even a numeric and statistical measure, but how can we know for certain which characters really are primitive and which ones really are derived? Relationships based on fossil morphologies can indeed be "testable" and a best-fit cladistic tree be constructed, but this is not a test of the validity of evolution. It's an internal test of the validity of cladistics. The reason this blindingly simple logical flaw is no problem in this science, is because evolution is already a fact (§1.5.2;§3;§20).

What's worse is that although already as woolly as a fur ball, cladistics is as empirical as evolution seeking paleontology can get. This is because to move from a cladogram to a phylogenetic tree, which is when the relationships between two fossils become specified, is to move to conjecture. Cladistics says particular fossils are related and by how much, but not how exactly. Ian Tattersall and his University of Pittsburgh anthropologist colleague Jeffrey Schwartz have pointed this out already by saying: "because it is impossible to demonstrate one kind of relationship as opposed to the other, trees are not susceptible to scientific testability."[33]

And yet having identified this quandary, and untestability in a science, they do not address the intractability of the problem nor suggest circumspection. Paleontology does not stop at evolutionary trees either. It hangs on explanations of trees. Our guide explains:

"Yet more complex than the [phylogenetic] tree is the "scenario". This is what you get when you add the really interesting stuff to the information already present in the tree. This added information includes everything you know about adaptation, ecology, behavior, and so forth, and it's certainly what makes the past come alive. But it does mean that the average scenario is a highly complex mishmash in which considerations of relationship, ancestry, time, ecology, adaptation, and a host of other things, are all inextricably intertwined, tending to feed back into each other. When you're out there selling such complex narratives, normal scientific testability just isn't an issue: how many of your colleagues or others buy your story depends principally on how convincing or forceful a story teller you are – and on how willing your audience is to believe the kind of thing you are saying (which brings us back to the importance of people's expectations)."

To summarize, there's an evolutionary bias brought to bear on all fossil interpretation that leads necessarily to the detection of evolution in the fossil record, to the formulation of evolutionary trees and to scenarios that cannot be disproved. Tattersall and Schwartz continue:

"Evolutionary relationships, then, are reflected in systems of resemblance that cannot simply be discovered, like fossils themselves, but that have to emerge from a process of morphological (anatomical) comparison and analysis."33

What "emerge" means here in the light of the above is clear despite the euphemism. Depending on the age of a fossil and its shape, a pattern of fossil species and its relationship to another is constructed on the basis of...Conjecture. Even if the cladogram demonstrates an excellent "fit", this is still no nearer at substantiating evolution. How do we know the relationships are correct? How do we know the fossils are related at all? *Nature* editor Henry Gee again:

"If we can never know for certain that any fossil we unearth is our direct ancestor, it is similarly invalid to pluck a string of fossils from Deep Time, arrange these fossils in chronological order, and assert that this arrangement represents a sequence of evolutionary ancestry and descent...such misleading tales are part of popular iconography: everyone has seen pictures in which a sequence of fossil hominids – members of the human family of species – are arranged in an orderly procession from primitive forms up to modern Man. To complicate matters further, such sequences are justified after the fact by tales of inevitable, progressive improvement. For example, the evolution of Man is said to have been driven by improvements in posture, brain size, and the coordination between hand and eye, which led to technological achievements such as fire, the manufacture of tools, and the use of language. But such scenarios are subjective. They can never be tested by experiment, and so they are unscientific. They rely for their currency not on scientific test, but on assertion and the authority of their presentation."34

§10.2.3 Cladistic Analysis Is Hardly Evidence

And so we find that the most robust method of analysis in paleontology is not even close to robust in showing evolution, the issue at hand. But we knew this already (§5.3.3). There is more though. Is the cladogram referred to with such confidence by Tattersall and Gee even a reliable method for generating phylogenetic hypotheses in the first place? Phylogenetic hypotheses are the starting point for any postulate about ancestry, and for any "scenario" that attempts to explain specific events that happened during human evolution. If phylogenetic hypotheses are unreliable, nothing can be considered steadfast in paleoanthropology except a bone's dimensions and its constituents.

It comes with some disquiet then, to discover phylogenetic relationships between the various species in the hominin fossil record remain unclear despite a century and a half of study since Huxley. Exactly which species are on the direct lineage to man and which are on side branches is unresolved, and of course only one phylogeny can be correct. The current muddle of conflicting phylogenies is shown in Fig.10.2. In the words of two of the leaders in this field, Bernard Wood at George Washington University and Mark Collard at University College London, "[d]espite, in paleontological terms, a relative abundance of fossil evidence, cladistic analyses of the hominins have so far yielded conflicting and weakly supported hypotheses of relationships."[35] In light of what we now know this comes as no surprise. It's not news to paleontologists either, but given the gusto with which human evolution is championed, it probably is to everyone else.

You see a relationship through ancestry is fundamental for the existence of evolution, but there are no convincing relationships between the fossils and yet human evolution is unquestioned in science, anywhere. Explanations for the failure of cladistics to resolve the evolutionary relationships have not been blamed on non-existent relationships either, but on faulty character selection, faulty analysis or taxonomic uncertainties instead.[35] This is the most interesting observation of all. How it happens exactly, is because of an *Emperor's New Clothes* illusion caused by the Law of the Emperor's New Clothes (§3.1,4;§10.3,4;§19).

It's not difficult to see how badly cladistics fails either. Simply choosing different fossil characters for cladistic analysis can generate different phylogenies. For the same species! ("Characters" are traits - the various particular measurements of a bone's shape like protuberance length, width, thickness, holes etc.). Which characters are chosen for analysis is arbitrary. Instead of being analyzed independently, fossil characters can also be clustered into "functional complexes" of a composite trait which are combined and analyzed as an already summated group. This is done in an attempt to have a uniform spread of characters for a particular fossil species, and reduce the chance of analyzing functionally related characters separately (which violates the assumption of character independence, implicit in all cladistic studies).[38,39] The clustering of traits into functional complexes to avoid "trait bias" belies the fact it's difficult to select characters that are truly independent. They may be related in unpredictable and unrealized ways, for instance through developmental constraints and pleiotropy (which is when distinct and seemingly unrelated characters are controlled by the same gene).[40] Be that as it may, different results are obtained depending on which cladistic method is chosen - independent analysis or clustering![39,41]

Another example of the problems of cladistic analysis is its assumption that the most parsimonious phylogenetic hypothesis is always the correct one. This would reject the hypothesis of Fig.10.2H for example, a tree that declares there was contemporaneous parallel evolution of *A. africanus* to *A. robustus* in Southern Africa and *P. aethiopicus* to *P. boisei* in East Africa. A plausible case has been made for this scenario, both from the fossil record and by informal non-cladistic analysis.[37] There seems no good reason to reject it offhand. How do we know the most parsimonious tree is really the one with biological reality? We don't. Now you're getting it.

That fossil characteristics are unreliable, or unable, to permit phylogenetic assumptions as these examples indicate, has only recently even been considered by the science itself. In light of the self-evidently obvious muddle of contradictions and its significance to the entire enterprise of uncovering human evolution, that oversight is extraordinary and worth seeing in print. Actually it was discovered scientifically, and by two of the most senior investigators in this science, Bernard Wood and Mark Collard. In 2000 they made the stunning declaration that despite nearly one and a half centuries of fossil study since Huxley:

"Cladistic analysis of cranial and dental evidence has been widely used to generate phylogenetic hypotheses about humans and their fossil relatives. However, the reliability of these hypotheses has never been subjected to external validation."[35]

That such a discovery could take so long to make is how powerful a hold the evolution paradigm has on science. If evolution is already "a fact", why would you test for its reality?

To perform their analysis Wood and Collard used the usual measurements of the cranium, mandible and teeth that are characters in

Evidence of human evolution I

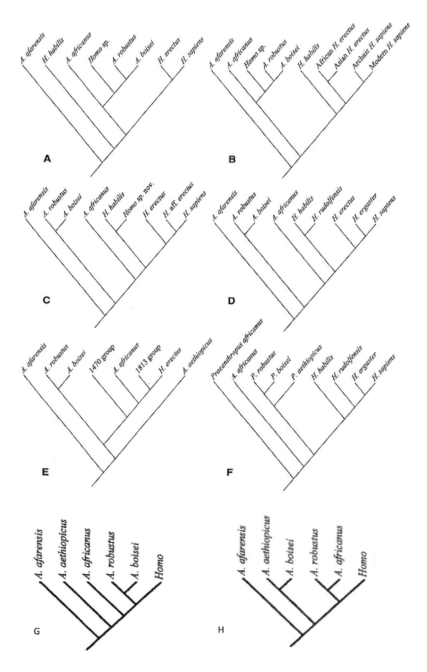

Fig.10.2 The proposed hominin phylogenies. Cladistic parsimony analysis was used for A-G and H is from informal analysis of morphology. (A-F from Wood and Collard, 1999[36] reprinted with permission from AAAS, G and H from Lockwood and Fleagle, 1999[37] reproduced with permission of John Wiley & Sons, Inc).

fossil analysis to generate a cladogram for two groups of extant primates - the hominids (*Homo sapiens, Pan, Gorilla* and *Pongo*), and the papionids (the Old World monkey tribe of baboons, mandrills, macaques and mangabeys). Next the most parsimonious fossil character cladograms were compared with the consensus molecular cladogram of each from the world literature. The molecular cladogram was taken as the reference standard for the phylogeny, consistent with the demands of evolutionary theory. They point out the reason why, in another demonstration of the evolutionary mindset and reference frame of this science:

"First, phylogenetic relationships are genetic relationships. Thus, in phylogenetics, morphology can never be more than a proxy for molecular data. Second, because osseous and other morphological characters can be highly influenced by external stimuli, such as the forces generated by habitual activities, they can be expected to provide misleading information about phylogeny more frequently than genetical characters."

Ignoring for purposes of argument our analysis of molecular homology that demonstrates this confidence wholly misplaced (§7.3;§8.3), their experimental design is as good an example of the blind leading the blind into a ditch as one could ever hope to find anywhere, much less in science. It speaks to the problem raised earlier for paleoanthropology. It has no foundation. Nothing can be certain but a bones' measurements. You'll still never guess what they found.

"We found that the phylogenetic hypotheses based on the craniodental data [skull and teeth fossils] were incompatible with the molecular phylogenies for the groups. Given the robustness of the molecular phylogenies, these results indicate that little confidence can be placed in phylogenies generated solely from higher primate craniodental evidence. The corollary of this is that existing phylogenetic hypotheses about human evolution are unlikely to be reliable...The results of the parsimony and bootstrap tests indicate that cladistic analyses based on standard craniodental characters cannot be relied on to reconstruct the phylogenetic relationships of the hominoids, papionins, and, by extension, the fossil hominins."[35]

The report was brief and made barely a ripple. At just 3 pages long, among 600 other pages in an issue of the *Proceedings of the National Academy of Sciences*, it was published without editorial comment too. Yet it is a devastating indictment of an entire discipline, at a stroke invalidating the main method of hominin fossil analysis of twenty-first century science, and it invalidates all the inferences made from phylogenetic analyses too In other words paleontologists have been telling us stories and calling it science, and this was finally found out scientifically. What can we be certain about in paleoanthropology if the only testable, objective and "scientific" aspect to the field, cladistic analysis, is so uncertain? There is still more to this science, and it is there that we find what it is that makes it so believable.

§10.3   THE SCIENCE OF PALEOANTHROPOLOGY: FINDING EVOLUTION PRESENT

§10.3.1   The Fact of Human Evolution Or Not

So much for the supposition of paleoanthropology, which is that evolution must be present. How is hands-on paleoanthropology conducted? There are two principal, and dare I say it, opposite hypotheses of human origins. The first is the linear or "tidy" model and the second is the bushy tree or "untidy" model.[42]

The former is the traditional view. It holds hominid features evolved only once and that evolutionary change occurred in a linear sequence of ancestor-descendant related species with no branching (cladogenesis) until after 3 million years ago.[43] This is the view of such notables as Berkeley's Tim White and Pennsylvania State University's Alan Walker.[44]

The alternative or "bushy" model holds there instead were a series of adaptive radiations, which resulted in mosaics of anatomic features in intermediate species between modern man and our shared ancestor with extant apes. It holds there are multiple branches in the evolutionary tree, only one of which reaches up to *Homo sapiens*, and that human features (like large brain, bipedalism and hand dexterity) likely evolved more than once by homoplasy.[42] This is the majority view now, and its supporters include Bernard Wood, Ian Tattersall, Jeffrey Schwartz, Maeve Leakey, Martin Pickford, Brigitte Senut and Stephen Jay Gould.

As regards the origin of *Homo sapiens* within either of these models, there are two further hypotheses. The first is called the multiregional or candelabra model. It posits modern man evolved independently at multiple places from an older worldwide species, most likely of African origin. The alternative and more popular theory is the single-origin "Out of Africa", "Eve" or "Noah's Ark" model (for some reason paleontologists love Bible terminology) which holds modern man originated in a single part of Africa and emigrated to repopulate the rest of the world around 200,000 years ago replacing previous populations.

The current state-of-the-art is reproduced in Fig.10.3.[42] How does paleontology convert the data on these figures into an evolutionary tree and show human evolution true? By joining the various boxes with connecting lines that indicate ancestry. But how? By inference. These inferences are based on cladistic analysis of fossil characters (i.e. by comparisons with other fossils and the bones of extant apes and humans) and by hypotheses of what characteristics a common ancestor must have had. But we have already seen that cladistic analysis on craniodental bone characters "cannot be relied on to reconstruct the phylogenetic relationships"[35] and that hominid cladograms are conflicting. There isn't even agreement over such basic questions as whether the *Homo* genus is mono- or paraphyletic.[36] There are no lines on the figure to help us either.

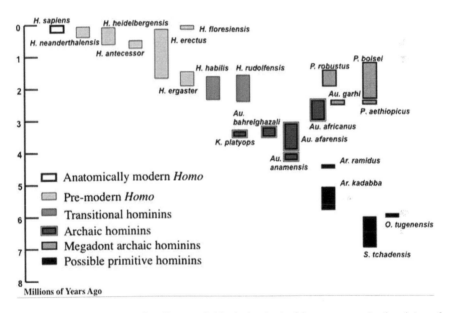

Fig.10.3 **The hominin fossil record**. Vertical extent of bars represents the dates of earliest and latest fossil discoveries. *Au. sediba* (published after this figure was made) would occupy a tiny box just above *Au. africanus* at 1.9 Mya. (Reproduced from Wood, 2010 with permission).[46]

Nonetheless the hominid fossil record is considered a robust confirmation of the progressive transformation of an ape-like ancestor to modern man, just as Darwin predicted. In fact, as you heard earlier, a sequence of fossils is supposedly seen: Ape fossils in the Miocene, both ape-like and human-like features in the Pliocene (though their equally important differences tend to be overlooked) and human fossils in the Pleistocene. This is the evidence of Human Evolution Argument #3 (§10.1). Here it is formally stated by the National Academy of Sciences:

> "today there is no significant doubt about the close evolutionary relationships among all primates, including humans...Not one but many connecting links – intermediate between and along various branches of the human family tree – have been found as fossils...They document the time and rate at which primate and human evolution occurred" (Table 3.1).

Evolutionary trends in the hominin lineage are thus claimed both for species and for individual traits such as locomotor behavior, dentition, brain size and tool artifacts. It sounds impressive all right. But where exactly are the "many connecting links" they speak of on Fig. 10.5? We will need to look more closely to see.

§10.3.2   Fossil Trait Analysis: Measurement But From No Beginning

The following statement would seem self-evident but perhaps it is not. It's that to demonstrate an evolutionary trend in a fossil lineage, there has to be a starting species and a finishing species over which to assess the change. Although we can accept human evolution ended with modern man, without a phylogeny none of man's assumed ancestors can serve as baselines for analysis (Fig. 10.2). Here is Harvard's Richard Lewontin on this matter:

> "Despite the excited and optimistic claims that have been made by some paleontologists, no fossil hominid species can be established as our direct ancestor...The earliest forms that are recognized as being hominid are the famous fossils, associated with primitive stone tools, that were found by Mary and Louis Leakey in the Olduvai gorge and elsewhere in Africa. These fossil hominids lived more than 1.5 million years ago and had brains half the size of ours. They were certainly not members of our own species, and we have no idea whether they were even in our direct ancestral line or only in a parallel line of descent resembling our direct ancestor."[47]

The point is that all the claims made by Human Evolution Argument #3 (§10.1) about fossil trends demonstrating evolution are all about assumed change. Either between known fossil species pre-arranged into an assumed lineage, or from the assumed shape of a phantom form - the common ancestor. This is hardly impressive science, whatever the result supposedly obtained.

Trend analysis between known fossil species is hamstrung. At the crudest level of comparison, which of the two species being compared is the ancestor and which is the descendant to confirm the polarity of the trend? Or are they instead sister clades evolving convergently without direct significance to evolutionary ancestry? Claims a trait trend is caused by evolution are circular for using the process one is attempting to prove (macroevolution), to prove the existence of an unproven relationship (ancestry). If a trend is detected between two fossils separated in time, the claim there is evolution between them cannot be disproved. Detecting a trait trend does make an ancestral relationship a possible hypothesis, but its validity depends entirely on whether or not the ancestral relationship is correct. This is because there can be only one true phylogeny. It's also another problem with the present paleoanthropological method, impugning its having any validity at all.

The next problem with fossil investigation is the assumption that the more alike a fossil ape trait is to a human, the more closely is that this creature is evolutionarily (genetically) related to human and vice versa. It's the similarity-equals-homology argument again. The problem and the point is not that the assumption is false, although it is false (§7.3-7). It's that the science takes it as true and then uses an assumed homology as the basis for still further evolutionary inference making. The net result of this compounding of false premises, is the assumption of paleoanthropology that what is being found in the fossil record are ape-man evolutionary intermediates of some kind. Brian Richmond at the Smithsonian Institution and Bernard Wood at George Washington University explain for us how this happens:.

How exactly does one interpret a fossil? For instance having made measurements of fossil's morphologic characteristics how does one distinguish it as being a chimpanzee ancestor, a human ancestor or an ancestor of both lineages?

"How, then, are we to tell a late Miocene/early Pliocene early hominin [human lineage] from the ancestors of *Pan* [chimpanzee], or from the lineage that provided the common ancestor of *Pan* and *Homo*? The presumption is that the common ancestor and the members of the *Pan* lineage would have had a locomotor system that is adapted for orthograde arboreality and climbing, and probably knuckle-walking as well. This would have been combined with projecting faces accommodating elongated jaws bearing relatively small chewing teeth, and large sexually-dimorphic, canine teeth with a honing system. Early hominins, on the other hand, would have been distinguished by at least some skeletal and other adaptation for a locomotor strategy that includes substantial bouts of bipedalism linked with a masticatory apparatus that combines relatively larger chewing teeth, and more modest-sized canines that do not project as far above the occlusal plane. These proposed distinctions between hominins, panins and their common ancestor are 'working hypotheses' that need to be reviewed and, if necessary, revised as the relevant fossil evidence is uncovered."[48] That's how you tell. By assuming chimpanzees and humans are related by evolution and assuming they have a common ancestor; that that much is already scientifically certain. It's just a matter now of finding the various linking creatures to demonstrate that fact. On the validity of the inference that an ape became a man the entire science of paleoanthropology rests also.

Another hindrance to truth in fossil interpretation is that despite intensive searching in some of the most inhospitable places on the planet, all the hominid fossil evidence available for what can be called a new species may be some teeth and bone fragments still requiring of a painstaking reassembly. How does the reassembly and the taxonomy happen?

The evolutionary "working hypotheses" identified by Wood and Collard to interpret a bone are what drive the field, not the other way around. It is hard to avoid the circularity, and it is the reason for the instability detailed by Ian Tattersall earlier (§10.2.2). Fossils cannot reveal their genetic relationships to uncover evolution directly. They live in the mind of the scientist. The paleontologist must first reassemble the bone fragments of a fossil, but reassemble them according to what plan? This can be a 3-D jigsaw puzzle without a picture key, that relies only on clues that the bones suggest or don't suggest. The reassembly of skull specimen KNM-ER 1470 is a notable example if this needs more explaining (§11.3.3).

These difficulties are not helped by a lack of clarity about species definitions either. For instance what exactly constitute the fossil characteristics of the genus *Homo*, much less its constituent species, are crucial concepts but these are still debated despite nearly a century and half of research! Bernard Wood and Mark Collard summarized these problems back in 1999 to discover that "conventional criteria for allocating fossil species to *Homo* are reviewed and are found to be either inappropriate or inoperable".[36] In another example Milford Wolpoff and Raymond Dart have determined that *Homo erectus* should be sunk within the *H. sapiens* taxon rather than be given separate species status, but the majority of their colleagues disagree. If the minority view is true then this wipes away much of the claim for human evolution made by Argument #3, arguably putting a human in the fossil record 1.8 million years ago (§11.2.4). So who's right and who's wrong? And how can human evolution still remain just as certain for both camps? No seriously. How? (§19)

To compound the difficulties, *Homo sapiens* does not even have a type specimen. This is a situation unique among hominins if not in all paleontology. A "type specimen" is a bone or set of bones that defines the species. The "type description" describes the similarities and differences of that species compared with others in the fossil record. A type specimen is in effect a yardstick used to tell one species from another. For a field directed toward discovering the bones that show how an ape changed into a human, knowing what a human is would seem to be important to establish at the outset. Reviewing the matter in 2000, Bernard Wood and Mark Richman gave the following analysis:

> "Paradoxically, it has proved easier to assemble information about the characteristic morphology of extinct hominin taxa than the only hominin species with living representatives, *H. sapiens*. Just what features of the cranium, jaws, dentition and the postcranial skeleton are specific to *H. sapiens*?...what are the 'boundaries' of living *H. sapiens* variation? How far beyond these boundaries, if at all, should we be prepared to go and still refer the fossil evidence to *H. sapiens*? These are simple questions, to which one would have thought there would be ready answers. However, the concept of 'modern humanness' has proved to be complex and difficult to express. Some researchers have made explicit suggestions that *H. sapiens* should be much more inclusive than just being limited to living and recent modern humans."[48]

Vague or absent boundaries allows evolution room between moving goal posts. Back in 1950 Ernst Mayr had argued for the "more inclusive" taxonomic view above rather than the speciose which is currently vogue.[36] He makes salient points about the biological relevance of hominin taxonomy employed in paleoanthropology:

> "...there is less agreement on the meaning of the categories species and genus in regard to man and the primates than perhaps in any other group of animals. Some anthropologists, in fact, imply that they use specific and generic names without giving them any biological meaning...The nomenclatorial difficulties of the anthropologists are chiefly due to two facts. The first one is a very intense occupation with a very small fraction of the animal kingdom which has resulted in standards that differ greatly from those applied in other fields of

zoology, and secondly, the attempt to express every difference of morphology, even the slightest one, by a different name."⁴⁹

An example is the separation made among the various supposed different *Homo* species – *sapiens, neanderthalensis, archaic sapiens, heidelbergensis, rudolfensis* and *erectus* (although unlike Dart and Wolpoff, Mayr felt *erectus* should retain separate species status) (§11.1). Mayr continues:

"This difference in standards becomes very apparent if we, for example, compare the classification of the hominids with that of the *Drosophila* flies. There are now about 600 species of Drosophila known, all included in a single genus. If individuals of these species were enlarged to the size of a man or of a gorilla, it would be apparent even to a lay person that they are probably more different from each other than are the various primates and certainly more than the species of the suborder Anthropoidea. What in the case of *Drosophila* is a genus has almost the rank of an order, or at least, suborder in the primates...Anthropologists should never lose sight of the fact that taxonomic categories are based on populations, not individuals. Different names should never be given to individuals that are presumably members of a single variable population."⁴⁹

So much for the history of paleoanthropology, its preconceptions and its problems. Setting these aside now, what evidence has been accumulated for human evolution since Huxley and Darwin? We need to inspect the data for Human Evolution Arguments #1 through #4. Here now is the evidence of human evolution. We take them in turn.

§10.4 INSPECTING THE EVIDENCE: BEFORE WE EVEN BEGIN - BEGGING THE QUESTION

Accepting evidence of human evolution in the fossil record requires making two assumptions before a fossil can even be picked up and examined, and each of them is taken to be incontrovertible. The first is Argument #1 that human evolution is a fact, and the second is Argument #2 that humans are ancestrally related to living apes because they look similar. If human evolution is indeed to be considered evidence of evolution (*LOE*#11), then this is a flaw of logic called begging the question - assuming the thing you are trying to prove. Take a look at the legitimacy of these two arguments all the same. Look at Argument #1 first.

474

§10.4.1 Inspecting Argument #1: Is Evolution A Fact?

The validity of human evolution rests entirely on the validity of macroevolution. This is obvious and yet as regards the evidence of human evolution, it could hardly be more important because if (macro)evolution is false then so is human evolution and the hominid fossil record meaningless in its current form. While the fact of evolution is taken as grounding the evidence for human evolution (Argument #1), on closer inspection it's much more than that. It has to be true for all of the other arguments #2-#4 to be true because they infer this process. The lines of evidence for human evolution are not independent after all.

We must accept evolution to be a fact if we are to proceed but it seems an opinion rather than a fact and one refuted by the evidence. The reason is because at least for all the other species we inspected earlier, instead of progressive evolutionary change in the fossil record there is sudden appearance, sudden extinction and stasis that not even the punctuated equilibrium model can render testable as definitively evolutionary (§4;§5).

Instead of evolutionary relationships between supposed homologous structures across species supposed homologous structures are not caused by homologous genes or homologous developmental pathways to permit the forces of evolution, mutation and natural selection, to explain them (§7). Instead of evolutionary trends in homologous molecules there is molecular equidistance across species and a molecular clock that can't keep time (§8).

Instead of a similar embryogenesis across species in a way consistent with inheritance, the similarities that do exist like homeobox genes are more striking for how similar genes cause different effects in different creatures, the opposite of an evolutionary prediction (§9).

In short, there is discontinuity in biological similarity across taxa whether it is of morphology or molecules or development. This is a demonstrable rejection of evolutionary relationships between taxa. Thus the evidence of biology declares the differences between organisms are more important than their similarities. It's only evolution that says that it's the similarities that are paramount.

We also find an uncoupling of the mechanism by which evolution works, mutation and natural selection, by the absence of empirical support for evolution in embryonic development (§9.7). Finally, instead of support from mathematics for a random process that could generate protein and DNA sequences, the order these molecules possess exceeds by even the remotest probability their arising by mere chance (§12.9). If evolution is false, Argument #1 is false and we need go no further.

Even if we ignore all these objections to human evolution another hurdle remains. Just as the idea of human evolution in the fossil record was conceived of before fossils were studied, so in reverse the attempt to use fossils to now prove the fact of human evolution, invokes a circular argument. Fossils can at best be consistent with the predictions of the theory. That would be confirmatory evidence for a scientific theory. Is there support for even this? The next chapter is an inspection of the evidence, its internal consistency and it's contradictions. But we have not examined the other necessary assumption, Argument #2. We must do that first.

§10.4.2 Inspecting Argument #2: Is Similarity Ancestral Or Typological?

The inference of Argument #2 is that humans must have descended from an ape-like ancestor because of their close morphologic and chemical similarities with the living apes, and chimpanzees in particular. On this view humans and chimpanzees are considered to have evolved from a common ancestor 4-8 million years ago. The morphology of the common ancestor is hypothetical, and thus the relevance of traits both exhibit now to that past evolution is speculative. Among the distinguishing features considered most useful by the experts to interpret the fossil record for evolution are a large brain, a retracted face as opposed to protruding "prognathic" one, a flexed cranial base, evidence of manual dexterity, a habitually upright posture and obligate bipedalism.[50] The question for you is whether the similarities and differences really are evidence of ancestry. Consider the following objections:

(1) Similarity can be explained as typological i.e. by design parsimony (§7.1). This needs to be tested before being rejected but it's not even considered. What if there were robust evidence for design? (*Part III*). There's also the paradox that although chimpanzees and humans are genetically 98.5% identical, they are not 98.5% identical in how that similarity is ultimately effected – say in behavior (frat houses and rugby players excluded), in technology, arts, worship, morality, language and cultural awareness. They are incomparable! So what if their DNA is similar? The human genome is 90% identical to the mouse,[51] 83% similar to bullfrogs and 62% similar to sunflowers, and nobody makes those comparisons (Fig.8.1). It's typological is the point, but the power of Argument #1 makes you see similarity in no other way than evolutionary.

(2) Morphologic similarity across taxa is not evidence of shared ancestry. This is why Argument #2 is false. We inspected the homology argument earlier to discover morphologic similarity across other animal

species is not caused by similar genes or similar development to allow inferences of evolution as cause (§7.3;§8.3;§9). We already know that fossil life (§5.5), extant life (§7.3), embryonic life (§9.3-7) and the molecules of life (§8.3) consistently and concordantly demonstrate discontinuity rather than links between taxa, removing any known natural mechanism to explain similarity across taxa as caused by ancestry. Why believe it's going to be any different for humans now?

(3) Skeletal similarities among extant apes and man do not correlate with genetic similarities to implicate ancestry as the cause. There is an assumption underlying the comparisons of fossil bones with those of extant apes and humans, and which is that greater morphologic similarity implies greater genetic similarity. But this is only an inference. Similarities surely speak to functional similarity as much as they do genetic, and from what we know from §7-§9, incomparably the more.

In other words the inference that similar morphology equals similar function, and which does not necessarily suggest an evolutionary relationship, is at least as likely in principle as similar morphology equals similar genetics which does suggest it but which is false (§7.3). Either way, an inference about the past is required that is hard to verify. It might seem intuitive that greater genetic similarity between two creatures means greater similarity of their fossil morphology. But is there evidence to support that reasoning used to interpret fossils? Actually the evidence indicates the inference is false, and in the very species at issue - man and chimpanzee. It's not new evidence either.

Charles Oxnard at the University of Chicago showed this in *Nature* in 1975 but the obvious anti-evolution significance of his results was not recognized. He made multivariate comparisons of the fore- and hind limbs of living primates with those of man. Comparative anatomy and comparative DNA sequencing both indicate man is most closely related to the chimpanzee, and to the chimpanzee and gorilla more than to orangutan (Fig 10.5). Was this relationship confirmed by his bone analysis? No. He said:

> "The results...do not present these classical pictures; man is not, in studies, especially close to the primates as a whole or to the chimpanzee in particular. Nor, however, can man be described as a mosaic of other forms. In almost all studies man lies quite separately from the spectra of non-human species although he is, of course, somewhat closer to particular species (for example, sifaka, saki monkey, orang-utang, chimpanzee) than to most others...The findings for man provide...further confirmation of the functional, more than any other biological, such as genetic, emphasis to these particular morphological

results."[52]

What he found was discontinuity across taxa then. Again! His interpretation thus becomes as interesting as his results for where could it have come from? Not his data set, which only leaves his mindset. It's the one-track mindset used for fossil analysis: evolution must be present even if it's not yet seen (§10.2). He continued:

> "This uniqueness for man does not contradict the findings from traditional morphology or from the molecular studies...This uniqueness presumably arises from his having a totally different behavioral repertoire from any other primates, and therefore unique functions in different anatomical regions. The lack of a mosaic description presumably results because functional parallels are not available from other living primates."[52]

There are no fossils in the extant ape lineage. If indeed man is descended from a common ancestor with the apes, a fossil lineage leading to the living apes is expected. Although there are multiple fossil species of supposed human ancestors (by 1976 there were 4,000 according to the *Catalogue of Fossil Hominids* published by the British Natural History Museum), and despite the supposed close evolutionary relationship of the chimpanzee, there is not one fossil for any of the great ape lineages! You need to hear this directly from an expert because it is extraordinary and disturbing. *Nature* editor Henry Gee again: "There are no known fossils, of any age, that might illuminate the ancestries of the extant African apes".[53] Anthropologist Jeffrey Schwartz at the University of Pittsburgh writes about this phenomenon in his book *Sudden Origins:* "When you put the whole picture together, you get a very unreal situation in nature - unreal when compared to virtually any other group of related organisms. Even though the three living great apes are more closely related to us than to any of the living primates, and one or two of these great apes are probably our closest living relatives, the actual, closest relatives of *H. sapiens*, and of each great ape, as well, are now extinct."[54]

The standard explanation for the "unreal situation" is not that it is unreal, but that it is what Penn State University evolutionary biologist Blair Hedges says: "The virtual lack of any fossil chimpanzees is most likely because chimps have lived in habitats – humid forests – where fossilization is rare." In other words evolution is already true and the evidence must be evolutionary in some way. Taken on its merits this explanation is the conjecture that presumes apes and humans though recently diverged, yet lived in totally different environments. The reverse would surely seem at least as likely, and discoveries of fossils like *S. tchadensis* in a woodland paleohabitat reject the notion. The disparity of the missing chimpanzee fossils and yet relative abundant ape-men fossils rather seems explained as evidence of an interpretation bias identified earlier (§10.2). Needless to say a fossil's standing, not to mention its scientist's standing, is boosted immeasurably more by being a human ancestor than an ape ancestor. Maybe this reality has something to do with the unreal situation also. The controversy over *Orrorin tugenensis* discovered by Martin Pickford and Brigit Senut, I suggest bears directly on this issue (§10.2.2).

Dr. Pickford was once a friend and colleague of Richard Leakey. They were at school together in Nairobi and "great buddies".[55] However they had a very public falling out in 1985 when Dr. Pickford co-authored a tell-all book called *Richard. E. Leakey: Master of Deceit.* Needless to say given that title the book upset Mr. Leakey a lot, and they have not seen eye-to-eye since. Dr. Pickford was subsequently barred from the National Museums of Kenya, an organization formerly directed by Richard Leakey who had "virtual veto power"[55] over fossil excavation permit applications in Kenya.

Not to be thwarted, Dr. Pickford managed to obtain permission to continue his research by way of a loophole at the Office of the President of Kenya which at the time had this authority as well. Adding further to the controversy, Yale researcher Andrew Hill disputes that and accuses Pickford of collecting the *Orrorin* material by illegally trespassing on land exclusively his own to research by authority of a permit issued to Hill by the same National Museums of Kenya. Pickford had been researching that territory for years before being displaced however. There is still more to this complicated web.

In March of 2000, just months before *Orrorin* was discovered, Dr. Pickford was arrested and detained for five days in a Kenyan jail on charges of collecting fossils without a permit. This occurred after Mr. Leakey filed a complaint to the police. Dr. Pickford has since sued Mr. Leakey for unlawful arrest and malicious harassment. The dispute has also spilled over into the wider paleontological community. The sensational aspects of the scandal are not what are of interest to us but rather how expert views became aligned after *Orrorin* was reported in the scientific literature. Data supposedly indicative of human evolution can be rejected for being meager and speculative by evolution scientists, if and when the need arises is the point.[55,56]

To show you what I mean, follow the story as it was reported in *The New York Times*. Dr. Pickford's team proposed their fossil species, dated at 6 million years, was bipedal and so a human-like ancestor, not merely another fossil ape. "But Dr. Alan Walker, an anatomist at Pennsylvania State University who has often excavated with the Leakeys, said he was "not impressed with their [discoverers, Pickford and Senut's] evidence for

479

bipedality of the species". Evidence from the lower end of the femur would have been more convincing, he said, but that was missing, and cited other evidence could have contradictory interpretations.

"If *Orrorin* was not bipedal then neither was Lucy, and we have to go back to the drawing board," Dr. Pickford responded. "In many ways, the *Orrorin* femur is closer to humans than they are to Lucy's, and they are quite different from those of chimps. Dr. Daniel E. Lieberman, a George Washington University paleontologist said, "I swear the fossils are so chimpanzee like that it's incredible. If it were a chimpanzee, it would be the first record we have of their early evolution."[57] The journal *Science* quoted Jeffrey Schwartz at the University of Pittsburgh on the same matter: "These conflicting views reflect the fact that experts lack a clear definition of a hominid".[56] In the context of what paleoanthropology claims to do, it's scarcely possible to think of a more damning self-admission by a scientist of his own field.

§10.5 IN CONCLUSION: HOW TO MAKE A MONKEY OUT OF YOU

In summary because of its requirement for two unproven assumptions, namely that human evolution is already a fact (Human Evolution Argument #1) and that humans descended from an ape because they look similar (Human Evolution Argument #2), paleoanthropology seeks to demonstrate what is already known – that man evolved from an ape-like animal by evolution. The fossils just need to be found and the precise pattern of the intermediate species discerned. History declares every fossil with the exception of the first one - the Neanderthals - has been sought, and interpreted, from this mindset (§10.6.2). We'll leave objections to the evidence for human evolution on those grounds here, but suffice it to say if Arguments #1 and #2 are false then human evolution is also false and there is no point us looking further.

But what if we ignore the evidence that cladistic analysis is unreliable, tolerate the explanatory inferences that are untestable, and accept the assumptions necessary for phylogenetics - namely that macroevolution is true and that man indeed is descended from an ape-like ancestor. Does the fossil data now fit the predictions of evolutionary theory? Fig. 10.3 shows the current hominid fossil record, the fruit of one-and-a-half-centuries of research looking really hard for evidence of macroevolution. There are no connecting lines on the drawing to indicate linkages, nor agreement over phylogenetic hypotheses (Fig.10.2). It's not enough to claim a trend because ape fossils have been found in the Miocene, ape-man fossils found in the Pliocene and human fossils in Pleistocene.[21] They need to be linked in a plausible, causal, ancestral manner for evolution to be demonstrated. A naturalistic solution that is scientific needs to be empiric, explicit and possess explanatory details of mechanism. The Miocene ape fossils have not been linked to the *Australopithecines* and may not be relevant. Nothing substantive of the evolutionary history of the extant great apes is known. Which of the Pliocene hominids link *Homo* and which of the Pleistocene forms link modern man are not known. Yet human evolution is still considered scientifically certain.

What if we reject these objections, assume the evolutionary viewpoint of Arguments #1 and #2 and look at the evidence now? The fossil data still appears equivocal at best and at worst incongruent with human evolution theory. This is a remarkable given the contrivances required to get this far, and it's not even because of the need for further reasoning by unproven inference. It's because of weak correlations and significant inconsistencies with evolution theory's own predictions. Counterpoint responses to Arguments #3 and #4 for human evolution are presented next, beginning with Argument #3 which will be rejected for five separate reasons, each contradictory to the predictions of evolution theory. Specifically: 1. There is no evolutionary trend in hominin fossil traits (§11.1). 2. There is abrupt and marked morphologic change (sudden appearance (§11.1.2). 3. Supposed evolutionary hominin fossil species are contemporaneous rather than in series (§11.1.3). Thereafter these and other objections will be inspected in more detail by inspecting the evolutionary credentials of the two hominin fossil species at the all-important interface of the clearly ape-like to clearly human-like transition, to discover 4. *Homo erectus* is not a transitional species (§11.2) and 5. *Homo habilis* is not an transitional species (§11.3). Take a look for yourself.

APPENDIX:
A BRIEF OVERVIEW OF THE HOMININ FOSSIL RECORD

§10.6.1　　　　　Hominid Taxonomy

The Linnaean taxonomy of man is as follows: Phylum Chordata, Class Mammalia, Order Primates and Family Hominidae. There are two classification schemes below the family level. The traditional version has the family Hominidae contain only "hominids". Hominids are the fossils of modern man (*Homo sapiens*) and extinct ancestral creatures considered more closely related to man than man's closest living relative the chimpanzee. The problem with this schema is that it doesn't reflect the closer genetic relationship of chimps to man than chimps are to the other great apes, and hence does not show the inferred closer evolutionary relationship of chimps and man.

To accommodate this idea, the alternative and current classification extends the family Hominidae to include the extant great apes (i.e. chimpanzee, gorilla and orangutan) as well as their evolutionary ancestors but has chimps and man grouped together in a subfamily called Homin<u>in</u>ae. Man and his ancestors are in a subcategory of that subfamily, the tribe Homin<u>ini</u>. Confusingly in this schema the word "hominid" refers to all the great apes, and therefore the term "homin<u>in</u>" is used to refer to man and his ancestors (with the same meaning as "hominid" in the older scheme).

The most widely accepted classification of the hominin series has seven genera in the tribe Hominini as shown in Fig. 10.4 and is the most speciose proposed.[36,42,44,48] Less speciose schemes simply collapse one or more of the above into each other. Those simpler taxonomies from our perspective of observing and inspecting the data for ourselves can be more difficult to follow therefore. At present there is no central authority to determine consensus criteria to resolve the differences of classification. As a result the fate of a fossil species depends purely on the opinion of influential individuals in the field. In general the older a fossil is and the closer it is to the evolutionary tree of man, so the more famous the fossil (and its finders) become.

How are bones sorted into species? By inference using two parameters, age and morphology (§5.3). The hominid species and their chronological sequence in the fossil record is shown in Figs. 10.3 and 10.5. The absolute duration of particular species in the fossil record differs somewhat depending on opinion as you will notice. The most productive way to meet these bones is to follow chronologically how we got to the current state of knowledge, in trace the progress of paleontology since Darwin. Come and meet the family, only hide the bananas, roll up the windows and remain in the vehicle at all times.

Genus	Species	Citation
Sahelanthropus	S. tchadensis	Brunet, Vignaud, 2002[4,5]
Orrorin	O. tugenensis	Pickford, 2001[59,60]
Ardipithecus	A. ramidis ramidis	White, 1994 & 1995[61,62]
	A. ramidis kadabba	Haile-Selassie, 2001[3]
Kenyanthropus	K. platops	Leakey, 2001[63]
Australopithecus	A. africanus	Dart, 1925[64]
	A. anamensis	Leakey, 1995[65]
	A. garghi	Asfaw, 1999[43]
	A. afarensis	Johanson, 1978[66]
	A. bahrelghazali	Brunet, 1996[67]
	A. sediba	Berger, 2010[68]
Paranthropus	P. aethiopicus	Arambourg, 1968[69]
	P. boisei	Leakey, 1959[70]
	P. robustus	Broom, 1938[71]
Homo	(A)H. habilis*	Leakey, 1964[72]
	(A)H. rudolfensis*	Alexeev, 1986[73]
	H. ergaster	Groves, 1975[74]
	H. erectus	Dubois, 1892[75]
	H. heidelbergensis	Schoetensack, 1908[76]
	H. neanderthalensis	King, 1964[77]
	H. floriensis	Brown, 2004[78]
	H. sapiens	Linnaeus, 1758[79]

Fig.10.4 Current hominin taxonomy. *Recent proposals recommend moving H. habilis and H rudolfensis into the Australopithecus genus.[36,58]

§10.6.2   A Brief Historical Overview Of The Hominin Fossil Record[11,48,82,83]

We will examine the fossils in the order they were found. Refer to Figs 10.3 and 10.5 as we go. The first hominin fossils were the **Neanderthals**.

Specimens were first discovered in 1829 in Belgium and later in Gibraltar in 1848 ,but they were not appreciated for what they were until those of better quality were found near Dusseldorf in the Neander Valley in 1856. Thal is valley in German, hence their name. Specimens were examined by the luminaries of the day, including pathologist Rudolf Virchow and at first explained as humans with rickets (vitamin D deficiency). It was William King at Queen's College in Ireland who distinguished the bones from those of modern humans and in 1864 gave them their present name of *Homo neanderthalensis*. The name came from the place where the holotype had been recovered. Holotype is the term used for the original fossil from which the description of a new species is made. Since then Neanderthal fossils from around 300 individuals have been recovered, by far exceeding any other hominin species. They have been recovered from over 70 different sites scattered throughout Europe, excepting Scandinavia, the Levant, West Asia and the Near East. Specimens date from 300,000 to 30,000 years old.

What makes a fossil bone a Neanderthal? The key characteristic is that they are "robust" meaning strong boned. Neanderthals have a larger skull case (and therefore large "endocranial capacity" which indicates brain size) than modern man, a thick brow ridge ("supraorbital torus"), a face that protrudes anteriorly in the midline ("prognathous"), a large nose, a projecting occipital bone to form a bun ("chignon"), a prominent occipital ridge ("torus"), a mastoid crest behind the auditory canal, large incisors and a stout body habitus with a barrel chest and prominent muscle insertions, consistent with muscular individuals).

In 1868 the first fossils indistinguishable anatomically from modern *Homo sapiens* were found in a rock shelter in the Cro-Magnon valley in southwest France. They are called **Cro-Magnon man** and date 30,000 years old. The word means "big cliff" and this name applies only to modern human bones found in this region of France even though the word has been since loosely used to describe all the upper Paleolithic fossil *Homo sapiens* recovered in Europe, Asia, Australasia and Africa since then. Specimens date around 100,000 years or less, although if more robust older specimens that have features generally considered evolutionarily more archaic (like a supraorbital torus) are included, the fossil record of anatomically modern humans dates back to 270-300,000 years.[84] Keep that date in mind because it is a challenge to Argument #4 for human evolution.

In 1891 the third fossil hominin was discovered. Today it is called **Homo erectus**. The first specimen was found by a physician in the Netherlands army called Eugene Dubois. Dubois was a disciple of Ernst Haeckel who went to Southeast Asia with the express intent of finding missing link fossils between ape and man. You will know how many links were missing when he started looking at two species there wasn't even a chain! In 1891 he found a low vaulted, thick walled skull cap (called a "calotte") with prominent brow ridges near Trinil in Java. The following year he found a femur identical to that of a modern human. He considered the two bones to have come from the same individual and called it *Pithecanthropus erectus* ("erect ape-man"). The name was an acknowledgement of the hypothetical genus Haeckel had coined anticipating the future discovery of the missing link between ape and man. This is another example of how evolution theory preceded or has correctly predicted depending on your view, fossil evidence of macroevolution (§5.5.4;§10.3-4). This specimen is now more popularly known as "Java man".

Between 1927 and 1937 over forty others were found near Zhoukoudian, China although at the time they were classified as *Sinanthropus pekinensis* ("Peking man"). In 1964 these and other examples found in Java were reclassified within the genus *Homo* by Oxford anatomist Wilfred Le Gros Clark (§5.1). In 1984 Alan Walker and Richard Leakey described the "Turkana boy" (Nariokatome boy) from West Turkana in Kenya (formally called WT 15000) and dated at 1.6 million years old. It is the most complete specimen of *Homo erectus* yet found and missing only hands and feet. It is an 11 or 12 year old boy projected to have grown to six feet as an adult and to an endocranial capacity of 910 cm^3. This stature was surprising and far taller than many had expected for a taxon considered more primitive than modern humans, an interesting matter we discuss later. Endocranial capacities of modern humans are variable with the low end ranges reported as low as 790cm^3 [85] and 850cm^3 [86] and an average of around 1200 cm^3.[87] Chimpanzees are around 400 cm^3 and gorillas 500 cm^3.

Beginning in 2000 a series of reports by a team lead by David Lorkipanidze has described fossil remains from Dmanisi in Georgia dated at 1.8 million years old, making them the oldest outside of Africa. The features are with strong similarities to the Nariokatome Boy, associated with Oldawan tools (§11.1.1.3) and currently assigned *H. erectus* although in 2002 proposed as a separate species called *H. georgicus* by the authors. These fossils put *erectus* in Asia at the same time as in Africa and so raise questions whether early *Homo* migrated out of or into Africa.[88]

Specimens classified as *H. erectus* have so far been found in Africa, Indonesia and China and have dated from 1.8 million years to 50,000 years old. Their morphologic features are of a long, low vaulted thick walled cranium that is widest near the base, an average endocranial capacity of around 1000 cm³, a large supraorbital torus, a variable saggital keel (a bony ridge over the top of the skull case in the midline), a sharply angled occiput (posterior part of the skull above the nape of the neck), robust mandible (jaw bone) and a skeleton more robust with flattening of the femoral shaft from front to back ("platymeria") and flattening of the tibial shaft from side to side (platycnemia) compared to modern humans.[89] A minority of paleontologists split African specimens classified as *H.erectus* by others into another species called **Homo ergaster** which they claim represents an earlier African form that eventually became *Homo erectus* in Asia.

In 1907 part of a mandible larger than that of modern Europeans was found at Mauer near Heidelberg, Germany. It dates to 500,000 years and it became the type specimen for what is today called **Homo heidelbergensis** (formerly "archaic Homo sapiens"). The taxon later fell into disfavor only to be revived again in the 1990s. Its validity as a taxon separate from modern man remains controversial. Specimens have been found in Europe, Africa and Asia and date from around 600,000 to 100,000 years old. Compared with *Homo sapiens* they generally have a larger body mass and more robust long bones, the skull is thickened posteriorly and it has supraorbital ridges that are separate (compared with *erectus* in which they are continuous), and the face is projecting. It is considered the earliest *Homo* species to have an endocranial capacity the size of modern man. Compared with *erectus* differences are that the occipital bone (posterior skull bone) is more rounded, and that the widest part of the skull is at the parietal region (above the ears) and not at the cranial base.

In 1924 Raymond Dart described the first **Australopithecus africanus**. The specimen was of a child and recovered at Taung in South Africa. He coined the name which means "Southern ape of Africa". The science community initially subjected him to intense criticism, regarding it as a chimpanzee. The accepted wisdom back then said Asia not Africa was the cradle of mankind's evolutionary development. As a juvenile skull (probably two or three years old), it is indeed difficult to interpret. However Dart's realization the skull was unlike any known ape and potentially capable of bipedal gait (because the foramen magnum - the hole through which the spinal cord exits the skull - was more anteriorly situated in a manner similar to man suggesting an erect spine) was later accepted as a highly perceptive observation. In 1936 the Scottish physician Robert Broom found the first adult *A. africanus* specimen at Sterkfontein, South Africa and specimens of this species today total around thirty-two partial or complete skulls, a hundred partial or complete jaws and thirty postcranial (the term for non-skull) bones.

All of them, like the original Taung skull, have been confined to cave sites in South Africa. As a result, dating is even more problematic and ages of between 2.4 and 3 million years have been inferred by comparing bones of animal fossils also found in the caves with similar looking but datable animal bones found far away in East Africa. There is contrary evidence that the Taung skull is less than a million years old however. If true it is the most recent Australopithecine species known, and by more than a million years. Needless to say, referring to the chronological sequence of the fossil record in Fig 10.3 and 10.5, this would completely upend the current evolutionary notions and Argument #3. T.C. Partridge obtained evidence for this notion in 1973 using uranium series dating. He demonstrated that the cave in which the Taung skull was found could not have been formed more than 0.87 million years ago.[90] The counter argument postulates uranium could have entered aprons after they formed resulting in a spuriously "young" age for the cave.[91] The true age of the specimen still remains uncertain eight decades after the discovery.[10]

As far as the consensus of scientific literature is concerned the Taung specimen is around 2 million years old. This is by curious logic. Namely because other *A. africanus* fossils from other cave sites in South Africa have dated older than 2 million years and because baboon fossils (of completely uncertain association) with morphology similar to those found in East Africa were dated older than two million years.[10]. One wonders what would have happened if some other creature, say a horseshoe crab, had been found in the cave. The endocranial capacity of *A. africanus* is 404-440 cm^3 and compared with *afarensis* (see below) the face is wider, with more prognathism, more robust mandibular bodies and smaller anterior and larger post canine teeth. The limbs are very similar to those of *A. afarensis*. It is considered a good arboreal climber but had a pincer grasp. It is considered a facultative biped with a gait different to humans because of an abductable hallux, curved lateral tibial condyle, different pelvic bone trabecular pattern (which reflects the biomechanical forces on the pelvis) and long lower back with six lumbar vertebrae compared with five in man (and three or four in the extant apes).

In 1938 Robert Broom described the first specimen of **Paranthropus robustus** at Kromdraai, South Africa. All specimens yet found of this species are from South Africa and all dated between 2-1.5 million years old. Only one specimen allows an assessment of endocranial capacity - 476 cm^3. Body size is a little larger than that of *A. africanus* with stature of 1.5-1.6 meters, and is more robust. Like *P. boisei*, it differs from *afarensis* (see

Evidence of human evolution I

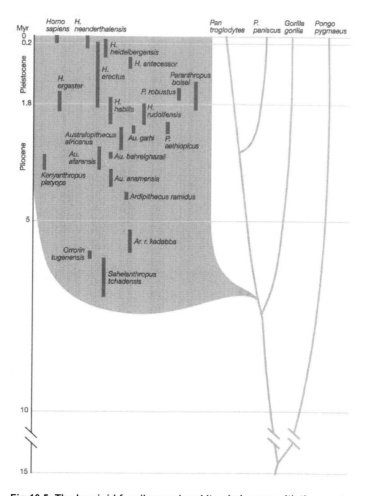

Fig.10.5. **The hominid fossil record and its phylogeny with the great apes.** There are no known links but the urge to draw them is irresistible! (§25.1). Phylogenies and links among the hominin fossils are not known. There are no ancestral fossils to the extant apes yet there are many in the human tree. Humans are the only living species of the genus *Homo* and according to the science, represent the last 3% of the time span of hominin evolution. (Reproduced from Carroll, S.B. *Nature* 2002 with permission).[80]

below) for similar reasons as *africanus*, but more marked. Incisors and canines are small but premolars and molars are large. The teeth have thick enamel like the other members of this genus. Compared with afarensis they have a more flexed cranial base. They have saggital crests on the skull for attachment of jaw muscle and marked postorbital constriction (depressions on the skull above the eye sockets). The distal humerus resembles that of man and is totally unlike the knuckle walking apes.

488

In 1959 Louis Leakey described a specimen Olduvai Hominid 5 (OH 5) from Olduvai Gorge of a species now called **Paranthropus boisei**. He called it *Zinjanthropus boisei* ("East Africa man") and regarded it as the maker of stone tools which were found in the same strata as the bones. Both of his views were later retracted. The name lives on however as the nickname of the type specimen OH5 "Zinj", although it has also been called "Nutcracker man" in reference to the prominent molars. *P. boisei* is considered an East African sister group to the South African robust australopithecine species. Specimens have been recovered from Tanzania, Ethiopia, Kenya and Malawi and date from around 2.3 million years to 1.2 million years. Its most distinctive feature is a large flat and wide face, and megadontia (large, broad based premolar and molars with thick enamel). The endocranial capacity is 450 cm³. There are no postcranial (non-skull) remains definitively linked to this taxon.

In 1964, a year after the description *Zinjanthropus*, Louis Leakey, John Napier and Phillip Tobias (the latter succeeded Raymond Dart at the University of the Witwatersrand, South Africa) proposed a new taxon called **Homo habilis**. The purpose was to allocate a variety of bones found at Olduvai Gorge (parietal skull bones and other skull fragments, foot and hand bones, teeth and a juvenile mandible). This species was considered to be on the direct line to modern man, have language and to have made stone tools also recovered at the site. *Habilis* translated means "handy man". The taxon has been controversial ever since, even though specimens have subsequently been added to it. It was not until the discovery of OH 62 by Donald Johanson, dated at 2 million years old and popularly dubbed "Lucy's child", that upper and lower limbs could firmly be associated with the same individual of the taxon. It had been hitherto been assumed that as a toolmaker it similarly had human-like limb proportions. It turned out OH 62 stood just 1 meter (3 feet) tall with arms almost as long as its legs, typical of an ape. So much for a supposed toolmaker, this was the smallest adult hominid ever discovered. It was a blow not credibly softened by inferring it a female and so from sexual dimorphism as some suggested. Another problem was the allocation of specimen KNMER1470 (<u>K</u>enya <u>N</u>ational <u>M</u>useum <u>E</u>ast <u>R</u>udolf) by Richard Leakey to the taxon even though it has much more human-like features. There is still wide disagreement over which specimens exactly are attributable to the taxon. Some suspect *habilis* represents more than one species. The most popular version of that view holds it to consist of **Homo rudolfensis**, a species originally proposed by Russian anthropologist Valery Alexeev in 1986 to accommodate ER 1470 and *habilis sensu stricto* (the residua). Others consider it a widely divergent but single species also called *habilis sensu lato*. Specimens are dated from 2.3 through to 1.6 million years and have been found in Tanzania, Kenya, Ethiopia and South Africa. Endocranial volume ranges widely (from around 500 to 800 cm^3) as does facial morphology and mandible robusticity. Since much of the post cranial specimens attributed to the taxon could belong to *P. boisei*, only OH 62 ("Lucy's child") and some foot bones from OH 7 are considered reliably to be of this taxon. The problem is compounded by the fact OH 62 is a collection of 302 bone fragments poorly preserved and that OH 7 is a juvenile. In the most recent development surrounding this taxon, Bernard Wood at George Washington University has proposed *habilis* be transferred to the Australopithecine genus from Homo.

In 1974 Donald Johanson and Tim White found specimen A.L.-288 at Hadar, Ethiopia, now known to the world as "Lucy". The specimen got its name from the Beatle's song "Lucy in the sky with diamonds" which was playing on the camp radio at the time. It represents a 40%-complete skeleton and dated at 3.2 million years. Along with other specimens from Hadar and Laetoli in Tanzania it became **Australopithecus afarensis** and is considered more ape-like ("primitive" in the evolutionary paradigm) than *A. africanus*. Political instability in the region prevented further excavations for almost 15 years. Since then potential members of this species have also been found in Kenya. Specimens are dated between 2.9 and 3.9 million years old. The "hypodigm" (the term referring to all the fossil evidence for a taxon) now consists of in addition to Lucy, a skull, several skull fragments, jaws and many limb bones.

A. afarensis stood just over 3 feet (1-1.5 m) tall and weighed 25 to 50 kg - a wide range that has been attributed to sexual dimorphism. It had an endocranial volume of 375-540 cm^3 and the skull is similar to chimpanzee with its small brain size and powerful jaw musculature. As a result Stanford anthropologist Richard Klein considers this "demonstrates beyond all doubt the essentially ape ancestry of the hominids." The canines are smaller and molars larger, more human like, and the enamel is thick. The overall body shape is apelike, and though the upper limb shorter than the great apes, much longer than man. The narrow apical tufts of the distal finger phalanges and less prominent thumb metacarpal are taken as indicative of poor manual dexterity. The ankle, knee, femur and pelvis structure suggest bipedal gait and just how human-like it was continues to be debated. In 2007 Yoel Rak at Tel Aviv University showed the *afarensis* jaw like a gorilla's (and unlike that of the humans, chimps and orangutangs which grouped together in contradiction to hominid phylogeny), a finding which he said "casts doubt on the role of Au afarensis as a modern human ancestor".[92]

In 1986 Alan Walker and Richard Leakey described an almost complete skull called KNM-WT 17000 from West Turkana, Kenya. It is better known as the "Black Skull" because of distinctive discoloration caused by mineral uptake during fossilization (§5.2). It has an endocranial capacity of is 410 cc and is considered "less derived" than *boisei* (i.e. the evolutionary inference of more ape-like) with more prognathism, bigger incisors and a less flexed cranial base with a prominent masticatory apparatus. Dated at 2.5 million years, it was also older than the existing specimens allocated to boisei at 2.3 million years. With an evolutionary preconception, it is not difficult to see what happened. Though no mandible was recovered with the skull, its features reawakened interest in an edentulous mandible (Omo 18) described back in 1967 by a French team led by Camille Arambourg and Yves Coppens. They had named it into a new taxon called *Paraustralopithecus aethiopicus* but both their discovery and its allocation had been ignored. This now became the holotype specimen for a new taxon now called **Paranthropus aethiopicus**, the hypodigm of which with the Black skull now includes another mandible (KNM –WT 16005), an anterior mandible fragment with a worn P4 crown dated at 2.7 million years not before assigned (L55s-33), and a fragmentary juvenile vault (Omo338y-6). Specimens are all from Kenya and Ethiopia. *P. aethiopicus* is considered an "early" member of a robust lineage that is later represented by *boisei*.

In 1994 Tim White at Berkeley, Gen Suwa at the University of Tokyo and Berhane Asfaw at the Ethiopian Ministry of Culture described seventeen specimens from Aramis in the Middle Awash region of Ethiopia of a new taxon they called *Australopithecus ramidis*. However 7 months later in a highly unusual move, the genus was changed in a note published in the back pages of the same journal (*Nature*) to **Ardipithecus ramidis**, and without reasons being given, by which it remains today. Dated at 4.4 million years, the specimens were older than any previous hominin discovered. This age together with its very ape like morphology (considered evolutionarily "primitive") the authors regarded to be a possible root species for the Hominidae. Its name reflects this view, translating literally as "ground man-root" in the Afar language of the region. The dentition is more ape-like than *A. afarensis* with larger canines, narrower and less complex molars and the teeth are capped with thin enamel like chimpanzee, gorilla and most monkeys, and unlike all more recent hominids. There is no evidence yet as to brain size nor any post cranial (non skull) evidence. The argument made for its hominid status is because compared with the extant apes, the canines are less prominent, the upper central incisors smaller, broad molar crowns and the foramen magnum located more anteriorly on the skull base (and from which bipedal gait has

491

been inferred). In 2001 Yohannes Haile-Selassie at Berkeley described 11 specimens dated at 5.2-5.8 million years and also recovered from the Middle Awash of Ethiopia. The bones were 4 teeth, a mandible, fragments of clavicle, humerus, ulna and fragments of single finger and toe phalanges. He considered them to represent a subspecies called *A. ramidus kadabba* because of differences in the molar and premolar teeth characteristics.

In 1995 Maeve Leakey and Alan Walker described **Australopithecus anamensis** as a new species dating from 3.9 to 4.2 million years ago and based on specimens consisting of 45 teeth, a mandible and mandible fragment, a tibia and a humerus fragment and two maxillary bones. They considered it to be a possible ancestor to *A. afarensis ramidus* because of its greater age and more ape-like characters and a sister species to *Ardipithecus*. The name "anam" means lake in the Turkana language because the specimens were found in sediments of an ancient lake near Lake Turkana, Kenya and have not been found outside that country. It weighed around 50 kg with features indicative of good climbing ability (curved fingers, long arm, reduced wrist extension similar to knuckle walking apes) and of bipedalism. Additional specimens to the taxon were added by these authors in 1998, confirming an intermediate age between *A. ramidus* and *A. afarensis*.

In 1999 Berhane Asfaw, Tim White and Gen Suwa described **Australopithecus garhi** which was discovered at Bouri, Middle Awash in Ethiopia. The species name means surprise in the Afar language. Dated at 2.5 million years its major effect was in filling the gap in East Africa at between 2 and 3 million years ago. They consider it a descendant of *afarensis* and a candidate ancestor for early Homo. The major features are small brain size (450 cm^3), primitive face (prognathic with marked post orbital constriction) and palate and larger post canine teeth than *afarensis* and as big as boisei but distinguished from Paranthropus by less enamel and large anterior teeth (which are small in all the Paranthropus species).

In 2000 twelve fossils dated at 6 million years from the Lukeino formation, Kenya were described by Bridget Senut and Martin Pickford to represent a new genus and species, **Orrorin tugenensis** ("Millenium man"). Orrorin means "original man" in the local dialect. The thick enamel on the molars, their square shape and small size (humans are small too) are claimed distinctive (Lucy's molars are larger and hence they consider it off the lineage to man), and the femur head is claimed to be indicative of bipedal gait.

In 2001 Maeve Leakey reported a Kenyan cranium, temporal bone, mandible and maxillary bone fragments and teeth as a new genus and species, **Kenyanthropus platops** contemporary with *A. afarensis* being dated at 3.5 million years old. The name means "flat-faced man of Kenya". This species introduced a "bushiness" into the East African lineage that had previously been considered simply anamensis to afarensis to later hominins. Furthermore, certain features like small molars that are shared by humans but not by afarensis and anamensis questioned the latter's presence in the direct lineage to man. The skull size is in the range of africanus and afarensis. The upper molar size is small (like man and *Orrorin*), and unlike Paranthropus in which they are large with thicker enamel. The midface is flat, unlike afarensis and Paranthropus.

In 2002 Michel Brunet and his team described **Sahelanthropus tchadensis** based on six bones discovered in Chad. It is the oldest fossil assemblage currently known, at between six and seven million years. This age was derived indirectly by estimates of the faunal age at the site. It is considered a hominid because of small canines, intermediate post canine enamel thickness, reduced subnasal prognathism, supraorbital torus and anterior positioning of the foramen magnum. It also had a small brain – estimated at 320-380 cm^3. As a result of its age and this mosaic of primitive and derived features, the authors consider it close to the last common ancestor of humans and chimpanzees.

In 2004 Michael Morwood and a joint Australian-Indonesian team reported **Homo floriensis**, nicknamed Hobbit. At least 9 specimens have now been found, all in a cave on the island of Flores in Indonesia. It is the smallest hominin yet reported, smaller in stature and brain size than a chimpanzee (106 cm 3'6" tall and endocranial size 385-417cm^3), yet evidence indicates they engaged in cooperative hunting and used fire and stone tools. Dated at 95-16,000 years they overlapped with humans. What exactly these fossils represent remains controversial with some insisting it is diseased human material from dwarfism or hypothyroidism or microcephaly, while others maintain it is represents a novel extinct hominin species as they do not show the skull case morphologic features associated with microcephaly, and have teeth, mandible wrist and foot features seen in early *Homo* fossils.

In 2010 Lee Berger, successor to Raymond Dart at the University of the Witwatersrand in Johannesburg, reported **Australopithecus sediba**. The specimens were found 15 miles away from Sterkfontein where the first adult *Au. africanus* was found. Dated at 1.9 million years with Australopithecine features of long arms, small body and small endocranial capacity (420 cm3), it also has mosaic features seen in Homo of longer legs, more erect pelvis, shorter fingers, smaller teeth, more gracile jaw and as such has been proposed as possibly occupying the stem of the *Homo* clade.

REFERENCES

1. Futuyma, D. J. *Science on trial*. 98 (Sinauer Associates Inc., 1995).
2. National Academy of Sciences. *Science and Creationism: A View from the National Academy of Sciences*. 2nd Ed edn, (The National Academies Press http://books.nap.edu/html/creationism/, 1999).
3. Haile-Selassie, Y. Late Miocene hominids from the Middle Awash, Ethiopia. *Nature* 412, 178-181 (2001).
4. Brunet, M., Guy, F., Pilbeam, D. & al, e. A new hominid from the Upper Miocene of Chad, Central Africa. *Nature* 418, 145-151 (2002).
5. Vignaud, P., Duringer, P., Mackaye, H. T., Likius, A. & Blondel, C. e. a. Geology and palaentology of the Upper Miocene Toros-Menalla hominid locality, Chad. *Nature* 418, 152-155 (2002).
6. Lemonick, M. D. & Dorfman, A. Father of us all? *Time*, 40-47 (2002).
7. Lemonick, M. D. & Dorfman, A. One giant step for mankind. *Time*, 52-61 (2001).
8. Spice, B. Paleoanthropologist Leakey speaking here on new fossils, debates human origins. *Post-gazette.com* Oct 14, http://www.post-gazette.com/healthscience/20021014leakey20021014p20021012.asp (2002).
9. Ridley, M. *Evolution*. (Blackwell Science, Inc., 1993).
10. Klein, R. G. *The Human Career*. (University of Chicago Press, 1989).
11. Tattersall, I. *The Fossil Trail*. (Oxford University Press, 1995).
12. Grant, V. *The evolutionary process. A critical review of evolutionary theory*. (Columbia Unversity Press, 1985).
13. Cowen, R. *History of Life*. (Blackwell Science, Inc., 1995).
14. Raven, P. H. & Johnson, G. B. *Biology* 477-492 (McGraw-Hill, 2002).
15. Futuyma, D. J. *Science on trial*. 100 (Sinauer Associates Inc., 1995).
16. King, M.-C. & Wilson, A. C. Evolution at two levels in humans and chimpanzees. *Science* 188, 107-116 (1975).
17. The Chimpanzee Sequencing and Analysis Consortium. Initial sequence of the chimpanzee genome and comparison with the human genome. *Nature* 437, 69-87 (2005).
18. Gibbons, A. Which of our genes makes us human? *Science* 281, 1432-1434 (1998).
19. Ruse, M. *Darwinism Defended*. 236-7 (Addison-Wesley Publishing Company, 1982).
20. Ruse, M. *Darwinism Defended*. 242 (Addison-Wesley Publishing Company, 1982).
21. Futuyma, D. J. *Science on trial*. 98-113 (Sinauer Associates Inc., 1995).
22. Simpson, G. G. *Fossils and the History of Life*. 211 (Scientific American Books, Inc, 1983).
23. Tattersall, I. *The Fossil Trail*. 4 (Oxford University Press, 1995).
24. Huxley, T. H. *Evidence as to Man's place in nature*. (University of Michigan Press, 1959 (1863)).

25 Darwin, C. R. *The descent of man, and selection in relation to sex.* 185 (John Murray, 1871).
26 Tattersall, I. *The Fossil Trail.* 51 (Oxford University Press, 1995).
27 Tattersall, I. *The Fossil Trail.* 57 (Oxford University Press, 1995).
28 Lewin, R. *Bones of Contention.* 21 (Simon & Shuster, 1987).
29 Johnson, P. E. *Darwin on trial.* 83 (InterVarsity Press, 1993).
30 Tattersall, I. *The Fossil Trail.* 165 (Oxford University Press, 1995).
31 Gee, H. *In Search of Deep Time.* 10 (The Free Press, 1999).
32 Tattersall, I. *The Fossil Trail.* 168-9 (Oxford University Press, 1995).
33 Tattersall, I. & Schwartz, J. H. *Extinct Humans.* 53 (Westview Press, 2000).
34 Gee, H. *In Search of Deep Time.* 5 (The Free Press, 1999).
35 Collard, M. & Wood, B. How reliable are human phylogenetic hypotheses? *Proceedings of the National Academy of Sciences USA* 97, 5003-5006 (2000).
36 Wood, B. & Collard, M. The Human Genus. *Science* 284, 65-71 (1999).
37 Lockwood, C. A. & Fleagle, J. G. The recognition and evaluation of homoplasy in primate and human evolution. *Yearbook of Physical Anthropology* 42, 189-232 (1999).
38 Skelton, R. R. & McHenry, H. M. Evolutionary relationships among early hominids. *Journal of Human Evolution* 23, 309-349 (1992).
39 Skelton, R. R. & McHenry, H. M. Trait list bias and a reappraisal of early hominid phylogeny. *Journal of Human Evolution* 34, 109-113 (1998).
40 Strait, D. S. & Grine, F. E. Trait list bias? A reply to Skelton and McHenry. *Journal of Human Evolution* 34, 115-118 (1998).
41 Strait, D. S., Grine, F. E. & Moniz, M. A. A reappraisal of early hominid phylogeny. *Journal of Human Evolution* 32, 17-82 (1997).
42 Wood, B. Hominid revelations from Chad. *Nature* 418, 133-134 (2002).
43 Asfaw, B. *et al.* Australopithecus garghi: a new species of early hominid from Ethiopia. *Science* 284, 629-635 (1999).
44 Gibbons, A. In search of the first hominids. *Science* 295, 1214-1219 (2002).
45 Wood, B. & Lonergan, N. The hominin fossil record: taxa, grades and clades *Journal of Anatomy* 212, 354-376 (2008).
46 Wood, B. Reconstructing human evolution: achievements, challenges and opportunities. *Proceedings of the National Academy of Sciences (USA)* 107 Supp 2, 8902-8909 (2010).
47 Lewontin, R. Human Diversity. *Scientific American*, 163 (1995).
48 Wood, B. & Richmond, B. G. Human evolution: taxonomy and paleobiology. *Journal of Anatomy* 196, 19-60 (2000).
49 Mayr, E. Taxonomic categories in fossil hominids. *Cold Spring Harbor Symposia in Quantitative Biology* 15, 109-118 (1950).
50 Elton, S., Bishop, L. C. & Wood, B. Comparative context of Plio-Pleistocene hominin brain evoution. *Journal of Human Evolution* 41, 1-27 (2001).
51 Hughes, A. L. Comparative genomics: genomes of mice and men. *Heredity* 90, 115-116 (2003).
52 Oxnard, C. The place of the Australopithecines in human evolution: grounds for doubt? *Nature* 258, 389-395 (1975).

53 Gee, H. *In Search of Deep Time.* 205 (The Free Press, 1999).
54 Schwartz, J. H. *Sudden Origins. Fossils, genes and the emergence of species.* 18-9 (John Wiley & Sons, 1999).
55 Balter, M. Paleontological rift in the Rift Valley. *Science* 292, 198-201 (2001).
56 Balter, M. Scientists spar over claims of earliest ancestor. *Science* 291, 1460-1461 (2001).
57 Wilford, J. N. On the trail of a few more ancestors. *The New York Times* (April 8 2001).
58 Wood, B. & Collard, M. The changing face of genus Homo. *Evolutionary Anthropology* 8, 195-207 (2000).
59 Senut, B. et al. First hominid from the Miocene (Lukeino Formation, Kenya). *Comptes Rendus de l'Academie des Sciences* 332, 137-144 (2001).
60 Pickford, M. & Senut, B. The geological and faunal context of Late Miocene hominid remains from Lukeino, Kenya. *Comptes Rendus de l'Academie des Sciences* 332, 145-152 (2001).
61 White, T. D. e. a. New discoveries of Australopithecus at Maka in Ethiopia. *Nature* 371, 306-312 (1994).
62 White, T. D., Suwa, G. & Asfaw, B. Corrigendum: Australopithecus ramidus, a new species of early hominid from Aramis, Ethiopia. *Nature* 375, 88 (1995).
63 Leakey, M. G., Spoor, F., Brown, F. H., Gathogo, P. N. & Kiarie, C. e. a. New hominin genus from eastern Africa show diverse middle Pliocene lineages. *Nature* 410, 433-440 (2001).
64 Dart, R. A. Australopithecus africanus: the man-ape of South Africa. *Nature* 115, 235-236 (1925).
65 Leakey, M. G., Feibel, C. S., McDougall, I., Ward, C. & Walker, A. C. New four-million-year-old hominid species from Kanapoi and Allia Bay, Kenya. *Nature* 376, 565-571 (1995).
66 Johanson, D. C., White, T. D. & Coppens, Y. A new species of the genus Australopiithecus (Primates: Hominidae) from the Pliocene of eastern Africa. *Kirtlandia* 28, 1-14 (1978).
67 Brunet, M. et al. Australopithecus bahrelghazali, une noubelle espece dHominde ancien de la region de Koror Toro (Tchad). *Comptes Rendus de l'Academie des Sciences* Serie IIa, 322, 907-913 (1996).
68 Berger, L. R. et al. Australopithecus sediba: A new species of Homo-like Australopith from South Africa. *Science* 328, 195-204 (2010).
69 Arambourg, C. & Coppens, Y. Decouverte d'un australopithecien nouveau dans les gisements de l'Omo (Ethiopie). *South African Journal of Science* 64, 58-59 (1968).
70 Leakey, L. S. B. A new fossil skull from Olduvai. *Nature* 184, 491-493 (1959).
71 Broom, R. The Pleistocene anthropoid apes of South Africa. *Nature* 142, 377-379 (1938).
72 Leakey, L. S. B., Tobias, P. V. & Napier, J. R. A New Species of the Genus Homo from Olduvai Gorge. *Nature* 202, 7-9 (1964).
73 Alexeev, V. *The origin of the human race.* (Progress Publishers, 1986).

74 Groves, C. P. & Mazak, V. An approach to the taxonomy of the Hominidae: gracile Villafranchian hominids from East and South Africa. *Casopis pro Mineralogii Geologii* 20, 225-247 (1975).
75 Dubois, E. Paleontologische and erzoekingen op Java. *Verslag van het Mijnwezen Batavia* 3, 10-14 (1892).
76 Schoetensack, O. *Der Unterkierfer des Homo heidelbergensis aus den Sanden von Mauer bei Heidelberg*. (W. Engelmann, 1908).
77 King, W. The reputed fossil man of the Neanderthal. *Quarterly Journal of Science* 1, 88-97 (1964).
78 Brown, P. et al. A new small-boded hominin from the Late Pleistocene of Flores, Indonesia. *Nature* 431, 1055-1061 (2004).
79 Linnaeus, C. *Systema Naturae*. (Laurentii Salvii, 1758).
80 Carroll, S. B. Genetics and the making of Homo. *Nature* 422, 850-857 (2002).
81 Klein, J. & Takahata, N. *Where do we come from? The molecular evidence for human descent*. (Springer-Verlag, 2002).
82 Lubenow, M. L. *Bones of Contention*. (Baker Books, 1992).
83 Klein, R. G. *The Human Career. Human Biological and Cultural Origins*. (University of Chicago Press, 1989).
84 Brauer, G., Yokoyama, Y., Falgueres, C. & Mbua, E. Modern human origins backdated. *Nature* 386, 337-338 (1997).
85 Schultz, A. H. in *Readings in Physical Anthropology* (ed T.W. McKern) (Prentice Hall Publishers, 1966).
86 Harris, M. *Culture, Man and Nature*. (Thomas Y. Crowell, 1971).
87 Ashton, E. H. & Spence, T. F. Age changes in cranial capacity adn foramen magnum of hominoids. *Proceedings of the Zoological Society of London* 130 (1958).
88 Wood, B. Did early Homo migrate "out of" or "in to" Africa? *Proceedings of the National Academy of Sciences (USA)* 108, 10375-10376 (2011).
89 Rightmire, G. P. *The evolution of Homo erectus*. (Cambridge University Press, 1990).
90 Partridge, T. Geomorphological Dating of Cave Openings at Makapansgat, Sterkfontein, Swartkrans and Taung. *Nature* 246, 75-79 (1973).
91 Vogel, J. in *Hominid evolution: past, present and future* (ed PV Tobias) 189-194 (Alan R. Liss, 1985).
92 Rak, Y., Ginzburg, A. & Geffen, E. Gorilla-like anatomy on Australopithecus afarensis mandibles suggests Au. afarensis like to robust australopths. *Proceedings of the National Academy of Sciences (USA)* 104, 6568-6572 (2007).

11. Finding ape men in the family album

§11.1 INSPECTING ARGUMENT #3: IS THERE AN EVOLUTIONARY FOSSIL RECORD?

§11.1.1 Stasis Instead Of Evolutionary Change

Argument #3 says there's a series of fossil intermediates extending from ape to human over geologic time in confirmation of human evolution (§10.1). Fossil traits claimed most demonstrative of this are gait, brain size and dentition but other traits, and even stone tools, non-living things mind you, are taken as evidence. Specifically changes in tool shape are taken evidence of an evolution of technology and as proxy evidence of brain evolution. The fossil trends are all so impressive that Douglas Futuyma says: "One 'species' blends into another as you go from younger to older fossils"[1] and Niles Eldredge says "the fossil record of human evolution is one of the very best, most complete, and ironclad documented examples of evolutionary history that we have assembled in the 200 years or so of active paleontological research."[2] This gradualism is exactly what Darwinism has always demanded of course (§2.2).

It might come as a surprise then, but depending on how cynical you've become by now perhaps not, to discover there is stasis instead of evolutionary change, sudden appearance instead of gradualism and simultaneous existence instead of anagenesis in the hominin fossil record. Just as for other organisms (§5.5). The next question is how such an opposite interpretation could come from the same fossil data set! We must first sort out which interpretation is true before we can sort out who is telling stories, how and why.

§11.1.1.1 Stasis In *Science*
In 1999 a quite remarkable review of the morphologic traits of the fossil hominins was published in *Science*.[3] It was remarkable not because it was written by two of the most senior voices in paleoanthropology, nor for its detail, nor for its high profile in that journal although it was all of these things. It was because across the hallowed pages a full-length article and review of the field was presented which documented an absence of human evolution all the while purporting its presence. Needless to say but equally remarkable because it demands an explanation of us how this could happen, neither the editors nor the readership or at least those who write letters to the editor, recognized it either.

The authors of the study were Bernard Wood at George Washington University and Mark Collard at University College London. They were attempting to clarify the criteria by which a fossil is allocable to the genus *Homo*. They declared the existing criteria "either inappropriate or inoperable" (§10.3.2). That such a situation could have persisted for more than a century and not impeded phylogenetic hypotheses showing human evolution "a fact" is breathtaking. Do we understand why? Precise species definitions are essential. How can you know human evolution is happening if you don't know what's changing?

In order to rectify this and develop verifiable criteria for diagnosing these fossils, the authors worked from what would seem a reasonable premise. Namely that the definition of a hominin genus, say *Homo* or *Australopithecus*, should refer to species whose members occupied a single "adaptive zone". Classification by shared "adaptive zone" is simply to group the creatures that in similar fashion fed, reproduced and otherwise behaved to maintain their homeostasis in the ecosystem or "zone" in which they lived. This attempts to give biological significance to an otherwise morphologically based taxonomy, and harks back to a plea made earlier by Ernst Mayr (§10.3.2).

But how do you decide such adaptive features from ancient bones of never before seen, extinct creatures? By inference. In another inference, the authors considered the traits most indicative of zone from fossil characters were 1) body size, 2) body shape, 3) locomotor apparatus, 4) rate and pattern of development, 5) teeth and jaws and 6) brain size. These six traits were compared across the hominin species with a view to stratifying them by evolutionary trends. You'll never guess what was found. We will take them in that order.

	Taxon	Mass (kg)
Ape-like	A. afarensis	37
	A. africanus	36
	P. aethiopicus	-
	H. rudolfensis	-
	P. boisei	44
	H. habilis	34
	P. robustus	36
Human-like	H. ergaster	58
	H. erectus	57
	H. heidelbergensis	62
	H.neanderthalensis	76
	H. sapiens	53

Fig.11.1. Estimates of mean hominin body mass. Species listed by dated age (Fig.10.3). Masses are estimated averages between male and female by arbitrary formulae using correlates with living apes and humans. The human mass given is low, so closer to ape, and "the average of the male and female values given for modern Africans". This is not racist (§11.5.3.3). Mean body mass of *Pan paniscus* is 35 kg, *Pan troglodytes* is 45 kg and *Gorilla gorilla* 105 kg.[4] There was insufficient data for analysis of *Ardipithecus*, *K. platops*, *O. tugenensis* and *S. tchadensis*. (Data from Wood & Collard, 1999 with permission).[3]

No evolutionary trend was found in body size or body shape. Instead there was a division into two discrete groups, those with distinct ape-like and those with distinct human-like characteristics (Fig. 11.1;Fig 11.4). This had been shown before. Five years earlier Henry McHenry at the University of California at Davis found "no evidence of a gradual trend of increased body weight through time, as might be expected from Cope's law."[4] No evolution going on here then. What's Cope's law? Edward Drinker Cope (1840-1897) was an American dinosaur paleontologist and child prodigy who began publishing scientific articles when still in his teens. Cope's law states that in general organisms tend to get bigger over time by evolution. The horse lineage is taken as demonstrative (§5.5.4.2).

What about the third fossil trait – gait? Among the goals of hominid fossil research is to show how locomotion evolved from an arboreal ancestral ape into an upright terrestrial obligate biped. In contrast to bipedal humans, chimps and gorillas are quadrupedal and orangutans use a mixed scrambling and climbing brachiator locomotion. Thus bipedal locomotion gets evolutionary significance because it distinguishes man from the extant apes and with which we supposedly share a common ancestor. The significance reaches still further.

Bipedalism is considered the defining characteristic of humanity in the fossil record![5] Where does that idea come from? Argument #2 resting on Argument #1 which rests on...? (§10.4.1). It's an extraordinary idea. How we walk would seem a poor correlate for how we think, which is surely the greatest difference between humans and apes and for that matter every other living thing including birds which also walk on two legs. Absent the preconception man evolved from an ape, bipedalism has no significance as an especially human trait. Henry Gee agrees: "There is therefore nothing special, advanced, or progressive about bipedality – only the fact that it is we who are bipedal, and it is we who are writing the book, make it so."[6] His is a minority view because the mainstream shouts, together with the sheep at Animal Farm, "Four legs good, two legs better!" In a non-evolutionary paradigm fossils of extinct hominids that walked on two legs evokes interest as to how they became extinct, not how they became human.

How did human evolution select for the adaptation to bipedalism from an ape-like gait? The hypotheses are as many as they are different. They include improved tool use, improved foraging by being able to reach higher for food, better carrying of food and tools and weapons, better sexual and territorial display, increased locomotor energy efficiency, greater vigilance to detect prey and predator over obstacles like tall grass, and improved thermoregulation as body exposure is reduced to the African sun when erect but increased to cooling breezes. All are speculations of course but you would never know from the seriousness with which they are advanced. In a rare moment of frankness from paleontology Maeve Leakey reviewed them for *Time* in 2001 and concluded: "There are all sorts of hypotheses, and they are all fairy tales really because you can't prove anything".[7] She continued:

"In the physical realm, any theory of human evolution must explain how it was that an apelike ancestor, equipped with powerful jaws and long dagger-like canine teeth and able to run at speed on all four limbs, became transformed into a slow, bipedal animal whose natural means of defense were at best puny. Add to this the powers of intellect, speech and morality, upon which we "stand upon a mountain top" as Huxley put it, and one has the complete challenge to evolutionary theory."[8]

Indeed.

Returning to Wood and Collard's review in *Science*, what changes in inferred locomotor behavior were found from the fossil record? Over the 4 million year datings of fossil evidence there was again no trend but a separation into two sharply divided groups of species. One with clear ape-like abilities that combined climbing and facultative terrestrial bipedalism (to a degree not seen in any extant apes) comprising *Australopithecus*, *Paranthropus* and *H. habilis*, and another with a terrestrial bipedalism indistinguishable from that of modern man comprising *H. erectus*, *H. ergaster*, *H. heidelbergensis* and *H. neanderthalensis*. There was also no evolutionary trend in the traits indicative of development but a dichotomous separation as before.

The next trait they examined was the time to develop to adulthood. This is almost twice as long in humans as in the great apes, and so becomes a discriminator as being more "advanced" and "human-like". There was no trend found but again dichotomy with an ape-like group of *Australopithecus, H. habilis* and *H rudolfensis* and a human-like of *H. neanderthalensis* (*erectus* and *heidelbergensis* were not assessed for this trait).

The fifth trait was masticatory apparatus. The results were the same with *sapiens, erectus/ergaster* and *neanderthalensis* in one group and the rest in the other (Fig 11.4). An obvious question raises itself in the light of the repeated similarities among the *erectus/ergaster, heidelbergensis* and *neanderthalensis* fossils. Are they really separate species, or just subspecies variants of *Homo sapiens*? Chihuahua and Irish wolfhounds are the same species. The answer is critical to the presence of human macroevolution or not. The final trait assessed was relative brain size, considered a surrogate for human brain evolution.

§11.1.1.2 Human Brain Evolution

They found the "four *Homo* species – *H. ergaster, H. habilis, H. heidelbergensis* and *H. rudolfensis* – are more similar to *Australopithecus* than they are to *H. sapiens*, and the relative brain size of *H. erectus* is intermediate between those of *A. africanus* and *H. sapiens*. The only fossil *Homo* species whose relative brain size is more similar to *H. sapiens* than it is to *A. africanus* is *H. neanderthalensis*."[3] Thus there is a trend of increasing brain size with *erectus/ergaster, heidelbergensis* and *rudolfensis* as intermediates (Fig 11.2). Is this finally the evidence of human evolution we have been looking for?

An evolutionary trend seems obvious. Brain sizes ascend as we look down the column on Fig 11.2, and it's even more apparent when the data is represented graphically, as in other analyses by Henry McHenry or Shannen Robson and Bernard Wood (Fig. 11.3).[4,9] What does the evidence look like when we look closer though?

The most significant problem, setting aside the assumption there is any evolutionary relationship at all, is that no phylogeny has been established to even align creatures for trend analysis (§10.3.2). A phylogeny based on chronology is just assumed, just as human evolution is assumed. Why are we looking at brain size data in this way, if not to explain the plausibility of an evolutionary phylogeny, which becomes "proved" in circular reasoning when one is seen?

	Absolute Brain Size (cm³)	Relative Brain Size
A. afarensis	-	-
A. africanus	457	2.66
P. aethiopicus	410	2.39
H. rudolfensis	752	2.76
P. boisei	513	2.40
H. habilis	552	2.72
P. robustus	-	-
H. ergaster	854	2.76
H. erectus	1016	2.87
H. heidelbergensis	1198	2.84
H. neanderthalensis	1512	3.06
H. sapiens	1355	3.08

Fig.11.2. Hominin brain volume. Taxa are listed in the order found in the fossil record. The brain volume of A. afarensis 384 cm³, Pan paniscus 343 cm³, Pan troglodytes 395 cm³ Gorilla 505 cm³).[4] The confounding effect of body size was normalized by the formula (cube root of the absolute brain size ÷ square root of the mean of the orbital area x 10). (Data from Wood & Collard,1999 with permission).[3]

Another observation not allowed to stand in the way of argument is also obvious on Fig.11.3. It's the large variation in brain size within normal *Homo sapiens*. In light of this, one asks again whether the distinctions between the various supposed species in the *Homo* genus and their arrangements, *erectus, heidelbergensis/archaic sapiens, neanderthalensis and sapiens* warrant species separation and might instead all be *sapiens*. If so we speak of microevolution and it's no big deal. We visit this matter latter focusing on *H. erectus*, the species on the graph (labeled E) looking intermediate between clearly human and clearly ape endocranial volumes. The problems with *rudolfensis* taxon (labeled F) will be reviewed later too, but in brief for now, *rudolfensis* was erected for "problematic" fossils in the *habilis* taxon which was erected on an evolutionary whim (§11.3). I assume you know by now what paleontologists mean when they call fossils "problematic".

The next question is how do we know "evolution" caused the shape changes claimed by Argument #3? - I mean apart from using Argument #2 or Argument #1 in circular reasoning. Without a mechanism or even one in theory, we hardly have warrant to conclude macroevolution. Yet "brain evolution" is taken as already incontrovertible from these data, and has been for more than a century. This is how evolution believers become even more certain of the truth of Argument #3. Every observation is more evidence (§10.2;§20). The reason is all observations are attributed to natural selection and mutation without offering details how this was done by them exactly. All the differences are caused by evolution and all the similarities are caused by evolution. On these terms evolution can deliver any kind of change imaginable.

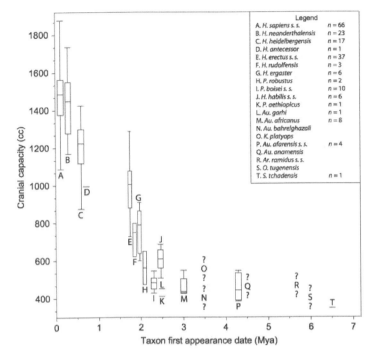

Fig.11.3. **An evolutionary trend in hominin brain size?** Brain size inferred from endocranial capacity plotted against first appearance date for the fossil hominin taxa. Box and whisker plots show median, lower and upper quartiles (box) and maximum and minimum values (whiskers). Note sample sizes in the legend. Brain volume of *Pan paniscus* is 343 cm³, *Pan troglodytes* is 395 cm³ and Gorilla 505 cm³).[4] *H. floriensis* is not shown. (Reproduced from Robson and Wood, 2008 with permission).[9]

One example of this magical thinking, I mean concluding evolution from an interpretation of an observation without a mechanism, is when Richard Dawkins explains brain evolution. This has a clarity that convinces until its simple dogmatism is challenged. He says:

"one of the fastest known evolutionary changes [is] the swelling of the human skull from an Australopithecus-like ancestor, with a brain volume of about 500 cubic centimeters (cc), to the modern *Homo sapiens's* average brain volume of about 1,400 cc. This increase of about 900 cc, nearly a tripling in brain volume, has been accomplished in no more than three million years. By evolutionary standards this is a rapid rate of change…But if we count up the number of generations in three million years (say about four per century), the average rate of evolution is less than a hundredth of a cubic centimeter per generation...a hundredth of a cubic centimeter is a tiny quantity in comparison to the range of brain sizes that we find among modern humans."[10]

This attempt to dilute dramatic change by slicing it up into (intuitively more plausible) change over single generations in a method he calls Climbing Mount Improbable does not change the crushing mathematics of the end result (§17.2). Small steps are taken randomly under the direction of natural selection he says, and yet brain size keeps going so dramatically and so wonderfully in one direction for a stochastic process though under the vagaries of natural selection for millions of years. Dr. Dawkins did not offer a mechanism for this wonder, so let's consider the magnitude of this problem for ourselves.

How much change is "a hundredth of a cubic centimeter" per generation that he calls "a tiny quantity". There are an estimated 10^{10}-10^{12} nerve cells in the human brain.[11] Knowing the complexity of neural connections, how does natural selection cause such a rapid, relentless change to a bigger and better brain? As a crude estimate, because nerve cells in the brain are not spread out uniformly but either layered over its surface or found in collections called "nuclei" deep within, a volume of this size represents about 700,000 nerve cells. The next question is whether during the evolution there was a concomitant increase in nerve cell number with the increase in brain volume or not. We don't know, but it makes no difference.

If we assume human brains have more nerve cells on a volume-for-volume basis than their ape-like small-brained ancestors, Dr. Dawkins would have us believe that on average there was an absolute increase of around 700,000 nerve cells per generation. On the other hand if we assume that the number of nerve cells remained the same through the hominin evolutionary series, we still have to believe that changes in brain size occurred with changes to permit synaptic connections between this number of new cells. It's not just a question of more cells. They have to be integrated into a highly complex previous primate generation's brain circuitry in way that is an improvement on the past to be selected. A typical neuron in the human cerebral cortex has several thousand synapses, and there are 1.5 billion (1.5 x 10^9) synapses per mm^3.[12] How the exquisite complexity of the integrated neocortical neural circuitry can be generated, much less how a near three- fold increase in neurons can be integrated with the precision required by randomly mutating genes and natural selection, beggars the mind. If no mechanism is offered we might as well believe it was done with a wand.

There is another more important illusion at work that is a final problem with the brain size change data. It's the assumption that a bigger brain means a smarter, more human-like brain and that what we're witnesses to in measuring hominin fossil endocranial volumes is really an evolution of neocortex and human intelligence. There is evidence to indicate the enterprise of brain size measurement by that inference, is misguided. The discovery of *H. floriensis*, a toolmaker who tamed fire but had a brain smaller than Lucy (*A. afarensis*) is one example, but consider the following (§10.6.2).

Intelligence is the greatest difference between man and all the other creatures (*sapiens* means wise), and demonstrating the evolution of intelligence goes a long way convincing observers that humans came from animals. Among the several assumptions made to reach that conclusion is the assumption that a bigger relative brain size means more braininess. It seems intuitively obvious. But what's the evidence for this assumption that paleoanthropology takes as a brute given? How do we go about investigating the question?

First we need to level the playing field between small and large animals. The simplest way to normalize for body size is as relative brain size, the ratio of brain mass to body mass, also called the encephalization quotient. What do these numbers show? It turns out, now for another human organ, that size is not important. Although brain to body mass ratios for humans are 1/40, chimps 1/129 and gorilla 1/200 in accord with human evolution theory, sampling other members of the animal kingdom we find small birds 1/12, mouse 1/40, frog 1/172, elephant 1/560 and horse 1/600. Thus relative brain size does not discriminate a birdbrain from a human brain. Perhaps there is a more discriminating measure we could use though.

The part of the brain principally responsible for mammalian intelligence is the neocortex, the mantle of neurons on the surface of the brain toward its anterior pole. This is folded into depressions (sulci) and protrusions (gyri) that increase the surface area to accommodate more nerve cells. Comparisons of primate neocortical surface area show there is a linear relationship between brain size and neocortical surface area and volume. Man has no more neocortical surface area than would be expected in any other primate of that brain size however. Again the measure is not discriminating an ape from a human, the most significant comparison to make in human evolution. Perhaps we need to refine the measure still further

Two ultrastructural measures of intelligence declare themselves. The first is the density of nerve cells in the neocortex, and the second is the number of synapses or connections they have to other nerve cells, something called "neural connectivity". This reasoning as measure of intelligence from brain structure remains entirely intuitive, but is proposed here in a good faith analysis of the evolutionary assumptions. Doing so, we discover that neuron density decreases with increasing brain size, but neural connectivity changes in the opposite direction i.e., there are more synapses when there are fewer nerve cells. Hence brain size as measure alone is still likely irrelevant to intelligence. Women have smaller brains than men but no less intelligence (Fig 11.3). Actually when adjusted for height, weight and intracranial volume, they have more neurons than men but let's keep that secret.[13] What's the point here?

The distinctions made between various *Homo* species by paleontologists trumpeting endocranial volume may not reflect biologically meaningful differences at all. This applies particularly to *Homo erectus* which differs from sapiens in only minor ways but brain size. If *erectus* is human, then there's been no macroevolution just stasis in 1.8 million years in the hominin lineage, and the gap from *Australopithecus* or *habilis* to *erectus* is as saltatory as any creation event you care to imagine (§11.2). That's the point.

In summary, while there are indeed differences in brain size between the most anciently dated and the most recently dated fossil species, an evolutionary continuum between them is rejected for reasons of implausibility, chief of which are the absence of a phylogeny to make that comparison and saltatory change rather than trend between a dichotomous pattern of ape-like and human-like species in the record. As regards the 5 other fossil traits deemed most significant in mapping for human evolution, instead of transitional forms there is stasis in the fossil record with fossil species mapping independently in a pattern across traits to clearly ape-like or clearly human-like species. They do not show intermediate "mosaic" features either, say combinations of various *Australopith* and *Homo* attributes, to prompt consideration of "missing links". Fig. 11.4 summarizes the fossil analysis in *Science* by Drs Wood and Collard and shows how the various species segregate sharply. The claim of Argument #3 that the fossil record shows a progressive gradation from ape-like skeletons into those that are human, is false.

What if we attack the problem of finding human evolution from the other end now, looking backwards for it from *Homo sapiens*? In other words, is there evidence of evolution by inspecting the various morphologies of modern humans? Harvard anthropologist William White Howells (1908-2005) performed 57 different measurements on 3 independent skull series for both sexes, each series consisting of 27-58 skulls taken from the populations of six different geographic areas (Europeans, sub-Saharan Africans, Far Easterners-Japanese & Chinese, Australo-Melanesians, Polynesians and Americans). The evolution he was

Species name	1	2	3	4	5	6
A.habilis	A	A	A	A	A	A
A.rudolfensis	?	?	?	A	A	A
H.ergaster	H	H	H	H	H	A
H.erectus	H	?	H	H	?	I
H.heidelbergensis	H	?	H	H	?	A
H.neanderthalensis	H	H	H	H	H	H

Fig.11.4. Summary of functional analyses of fossil *Homo* species.[3] Traits (top row) are 1) body size, 2) body shape, 3) locomotion, 4) jaws and teeth, 5) development, 6) brain size. (H=humanlike, A=australopithecine like, I=intermediate, ?=no data. (Data from Wood & Collard, 1999 with permission).

searching for was postulated in 1962 by University of Pennsylvania anthropologist Carleton Cool in *The Origin of Races*. Namely that after evolving out of *Homo erectus*, modern man evolved in parallel as five separate subspecies called Caucasoid, Congoid, Capoid, Mongoloid and Australoid. Dr. Howells found cranial variation was "highly limited" between populations however, and none appeared diverged from any other. His conclusions simply stated were a whole bunch of evolutionary no's. If you want to hear it directly though, he says:

"Statistically, modern cranial populations are a unit according to criteria of shape...It is notable that, if anything, intraregional heterogeneity is greatest in Polynesia and the Americas, the two regions we can certify as the latest to be occupied. This goes counter to any expectation that such recency would be expressed in cranial homogeneity. Altogether, the whole forgoing analysis must be recognized as...one giving largely negative signals. Such negatives are as follows: No support for lineages deriving separately from Homo erectus. No support for a special eastern common ancestry for East Asiatics and Australians. No support for a sub-Saharan first source for anatomical moderns, i.e., *Homo sapiens sapiens*..."Analysis also suggested that recent humanity is relatively homogeneous in cranial shape. Within the limits of the evidence no support is apparent for tracing the perceived distinctions any considerable distance into the past (i.e., to a Homo erectus level), and no distinctions of sub-Saharan Africans suggest that they are parental to any modern populations".

Thus both models of human evolution, the multiregional evolution pattern which predicts there will be interregional differences, and the late radiation of a recent sapiens "Out of Africa" which predicts there won't, (but that differences will be greatest between evolutionary older Africans and descendant non-Africans), were rejected (§10.4.1). No evolution going on here either.

§11.1.1.3 Ape-Men At Work? – The Evolution of Stone Tools

What about human evolution being demonstrated by tool artifacts? (§10.1) There's been an incomparable advance from the hand tool artifacts found in the fossil record to the electronic and mechanical tools of today. This might be called an evolution of human technology, but it was not the result of random mutation and natural selection and which is the biological meaning of the word. Yet this is the claim being made for stone tools.[1,14] Specifically it is that observed changes in stone tool shape are surrogate evidence of the evolution of intelligence from ape-men into human beings. Douglas Futuyma tells us what this evidence is:

> "[A.] *Africanus*...is found in association with "pebble tools" that it apparently made by knocking chips off larger stones...*Habilis* is associated with an extensive industry of stone tools, which become increasingly sophisticated...The next stage in human evolution is...now known as *Homo erectus*...is essentially the same as *habilis*, except for its larger brain, and is larger in more recent than in older specimens. It also is associated with more sophisticated tools than *habilis*, including stone hand axes...By the late Pleistocene, 200,000 to 100,000 years ago, humans of almost fully modern form had expanded into Europe...Whether fossils from this period should be considered *Homo erectus* or *Homo sapiens* is entirely arbitrary...By the end of the Pleistocene...the brain had reached its modern level...in Western Europe they had even larger brains. These western populations, known as Neanderthals...had a sophisticated industry of stone tools and doubtless were thoroughly intelligent people"[1]

Evolution by proxy of stool tool shape is an illusion read into the record by anthropologists because they know an evolution of ape-men exists already. Like all illusions this one is very apparent when we look for its details and it begs the question again, why we see it but experts seeing human evolution can't. A little background information is necessary first.

Tool artifacts recovered from a single excavation site are called an assemblage and similar assemblages from the same time or place are called an 'industry'. Related industries are grouped into the same industrial complex or 'tradition'. The best type of rock for stone tools is hard and smooth of uniform consistency and fractures easily like flint, which is common throughout Europe. In Africa where flint is rare, quartzites and volcanic rocks were used. The oldest known stone tools are called the Oldowan Industrial Complex, a term coined from the site where they were first found at Olduvai Gorge by Mary Leakey in the 1930s. They are dated at 2.5 million years old.[15] Her husband Louis ascribed these tools to a new species he called *Zinjanthropus*. However when *Zinjanthropus* was discovered to be just another specimen of *P. boisei,* tool ownership was quickly transferred to a new taxon he erected especially for them called *H. habilis,* and with whom they have remained.[16,17] Oldowan tools are also called choppers. They consist of stones which have had either one (unifacial) or both (bifacial) convergent surfaces flaked off to form a sharp edge. The removed fragments are called flakes and the remaining rock chopper is called the core. When bifacial flaking is extended around the entire periphery of the core, they are called bifaces. Bifaces are of two types: hand axes, in which one pole is sharpened, and cleavers where one pole is straight like a guillotine. The appearance of bifaces, dated around 1.4 million years ago, characterizes the transition from the Oldowan to the Acheulean Industrial Tradition, and named after St. Acheul in northern France where they were first found. Extensive shaping is called a Levallois core named after the Paris suburb of that name, and found to within 400,000 years ago. Bifaces are ascribed to *H. erectus* and Mousterian tools to Neanderthals. Simple tools are equated with a more primitive state and coupled to species chronologically separated in the fossil record you get tool evolution and intelligence evolution by toolmakers. What's the problem?

The first point is the shape changes are hardly overwhelming. Look at the evolutionary "leap" from a chopper to a hand axe.

The second point is that for the all the assumed certainty stone tools imply evolution, exactly how they were even used is not known. Without a knowledge of their purpose, how is it reasonable to infer their progress? There is no evidence for how exactly the 'hand ax" was used for instance. The image we have of them strapped by twine to a handle in the way Fred Flintstone does it, is only comic strip supposition. Alternative uses have been advanced , the most intriguing of which is the idea they were missiles. This notion is supported by replicas landing point-first, like darts, most of the time when thrown.

The third point is what is supposedly simple and primitive is not. This scuttles the basis for inferring evolution. Stanford anthropologist Richard Klein describes the observation and the evidence for Argument #3 for us. "From a typological and technical perspective, Acheulean hand axes and other bifaces are a logical development from preceding Oldowan bifacial choppers".[18] The problem is while the Acheulean might seem of greater "complexity" than the Oldowan to the eye, Oldowan tools were manufactured throughout the Stone Age rather than becoming obsolete. Lawrence Robbins writes about this in *Science*. "In Turkana, pebble tools of the Oldowan type can be found on the surface in many localities, but the age of these tools is open to question. In fact, some of these implements have been found in Late Stone Age, post Pleistocene contexts in association with pottery. It is interesting that these oldest of technological items were among the most successful inventions, for they continued to be manufactured throughout the entire Stone Age."[19]

In fact they are still used. Mary Leakey, daughter-in-law to Mary and Louis, and the wife of Richard Leakey, reported in *Nature* that Oldowan tools are used by rural Kenyans today. Specifically "a crude form of stone chopper is used in the present time by the more remote Turkana tribesmen in order to break open the nuts of the doum palm."[20] And so we discover that Oldowan tools are in the hands of humans. So are Acheulean tools. Here is Richard Klein: "From a strictly archeological perspective, there is little to distinguish early *Homo* sapiens from *H. erectus*...In Africa and Europe, the Acheulean Tradition persisted after the emergence of *Homo sapiens* and is well represented at some of the same sites that contain early *H. sapiens* fossils."[21] No evolution going here either then.

Why aren't Oldowan tools just those of ancient humans rather than *Homo habilis* the ape-man? Is there any support for that? Yes, coming up. Still more bizarre is whether *habilis* even exists (§11.3). The all-important question is who made these tools. If we believe Fig.10.4 there is no single species that spans the time range of 2.5 to 1.5 million years to ascribe the Oldowan tools to. Bernard Wood postulates it is Paranthropus.[22] You can take a guess too.

Another problem is over a period of at least a million years in this record there is more evidence for stasis than evolution. The oldest known Oldowan tools have been recovered at Gona, Ethiopia and dated 2.5-2.6 million years old. Yet they are not significantly different a million years later. This is what the authors wrote about them: "The artifacts show surprisingly sophisticated control of stone fracture mechanics, equivalent to much younger Oldowan assemblages...The very early age of the Gona artifacts shows a technological stasis in the Oldowan industrial complex for over a period of 1 million years. The Acheulean appears abruptly at about 1.6-1.5 Myr with large bifacial tool forms such as handaxes and cleavers."[15]

And as regard the Acheulean, Richard Klein writes: "The total time span of the Acheulean in Africa remains to be established, but it certainly exceeded 1 million years, from 1.5-1.4 to perhaps 200,000 years ago. Although many crucial sites remain poorly dated within this long interval, it nonetheless appears that Acheulean artifacts changed remarkably little through time...it is difficult if not impossible to date an Acheulean assemblage on typological or technical grounds alone...the overall impression is one of remarkable behavioral conservatism through time and space...it seems strange that there was no significant artifactual change when early *H. sapiens* appeared. From the artifactual evidence it could in fact be argued that the transition to *H. sapiens* was very gradual...the European Acheulean still spanned at least 300,000 years; yet, as in Africa, there appears to have been little artifactual change through time or cultural differentiation through space...Overall, the European artifactual data, like those from Africa and southwest Asia, suggest remarkable behavioral conservatism over long periods.[23] Not much evolution going on here either.

§11.1.1.4 Are Ape-Man Mosaics Really Evolutionary? A Look At The Man-Ape From Chad
Although we have failed to find convincing mosaics, macroevolution implies their existence and they remain an important principle for interpreting the hominin fossil record. It was with this expectation that *Sehalanthropus tchadensis* (§10.2.2) was presented and greeted because this species has mixed ape and human fossil characteristics. Is this finally an evolutionary mosaic? Incidentally the entire record of this creature is a skull, two jaw fragments and three teeth.[24,25] The skull is chimp-sized in capacity, with large incisors and wide-spaced eyes like those of a gorilla. On the other hand the lower face is flattened, instead of protruding like extant apes. The canines are small, apically worn, and without a space between canine and incisor. Such traits are considered evolutionarily more "advanced" because these traits are also found in man.

The fossils were dated at 6-7 million years old, three million years older than any previously discovered, and arguably more human-like than Lucy (*A. afarensis*), dated at 3.4 million years. As a result this skull sitting at the roots of the evolutionary tree has come as a big surprise. Sexual dimorphism could be an important factor in the analysis of the traits, but how do the *S. tchadensis* fossils show evidence for evolution exactly, and do they speak to the concept of mosaicism being evidence of evolution from and ape into man? [26]

Whichever model of human evolution one subscribes to, linear or bushy, *S. tchadensis* at the base of the human clade is a problem. Bernard Wood says the skull: "plays havoc with the tidy model of human origins. Quite simply, a hominid of this age should only just be beginning to show signs of being a hominid [revealing his evolutionary preconception bias]. It certainly should not have the face of a hominid less than one-third of its geological age. Also, if it is accepted as a stem hominid, under the tidy model the principle of parsimony dictates that all creatures with more primitive faces (and that is a very long list) would, perforce, have to be excluded from the ancestry of humans."[25] On the other hand he considers the untidy model able to accommodate the fossils well: "it would predict that at 6-7 million years ago we are likely to find evidence of creatures with hitherto unknown combinations of hominid, chimp and even novel features...it will prove telling evidence of the adaptive radiation of fossil ape-like creatures that included the common ancestor of modern humans and the chimpanzee."

With only one skull and three teeth as evidence, his view is more spin than science. At present his explanation only attempts to deflect criticism of the untidy model to preserve the hominid evolution paradigm. The stunning differences between the skulls' composite features in light of its age, and that of the entire rest of the hominid fossil record, remain unaccounted for. The major problem with this specimen of course, and something he does not mention, is it challenges the preconception of Argument #2 that mosaic fossil features of ape and human characters represent evolutionary intermediates between ape ancestor and man. *S. tchadensis* is a mosaic of features found in ape and man, but being among the oldest fossil species in the record, hardly an evolutionary mosaic if it sits at the base of the evolutionary tree!

§11.1.1.5 Evolution Going Backwards – Reversal In The Hominin Series
Evolution represents change in individuals in a population, by the effect of natural selection on new mutations and new gene recombinations, spreading to populations of species, then to higher taxa (§2.2). It's not expected to be reversible to any significant degree therefore. This prediction is called "Dollo's law", named after the Belgian paleontologist Louis Dollo (1857-1931). Dollo's law says: "An organism is unable to return, even partially, to a previous stage already realized in the ranks of its ancestors". This is because the precision required for back mutations occurs far more rarely than the randomly located new mutations which accumulate with every generation (about 6×10^{-9} reversal mutations compared to 0.3 new mutations per chromosome set per generation respectively §2.3.1.4). The likelihood of reversal is even less than this, because we have to account for the odds of the reversed trait now being positively selected by natural selection, when before it was lost by being subjected to the opposite pressure. In summary, trait reversal requires that genes or developmental pathways silenced by mutation and released from the selective pressure on a trait, overcome overwhelming odds against concurrent reversal mutation and selection for the trait.

There is no theoretical reason why this could not happen though. Charles Marshall and Rudolf Raff have shown over timescales of less than 6 million years this is possible, although the actual mechanism by which their examples occurred was not given.[27] The problem with evolutionary reversals is not that they cannot occur, it's that by having to reach statistically very unlikely possibilities, questions of plausibility are raised, particularly if it is with frequency.

It is therefore a challenge to the human evolution paradigm to discover several examples of evolutionary reversal in the hominin series. Some of these trait reversals, for what that's worth, are hard to reconcile intuitively with selective pressure. There is a reversal over time of molar size which is small for *Pan, Ardipithecus* and *Orrorin*, large for Australopiths and then small again for *H. erectus*. There is reversal of skull characteristics. For *A. africanus* the cranium is gracile, high domed and thin, for *H. erectus* it is thick and low domed and for *H. sapiens* it is gracile, thin and high domed again. Another reversal is of brow ridges which are minimal in Australopithecines, prominent in *erectus* and back to minimal again in *sapiens*.

The logical imperative of these changes contributed to Louis Leakey's insistence that *erectus* was not on the lineage to man, and with no other taxon available, his proposal that *H. habilis* was.[28] Despite his designation and description of *habilis*, Leakey's problem remains unresolved because so far there are no fossil species except *erectus* in the time gap between *habilis* and *sapiens*, and erectus remains firmly in the lineage to man. What do the experts say about this problem of reversal in the face of the supposed fact of human evolution? Here is paleoanthropologist Richard Klein at Stanford University:

"While it is widely agreed that *H. erectus* was ancestral to *H. sapiens*, debate continues about whether the former evolved toward *H. sapiens* gradually or in a burst after a long period of morphological stasis. An even more crucial question is how and why the long, low-vaulted, thick-walled cranium of *H. erectus,* with its sharply angled occiput, changed into the higher, shorter, thinner-walled cranium of *H. sapiens*, with its much more rounded occiput. The issue is particularly vexing, because in some aspects of cranial form *H. sapiens* is actually more like *H. habilis*, - and thus the evolution of *H. sapiens* from *H. erectus* would appear to require an evolutionary reversal. On the basis of cranial morphology some have even argued that H. sapiens was derived directly from *H. habilis* and that *H. erectus* was an evolutionary side branch and dead end, but, if this is true, it is puzzling that, so far, the only human fossils in the time gap between *H. habilis* and *H. sapiens* come from *H. erectus*."[14]

The interesting thing is not the observation he makes because the observation is obvious, even though we don't hear it when Argument #3 is being made. The interesting thing is what he does with this observation when evolution is already a fact by Arguments #1 and #2. You know what happens:

> "Perhaps future discoveries will show that *H. erectus* was accompanied by a contemporaneous, more sapiens-like species in Africa or Europe [Darwin's imperfect fossil record argument rejected in §6.2.], but for the moment it seems more likely that the distinctive cranial features of *H. erectus* were mechanically determined by its brain size (or form) and by the way it used its jaws and that it evolved into *H. sapiens* when the mechanical constraints were altered or relaxed by changes in behavior and brain size."[14]

The observation is the evidence. Not much evolution going on here either then.

§11.1.2   Sudden Appearance Instead Of Gradualism

Rather than progressive change as claimed by Argument 3#, the most striking feature of the hominin fossil record is sudden appearance specifically as gap between clearly ape-like and clearly human-like fossil species (Fig 10.3;Fig 11.4). What about evolutionary trends within individual species? *Homo erectus* is the species with the longest duration in the record but it demonstrates stasis for 1.8 million years (§11.2). As regards change within other hominin species, Henry McHenry wrote a review assessing the tempo and mode of human evolution and concluded: "It is interesting, however, how little change occurs within most hominid species through time"[4]. The pattern in the hominin fossil record is stasis rather than change, just as was found for other fossil creatures (§5.4.3-4).

§11.1.3   Simultaneous Existence Rather Than Anagenesis

Also evident is the pattern of contemporaneous existence of species and not just of *Homo erectus/ergaster*, *Homo habilis* and *Paranthropus*, but *erectus*, *sapiens*, *neanderthalensis* and *heidelbergensis* (Fig. 10.3;Fig 10.5). Argument #3 for human evolution holds that the fossil record shows successive ancestor-descendant species over time. Rather than tediously inspecting each hominid species for its evolutionary credentials, the two species in the series where we should see it most, at the interface between the apelike and humanlike characteristics, will be investigated as evolutionary transitional species. They are *Homo erectus* and *Homo habilis*.

§11.2   INSPECTING ARGUMENT #3:
          HOMO ERECTUS IS NOT A TRANSITIONAL SPECIES

A detailed study of this taxon alone provides evidence to refute human evolution. E*rectus* is a taxon critical to Argument #3 because of its long duration in the fossil record, it is the first indisputably human-like species, and because it sits at the interface of the clearly ape-like to human-like transition of the record (Fig.10.3;Fig.10.5). *Erectus* is even more important than these diagrams indicate because most experts say *H. ergaster* is not a separate species from *erectus*. Bernard Wood disagrees as Fig. 10.3 shows, but he concedes "the majority of researchers do not regard the *H. ergaster* hypodigm as worthy of a separate species".[29] Separating them as he does into an ancestral *ergaster* and descendant *erectus* is to have found human evolution already, needless to say. In another demonstration of the power of personality in paleoanthropology his minority view which is much more appetizing if you like the taste of evolution, was nevertheless presented to the general science readership of *Nature* as "[t]he known fossil record".[25]

Wood's view is that *ergaster* fossils represent an earlier African form in the hominin series, and *erectus* is the later evolutionary Asian lineage (clade). This is because he considers various skull features (frontal keeling, an angular torus on the parietal bone, thicker vault and other characters of the temporal bone) as sufficiently different from the African specimens to be Asian autapomorphic features. He acknowledges he holds a "splitting" and so necessarily evolutionary trend favoring, rather than a "lumping" taxonomic interpretation.[3] We assess the validity of a splitting approach for the other *Homo* members later but as regards *ergaster*, Berhane Asfaw and Tim White at Berkeley reviewed and refuted this idea in the same journal back in 2002 with their description of the Ethiopian "Daka" *Homo erectus* calvarium (BOU-VP-2/66). They found:

> "the cladistic method…fails to support the division of *H. erectus* into Asian and African clades. Whether viewed metrically or morphologically, the Daka cranium confirms previous suggestions that geographic subdivision of early *H. erectus* into separate species lineages is biologically misleading, artificially inflating early Pleistocene species diversity.[30]

Philip Rightmire at the State University of New York found the same thing, more than a decade before Asfaw and White.

> "Differences in cranial morphology are of the same order and should be expected within a species that is geographically widespread…One has to check quite carefully for signs of consistent regional change, and the

evidence for overall similarity of African and Asian populations is much more striking."[31]

Making comparisons of facial anatomy between African and Asian *H. erectus* skull specimens, he concluded:

"Neither detailed anatomical comparisons nor measurements bring to light any consistent patterns in facial morphology which set the African hominids apart from Asian *erectus*."[32]

Thus the *ergaster* taxon and the box representing it on Fig. 10.3 should, on this evidence, be sunk into *erectus*. This removes what would be an evolutionary intermediate, surely the only reason why *ergaster* still lives.

Either way, along with the *H. ergaster*-ascribed specimens with which it has subtle differences at best, *H. erectus* represents the first fossil hominid all agree is human-like. It is the link backwards to those that all agree are ape-like. The importance of *erectus* to theories of human evolution cannot be overstated then, and the consensus in the science is that *Homo erectus* is a transitional species on the direct lineage to modern man (Fig.10.3).[30,33-35] Richard Klein spells all this out in his book about human origins called *The Human Career*:

"About 1.8-1.7 million years ago (mya), *Homo habilis* (or one of the species into which it may ultimately be split) evolved into the primitive human species, *H. erectus*...Nearly all specialists agree that *Homo erectus* was directly ancestral to the modern human species, *H. sapiens* (broadly understood), which appeared in Europe and Africa 500,000-400,000 years ago."[34]

The question for us then is, is this true?

The features of *erectus* suggest the absence rather than the presence of evolutionary change and no matter whether phylogenies go through this taxon to *sapiens* or they don't. There are four main reasons why. *Erectus* fossils show: (1) stasis rather than evolutionary change; (2) sudden human-like appearance from other earlier ape-like forms rather than evolutionary transition; (3) contemporaneous existence with both supposed ancestor and descendant species; and because of (4) behavior not significantly different from *Homo sapiens*. See if you agree.

§11.2.1   *Erectus* demonstrates stasis, not evolutionary change

As the species with the longest duration in the hominid fossil record *erectus* would seem the best indicator of hominin evolutionary change. Instead it shows remarkable morphologic stability for at least 1.8 million years. Claims for an evolutionary trend in increasing endocranial volume have been made by Milford Wolpoff amongst others however. This required lumping the crania constituting this species into "early" and "late" groups by their fossil dating, then comparing the means of the two assigned groups. By Dr. Wolpoff's read there was a progressive increase in brain size to the extent of a 30% rise in the mean of the Late Pleistocene "late" group compared with the Early Pleistocene "early" assemblage.[36] However the exclusion of just three of his specimens from the analysis, and for good reason, causes the trend to evaporate back into stasis.[37]

Philip Rightmire reviewed the matter in his book *The Evolution of Homo Erectus*. His linear regression analysis of the 23 skull specimens shows no trend when the so-called Ngadong specimens are excluded because of concerns their recent age dating is spurious (§11.2.3). When the Ngadong specimens remain, there is a trend of increasing volume at a rate of 181 ml per million years. However the 95% confidence limits are -181.5 ± 113.9 ml and the range of the *erectus* skull volumes is 524 cm^3 by comparison! His conclusion that there is no evolution was still very hard for him admit. Instead he said: "later members of this species show an increase in cranial capacity in comparison to earlier ones, although the trend is not significant...Neither least-squares nor other approaches to the data...support unequivocally a claim that there is a continuous expansion of the vault within *Homo erectus*."[38] As regards trend analyses of other skull measurements and teeth looking for evolution, they can be given curt conclusion in Rightmire's own words: "Other characters change slowly or not at all."[37]

Modest microevolutionary change by *erectus* (or any other species for that matter) is not at issue. What is at issue is macroevolution meaning new paleospecies that are the claims of Argument #3. Even the differences between *erectus* and *sapiens* are not the main issue. The main issue is the change into *erectus* from ape-like species to man-like, which is the core claim of human evolution. The stasis of *erectus* fossils is thus the same as for other creatures (§5.5.3), and while it does not disprove evolution since it could have happened as punctuated equilibrium, it does provide objective evidence that at least evolution as is claimed by Argument #3 has not. Believing evolution by punctuated equilibrium is to hold to a radical saltational change from an ape-like Australopithecine into this species, and to hold to a theory for which there is no direct evidence nor can there ever be, except the absence of evidence - and which is stasis! (§6.3) Over a much shorter time span than the stasis of *erectus*, an ape is claimed to have evolved into a human and from that same ancestor, one daughter species evolved so well as to land on the Moon while the other became a hooting jungle beast. Why should we believe this again?

§11.2.2 *H. erectus* is a sudden appearance from earlier hominin forms

The argument has been made already. There is a saltationary change to this species by almost every measure with any fossil taxon before it in the record (§11.1). To use two of the most complete skeletons to demonstrate the point, there is a gap from the *habilis* morphology of fossil Olduvai Hominid 62, dated at 1.8-2 million years old standing 3 feet tall with arms almost as long as its legs, to the *erectus* morphology of the Turkana boy (WT 15000), dated at 1.6 million years and standing 6 feet tall as adult. Behane Asfaw and Tim White make this admission:

> "The origins of the widespread, polymorphic, Early Pleistocene *H. erectus* lineage remain elusive. The marked contrasts between any potential ancestor (*Homo habilis* or other) and the earliest known *H. erectus* might signal an abrupt evolutionary emergence some time before its first known appearance in Africa at ~1.78 Myr. Uncertainties surrounding the taxon's appearance in Eurasia and southeast Asia make it impossible to establish accurately the time or place of origin for *H. erectus*."[30]

Ian Tattersall has alluded to this problem too, but he provides a post hoc untestable, adaptationist explanation to account for this massive change. The observation is the evidence of his claim:

> "...accommodation to the environment is something which the history of the differentiation of *Homo sapiens* (with populations as distinctive as Dinkas and Eskimos) shows need not take too long: certainly well under 100 kyr. It's less obvious, however, that a radical shift in body design such as that exemplified between Olduvai Hominid 62 (1.8 myr old) and the Turkana Boy (1.6 myr old) is something that one might expect to happen (in evolutionary terms) overnight. But unless the known fossil record is substantially biased, it's nonetheless apparent that the acquisition of essentially modern human body build and proportions was a relatively rapid process. The acceleration of this process was probably due to strong competition between differentiating hominid populations (species?) as they strove for survival in increasingly severe environmental conditions."[39]

If you can believe that amount of transition with that amount evidence you will have absolutely no trouble believing any of the other evidence of human evolution. Whether that kind of faith is reasonable or not we may disagree, but at least we can agree that at this crucial step in the fossil record when an ape-like creature became man-like, there's no support for evolutionary transitional forms. Even invoking punctuated equilibrium to escape this impasse demands it explain a change so large it is a saltation requiring all of the miraculous prowess of Goldschmidt's "hopeful monsters" (§2.2.1).

In summary the evidence rejects the claim of Argument #3 that the hominin series shows that "[one] 'species' blends into another as you go from younger to older fossils".[1] The bones demonstrate a cosmic disconnect between what Darwin, Huxley and evolution theory predicted they would show , what scientists say they show, and what they actually show. How and why this disconnect can continue to persist, in science, is the question I leave with you to ponder for now.

§11.2.3   *H. erectus* is contemporaneous with ancestral and descendant species

Richard Klein and Bernard Wood consider *erectus* to extend from around 1.8 million to 500,000 years[25,40], and elsewhere (e.g. Fig.10.3) like others up to 200,000 years ago (Fig.10.5).[3] Ian Tattersall and Henry McHenry date *erectus* from 1.8 million to 300,000 years,[4,41] and Richard Cowen from 1.7 million to 300,000 years ago.[33] These dates all fit with the notion of an evolutionary intermediate status between the ape-like state of Australopiths and the group afterward which are all human-like (§11.1;Fig.11.4). The trouble is these date ranges do not square with the data. We begin by looking at the recent end of the range and then the oldest end.

In 1996 the journal *Science* published evidence demonstrating *Homo erectus* was living in Java as recently as 27,000 - 53,000 years ago. In other words some 200,000 years after it had been presumed extinct.[42,43] If this is true, all of the above listed ranges need revision and further fancy footwork will be needed to keep *erectus* in the evolutionary tree. Lead author Carl Swisher of the Berkeley Geochronology Center used two independent dating techniques called electron spin resonance and mass spectrometric uranium-thorium dating. He studied the large *H. erectus* fossil assemblage at Ngadong on the Solo River in Java that includes 12 crania. The specimens were discovered between 1931 and 1933, just forty years after the *H. erectus* type specimen called "Java man" was recovered by Eugene Dubois further downstream on the Solo River at Trinil (§10.6). Direct dating of the Ngadong fossil material was not permitted by the keeper of the fossils, nor was there volcanic material for indirect analysis (§5.3.1).[43] Age determination was therefore done by showing that two independent dating methods gave similar results on bovid teeth collected with the hominid fossils originally, on water buffalo teeth recovered from the fossil bearing layer of the site, and on bovid teeth from Sambungmacan, a nearby site of the *H. erectus* fossils upstream. Fluorine analysis of the bovid teeth and an *erectus* leg bone (tibia) from Sambungmacan was taken as confirming the fossils of similar age.

Their results showing a modern times existence for *erectus* play havoc with the prevailing notion of human evolution. They show *erectus* and *sapiens* were living contemporaneously in Southeast Asia, in the same way that Neanderthals and *sapiens* did in Europe. Needless to say these dates are disputed,[44] and clearly Bernard Wood does not accept them either given the truncated upper limit of the erectus taxon from what the Ngadong fossils warrant. Incidentally the dates on his 2002 figure also represent a departure from his view three years before when he reported the range for erectus to extend to 200,000 years ago.[3] This kind of blurring of taxon ranges makes evolution appear and disappear.

To return to the Ngadong fossils, their dating will likely remain controversial until direct analysis of the bones can be performed. Is there any other evidence of *erectus* fossils with this recent age to substantiate that unexpected result however, at least unexpected if you believe in human evolution? Well, yes.

It's not new data either. In 1972 Alan Thorne at the Australian National University in Canberra described an assemblage of at least 40 *erectus* individuals at Kow Swamp, Australia almost two and a half decades earlier. He wrote in *Nature* that the assemblage "reveals the survival of *Homo erectus* features in Australia as recently as 10,000 years ago".[45] The repercussions of this surprising result were likely contained by his speculation "the Kow Swamp series represents an isolated and remnant population," presumably meaning they were a Jurassic Park-like enclave of ancestors, alienated from the evolutionary mainstream that managed to survive out to modern times.

The subsequent discovery of the so-called Cossack skull on the opposite (western) side of the Australian continent in 1979 dated at between 2,000-6,500 years old,[46] and the Solo fossils in Java dated by Swisher above all with *erectus* features, reject this his explanation.[47,48] *Erectus* existed over a whole continent, and to comparatively modern times. It turns out that there are plenty more fossils of similar robust *erectus*-like morphology, and similar recent age. Specimens have been found both in Southeastern Australia (Lake Mungo dated at 30,000 years, Mossgiel, Talgai, Broadbeach, Willandra Lakes, Coobool Crossing) and Western Australia (Devils Lair at 25,000 years). In summary the evidence shows *Homo erectus* lived alongside *Homo sapiens*.

The interesting thing is that the recent dates Carl Swisher obtained for the Javan Solo fossils have not impacted the accepted survival times for *H. erectus* by the major authorities in the field. If these fossils are not explained as being what they at least appear to be, evidence of *Homo erectus* living contemporaneously with modern man, how has this conclusion been rejected? They are taken to be anatomically modern humans and the reason is they are too recent to be *erectus*. Three observations can be made.

First, the evolution paradigm is seen to be driving fossil interpretation not the fossil data. Just as it always did and just as always has. Second, evolution science has no problem accepting differences between *erectus* fossils and *sapiens* as variability within a single species called *Homo sapiens* when human evolution is not jeopardized. Third, macroevolution is not considered necessary by the experts to explain the morphologic differences between erectus and sapiens.

How are the morphologic differences between the contemporaneous *erectus*-like fossils and *sapiens*-like fossils dispensed with? Richard Klein presents two alternatives. Either "they inherited their rugged skulls more or less directly from Southeast Asian *Homo erectus*," so conflating the obvious, or else "it is at least as likely that...their "archaic" cranial features are due wholly or partly to artificial cranial deformation," namely head binding [49]. Wrapping the head up in bandages to intentionally deform it is what he means. It's an extraordinary idea but invoked here at this time and only on these fossils, it at least appears contrived.[48] It also flies in the face of a specific rejection of head binding by the authors of the Cossack skull description who determined the parietal index, a discriminatory measure of the procedure.

Klein wrestles even more with the problem these fossils pose for evolution, considering it "a puzzle that no available theory of modern human origins parsimoniously accommodates,"[49] begging the question why he believes it at all. He can sees the differences at least. "Curiously, compared with the historic Australian aborigines, the Lake Mungo people, together with broadly contemporaneous or somewhat younger ones from Keilor (Victoria) and Lake Tandou (New South Wales), possessed relatively high vaulted, thin walled, smooth-browed, spherical skulls with relatively flat faces [like sapiens], while the people who lived at Kow Swamp (northern Victoria) and Talgai (Queensland) had exceptionally rugged skulls with relatively low vaults, thick walls, flat and receding foreheads, strong browridges, and projecting faces [like erectus]. The range of variation is extraordinary."[49] The chronological demands on the fossils made by the paradigm of evolution theory are more compelling than the logic of the observations he makes. The fossils with *erectus* morphology are summarily stamped anatomically modern human, and dumped into that taxon instead!

These Australian and Javan fossils as well as others from China are used to support the multiregional model of hominid evolution (§10.3.1). The principle proponent of the model is Milford Wolpoff at the University of Michigan.[50] He posits *erectus* dispersed widely out of Africa initially, and subsequently evolved into *Homo sapiens* concurrently in different regions of the Old World through morphologically distinct populations as a result of genetic drift and natural selection. In this theory it is posited that there was sufficient gene flow between them to preserve the species from evolving into another one. The counter arguments to this model are that the multiregional model first "ultimately requires the discovery of unequivocal transitional fossils between archaic and modern forms of Homo; but so far these are as conspicuously lacking in the Far East as they are in Europe",[51] and second how the model can accommodate the variability seen between erectus-like and sapiens-like skulls in these regions. Further, as Klein points out, "[t]here is also the problem that some of the features that some early modern Australians share with archaic Javan *Homo* are also shared with archaic *Homo* elsewhere, including Africa. They therefore do not demonstrate a unique local lineage."[51] I take Dr Klein to mean what the fossil morphology declares (and the corollary he will not make): the Australian and Javan fossils look like African *erectus* or *ergaster*. Ergo *erectus* and *sapiens* were living together, on the Australasian continent, only a few tens of thousands of years ago.

While *erectus(/ergaster)* was contemporaneous with *sapiens* at the front end, at the back end it was contemporaneous with *habilis* over some (Fig. 10.4), or more (Fig.10.7), or all (Fig. 11.7) of the fossil material assigned to that taxon. The oldest *erectus* fossils are the Damiao mandible and teeth (2 Myr), the KNM-ER 2598 skull fragment (1.85-1.95 Myr), the KNMER-3228 hip bone (1.9-2.0 Myr) and the Java/Djetis teeth and cranium (1.9 Myr) and the Modjokerto child (1.8-1.9 Myr).

The evolutionary origins of *erectus* have been summarized by Philip Rightmire as follows: "How such [ancestral *Homo habilis* fossil] species may be related to one another or to *Homo erectus* must be decided on anatomical grounds...At the moment, there is no consensus on this issue. A complicating factor is that some early *Homo* specimens overlap in time with *Homo erectus*".[37]

Thus we see at first hand, (1) how evolution is read into the fossil record ("must be decided") not the other way around; (2) how there is "no consensus" about *erectus* evolution though it is a fact in the science nonetheless, and (3) that he yet recognizes the problem of an "overlap" of supposed ancestral and descendant species. In summary the *erectus* taxon is not evolutionary but fossil discoveries continue to be applied in back to front reasoning because of the assumption that human evolution is already certain.

§11.2.4   Could E*rectus* Be Just An Ancient Human?

According to evolution theory *erectus* is more ape-like and a separate transitional species in the lineage to *sapiens*, but what's the evidence *erectus* is different from a human? Could the differences be mere microevolutionary variations within the *sapiens* species in the way *ergaster* is considered a separate species from *erectus*? If so, this alone would cripple the concept of human evolution because it creates an insurmountable separation of the two groups of hominid fossil species, establishing one as all human and the others as all ape-like going back at least 1.8 million years in the fossil record (§11.1;Fig 11.4). Yet precisely such a case can be made, and for three separate reasons.

First, *erectus* is not morphologically different from *sapiens* to warrant a separate species classification. Second, there is human behavior by *erectus* in the fossil record, and third *erectus* is lived contemporaneously with humans (§11.2.3). A handful of evolutionist paleontologists have also said *erectus* should be sunk within *sapiens*. The most prominent of them is Milford Wolpoff, but Phillip Tobias, Alan Thorne and Franz Weidenreich agree. Let's look for ourselves. What exactly are the differences between *erectus* and *sapiens*?

The first thing we need to know is what the morphology of *Homo sapiens* is. This would seem self-evident but it is not as we know (§10.3.2). There is no "type" specimen and yet evolution from an ape-man into *sapiens* is said to be a fact. How can the science tell? Type specimens serve as a standard reference for determining what constitutes the definition of the species, and as a measure to assess the eligibility of other specimens for inclusion into the taxon. Bernard Wood looks at this shifting goal post, sees it gyrate and even describes the contradiction, but then in another triumph of philosophy over fossils by paleontology concludes it just "a paradox", as if that settles it. You have to hear it to believe it:

> "It is a paradox that of all the hominid species the one that is the least well-defined is that to which the reader and the writer belong. Whereas all other hominid species have a 'type specimen' to which the species name is irrevocably attached, there is no designated type specimen of *H. sapiens*. In addition, because it is 'polytypic', that is it incorporates a

relatively large range of variation, a spectrum of skull shapes and limb proportions have to be taken into account when considering whether fossils can be included within the species category *Homo sapiens*."[52]

The only reason this absurdity is not ridiculous to him also, is because human evolution is a fact (§3.1;§19). *Homo sapiens* has a high morphologic variability. Adult human brain size varies by over 50% even among the normal non-elderly (975cm^3-1499cm^3).[53] Criteria used to define other species in the hominid fossil record at the expense of *sapiens* are too restrictive is the point, and specifically the *sapiens* taxon is surely inclusive of the supposed (macroevolutionary) separate *Homo* (chrono)species. Wood and Collard are deliberate in proposing an overspeciose taxonomy for theoretical reasons, but it sustains human evolution for those who already have evolution in mind, and so it stays.[3]

With silence on the *sapiens* end of the human evolutionary tree as regards differences, what are the fossil features of *erectus* to look at the other end? Philip Rightmire has summarized them for us. Compared with *sapiens*, the cranium is long, relatively low and sharply angled at the occipital pole. The basicranial axis is flattened compared with the more flexed base of *sapiens*. There are heavy brows (supraorbital tori) and transverse occipital torus, with an average endocranial capacity of 1000 cm^3 (about 80% of the average for modern man), an average mass of 48 kg, a robust facial skeleton with subnasal prognathism (although few specimens), the frontal squama flattened and may have a midline keel, a short parietal chord compared to later *Homo*, sharp flexion of rear of the vault, crests associated with the mastoid process prominent, glenoid cavity narrowed, sphenoid spine not developed, the inferior tympanic plate thick, the mandibular articular tubercle is not prominent, mandible robust and thick with broad ascending ramus and minimal indication of chin. The thickened brow with flattened supratoral shelf, frontal keeling and parietal angular torus are considered derived features. Every other bone apart from the skull is the same. Indeed the height of WT 15000 is estimated to have been 6 feet as adul, and "by earliest *H. erectus* times body size was essentially modern".

Here is Franz Weidenreich back in 1943 speaking about Peking man *H. erectus*: "It would be best to call it *Homo sapiens erectus pekinensis*. Otherwise it would appear as a proper 'species', different from '*Homo sapiens*' which remains doubtful to say the least."[54] Here is Douglas Futuyma: "The "Java man" [*Homo erectus*], discovered by Dubois in the 1890's...was essentially a modern human in every aspect except its brain size, which ranged from 750 to 900 cm^3"[1] And here is Donald Johanson discoverer of "Lucy", deep in evolutionary thought: "It would be interesting to know if a modern man and a million-year-old *Homo erectus* woman could together produce a fertile child." Recalling Mayr's definition of species earlier this would solve the question definitively. You'll never guess his answer then. "The strong hunch is that they could; such evolution as has taken place is probably not of the kind that would prevent successful mating".55

How does he reconcile the anomaly for human evolution though? "But that does not flaw the validity of the species definition...because the two cannot mate. They are reproductively isolated by time." This is the logic that makes your great-great-great-great grandmother a different species from you. It's logical only if macroevolution is logical. Now what is it that makes it logical again?

Not only is *erectus* morphologically within the range of modern human, it displays human behavior. The first is in tool use. Summarizing the evidence Richard Klein tells us:

"From a strictly archeological perspective, there is little to distinguish early *Homo sapiens* from *H. erectus*...In Africa and Europe, the Acheulean Tradition persisted after the emergence of *Homo sapiens* and is well represented at some of the same sites that contain early *H. sapiens* fossils."21

There is no evidence of geographic (evolutionary) difference between *Homo erectus* in those tool technologies either. To quote Huang Weiwen at the Institute of Vertebrate Paleontology in Beijing, the leader of the excavation which found stone tools in South China akin to Acheulean technology in Africa,56 "There is no essential difference in the biology and culture between the early human groups of the East and West".57

We also read in *Nature* that "sometime between 800,000 and 900,000 years ago *H. erectus*...had acquired the capacity to make water crossings" from mainland Southeast Asia to the island of Flores, east of Java. The authors inform us "*Homo erectus* in this region was capable of repeated water crossings using water craft".58

Elsewhere in *Nature* is the report of wooden throwing spears dating at 400,000 years ago from Schoningen, Germany. Specifically "[f]ound in association with stone tools and the butchered remains of more than ten horses, the spears strongly suggest that systematic hunting, involving foresight, planning and the use of appropriate technology, was part of the behavioral repertoire of pre-modern hominids. The use of sophisticated spears as early as the Middle Pleistocene may mean that many current theories on early human evolution and culture must be revised."59

Erectus would therefore seem by every measure to be an ancient *Homo sapiens*. Not much human evolution going on here either. Argument #3 is rejected again. We turn next to look at the credentials of *Homo habilis*, more recently called *Australopithecus habilis*.

§11.3 INSPECTING ARGUMENT #3: *HABILIS* IS NOT A TRANSITIONAL SPECIES

Homo habilis is a fossil species of great significance to human evolution for at least three reasons. First it's in the direct lineage to man in most phylogenetic analyses (Fig.10.3). Second it occupies the fossil record at a time of marked transition from fossils clearly ape-like to clearly man-like and specifically it is on the ape-like side of a divide that has *erectus* on the other (Fig.10.4 & Fig.11.4). Third, *habilis* has been distinguished as the first member of the human genus to appear in the fossil record. It is an ideal species to inspect for human evolution therefore.

§11.3.1 A Birth By Handy Intuition

The history of the rise and fall of *Homo habilis* and the way fossil specimens have been shuffled in and out of this hypodigm (the term referring to the fossil record of a particular taxon) by reason of personal opinion not objective protocol, is an insight not only into the fossil evidence of human evolution but how its minders manage it. A look at *habilis* is to see the power of an evolutionary mindset operating by Argument #1 with a subjectivity uncomfortably high for a science. The shambles that is this taxon hinted at earlier rejects the claim of Argument #3 that *habilis* is a transitional species, and the reason is because *habilis* is an artificial construct that at best contains two different species, and at worst none at all. Ridiculous you say? Take a look at the evidence.

First, where did *Homo habilis* come from? It was a creation in 1964 of the collective will of three of the most important men in anthropology at that time, Louis Leakey, Phillip Tobias and John Napier.[16] A force of additional wills turned out to be needed though, because its announcement in the pages of *Nature* was still met with widespread skepticism. This criticism was, and still is, founded on three principal concerns.

The first is the fossils presented as the evidence for the new species were variable, fragmentary and juvenile. There was no skull, no face, nor even a complete brain case for instance, and the amount of juvenile material was substantial. Even the type specimen was juvenile. Second, the material presented was not significantly different from either *A. africanus* or *H. erectus* to justify a new species designation. It was arguably instead a potpourri of their miscellaneous remains, with *A. africanus* representing the older dated specimens and *H. erectus* the recent.[60,61] Third, experts objected there was insufficient "morphogenic space" to squeeze another evolutionary taxon into the accepted (though unproven) paradigm of a linear anagenic evolution from *A. africanus* to *H. erectus* in the series to man. The latter objection was of course a purely philosophical one, although that did not diminish its force, and is another example of how evolution theory drives the fossil data not the other way around (§10.2). Though powerful at the time, the last objection has since evaporated because human evolution theory now has been revised to accommodate the seeming contradiction.

Even more controversially, Leakey and his co-authors proposed that *habilis* was the first member of the *Homo* genus. But with a mean cranial capacity estimated 673-680 cm^3, it fell under the accepted definition for *Homo* set by Oxford's Wilfred Le Gros Clark of 700 cm^3.[16,62] This criterion had stood as a nomenclatural standard for the genus since 1955. How was this obstruction overcome? By changing the definition of course. What formed the basis for this scientific reasoning? Opinion. Opinion about stone tools, opinion there was an undiscovered ancient evolutionary species of *Homo* in the African sand, and opinion *H. erectus* was not on the direct lineage to man. Let me explain.

The original *habilis*-allocated fossils were found at Olduvai Gorge. They were dated at 1.75 million years and found in association with Oldowan stone tools (§11.1.1.3). Tool making is a human characteristic, it was argued, thus warranting a genus designation to *Homo* despite its ape-sized brain. Indeed the name *habilis* means "handy man". But how valid was this reasoning really? That the authors could neither prove tool ownership by these fossils nor refute ownership by other contemporaneous species in the fossil record did not seem to matter.

As far as Leakey was concerned, *Australopithecus* and *Paranthropus* could not make tools, *Homo erectus* was not ancestral to man and so could not perform this human activity, and the fossils could not be australopithecine because of a larger brain size and different proportions of anterior teeth and premolars. But this was not enough evidence for creating a new taxon, either back then or today. In an added inference justifying an allocation to *Homo*, Tobias argued *habilis* had another quintessentially human ability: speech. What was the evidence?

An impression left by the "language" area on the inside wall of the *habilis* skull. An amazingly fanciful idea realizing concerns raised in

§10.2.2, and although once held in all seriousness, is now rejected even within paleoanthropology.

§11.3.2 Paleontology By Personality

What is less known is Dr. Leakey four years earlier had allocated these same tools with just as much confidence to another species called *Zinjanthropus* (East Africa Man) in a report that was also published in *Nature*.[63] He trumpeted *Zinjanthropus* as the world's earliest "true man,"[63] who was "quite obviously human".[64] Subsequent investigation showed it just another specimen of *P. boisei* and so quite obviously not. This matter, like his earlier false claims about "Oldaway man" being "almost beyond question...the oldest known authentic skeleton of *Homo sapiens*", and *Homo kanamensis* being "the direct ancestor of Homo sapiens",[65] were forgotten as if they had never happened. The importance of personality in this field cannot be underestimated as you can see (§10.2). Ian Tattersall says
> "it is possible in retrospect to see that, largely through the efforts of the Leakeys, the human fossil record had begun, by the mid-1960s, to take on the outline that is familiar today."[66]

Is this unfair and personal criticism of a household name in anthropology and to most, one of the fathers of hominid paleontology? I submit not but rather another indication of how individual genius rather than data directs this rudderless science.

Perhaps in some measure the facility with which these grand claims were made could have had to do with finding benefactors. Leakey used the publicity over *Zinjanthropus* to obtain financial support for his research, and his long relationship with the National Geographic Society began at this time.[67] However because equally extravagant evolutionary claims continue to be made by investigators with disputable data today, the root cause certainly lies elsewhere. It is testimony rather to the fundamental problem inherent to paleoanthropology discussed earlier (§10.2). In other branches of science the ability to test hypotheses restrains investigator enthusiasm and bias, but in this historically based discipline about a singular historical event driven in the end only by the falsehood of Argument #1 (*LOE#*3), there are no objective standards. It is inevitable the power of personality will loom large.

In this regard, Roger Lewin cites Duke University primatologist Elwyn Simons' assessment of Leakey's *H. kanamensis* embarrassment as: "a fine example of the Louis Leakey syndrome, which is a subsidiary, but highly prominent, component of Leakey's pursuit of ancient *Homo*. Leakey's modus, says Simons, was that "The fossils I find are the important ones and are on the direct line to man, preferably bearing names I have coined, whereas the fossils you find are of lesser importance and are all on side branches of the tree."[65]

With *Zinjanthropus* disqualified, another owner for the stone tools needed to be found though. Leakey did not consider *H. erectus* or *Australopithecus* on the lineage to man and therefore as far as he was concerned, there had to be an ancient *Homo* still out there. It just needed to be found. Roger Lewin says "the great antiquity of man – so dominated Leakey's view of the past that it would lead him repeatedly to see in fossils what he wanted to see."[65] How could bias be avoided, knowing what we know (§10.2)? Even Ian Tattersall grudgingly concedes the obvious:

"In reading their description of *Homo habilis*[16] it is hard to avoid the conclusion that Leakey and his colleagues were swayed principally by the concept of "man the toolmaker," an ancient notion which had been gaining ground since Kenneth Oakley had published the first edition of his booklet of that name in 1949."[68]

Welcome now to the halls of the science of paleoanthropology. Turn your eyes now, to the exhibit standing right in front of you.

§11.3.3   A Bone To Pick

What was the original fossil evidence used to show *habilis* had biological reality? OH 7 was the type specimen. It consists of fragments of both parietal bones (parts of the skull cap), a partial mandible with teeth, and at least thirteen hand bones of a juvenile. Six other specimens were also presented: OH 13 (an incomplete adolescent skull), OH 14 (juvenile skull fragments), OH 8 (a partial adult foot), and OH 4 and OH 6 (skull fragments and teeth). A lot of fragmentary and juvenile material is the point. Ian Tattersall and Jeffrey Schwartz are more blunt: "it has to be admitted that the new hominid itself consisted of no more than a couple of handfuls of fragments that could be interpreted in several different ways."[69]

How this material could all belong to the same taxon without a more complete single skeleton was left unanswered back then and remains unanswered. While Tobias continues to offer a dissenting view,[70] it is fair to ask what is definite and objective about hominid fossil classification if such arbitrary proposals and ad hoc adjustments can be made to consensus definition criteria, and a whole new taxon be formulated using data of this quality and quantity. It does nothing to refute the impression *habilis* was erected by a manifestly subjective and evolutionary judgment on an intuition by a few dominant individuals in the field. After whom the rest have followed by making further fossil allocations in a case of a tale wagging dog. Bernard Wood has summarized this history by saying:

> "Few, if any, of the initial interpretations of *H. habilis* have survived closer scrutiny, or have been supported by subsequent additions to the hypodigm",[29]

and yet he does not see any cause for concern. Ironically the rule changing has now come a full circle with the proposal the *habilis* material be transferred, although for some of the fossils merely returned, to *Australopithecus* thirty-five years after *Homo habilis* was created.[3]

The next major development to the *habilis* hypodigm was the addition of OH 24 in 1968. Nicknamed "Twiggy", and also found at Olduvai Gorge, it consisted of a totally fractured cranium and seven teeth. In the reconstruction over a hundred fragments could not be replaced and it is highly distorted. The skull was considered to represent *habilis* rather than another australopithecine because of its larger brain size (estimated at 590 cm^3), less postorbital constriction, a maxillary notch, an anteriorly positioned foramen magnum and the absence of post canine megadontia.

Given the distortion of the specimen one wonders whether all these assessments are valid. The endocranial capacity in particular is considered by some as spuriously elevated. OH 24 nevertheless made an important contribution to the taxon. This was due to its age given as 1.85 million years, making it the oldest *habilis* specimen ever recovered from Olduvai Gorge. This fact was used to counter the argument *habilis* was simply a phantom amalgamation of *africanus* specimens from Bed I and *erectus* specimens from Bed II. With this age, OH 24 was a fossil with the more "derived" erectus-like features of those from Bed II (like OH13), but which actually derived from Bed I.[71] It turns out however, that the "fact" of its great age was really an inference. The fossil had actually been found on the surface and was assigned to Lower Bed I.[71]

Habilis only gained wider acceptance in 1972 with the discovery of the skull ER 1470 at Koobi Fora in Kenya. This was an event of considerable irony too, and for more than one reason.

First, as fate would have it Louis Leakey's son Richard discovered this specimen which has preserved the taxon his father helped birth. Unfortunately only days after the discovery, Louis unexpectedly died from a myocardial infarction.

Second, the skull that secured the taxon was incongruously unlike material originally presented for the species! The endocranial capacity was of far larger volume (around 800 cm^3) and of far greater age, "securely dated," he said in his announcement in *Nature*. It was excavated below rock dated at 2.6 million years, above rock dated at 3.2 million years and estimated to be 2.9 million years old.[72]

Third, this date which was determined before the skull was discovered, upended notions of human evolution. The number was subsequently revised downward, ultimately to 1.9 million years, in a decade long debacle that had less to do with fossil dating and everything to do with supposed compatibility with evolutionary scenarios. The revised date would reject earlier concordant results obtained from four different methods. It would ultimately rest, you'll never guess, on the fact that pig fossils at the site were similar to pig fossils further away at Omo and Olduvai, which had been dated as younger.

In added irony, this choice of pig fossils was at the expense of elephant fossils also present that supported the older date. So the pigs won the day in another paleoanthropology rule right out of *Animal Farm*: "All animals are equal, but some animals are more equal than others". The events of the ER 1470 dating controversy have been recounted by Max Lubenow in a withering critique.[73]

But at least in 1470 there was now a skull of a size that appeared intuitively capable of making tools, and that after all, is what had supposedly characterized *habilis* from its beginnings. Indeed the skull appeared more modern than *erectus*! Ian Tattersall recounts the events:

"it's pretty ironic that ER 1470 was, rightly or wrongly, ultimately responsible for the general acknowledgement of *Homo habilis* as a valid species of early human. What was pretty clear from the start was that 1470 didn't fit at all comfortably into any of the generally accepted early human species; but it certainly could not be ignored, and the sheer lack of definition of *Homo habilis* made it a useful slot into which this remarkable specimen could be squeezed. Suddenly the new species had a purpose...[and] hastened the acceptance of Homo habilis as a real biological entity." [74]

Interestingly a complete femur (ER 1481) was also discovered with the skull. The femur was not significantly different from that of modern man. This suggests an obvious question. Why were 1470 and 1481 not the remains of an ancient human? The reason is found with even the most cursory inspection of the 1470 skull. The protrusion of the lower face is apelike. How real is that protrusion is the obvious next question.

As frequently happens in paleontology, it turns out the skull was not found intact but had to be reassembled from hundreds of fragments. It also turns out that pro-evolutionary bias entered into the reconstruction.[75-77] Paleontologist Roger Lewin, who has worked with Richard Leakey, drew attention to this back in 1987. He recounts the circumstances relating to the reconstruction of the 1470:

> "As might be imagined, there was a great deal of discussion around the Koobi Fora camp and back at the museum in Nairobi about what 1470 might be, for it was not unequivocally any known species. One point of uncertainty was the angle at which the face attached to the cranium. Alan Walker remembers an occasion when he, Michael Day, and Richard Leakey were studying the two sections of the skull. "You could hold the maxilla forward, and give it a long face, or you could tuck it in, making the face short", he recalls. "How you held it really depended on your preconceptions. It was very interesting watching what people did with it." Leakey remembers the incident too: "Yes. If you held it one way, it looked like one thing; if you held it another, it looked like something else. But there was never any doubt that it was different. The question was, was it sufficiently different from everything else to warrant being called something new."[75]

Subsequently the cranial fragments of a juvenile of similarly sized endocranial capacity (ER 1590) were added to the *habilis* hypodigm, along with seven teeth larger than those of *erectus*, and displaying a sequence of development different to that of sapiens (Wood 1991). But how are fossils allocated to a taxon?

Using the pro-evolutionary preconception. For instance in 1973 the ER 1813 cranium was found with a volume around 500 cm^3. Richard Leakey described it in 1974 as having resemblances to *A. africanus*, although others have since disputed this for facial morphologic differences. Eventually the taxonomy was settled by Clark Howell, who suggested it represented a *habilis* female. This post hoc evolutionary explanation of sexual dimorphism permitted the small brain to cheat the definition for *Homo* (again!), it since having been reset at 600 cm^3, and to fit into the (tool making) *habilis* hypodigm. Tattersall is frank about the fossil allocation:

> "This assignment [of 1813] more than anything else reflects the usefulness of having around a basket called *Homo habilis* into which paleoanthropologists could sweep a lot of fossil loose ends. And of course, the more this basket swelled, the less biological meaning it possessed." [78]

The fact is h*abilis* has remained controversial ever since it was erected. The specimens comprising the hypodigm are dated from 2.3 million years (A.L. 666-1) to 1.6 million years (OH 13) and found in Tanzania, Kenya, Ethiopia and South Africa. Endocranial volumes range widely too, from just under 500 cm^3 to around 800 cm^3 and there is a similar large variability of the face and mandible. There seems to be a dichotomy in the hypodigm with human-like skulls (e.g. ER1470, ER1590) and femur (ER 1481) on the one hand, and fossils like ER 1805, ER 1813, OH 24 and OH 62 on the other with features reminiscent of *Australopithecus africanus* or *afarensis*. While many suspect habilis does indeed represent more than one species, others do not. And even within this polarized opinion there is disagreement over exactly which specimens belong in the taxon. Richard Leakey has even stated that of the specimens attributable to *habilis* "at least half probably don't. But there is no consensus as to which fifty percent should be excluded." [79]

We know how this happened and still happens, but how can we sort through this nebulousness now? Bernard Wood recently reviewed the hominin fossil record. It comes as a surprise then to find he informs us:

"Knowledge of the postcranial skeleton has traditionally come from the remains of Bed I at Olduvai Gorge, but although these were allocated to *H. habilis* it is by no means certain that one can exclude their allocation to *P. boisei*. The only postcranial evidence that can, with confidence, be allocated to *H. habilis* is the associated skeleton OH 62, and some of the postcranial bones associated with the OH 7 type specimen...In sum, there is little to distinguish the postcranial skeleton of *H. habilis* from that of *Australopithecus* and *Paranthropus*."[29]

In light of the claims made for the status of this taxon in the hominin evolutionary tree, it is an amazing admission of uncertainty. *Habilis* based on this amount of evidence looks more like a phantom than a genuine extinct species. The fact that OH 62 is a collection of 302 bone fragments poorly preserved and that OH 7 is juvenile, does little to alleviate the concern either. But there is even more to the story of OH 62 however.

§11.3.4 OH 62 – A Case of Standing Tall or Standing Down

It was not until 1986, twenty-two years after the *habilis* taxon had been in use, that upper and lower limb fossils could finally be firmly assigned to the same individual. This was the discovery of OH 62. Dated at 2 million years old it is also called "Lucy's child" and like Lucy, was discovered by Donald Johanson's team [80]

As a toolmaker, it had been assumed that *habilis* not only had human-like limb proportions, but was fully bipedal with a "hindlimb skeleton...adapted to habitual erect posture," as Leakey had defined it.[16] Imagine the surprise when it turned out OH 62 stood only 3 feet tall and had arms almost as long as its legs, typical of an ape. *Habilis* had a

535

humerofemoral index (humerus/femur length) of 95 compared to 100 for chimpanzee, 85 for Lucy and 70 for humans. As for its bipedalism, this was just like that of the australopiths and an analysis of the hand indicates it is more capable of grasping than any other in the hominin record.[29] So much for a human-like toolmaker, this was the smallest adult hominin yet found.

The more extensive partial skeleton also finally provided the opportunity to test the hypothesis that *habilis* was an evolutionary transitional species between *Australopithecus (afarensis)* and *Homo (erectus)*. In 1991 Sigrid Hartwig-Scherer and R.D. Martin performed pairwise comparison of the fossils OH 62 ("Lucys child") and AL 288-1 ("Lucy"), with African ape and human bone measurements to this end. In 24 out of 28 measurements OH 62 was closer to an ape than Lucy, one was the reverse condition and in three they were equally distant.[16] Thus OH 62 is even more primitive than *afarensis*, making "Lucy" more human-like than "Lucy's child". Though more than a million years older.[81]

Analysis of the dentition of OH 62 also rejected the idea of *habilis* as a transitional species. To quote Donald Johanson, "the degree of megadontia in *Homo habilis* may, in fact, be little changed from the *Australopithecus* condition."[80] Which incidentally agreed with other habilis tooth , such as OH16[82] and Stw 151 (e.g."the dental development pattern of Stw 151 does not seem to differ much from those of *A. africanus* specimens.")[83] .This is a blow to the evolution paradigm not credibly softened by the post-hoc supposition that OH 62 represents the female of a species with a wide range of sexual dimorphism, and others have shown tooth development of *habilis* specimen OH16 is more similar to great apes than human too. Suffice it to say, *habilis* is no transitional fossil.

§11.3.5   The Monkey Business of *Homo habilis*

What is the current status of *habilis*? There is no consensus yet but the various opinions can be boiled down to variations on one of two themes.

The first is that *habilis* is a single valid species ancestral to *H. erectus*, also called "early *Homo*" or *habilis sensu lato*.[70,80] This is the view of Phillip Tobias, Donald Johanson, Tim White and Gen Suwa among others. It requires having to explain the wide range in morphology and some similarities of fossils in the hypodigm to already defined species like *Australopithecus*. Even more difficult is the requirement to reconcile the following. Evolutionary stasis for over a million years from Lucy (*A. afarensis*) at 3.2 million years standing around 3 foot tall, through to OH 62 of similar proportions at 1.8 million years ago, followed by a doubling in height in just 200,000 years into *H. erectus* by 1.6 million years (as demonstrated by WT 15000 at 6' and ER 1808 at 5'9"). It also requires dealing with the evolutionary reversal problem (§11.1.5) and the contemporaneity of *habilis* and *erectus* (Fig.10.4). An alternative single species view holds *habilis* to be no more than a highly variable Australopithecine species.[84]

The second view concedes the variation within the taxon is too much to be plausibly accommodated within a single species. The most popular version is represented by that of Bernard Wood.[29,71] He proposes that *habilis* and *rudolfensis* be moved from *Homo* into *Australopithecus* on the basis of information presented in §11.1.1. He sorts all the Olduvai material and the more apelike fossils from Koobi Fora into *habilis sensu stricto* which he simply calls *habilis*. The remainder from Koobi Fora he considers *Australopithecus rudolfensis*. *Habilis* then represents a creature with ape-like characteristics (a small brain, arboreal adaptation and megadontia), while *rudolfensis* is distinguished by a larger cranial capacity, a face widest in the mid-section (not the upper face) and more complex premolar teeth. Of some interest is Wood's rejection of 1481 and others, considering this taxon to have as yet no postcranial specimens whatever. Alan Walker and Richard Leakey have the converse view, considering *habilis* to represent the more human-like of the hypodigm and the rest to represent a late surviving *A. africanus*.

In summary, diametrically opposite views of *habilis* are held yet none are rendered illogical by the science. That's the important observation we make. Although there can be only one correct phylogeny, human evolution theory is so flexible it can accommodate both viewpoints without ever being inconsistent. In light of this uncertainty, how can there be unswerving, emphatic and universal consensus in anthropology extending back almost four decades that *habilis* demonstrates evolution? Because human evolution is a philosophical viewnot an empirical view. *Habilis* was constructed out of an evolutionary opinion that considered it a necessity and its erection flouted the taxonomic rules of its own science. It persists as such, but the situation is now even worse.

Exactly what it is, depends on the eye of the beholder. Yet what little evidence there is indicates only two partial specimens unequivocally represent the species at Olduvai Gorge, and none of the postcranial evidence can be well discriminated from *Paranthropus* and *Australopithecus*.[29] These are more the characteristics of a phantom. Far from an evolutionary species, *habilis* likely does not exist at all.

§11.4   INSPECTING ARGUMENT #4: IS THERE EVIDENCE FOR
         ANCIENT HUMANS IN THE FOSSIL RECORD?

If to the contrary of what Argument #4 claims there were a fossil record of humans living at the bottom of the hominin fossil series, then human evolution in any of its formulations would be rejected. Is there any evidence of this? Consider the following four fossils and artifacts. Time will tell if they turn out to be anything more, but I show you them to make different point. It is to show how the paleontological community has responded to this data and which is far more interesting observation than even the fossils themselves. We already know human evolution is bankrupt from our inspection of Arguments #1-#3. The more interesting question is why human evolution can still be believed as a scientific truth. How these four fossils are handled is to see how paleontology does business and how it makes its discoveries. Let's watch.

§11.4.1   No Elbow Room: KNM-KP 271

In 1965 the fossilized distal end of a humerus (arm bone) was found at Kanapoi, 45 miles southwest of Lake Turkana in Kenya.[85],[86] It was in an excellent state of preservation and was dated as being 4.5 million years old. The fossil is now called KNM-KP 271 (the abbreviation of Kenya National Museum – KanaPoi, specimen 271). The problems this fossil poses for human evolution were pointed out by Marvin Lubenow more than a decade ago.[87] The fossil gets hardly a mention in books on human evolution now, and yet the original description of the fossil was made very conspicuously in the journal *Science*. In this announcement the specimen was compared with the bones of *Paranthropus robustus*, the chimpanzee and the human. The authors William Howells and Bryan Patterson found KP 271 "strikingly close to the means of the human sample". However in a curious departure from that result, they came to the conclusion the fossil "may prove to be *Australopithecus*".[85] What was the evidence for that conclusion?

It was because the proximal humerus of *A. africanus* was similar to that of humans. There was no evidence this applied to the distal humerus they had examined and that was KP271 however, and of course it also left their result unexplained. Their conclusion, published by *Science* to the world, was no more than a wild speculation made at the expense of avoiding the obvious. A fossil dated 4.5 million years old with the same characteristics as man had been discovered. But this was not even considered, much less concluded.

Five years later the bone got a backhanded and ironic redress, when University of California Davis paleontologist Henry McHenry who had performed his own multivariate analysis of the fossil earlier,[88] wrote in a review also published in *Science* that: "The results show that the Kanapoi specimen, which is 4 to 4.5 million years old, is indistinguishable from modern *Homo sapiens*".[89] The comment was made without further elaboration. Its startling significance, and inconsistency with the rest of his paper (about the good evidence for the fact of human evolution), not to mention the inconsistency with what the fossils' finders had said it was, was not addressed. The specimen was accepted to be *A. africanus* and for the same reasons as before – it was too old to be human. You don't believe me? Here is William Howells to say so:

"The humeral fragment from Kanapoi, with a date of about 4.4 million, could not be distinguished from *Homo sapiens* morphologically or by multivariate analysis by Patterson and myself in 1967 (or by much more searching analysis by others since then). We suggested that it might represent *Australopithecus* because at that time allocation to *Homo* seemed preposterous, although it would be the correct one without the time element."[90]

The story of KP 271 took a new turn in 1995, when Maeve Leakey appropriated it into a description of a whole new species she called *A. anamensis*. Her species was proposed as the ancestor to *A. afarensis*. The hypodigm consisted of eight other fossils, also found at Kanapoi (a full 39 years after KP 271), together with twelve others found between 1982 and 1995 at Allia Bay in Kenya.[91] Still more specimens were added 3 years later.[92] There was no humerus among any of the bones however. So how do we know KP 271 belongs there at all, you ask? Objectively speaking we don't. Decisions as to what constitutes a species when it is constructed from multiple bone fragments, separately discovered in time and space, at least appears to be an arbitrary construct and the more so when Dr. Leakey and her co-authors conceded, in tortured and circular reasoning "The distal humerus, KNM-KP 271, was originally seen to be human-like, and it does show many derived hominid features".

The most recent and the prevailing word on the matter came in 1996 with the publication of yet another analysis. The authors concluded that contrary to the findings of five independent groups of investigators which had examined the specimen over three decades before them,[85,89,91-95] "Given its great age, it is much more likely that the specimen [KP 271] is representative of *Australopithecus* than early *Homo*. The specimen is therefore reasonably attributable to *A. anamensis*, although the results of this study indicate that the Kanapoi specimen is not much more "human like" than any of the other australopithecine fossils, despite prior conclusions to the contrary." [96] So the age of the fossil was now the decisive factor, not just in what it was, but in what it looked like.

Among those five groups of investigators reporting its similarity to modern man were determinations made by Brigitte Senut (and her views about another distal humerus Gomboré IB 7594 did also).[94] How can such irreconcilable interpretations of the same dataset be a science? We know how it happens (§10.2). In summary the Kanapoi elbow fossil at least demonstrates illogical thinking in paleontology, the power of opinion over protocol, and how the philosophy of an obligate evolutionary preconception drives fossil interpretation not the other way around. And perhaps more.

§11.4.2   Following The Trail of Evidence At Laetoli

In 1978 Mary Leakey's team made a discovery at Laetoli, which is about 30 miles south of Olduvai Gorge in Tanzania, that remains one of the most remarkable finds in all of paleontology. It's a trail of 70 fossil footprints, preserved in volcanic ashfall and extending over a distance of 27m (89 ft.) (Fig. 11.6).[97-101] The footprints were made by 3 individuals researchers have named G1, G2 and G3. Potassium-Argon dating has demonstrated the age of the stratum above the prints to be 3.6 million years old and the age below the stratum to be 3.8 million years old. The significance of the discovery is that despite this supposed great age, undisputed analysis has confirmed Mary Leakey's original assessment that the footprints are "remarkably similar to those of modern man".[97,102-109] In the words of University of Chicago's Richard Tuttle who has also studied this fossil, "In all discernible morphological features, the feet of the individuals that made the G trails are indistinguishable from those of modern humans."[101]

While there is general uniformity in that sentiment, it is worth watching this science at work because don't for a moment think anybody actually attributes them to humans. First, listen to two more of the experts:
"The whole pattern of weight transmission and force distribution on the substrate is in essence that seen in imprints made on wet sand by modern man...That the stride is human in form we agree entirely...the patterns thus far reported from the ashfall tuff of Laetoli are patterns consistent with what humans call 'strolling'..."[109]; and
"What these footprints, and their photogrammetric analysis shows, it that bipedalism of an apparently human kind was established 3.6

million years ago. The mechanism of weight and force transmission through the foot is extraordinarily close to that of modern man. The spread of the forefoot as shown by the 'barefoot gap', and the varus position of the hallux and the presence of short toes, have close similarities with the anatomy of the feet of the modern human habitually unshod; arguably the normal human condition."[105]

Fig.11.5 **The footprints at Laetoli.** Right panel is higher power magnification. (Reproduced from Leakey & Hay, 1979 with permission).[113]

The individuals walked with similar step and stride lengths. Their stature has been estimated (by such methods as using foot length-to-stature ratios of 15% and footprint length/stature correlation tables using contemporary American subjects) at around 5'9" and 5'0" for G2 and G3 respectively, and 3'10' for G1. G1 is below the range of the normal adult human but within the range of a 3-14 year old.[101,110] The question of course is whose footprints are they?

The obvious answer is barefoot humans. But this interpretation dated at 3.7 million years old rejects every notion of human evolution and so is not considered. As a result this science gets really interesting to onlookers because all manner of unwarranted assumptions must be made to avoid this logic of the data. So if you like gymnastics or plasticine, come and look.

Having made the observation quoted above, Russell Tuttle invokes an as yet undiscovered advanced Pliocene hominid as the strollers, though none of the more than 5000 vertebrate remains and 26 hominids uncovered at Laetoli provide any evidence this faith is objectively grounded, or any reason for an optimism that it will be.[106] You can see for yourself what is fact and was evidence in his mind, because he explains the reasoning clearly: "If the G footprints were not known to be so old, we would readily conclude that they were made by a member of our genus, *Homo.*"[111]

It gets more ridiculous. Ronald Clarke suggests *P. africanus* made the footprints. On the basis of skull data, mind you.[112] The most popular opinion today is that they were made by *A. afarensis*. This notion was first proposed by Tim White at Berkeley and Gen Suwa at the University of Tokyo. What is the evidence for that idea? You should know no complete set of *A. afarensis* foot fossils has ever been found and the most complete example, from Hadar (AL333-115), is incompatible because of its long curved phalanges.[106,108] As Russell Tuttle has responded: "It is difficult to imagine a foot with such markedly curved phalanges fitting neatly into the footprints at Laetoli."[106]. The authors counter by speculating the AL-333-115 specimen is male and that sexual dimorphism would permit females like Lucy to make these prints nevertheless. This supposition is based on a reconstruction of *Preanthropus* foot using the only 3 foot bones available from the Lucy skeleton (AL288-1) (talus, a proximal phalanx and an intermediate phalanx) and the remainder of the foot from OH8 (classified *H. habilis*), selected because it was "the smallest normal adult hominid foot skeleton represented by tarsals (including talus) and metatarsals of which we were aware." [108] This is not convincing.

Believing *A. afarensis* made the prints requires leaping to a belief that despite the obvious existing evidence demonstrating the differences between *Australopithecus* and *Homo sapiens* over the rest of their skeletons (§11.1), unknown evidence exists to show their feet were really identical. The academic establishment finds this easier to believe than the utterly obvious. In summary, there is no fossil hominin candidate for the Laetoli footprints to date, and no evidence refuting the argument anatomically modern humans made them.

§11.4.3   The Olduvai Stone Structure

A circular stone artifact of unclear function measuring 14 feet in diameter was found by Mary Leakey at Olduvai'in 1961. It is dated at 1.8 Myr (by K-

Ar) and 2 Myr (by ^{40}Ar-^{39}Ar),[114] which is also the time *Homo erectus* fossils have been found. The structure consists of several hundred lava rocks that not are indigenous to the area. The nearest are miles away. The stones have been systematically placed on one another too, and the structure was associated with Oldowan tools. As such it would least appear to be the product of activity we associate with humans. What's the take home point?

Homo species (*rudolfensis* and *habilis* excepted), are within the range of variation seen in modern humans. If those species that are clearly human-like (*erectus, heidelbergensis, neanderthalensis*) are nothing more than microevolutionary differences within *sapiens*, it brings the separation between the ape-like and human-like groupings (Fig.11.4) into sharper relief and puts humans in the fossil record, at least according to the fossil dating methods, millions of years ago and at the same time as the ape-like fossils. Human evolution is then crippled. Consider these three species in turn.

As regards erectus, most in paleoanthropology claim it was a chronospecies to *sapiens* that was more ape-like in morphologic features and in its behavior. Yet skull specimens KNM-ER 1470, 1481, 1590 are of human morphology, they are dated by the science at 1.9 million years ago, *erectus* lived contemporaneously with its supposed ancestral and descendant species, and exhibited human behavior (§11.3.3-4). We find fossil evidence to indicate human behavior dated to 4 million years ago (§11.4.1-2).

As regards *heidelbergensis*, the differences from *sapiens* are minor and there is no type specimen for *sapiens* to discriminate it objectively (§10.2.2). Besides we know that if dogs are anything to go by, relying on bone size and shape to determine separate species is misleading.

As regards *neanderthalensis* and its supposed differences from *sapiens*, here is Bernard Wood:

> "The numerous associated skeletons of *H. neanderthalensis* indicate that their body shape was within the range of variation seen in modern humans...Contrary to the suggestions of early commentators who depicted *H. neanderthalensis* as ape-like, it is now clear that their posture, foot structure, and limb and muscle function were essentially the same as those of modern humans."[3] and

> "Detailed comparisons of Neanderthal skeletal remains with those of modern humans have shown that there is nothing in Neanderthal anatomy that conclusively indicates locomotor, manipulative, intellectual or linguistic abilities inferior to those of modern humans."[115]

We will stop here and I leave you with the evidence to consider further. Next we switch tack to consider what all this human evolution means.

§11.5 THE PHILOSOPHICAL IMPLICATIONS OF
HUMAN EVOLUTION

In closing we must consider an observation made earlier that never went explored. Namely that the answer to the question of the origin of man has profound philosophical implications no matter whether it's evolution or design (§10.1). We know design demonstrates the existence of God and what that means, but what are the philosophical implications if it was by evolution instead? With all the attention evolution gets as the only rational solution to the origins of biology, what macroevolution necessarily means for rational much less livable human existence is seldom given remotely equivalent disclosure. It seems evolution educators are only too happy to tell us of our hairy ancestors and how far we've come since our pond dwelling days, but not what this necessarily means for living out lives in coherence to such origins. They're not difficult to see if you look however, and you might be surprised at what you find. They have no bearing on whether human evolution is true or not. They're just the corollaries if it were.

So why bother mention them? Because it's amazing how little those who wax lyrical about an evolutionary worldview know, or if they know let on, about how livable that world would be. Teaching false philosophy in science class is bad enough but calling it a good idea is perverse. The reason is this idea is toxic, and ideas have consequences. If that seems ridiculous, then look at what macroevolution means.

For a start, what difference did Darwinism make to philosophies of human existence? For centuries Greek philosophers had said man was an animal and Linnaeus had classified humans in the primate order (§1.1). What was so special? It was to declare entirely natural processes accounted for all of human nature and spiritual thinking. This might seem innocuous, but it overturned a whole world order. Darwin was insistent. "My object," he said in *Descent of Man*, "is to show that there is no fundamental difference between man and the higher animal in their mental faculties...the difference in mind between man and the higher animals, great as it is, certainly is one of degree and not of kind." This extended to morality and religion. Specifically, "any animal whatever, endowed with well-marked social instincts, the parental and filial affections being here included, would inevitably acquire a moral sense or conscience, as soon as its intellectual powers had become as well, or nearly as well developed, as in man."

This is the view most everywhere in intellectual circles now. The

National Academy of Sciences for instance says "scientific investigations have concluded that the same evolution of all other life forms on Earth can account for the evolution of human beings."[116] On this view the things that set humans apart from other animals, the things that make a chimpanzee 98.8% different from a human although their DNA is 98.8% the same, namely intellect and morality and spirituality, these things are not particularly special or particularly different from the brutes after all. It's a stunning claim and it has profound consequences for living. Ernst Mayr explains why:

"But perhaps the most important consequence of the theory of common descent was the change in the position of man. For theologians and philosophers alike, man was a creature apart from the rest of life. Aristotle, Descartes, and Kant agreed in this, no matter how much they disagreed in other aspects of their philosophies...But Ernst Haeckel, T.H. Huxley and in 1871 Darwin himself demonstrated conclusively that humans must have evolved from an ape-like ancestor, thus putting them right into the phylogenetic tree of the animal kingdom. This was the end of the traditional anthropocentrism of the Bible and of the philosophers...today this derivation of man is not only remarkably well substantiated by the fossil record, but the biochemical and chromosomal similarity of man and the African apes is so great that it is puzzling why they are so relatively different in morphology and brain development."[117]

If you're still nonplussed, think what this "change in position of man" means. What does it mean? Three things:

(1) Man is just another animal, albeit more intelligent.
(2) Humans arose from non-ordained, purposeless natural processes.
(3) Humans are stratified by evolutionary forces, like all living things.

They might seem as innocuous as they are straightforward, but look more closely. At least one thing is clear if you haven't recognized already. Evolution is emphatically a philosophical "religious" worldview no matter how "scientific" it also might be (§1.4;§3.4-8;§19;§26). It's an inherently atheistic worldview too, because only what is natural is real and what's supernatural cannot ever become real.

A large number of church and synagogue goers now throw up their hands in protest, saying you most certainly can believe both in God and in evolution at the same time and that they are even living proof of it. This is a belief called theistic evolution (§1.3;§26.1). The point they miss, is their beliefs in a God behind the working of evolution is no more than a private conviction because even the brightest of these believers in biology say this has no evidence in science (§1.3). Anyway the entirely opposite conclusion about the world is made by atheists, with whom theistic evolutionists agree about evolution. Leaving that nonsensical if nonetheless hugely popular religious mindset for analysis later, we should unpack each of those three philosophical implications of human evolution to see what they mean for living out human lives (§26.3.3). Distinguished faculty are already here to be our guides.

§11.5.1 Man Is Just Another Animal...

A whole research field has emerged to uncover the evolutionary implications of similarities and parallels between apes and humans. Evolution demands this thinking that uncovers continuities across taxa in accord with common ancestry. Molecular clock pioneer Emile Zuckerkandl for instance says that since only 1 of 287 amino acids in hemoglobin are different between man and gorilla, "from the point of view of hemoglobin structure, it appears that gorilla is just an abnormal human, or man an abnormal gorilla, and the two species form actually one continuous population."[118] (§8.4) UCLA physiologist and Pulitzer Prize winning author Jared Diamond makes similar arguments to declare human beings another chimpanzee,[119] and in a summary of this field, molecular biologist Morris Goodman at Wayne State University concludes: "Behaviorally, the separation between chimpanzees and humans is much smaller than once thought...In agreement with the newer information on the social lives and intelligence of chimpanzees and other apes, the results of molecular studies of primate phylogeny challenge the traditional anthropocentric view that humans are very different from other animals."[120] This is not disturbing? Keep reading.

The more that differences between man and beast are blurred, the more plausible are explanations of ape-still-in-man of human behavior that otherwise are just a rank deviancy from what consciences plainly declare. Absent an *a priori* commitment to evolution the idea of a man being a victim to a brute for his behavior is ridiculous. The premise grounding that reasoning, that similarity must mean common ancestry, we know is false (§10.4.2). Man was made and therefore is responsible to his Maker. Any other view is just a clenched fist, shaken in the face of God.

There is a second objection to the notion man is just another animal, and it's one that turns the man-ape similarity claim on its head. It's what Ernst Mayr found "puzzling" above yet dismissed without engaging because it contradicted his claim. It's to point to the differences between apes and man in spite of their similarities. Similarities do nothing to change what so obviously separates a human from an ape - human intellectual, moral and spiritual capacities which surely are not only 1.2 percent different! Well they're not is the argument.

Among the evidence is a claim animals share the human trait of tool usage. Twig use by chimps to extract termites from their nests is the observation, and it was first made in 1960 by a young English woman only just hired by Louis Leakey to study chimpanzees, called Jane Goodall. Famous all over the world now, her finding is still taken as illuminative of how tool making might have evolved in humans and why humans are not so different after all. This mindset is perhaps best demonstrated by Louis Leakey himself, who when told of her discovery said: "Now we must redefine 'tool', redefine 'man', or accept chimpanzees as humans." Another argument used to show apes are like humans and vice versa, is to emphasize ape "language".[120] The idea is dubious at best but whatever language they have, we can surely agree it cannot compare with Dante, Shakespeare or Keats but these differences are what are at issue. Apes do not have the arts or technology. They do not worship. They do not have the moral sensitivities of humans. Absent these man is indeed an ape, no matter what his DNA may say.

Even co-discoverer with Darwin of natural selection Alfred Russell Wallace, baulked at how evolution could explain the ascent of an ape to a human (§1.3). Wallace eventually concluded humans were created, and in 1870 said: "a superior intelligence has guided the development of man in a definite direction, and for a special purpose, just as man guides the development of many animal and vegetable forms."[121] Darwin scoffed at his retreat and wrote *Descent of Man* in response. It was all natural selection Darwin said, in an explanation that was also the evidence.

There was another wrinkle. Darwin observed: "in the rudest state of society, the individuals who were the most sagacious, who invented and used the best weapons or traps, and who were best able to defend themselves, would rear the greatest number of offspring." He could not explain how natural selection could cause that common consequence of the human conscience called altruistic behavior, and admitted:

"It is extremely doubtful whether the offspring of the more sympathetic and benevolent parents, or of those which were the most faithful to their comrades, would be reared in greater number than the children of selfish and treacherous parents of the same tribe. He who was willing to sacrifice his own life, as many a savage has been, rather than betray his comrades, would often leave no offspring to inherit his noble nature. The bravest men, who were always willing to come to the front in war, and who freely risked their lives for others, would on average perish in larger number than the other men."

He had no solution. Today the phenomenon is called "reciprocal altruism" in evolution theory and it requires just as much hand waving.

Morality was and still is a direct rebuttal of human evolution. Moral rules are immaterial and yet they are real. They are discovered not declared and they are universal. That's a disproof of materialism right there. Consider their incumbency or "oughtness". Where does that come from? It's a compulsion before it is ever an activity and so is a communication. But only a mind can communicate to a mind. Consider the power of moral laws. Transgression causes guilt. Where does that authority come from? If it is all from chance purposeless processes it could not be rational. But it is rational. If it's just a neurosis, the self deprecating the self, it's better called a denial because we each know our consciences are real and our guilt is deserved. Every human finds herself under the constraints of a Moral Law she did not make, which resides outside of her material self, and is shared by all peoples and cultures. But we stray from the topic. The topic is to answer what the implications of a human being, being just another animal are. What are they?

We should treat each other as we would any other animal. On this view there's no such thing as intrinsic inalienable human dignity or human rights, any more than any animal has intrinsic dignity or inalienable rights. Animals have rights as much as humans have rights and all rights are conferred. They're not inalienable. They're conferred to whom and to what we say, if and when we say, and for as short or as long as we say.

§11.5.2 Man Is Just Another Animal That Became A Human By Chance...

Experts say this in different ways but they mean the same thing. George Gaylord Simpson (1902-1984) for instance says: "Man is the result of a purposeless and natural process that did not have him in mind".[122] Nobel laureate Bertrand Russell (1872-1970) says: "Man, as a curious accident in a backwater, is unintelligible; his mixture of virtues and vices is such as must be expected to result from a fortuitous origin."[123] Another Nobel prize winner, molecular biologist Jacques Monod (1910-1976) says it like this: "With that [the discovery of the genetic code], and the understanding of the random physical basis of mutation that molecular biology has also provided, the mechanism of Darwinism is at last securely founded...man has to understand he is a mere accident. Not only is man not the center of creation; he is not even the heir to a sort of predetermined evolution that would have produced either man or something like him in any case." [124]

Monod was much more blunt in his book *Chance and Necessity*. He said: "Pure chance, absolutely free but blind, [is] at the very root of the stupendous edifice of evolution...The universe was not pregnant with life nor the biosphere with man. Our number came up in the Monte Carlo game."[125] Consider what the consequences of this are.

Being no more than an animal, man is a victim of base animal instinct. As an animal that arose randomly by natural processes, he is as purposeless as the processes that created him. Any particular human individual has no special significance or inalienable purpose. A human is an animal that arose by chance, lives like an animal for a while and then dies when no longer able to compete for its survival. Since there is no ultimate purpose or meaning humans are not accountable for their behavior or the directives of their consciences to any other authority than themselves, because man is the highest authority there is. The view is captured in few places better than William Ernest Henley's (1849-1903) poem *Invictus* which concludes: "I am the master of my fate, I am the captain of my soul". Monod agrees. He says: "The ancient covenant is in pieces: man at last knows that he is alone in the unfeeling immensity of the universe, out of which he had emerged only by chance. Neither his destiny nor his duty have been written down."[126]

Theodosius Dobzhansky reviewed Dr. Monod's book for *Science*. He said Monod "has stated with admirable clarity, and eloquence often verging on pathos, the mechanistic materialistic philosophy shared by most of the present "establishment" in the biological sciences. But while many see this philosophy through a glass darkly Monod makes it crystal clear."[127] What he had just made clear, including to Dr. Dobzhansky, was the inevitable consequence of the philosophy that his colleagues aver for human existence. There are three consequences and they follow inescapably from one other: 1) There are no moral absolutes therefore 2) Life is ultimately meaningless therefore 3) Living is ultimately hopeless. Consider them each in turn.

First, since there is no God who made man or at least no objective reason to believe that if evolution is a fact, there's no moral absolute for human living. How can there be normative rules for all mankind, much less for all time when morals are the products of minds purposelessly and naturalistically evolved by chance? Richard Dawkins explains: "Our genes have survived...in a highly competitive world. This entitles us to expect certain qualities in our genes; a predominant one being ruthless selfishness...Much as we might want to believe otherwise, universal love and the welfare of the species as a whole are concepts which simply do not make evolutionary sense...Be warned that if you wish, as I do, to build a society in which individuals cooperate generously and unselfishly towards a common good, you can expect little help from biological nature."[128]

The point is not just that human nature is animal nature, it's that it's victim to nature and so not morally responsible for acting out the nature. Without a Creator morals are relative, they are intangible and they are without incumbency. Paul Churchland at the University of California San Diego explains. "The important part about the standard evolutionary story is that the human species and all of its features are the wholly physical outcome of a purely physical process...If this is the correct account of our origins, then there seems neither need, nor room, to fit any non-physical substances or properties into our theoretical account of ourselves. We are creatures of matter. And we should learn to live with that fact."[129]

Moral imperatives by definition are religious imperatives. All religions, at least all except the one that begins with an "a", go out the window on this view. Thomas Huxley spelt out long ago what religion means when evolution is a fact. "Extinguished theologians lie about the cradle of every science as the strangled snakes beside that of Hercules; and history records that whenever science and orthodoxy have been fairly opposed, the latter has been forced to retire from the lists, bleeding and crushed if not annihilated."[130].

If everything in the universe happens by processes ultimately meaningless, if humans evolved from slime by natural selection, and if matter became mind by chance, then the thoughts of man are the product of natural selection also. If this is so then there's no way of knowing if human reasoning can ever have truth beyond what is expedient or functional or what promotes survival. How can man trust the conclusions of his own reason? This impasse is engaged by Richard Dawkins, who says we must yet construct a world view "from our own wisdom and ethical sense". But where do ethics come from? And what exactly is "wisdom"? He does not say. He has no plank to pontificate from is the point, and the one he stands on now he took from a theist. The reason is because by his worldview there is no higher authority. Values are relative and all of them, objectively speaking, equally valid. He borrows from a world with absolute truth to make claims about a world he makes with his mind but it's one as unlivable as it's illogical. In his world ethics in the end are made by might. Rex Lex. This is the illogic, and the anarchy, of moral relativism.

The contradiction of trusting in an evolved mind was evident at least to Charles Darwin. He discovered his certainty about the truth of evolution might even be self-defeating. He said: "But then arises the doubt, can the mind of man, which has, as I fully believe, been developed from a mind as low as that possessed by the lowest animal, be trusted when it draws such grand conclusions?" Don't for a moment think he was serious. He was certain evolution was true and this is what's interesting. His observation about the grounding of evolutionary convictions applies to its reasoning. There are no universal moral values in an evolutionary world view by moral relativism. Abortion and same sex marriage and euthanasia are matters of personal choice because morals are made and morals are chosen. Thomas Huxley spelt this out long ago. "Cosmic evolution may teach us how the good and the evil tendencies of man may have come about; but, in itself, it is incompetent to furnish any better reason why what we call good is preferable to what we call evil than we had before."[131] What else does this mean?

It means the evolutionist has no business judging Adolf Hitler any more than Mahatma Gandhi. Absent absolute truth there's no way to tell who's "right" and who's "wrong", even if we might prefer to be ruled by the one than the other. What makes moral absolutes and makes meaning and makes purpose to living? God. What refutes atheism and moral relativism? A world created by God who defines everything.

The second consequence is in corollary to an absence of moral absolutes. Life is ultimately meaningless. Jacques Monod again: "[M]an has to understand he is a mere accident. Not only is man not the center of creation; he is not even the heir to a sort of predetermined evolution that would have produced either man or something like him in any case."[124] Here is Douglas Futuyma: "Some shrink from the conclusion that the human species was not designed, has no purpose, and is the product of mere mechanical mechanisms – but this seems to be the message of evolution." Stephen Jay Gould joins the panel to sum up: "We are here because one odd group of fishes had a peculiar fin anatomy that could transform into legs for terrestrial creatures; because comets struck the earth and wiped out dinosaurs, thereby giving mammals a chance not otherwise available (so thank your lucky stars in a literal sense); because the earth never froze entirely during an ice age; because a small and tenuous species, arising in Africa a quarter of a million years ago, has managed, so far, to survive by hook and by crook. We may yearn for a "higher" answer – but none exists. This explanation, though superficially troubling, if not terrifying, is ultimately liberating and exhilarating."[132]

And this is the amazing thing about it all. The thinking is not anywhere in elite intellectual or academic or education circles taken as ominous. Actually it's welcomed as here as by Gould, as liberating, as if bursting bonds and casting off cords. The compounded irrationality would be hilarious if it were not also so false, so perilous and so sick.

Third, and in corollary to the discovery that life meaningless, life is pointless and so ultimately hopeless. There is no ultimate meaning to living or suffering or justice or injustice, to happiness or sadness or to anything. Bertrand Russell speaks now: "[That man's] origin, his growth, his hopes and fears, his loves and beliefs, are but the outcome of accidental collocations of atoms; no fire, no heroism, no intensity of thought and feeling...and that whole temple of Man's achievement must inevitably be buried beneath the debris of a universe in ruins...Only within the scaffolding of these truths, and on the firm foundations of unyielding despair, can the soul's habitation henceforth be safely built." Richard Dawkins has the last word: "In a universe of electrons and selfish genes, blind physical forces and genetic replication, some people are going to get hurt, other people are going to get lucky, and you won't find any rhyme or reason in it, nor any justice. The universe that we observe has precisely the properties we should expect if there is, at bottom, no design, no purpose, no evil, no good, nothing but pitiless indifference."[133]

There is one last implication for *Homo sapiens* from Darwinism and it's one that not only goes unmentioned, it's even denied. It's that human evolution is inherently racist.

§11.5.3.3 Man Is Just Another Animal That Became A Human By Chance By Better Survival

Look how natural selection works in other species if you can't see it working here. The domination of the strong over the weak is not a perversion of natural selection, that is natural selection. The writings of Nietzsche and the actions of Hitler and Stalin are its logical outworking. If you won't hear it from me, here's historian Paul Johnson: "Darwin's notion of the survival of the fittest was a key element both in the Marxist concept of class warfare and of the racial philosophies which shaped Hitlerism."[134] There are necessarily some people closer to our ape-like ancestors than others or else human evolution is not happening and never happened. In the name of evolution people were once sterilized, incarcerated in zoos, hunted and exterminated. Ota Benga was a Congolese Pygmy put on display in the monkey house at the Bronx Zoo in 1906. He later committed suicide. Again there are arguments why human evolution should not be interpreted in this way but why should we believe those arguments? If evolution is a fact there is no evidence from biology to compel that thinking, and no moral imperative either. This is what human evolution means and what our children learn in science class. Now watch as they start living out their lessons.

§11.6 IN CONCLUSION: THE SCIENCE OF TURNING FROGS INTO PRINCES

In the journey just made the methods and the logic used by paleoanthropology to prove human evolution were as much the focus of inquiry as the evidence itself. We discover that accepting human evolution demands two preconditions. The first is that evolution is already a fact and the second is that man is descended from an apelike ancestor because of structural, chemical and genetic similarities with extant apes (Arguments #1 and #2). Science everywhere considers them today as undisputable as they were to Darwin and Huxley, which was before there was any fossil evidence for either idea. All the hominid fossils have been interpreted in this evolutionary light, and all show evolution in self-serving circular reasoning.

The science of paleoanthropology seeks to show how human evolution happened by searching for fossils of ape-men intermediates that must be in the ground. Whether it happened is not a consideration and the reason is because evolution is already a fact in a flaw of logic called begging the question. The problem with Argument #1 is that evolution is not a fact but an inference, and one the evidence of biology rejects as false (§10.4.1;Table 17.1;§17.3). Since Argument #1 is a precondition for the three other arguments for human evolution if is false, they collapse too.

The problem with the second precondition, Argument #2, is that an observation of similarity is not proof of causation by ancestry. That is an inference resting on the validity of Argument #1, and on evidence for homology which is absent at least in other species (§7.3, §8.3). Yet on these two the entire enterprise of paleoanthropology and the supposed truth of human evolution rests.

There are two more problems with the reasoning employed to interpret hominin fossils as well. The first is that multivariate analysis of human and ape bone characters does not confirm the accepted phylogenetic relationships of man to the great apes, and the second is that cladistic analysis is unreliable in formulating phylogenetic hypotheses. Yet these are the principal methods of research used to demonstrate human evolution.

Evolution predicts that there will be progressive change both within and between species over time. Argument #3 is the claim this evidence exists (§10.1). Claims of an evolutionary trend without a demonstrated phylogeny are spurious and yet this is what Argument #3 represents. Anyway instead of transitions as is claimed the overwhelming feature of the fossil record is a saltatory change around two million years ago between fossils that are clearly ape-like and fossils that are clearly human-like. Why were we not told about this? Darwin predicted anagenetic change would be found in the hominin fossil record. Instead we find supposed ancestral and descendant species co-existing. A retreat to punctuated equilibrium to explain this is to retreat to a theory that cannot be disproved. All those human evolution claims listed in §10.1 and Table 3.1 are not more evident but more intangible, when we try to look for them ourselves.

In the same way a detailed inspection of two key species in the hominin fossil record, *Homo erectus* and *Homo habilis*, fails to confirm their claimed credentials as evolutionary species. In contradiction of the predictions of evolution theory *erectus* is an abrupt and radical change from older hominin species, it shows no evolutionary trend over the 1.8 million years of its existence, is contemporaneous with supposed ancestral and descendant species, and displays overtly human behavior.

Homo habilis on the other hand is a species born of the necessities of an evolutionary prejudice and is maintained as such, and for the same reason. In a field directed toward demonstrating change in species, this is a taxon that changes depending on the opinion of the viewer. *Habilis* is more illusion than reality and its real significance is to expose the emptiness of a supposed science called paleoanthropology that is so bereft of objectivity it cannot discover this for itself.

Argument #4 argues for human evolution on grounds that the fossil record of humans is recent but since the fossil record supposedly dates back millions of years, humans must have evolved out of another creature. There are suggestions to the contrary however, even setting aside any objection with the fossil dating methodologies deployed. The point made here is another one anyway, and it is an assessment of paleoanthropology's scientific credentials and objectivity. We discover that this suggestive data is either deflected or rejected off hand, and on philosophical not empirical grounds.

The end result of this science is the extraordinary situation we find today, that when a child hears a frog became a man at school it's called science but when she hears it at bedtime it's called a fairy tale.

There is one more line of evidence for macroevolution still to consider. It's evidence that goes all the way back to *The Origin* and remarkably has remained essentially unchanged since.[135] Actually it was one of the four principal categories of evidence for evolution for Charles Darwin, along with paleontology, homology and embryology. It's the evidence of biogeography (*LOE#12*). Take a look at it again on Table 3.1.

§11.7 THE EVIDENCE OF EVOLUTION FROM BIOGEOGRAPHY OR NOT

Biogeography is the science that seeks to explain why and how living things are distributed over the Earth. Although compatible ecological factors between an organism and its environment are a prerequisite, distribution cannot be explained by these factors alone. Take for example those immigrant English rabbits living it up down in Australia (§1.2.2). Despite the favorable ecology, there were no rabbits in Australia until humans artificially introduced them. Thus while ecological factors limit the distribution of an organism, it is historical factors - the history of the region and the history of the species - that have the final word. It is in regard to this biological history that evolution fits in.

The claim made for evolution by *LOE*#12 in Ernst Mayr's words is that evolution "helped explain the reasons for the geographic distribution of animals and plants"[136] In Charles Darwin's words, "all the grand leading facts of geographical distribution are explicable on the theory of migration, together with subsequent modification and the multiplication of new forms" (p.547). In other words it is two things, the migration of species and their evolution through natural selection that explains biogeography.

If this is true then it is evidence of evolution indeed. Ernst Mayr summarizes the view of science for us: "Darwin...showed that the present distribution of animals and plants is due to the history of their dispersal from their original points of origin. ...For a creationist there is no rational explanation for distributional irregularities, but they are completely compatible with a historical evolutionary explanation".[137] There would seem nothing more to discuss. If only we could believe him. Take a closer look.

What were the observations that Darwin pointed to about biogeography? There were three and the first was just mentioned. It's that patterns of biodiversity cannot be wholly explained by climatic and other environmental conditions. The reason is because similar conditions exist on different continents, yet very different species live there. Take the environmental conditions between 25° and 35° latitude shared by Australia, South Africa and western South America. They are essentially the same and yet as Darwin pointed out "it would not be possible to point out three faunas and floras more utterly dissimilar" (p.483). Why should that be?

The second observation was his claim that "barriers of any kind, or obstacles to free migration" (p.483) of species between different regions of the Earth "are related in a close and important manner to the differences" between species. Examples of such barriers are oceans, deserts, mountains and rivers. In this he was presenting the first argument for allopatric speciation of course (§6.3). Thus the barrier to species migration provided by the ocean-separated continents of Australia, South Africa and western South America is taken to answer why their fauna and flora are so different. They were left to evolve independently without interbreeding.

His third observation was the inverse of the second observation. Namely that the similarities (which he called "affinities") between species in neighboring regions is explained by barriers to migration being weak. Examples for this are the similar morphologies of island life to that on adjacent continental mainland, the dozens of different species of honey creeper in Hawaii, and the thirteen endemic species of Galapagos finch (§4.5). In summary, Darwin claimed that both the similarities and the differences among species in different geographical locations are explained by their past migration and their past evolution through natural selection. Still think it's good evidence? Look a little more.

On the one hand the similarities are by evolution, and on the other hand the differences are by evolution. As for the similarities, they happen either because they are inherited traits from a common ancestor genealogically and geographically close (divergent evolution); or because they are not inherited traits at all but the similarity arose independently because natural selection favored parallel adaptations in similar environments (convergent or parallel evolution). How can we tell which of them it is? All bases are irrefutably covered and yet *LOE#12* is used as if it's still serious.

One of the most popular arguments made by *LOE#12* is about marsupials. These creatures dominate in Australia unlike in the rest of the world. In fact there are no Australian placentals at all (apart from bats and a few colonizing rodents) to compete with them for ecological niches. As a result, it is claimed, although marsupials nurture their newborn in a pouch while mammals do it by a placenta, the Australian marsupials to
"an astonishing degree resemble the placental mammals living today on the other continents. The similarity...argues strongly that they are the result of convergent evolution, similar forms having evolved in different, isolated areas because of similar selective pressures in similar environments".[138] If similarities can be caused by natural selection as well by ancestry as not by ancestry, then there is no effect that evolution cannot do.

This is great evidence because it is also irrefutable evidence. It requires faith nonetheless, a lot of it actually, and the reason is because the similarities that supposedly arose convergently are very impressive. The Tasmanian wolf for example looks very much like a North American wolf,

556

the marsupial mouse like the placental mouse, the marsupial mole like the placental mole, the Tasmanian tiger cat the placental cats, the flying phalanger to the flying squirrel, the cuscus to the lemur and the numbat to the anteater. It would rather seem that the creatures are more complicated than the system of human classification that tries to categorize them , not to mention these attempts to explain how those categories developed. Taxonomy is built up on similarities and yet we find that despite the best efforts of the best minds in biology over centuries they still vexingly conflict. Which are we to choose - that the Tasmanian wolf cares for its young like a kangaroo, or that in all other respects it is like a wolf?

There is another matter to consider that usually goes unmentioned when marsupial evolution is argued. Where did they come from if they are such good evolution evidence? Living marsupials are restricted to Australia and South America (the opossums of North America are recent immigrants). This is explained on grounds Australia and South America were once part of a single supercontinent called Gondwanaland representing today's Southern Hemisphere and India before continental drift moved them apart. The problem is that fossils taken to be marsupial ancestors are found only on Laurasia, the other supercontinent that represents today's Northern Hemisphere. The bottom line?

> "This geographical switch remains unexplained. The timing of the split between eutherians [placentals living and fossil] and metatherians [marsupials living and fossil] is also controversial. To date, the geographical record has yielded few fossils that bear directly on the origin of marsupials."[139]

The differences between species are explained just as well by *LOE*#12. It is by natural selection operating in different environments. Evolution explains the similarities and it explains the differences. Darwin freely admitted that evolutionary biogeography was descriptive and not predictive. He said: "neither migration nor isolation in themselves effect anything. These principles come into play only by bringing organisms into new relations with each other and in a lesser degree with the surrounding physical conditions" (p 487). Evolution seen by biogeography can explain anything, and so it explains nothing.

In the face of these uncertainties the real question is why it was, and still is, yet so compelling? Darwin tells us. It was not the objective evidence that was so compelling, it was his faith in the Naturalistic Premise that was so compelling (§3.2). "Undoubtedly there are many cases of extreme difficulty in understanding how the same species could possibly have migrated from some one point to the several distant and isolated points, where now found. Nevertheless the simplicity of the view that each species was first produced within a single region captivates the mind. He who rejects it...calls in the agency of a miracle" (p. 488).

Now we see it. But what if ID and migration were the true explanation of biological history? It's the only alternative after all. Rejecting this off hand beforehand for emotional reasons would seem willful blindness. The real problem with evolutionary biogeography is not just that it is "for the most part, a comparative observational science rather than an experimental one",[140] and thus a descriptive enterprise of history past, nor even that its explanations are the only evidence of the explanation. It's that evolutionary biogeography rests on intuitive notions of biological similarity as being ancestral which we know to be false (§7,§8,§9). The evidence of macroevolution from biogeography is not only circular, it's empty.

Since Darwin's day when migration was considered the only means of organism dispersal, today continental drift (§13.6.3) has become accepted as another second method, also called "vicariance". The Earth was once a single supercontinent called Pangea which broke up into Laurasia and Gondwanaland before separating into the five continents we now see. Thus the organisms themselves can move to distribute themselves over the Earth or the continents can move to distribute them. These cannot be discriminated however "no method is able to 'choose objectively' between either mechanism. The interpretation of dispersal and vicariance events is pure hypothesis rather than observation."[141] And yet the biogeographical evidence of evolution is considered to be just as strong as ever it was. For instance we read in *Science* that "[b]iogeography – the study of what creatures live where and why - provides a heap of compelling evidence for evolution and natural selection"[142] and indeed it is cited everywhere as evidence by experts[135,136,143-145] (though with rare exception not[146]). The point that is missed is biogeography does not discriminate between evolution and design.

Migration and vicariance dispersal of organisms is not disputed. Evidence of the former is how at least 3 species of birds from the South American mainland were blown to the Galapagos Archipelago, a distance of 1000 km, during the 1982-3 *El Nino*.[147] (§4.7) Darwin caught a locust 370 miles off the coast of Africa (p. 498). The Indonesian island of Krakatau was repopulated within 50 years after a volcano destroyed its plant and animal life in 1883. The nearest land masses are Java and Sumatra, 40 and 80 km away respectively.

As regards evidence for variance, among the most dramatic data for continental drift are the faunal differences that Alfred Russell Wallace (§2.2) first noted between northern and southern Indonesia, a division also now called "Wallace's line". He explained them another way, of course. The reason why terrestrial mammals are not found on islands or continents like Australia is indeed simply because they could not migrate there, unlike flying mammals (bats) that could. The potential of natural selection to produce morphologic differences and the high proportion of endemics, as is claimed for island biodiversity, is not disputed either. Microevolution is not in dispute. Dog breeds as evidence for artificial selection is evidence enough. That finches in the Galapagos (§4.7) appear similar to finches on the South American mainland because they are descendants is further evidence of microevolution. To claim there are 13 separate species of finch is to overstate the facts however. These "species" are better considered varieties or subspecies. They do not begin to approach the morphologic differences among dog breeds or even among the races of humans. They interbreed and field work shows one species even trending in morphology to another (§4.5).

In summary the lines of evidence for *LOE#12* are evidence of microevolution not macroevolution no matter what the textbooks, *National Geographic*[135] or the National Academy of Sciences[116] say (Table 3.1). The problem we need to solve is not the spreading of organisms over the Earth, it's how major macroevolutionary adaptations and separate species emerged in the first place, and whether natural forces could do it alone. On these questions *LOE#12* is bankrupt.

REFERENCES

1. Futuyma, D. J. *Science on trial*. 98-113 (Sinauer Associates Inc., 1995).
2. Eldredge, N. *The Triumph of Evolution*. 60 (W.H. Freeman & Co., 2001).
3. Wood, B. & Collard, M. The Human Genus. *Science* 284, 65-71 (1999).
4. McHenry, H. M. Tempo and mode in human evolution. *Proceedings of the National Academy of Sciences USA* 91, 6780-6786 (1994).
5. Tattersall, I. *The Fossil Trail*. 115 (Oxford University Press, 1995).
6. Gee, H. *In Search of Deep Time*. 214 (The Free Press, 1999).
7. Lemonick, M. D. & Dorfman, A. One giant step for mankind. *Time*, 52-61 (2001).
8. Lewin, R. *Bones of Contention*. 312-3 (Simon & Shuster, 1987).
9. Robson, S. L. & Wood, B. Hominin life history: reconstruction and evolution. *Journal of Anatomy* 212, 394-425 (2008).
10. Dawkins, R. *The Blind Watchmaker*. 228 (W.W. Norton & Company, 1996).
11. Williams, R. W. & Herrup, K. The control of neuron number. *Annual Review of Neuroscience* 11, 423-453 (1988).
12. De Felipe, J., Marco, P., Busturia, I. & Merchan-Perez, A. Estimation of the number of synapses in the cerebral cortex: methodological considerations. *Cerebral Cortex* 9, 722-732 (1999).
13. Gur, R. *et al.* Sex differences in brain gray and white matter in healthy young adults: Correlations with cognitive performance. *Journal of Neuroscience* 19, 4065-4072 (1999).
14. Klein, R. G. *The Human Career*. (University of Chicago Press, 1989).
15. Semaw, S. *et al.* 2.5-million-year-old stone tools from Gona, Ethiopia. *Nature* 385, 333-336 (1997).
16. Leakey, L. S. B., Tobias, P. V. & Napier, J. R. A New Species of the Genus Homo from Olduvai Gorge. *Nature* 202, 7-9 (1964).
17. Elton, S., Bishop, L. C. & Wood, B. Comparative context of Plio-Pleistocene hominin brain evoution. *Journal of Human Evolution* 41, 1-27 (2001).
18. Klein, R. G. *The Human Career. Human Biological and Cultural Origins*. 222-3 (University of Chicago Press, 1989).
19. Robbins, L. H. Archeology in the Turkana District, Kenya. *Science* 176, 359-366 (1972).
20. Leakey, M. Primitive artifacts from Kanapoi Valley. *Nature* 212, 579-581 (1966).
21. Klein, R. G. *The Human Career*. 251 (University of Chicago Press, 1989).
22. Wood, B. The oldest whodunnit in the world. *Nature* 385, 292-293 (1997).
23. Klein, R. G. *The Human Career*. 211-5 (University of Chicago Press, 1989).
24. Brunet, M., Guy, F., Pilbeam, D. & al, e. A new hominid from the Upper Miocene of Chad, Central Africa. *Nature* 418, 145-151 (2002).
25. Wood, B. Hominid revelations from Chad. *Nature* 418, 133-134 (2002).
26. Gibbons, A. First member of human family uncovered. *Science* 297, 171-172 (2001).

27 Marshall, C. R., Raff, E. C. & Raff, R. A. Dollo's law and the death and resurrection of genes. *Proceedings of the American Philosophical Society* 91, 12283-12287 (1994).
28 Leakey, L. B. Homo habilis, Homo erectus and the australopithecines. *Nature* 209 (1966).
29 Wood, B. & Richmond, B. G. Human evolution: taxonomy and paleobiology. *Journal of Anatomy* 196, 19-60 (2000).
30 Asfaw, B. *et al.* Remains from Homo erectus from Bouri, Middle Awash, Ethiopia. *Nature* 416, 317-320 (2002).
31 Rightmire, G. P. *The evolution of Homo erectus*. (Cambridge University Press, 1990).
32 Rightmire, G. P. Evidence from facial morphology for similarity of Asian and African representatives of Homo erectus. *American Journal of Physical Anthropology* 106, 61-85 (1998).
33 Cowen, R. *History of Life*. 417 (Blackwell Science, Inc., 1995).
34 Klein, R. G. *The Human Career*. 183 (University of Chicago Press, 1989).
35 Stringer, C. B. & Andrews, P. Genetic and fossil evidence of the origin of modern humans. *Science* 239, 1263-1268 (1988).
36 Wolpoff, M. H. Evolution in Homo erectus: the question of stasis. *Paleobiology* 10, 389-406 (1984).
37 Rightmire, G. P. *The evolution of Homo erectus*. 180-203 (Cambridge University Press, 1990).
38 Rightmire, G. P. *The evolution of Homo erectus*. 193,197 (Cambridge University Press, 1990).
39 Tattersall, I. *The Fossil Trail*. 238-9 (Oxford University Press, 1995).
40 Klein, R. G. *The Human Career*. 191 (University of Chicago Press, 1989).
41 Tattersall, I. *The Fossil Trail*. (Oxford University Press, 1995).
42 Swisher, C. C. *et al.* Latest Homo erectus of Java: potential contemporaneity with Homo sapiens in Southeast Asia. *Science* 274, 1870 (1996).
43 Gibbons, A. Human origins: Homo erectus in Java: a 250,000-year anachronism. *Science* 274, 1841 (1996).
44 Grun, R. & Thorne, A. Dating the Ngandong humans. *Science* 276, 1575-1576 (1997).
45 Thorne, A. & Macumber, P. G. Discoveries of Late Pleistocene man at Kow swamp, Australia. *Nature* 238, 316-319 (1972).
46 Freedman, L. & Lofgren, M. Human skeletal remains from Cossack, Western Australia. *Journal of Human Evolution* 8, 283-299 (1979).
47 Freedman, L. & Lofgren, M. The Cossack skull and a diybrid origin of the Ausralian Aboringies. *Nature* 282, 299 (1979).
48 Lubenow, M. L. *Bones of Contention*. 144 156 (Baker Books, 1992).
49 Klein, R. G. *The Human Career*. 396 (University of Chicago Press, 1989).
50 Wolpoff, M. H., Hawks, J., Frayer, D. W. & Hunley, K. Modern human ancestry at the peripheries: a test of the replacement theory. *Science* 291, 293-297 (2001).
51 Klein, R. G. *The Human Career*. 355-6 (University of Chicago Press, 1989).

52 Wood, B. Human Evolution. *BioEssays* 18, 945-954 (1996).
53 Allen, J. S., Damasio, H. & Grabowski, T. J. Normal neuroanatomical variation in the human brain: an MRI-volumetric study. *American Journal of Physical Anthropology* 118, 341-358 (2002).
54 Laughlin, W. S. Eskimos and Aleuts: Their origins and evolution. *Science* 142, 633-645 (1963).
55 Johanson, D. C. & Edey, M. A. *Lucy: The beginnings of humankind.* (Simon and Schuster, 1981).
56 Yamei, H. et al. Mid-Pleistocene Acheulean-like stone technology of the Bose Basin, South China. *Science* 287, 1622-1626 (2000).
57 Gibbons, A. Chinese stone tools reveal high-tech Homo erectus. *Science* 287, 1566 (2000).
58 Morwood, M. J., O'Sullivan, P. B., Aziz, R. & Raza, A. Fission-track ages of stone tools and fossils on the east Indonesian island of Flores. *Nature* 392, 173-176 (1998).
59 Theime, H. Lower Palaeolithic hunting spears from Germany. *Nature* 385, 807-810 (1997).
60 Robinson, J. T. "Homo habilis" and the australopithecines. *Nature* 205, 121-124 (1965).
61 Robinson, J. T. Reply to Tobias. *Nature* 209, 957-960 (1966).
62 Tobias, P. V. The Olduvai Bed I Hominine with special reference to its cranial capacity. *Nature* 202, 7-9 (1964).
63 Leakey, L. S. B. A new fossil skull from Olduvai. *Nature* 184, 491-493 (1959).
64 Leakey, L. B. Finding the world's earliest man. *National Geographic*, 421 (1960).
65 Lewin, R. *Bones of Contention.* 128-151 (Simon & Shuster, 1987).
66 Tattersall, I. *The Fossil Trail.* 230 (Oxford University Press, 1995).
67 Tattersall, I. *The Fossil Trail.* 107 (Oxford University Press, 1995).
68 Tattersall, I. *The Fossil Trail.* 114 (Oxford University Press, 1995).
69 Tattersall, I. & Schwartz, J. H. *Extinct Humans.* 108 (Westview Press, 2000).
70 Tobias, P. V. *Olduvai Gorge IV: The skulls, teeth and endocasts of Homo habilis.* (Cambridge University Press, 1991).
71 Wood, B. Origin and evolution of the genus Homo. *Nature* 355, 783-790 (1992).
72 Leakey, R. E. Evidence for an Advanced Plio-Pleistocene Hominid from East Rudolf, Kenya. *Nature* 242, 447-450 (1973).
73 Lubenow, M. L. *Bones of Contention.* (Baker Books, 1992).
74 Tattersall, I. *The Fossil Trail.* 134,136 (Oxford University Press, 1995).
75 Lewin, R. *Bones of Contention.* 160 (Simon & Shuster, 1987).
76 Lubenow, M. L. *Bones of Contention.* 163 (Baker Books, 1992).
77 Tattersall, I. *The Fossil Trail.* 133 (Oxford University Press, 1995).
78 Tattersall, I. *The Fossil Trail.* 135 (Oxford University Press, 1995).
79 Leakey, R. E. & Lewin, R. *Origins Reconsidered.* 132 (Doubleday, 1992).
80 Johanson, D. C. et al. New Partial Skeleton of Homo habilis from Olduvai Gorge, Tanzania. *Nature* 327, 205-209 (1987).

81 Hartwig-Scherer, S. & Martin, S. D. Was "Lucy" more human than her "child"? Observations on eraly hominid postcranial skeleton. *Journal of Human Evolution* 21, 439-450 (1991).
82 Dean, M. C. in *Aspects of Dental Biology: Paleontology, Antrhopology and Evolution* (ed J. Moggi-Cecchi) 239-265 (International Institute for the study of Man, 1995).
83 Moggi-Cecchi, J., Tobias, P. V. & Beynon, A. D. The mixed dentition and associated skull fragments of a juvenile fossil hominid from Sterkfontein, South Africa. *American Journal of Physical Anthropology* 108, 425-465 (1998).
84 Brace, C. L., Mahler, P. E. & Rosen, R. B. Tooth measurement and the rejection of the taxon "Homo habilis". *Yearbook of Physical Anthropology* 16 (1972).
85 Patterson, B. & Howells, W. W. Hominid humeral fragment from Early Pleistocene of Northwestern Kenya. *Science* 156, 64-66 (1967).
86 Patterson, B., Behrensmeyer, A. K. & Sill, W. D. Geology and fauna of a new Pliocene locality in North-western Kenya. *Nature* 226, 918-921 (1970).
87 Lubenow, M. L. *Bones of Contention*. 52-58 (Baker Books, 1992).
88 McHenry, H. M. & Corruccini, R. S. Distal humerus in hominoid evolution. *Folia Primatol (Basel)* 23, 117-144 (1975).
89 McHenry, H. M. Fossils and the mosaic nature of human evolution. *Science* 190, 425-431 (1975).
90 Howells, W. in *Homo erectus: Papers in honor of Davidson Black* (eds BA Sigmon & JS Cybulski) 79-80 (University of Toronto Press, 1981).
91 Leakey, M. G., Feibel, C. S., McDougall, I., Ward, C. & Walker, A. C. New four-million-year-old hominid species from Kanapoi and Allia Bay, Kenya. *Nature* 376, 565-571 (1995).
92 Leakey, M. G., Feibel, C. S., McDougall, I., Ward, C. & Walker, A. C. New specimens and confirmation of an early age for Australopithecus anamensis. *Nature* 393, 62-66 (1998).
93 Day, M. H. in *Early hominids of Africa* (ed C.J. Jolly) 311-345 (Duckworth, 1978).
94 Senut, B. Humeral outlines in some hominoid primates and in Plio-Pleistocene hominids. *American Journal of Physical Anthropology* 56, 275-283 (1981).
95 Senut, B. & Tardieu, C. in *Ancestors: the hard evidence* (ed E. Delson) 193-201 (Alan R. Liss, 1985).
96 Lague, M. R. & Jungers, W. L. Morphometric variation in Plio-Pleistocene hominid distal humeri. *American Journal of Physical Anthropology*, 401-427 (1996).
97 Leakey, M. & Hay, R. Pliocene footprints in the Laetolil beds at Laetoli, northern Tanzania. *Nature* 278, 317-323 (1979).
98 Hay, R. & Leakey, M. The fossil footprints of Laetoli. *Scientific American* 246, 50-57 (1982).
99 Leakey, M. & Harris, J. *Laetoli: a Pliocene site in northern Tanzania*. (Clarendon, 1987).
100 Leakey, M. Tracks and tools. *Philosophical Transactions of the Royal Society of London* B292, 95-102 (1981).

101 Tuttle, R. E. in *Laetoli. A Pliocene site in Northern Tanzania* (eds M.D. Leakey & J.M. Harris) 503-520 (Oxford University Press, 1987).
102 Leakey, M. Footprints in the ashes of time. *National Geographic* April, 446 (1979).
103 Hay, R. & Leakey, L. The fossil footprints of Laetoli. *Scientific American* 246, 50-57 (1982).
104 Day, M. & Wickens, E. Laetoli Pliocene hominid footprints and bipedalism. *Nature* 286, 385-387 (1980).
105 Day, M. H. in *Hominid Evolution: Past, Present and Future* (ed PV Tobias) 115-127 (Alan R. Liss, 1985).
106 Tuttle, R. E. Evolution of hominid bipedalism and prehensile capabilities. *Philosophical Transactions of the Royal Society of London* B292, 89-94 (1981).
107 Tuttle, R. H. in *Hominid Evolution: Past, Present and Future* (ed PV Tobias) 129-133 (Alan R. Liss, 1985).
108 White, T. & Suwa, G. Hominid footprints at Laetoli:facts and interpretations. *American Journal of Physical Anthropology* 72, 485-514 (1987).
109 Charteris, J., Wall, J. & Nottrodt, J. Pliocene hominid gait: new interpretations based on available footprint data from Laetoli. *American Journal of Physical Anthropology* 58, 133-144 (1982).
110 Robbins, L. M. in *Laetoli. A Pliocene site in Northern Tanzania* (eds M.D. Leakey & J.M. Harris) 497-502 (Oxford University Press, 1987).
111 Tuttle, R. E. The pitted pattern of Laetoli feet. *Natural History*, 60-65 (1990).
112 Clarke, R. Early hominid footprints from Tanzania. *South African Journal of Science* 75, 148-149 (1979).
113 Leakey, M. D. & Hay, R. L. Pliocene footprints in the Laetolil Beds and Laetoli, northern Tanzania. *Nature* 278 (1979).
114 Walter, R., Manega, P., Hay, R., Drake, R. & Curtis, G. Laser fusion 40Ar-39Ar dating of Bed 1, Olduvai Gorge, Tanzania. *Nature* 354, 145-149 (1991).
115 Trinkaus, E. Hard Times Among the Neanderthals. *Natural History* 87, 58 (1978).
116 National Academy of Sciences. *Science and Creationism: A View from the National Academy of Sciences*. 2nd Ed edn, (The National Academies Press http://books.nap.edu/html/creationism/, 1999).
117 Mayr, E. *One Long Argument*. (Harvard University Press, 1991).
118 Zuckerkandl, E. in *Classification and human evolution* (ed S.I. Washburn) (Aldine de Gruyter, 1963).
119 Diamond, J. M. *The Third Chimpanzee*. (Perennial, 1992).
120 Goodman, M. The genomic record of humankind's evolutionary roots. *American Journal of Human Genetics* 64, 31-39 (1999).
121 Wallace, A. R. in *Contributions to the Theory of Natural Selection* (Macmillan, 1870).
122 Simpson, G. G. *The Meaning Of Evolution*. 345 (New American Library Mentor Book, 1953).
123 Russell, B. *Religion and Science*. 233 (Oxford University Press, 1935).
124 Judson, H. *The eighth day of creation*. 217 (Simon & Shuster, 1979).
125 Monod, J. *Chance and Necessity*. 112,145-6 (Collins, 1972).

126 Monod, J. *Chance and Necessity.* 167 (Collins, 1972).
127 Dobzhansky, T. A Biologists world view. *Science* 175, 49-50 (1972).
128 Dawkins, R. *The Selfish Gene.* 3 (Oxford University Press, 1989).
129 Churchland, P. *Matter and Conciousness: A contemporary introduction to the philosophy of mind.* 12 (M.I.T. Press, 1984).
130 Huxley, T. H. *Westminster Review* 17, 541-570 (1860).
131 Huxley, T. H. *Evolution and Ethics, and Other Essays.* (Macmillan, 1901).
132 Friend, D. *The Meaning of Life* in *Life* 33 (Little, Brown and Co. 1991).
133 Dawkins, R. God's utility function. *Scientific American*, 85 (1995).
134 Johnson, P. *A History of the Modern World.* 5 (Weidenfeld & Nicolson, 1983).
135 Quammen, D. Was Darwin Wrong? *National Geographic*, 4-35 (2004).
136 Mayr, E. *What Evolution Is.* 12-39 (Basic Books, 2001).
137 Mayr, E. *What Evolution Is.* 31 (Basic Books, 2001).
138 Raven, P. H. & Johnson, G. B. *Biology.* 453-4 (McGraw-Hill, 2002).
139 Cifelli, R. L. & Davis, B. M. Marsupial Origins. *Science* 302, 1899-1990 (2003).
140 Brown, J. H. & Lomolino, M. V. *Biogeography.* (Sinauer Associates, Inc., 1998).
141 Ebach, M. C. & Humphries, C. J. Cladistic biogeography and the art of discovery. *Journal of Biogeography* 29, 427-444 (2002).
142 Leslie, M. Netwatch: Birth of biogeography. *Science* 296, 1771 (2002).
143 Futuyma, D. J. *Science on trial.* 197-207 (Sinauer Associates Inc., 1995).
144 Grant, V. *The evolutionary process. A critical review of evolutionary theory.* (Columbia Unversity Press, 1985).
145 Freeman, S. & Herron, J. C. *Evolutionary Analysis.* 2nd edn, (Prentice-Hall, Inc., 2001).
146 Ridley, M. *Evolution.* 40-68 (Blackwell Science, Inc., 1996).
147 Patterson, C. *Evolution.* 2nd edn, (Cornell University Press, 1999).

12. The origin of species

§12.1   BACK TO THE FUTURE

The inquiry of the last 8 chapters has been a search of evidence for the claim that the diversity, unity, complexity, adaptiveness and similarities of living things which is the wonder and the panorama of life, arose solely by natural forces (*LOE*#1-12). We switch gears now to focus on another matter, albeit one intimately related, the origin of the very first life. While evolution is all about natural forces causing new biological complexity, it requires some pre-existing complexity to work. It requires a self-replicating genetic system by which offspring are generated. Where did that come from? What came before mutation and natural selection? How did life begin?

    In case there's any doubt in your mind about the answer, rest assured there's no doubt in the mind of science. Like the cause of the wonder and the panorama of life this happened entirely naturalistically too. It's worth making sure of this before going further. Here's the view of the National Academy of Science for instance:

> "For those who are studying the origin of life, the question is no longer whether life could have originated by chemical processes involving nonbiological components. The question instead has become which of many pathways might have been followed to produce the first cells."[1]

Here is Douglas Futuyma:

> "We will almost certainly never have direct fossil evidence that living molecular structures evolved from nonliving precursors. Such molecules surely could not have been preserved without degradation.

But a combination of geochemical evidence and laboratory experiment shows that such evolution is not only plausible, but almost undeniable."[2]

Here is Michael Ruse:

"It can therefore be stated with some confidence that, at least in outline, we know how life evolved from its earliest beginnings."[3] "Origins are no longer the mystery that they were once."[4]

The list of voices could go on (e.g.)[5-9] but Nobel laureate and Rockefeller University scientist Christian de Duve gets the last word for now:

"But one affirmation can safely be made, regardless of the actual nature of the processes that generated life. These processes must have been highly deterministic. In other words, these processes were inevitable under the conditions that existed on the prebiotic earth...Life is a cosmic imperative."[10]

There should be very good scientific evidence then, given consensus about such a monumental claim. And yet this is what Richard Cowen at the University of California at Davis says, and he is an evolution advocate[11]: "Life on Earth is a fact, yet we don't know where and how it began."[12] For science to be so certain his voice comes as a little unsettling. Somebody is not telling the truth, that much we can tell already. The question is whether it's our first whiff of another scientific disconnect. We will need to look at the evidence ourselves to tell, but to cut to the conclusion and in a theme repeated in every chapter so far, we will discover although the primary data about the beginning of life is interesting, what is even more interesting is how science interprets that data, sees an evolutionary solution there, and stays certain it cannot be design despite contradiction. There's a scientific disconnect all right. Let me show you how big it is.

The reason for the disconnect? Methodological naturalism (§1.6;§3.3-4). The origin of life must have a naturalistic cause or the solution is not scientific or rational (§19;§21;§22). The spontaneous generation of life is true already, whatever the evidence might say. (§20). If that sounds crazy that's because it is crazy. Take a look.

It might come as a surprise to learn Charles Darwin never addressed the question of how life began in *The Origin of the Species*. Given the claim of his title the omission is at least ironic. The subject could hardly be more important to his thesis. What's the point of proposing an entirely naturalistic, non-saltatory mechanism for the diversity and complexity and order of life if it still demanded a fantastic supernatural creation to start it off? For if there is one supernatural intervention like that, why not more than one? And if there is even one the boasts of Darwinism are drained of their philosophical power that reside in the self-sufficiency of naturalism.

Darwin made only a passing allusion to ultimate origins in *The Origin,* and it was in the very last sentence of the book. There he concluded life was:

"originally breathed by the Creator into a few forms or into one...from so simple a beginning endless forms most beautiful and most wonderful have been, and are being evolved." (p.649).

If we did not know better we might suspect he was a creationist from this! It's much more likely he had a naturalistic opinion of life's origin given his subsequent writings. In 1871 he wrote:

"It has often been said that all the conditions for the first production of a living organism are now present which could ever have been present. But if - (and Oh! what a big if!) - we could conceive in some warm little pond, with all sorts of ammonia and phosphoric salts, light, heat, electricity, etc., present, that a protein compound was chemically formed ready to undergo still more complex changes, at the present day such matter would be instantly devoured or absorbed, which would not have been the case before living creatures were formed."[13]

The hypothesis that Darwin so speculatively advanced here, more than one and a half centuries ago, is a major paradigm for origin-of-life research still. There are other research paradigms we will also hear about, but all of them emphatically agree about this: living things arose, spontaneously, from chemicals.

It's an extraordinary idea come to think of it. It's not significantly removed from the widely accepted nineteenth century notion of "spontaneous generation". Spontaneous generation is the process by which dead things give rise to living things. Examples offered as the proof were microorganisms and maggots that appeared from nowhere in the food or drink left out on a laboratory bench.

Louis Pasteur (1822-1895) finally debunked the myth in 1859, in a public competition convened to solve the mystery. By comparing what happened to a broth after dividing it into two flasks, one with a narrow S-shaped neck with an opening pointing down, and another with a wide neck pointing up, he showed it was atmospheric contamination by bacteria and fungal spores and not inanimate chemicals that was the source and cause of that new pungent life. Life came from pre-existing life not from spontaneous generation. "Never will the doctrine of spontaneous generation recover from the mortal blow of this simple experiment!," Pasteur exclaimed.[14] *The Origin* was published that year.

Pasteur he saw no contradiction between science and creationism and rejected Darwinism. He said "science brings men nearer to God"[15] and "The more I study nature, the more I stand amazed at the work of the Creator".[16]

His demonstration contradicted claims of a natural origin of life. Darwin was frank about the problem for his theory. "I have met with no evidence that seems in the least trustworthy, in favor of so-called Spontaneous Generation."[17] It took another foe of Darwinism, physicist William Thompson (also called Lord Kelvin 1824-1907), to point out the common sense to everyone: "Dead matter cannot become living without coming under the influence of matter previously alive". Has the evidence changed? We expect so because the consensus of science has. We need to review the state of the science of the origin of life so buckle up. Keep your wits about you. Science by methodological naturalism can be a wild ride.

In Darwin's day it was assumed without any evidence mind you, that the cell was a very simple structure. As a result it seemed reasonable to believe it could arise entirely by chance processes. Ernst Haeckel called the cell a mere "homogeneous globule of plasm" and a "simple little lump of albuminous combination of carbon".[18] We know it's the opposite. Editor of the journal *Science* and past President of the National Academy of Sciences Bruce Alberts has this assessment:

> "The entire cell can be viewed as a factory that contains an elaborate network of interlocking assembly lines, each of which is composed of large protein machines...Why do we call the large protein assemblies that underlie cell function *machines*? Precisely because, like the machines invented by humans...these protein assemblies contain highly coordinated moving parts."[19]

Setting aside for now the question of how the intricate harmony and complexity of the cell could also become a "factory" possessing "machines" and so demonstrate purpose beyond its own brute structure (§15.3), how did the simplest life arise in the first place? The "simplest" forms of life in the fossil record - bacteria, look just like modern day bacteria without simpler forms before them (§5.5.2). (While viruses are indeed "simpler" than bacteria, they require the preexistence of a bacterial or eukaryotic host cell and therefore "[e]volutionarily speaking...are late arrivals on the Darwinian stage").[20] To answer, the naturalistic hypothesis has three requirements which must be satisfied to get a self-replicating biological system going to permit descent with modification and thereby evolution make nature. These requirements can be considered the steps needed to kick-start biological evolution, from chemicals, spontaneously.

First we need the building blocks for a self-replicating system. Next we need monomers to become polymers which self-replicate with modifications. The third step is when replication with mutation generates so much complexity as to eventually cause life and then eventually all of the diversity of life. This is standard, non-controversial thinking in the science now. We therefore ask the following questions to inspect the evidence for the idea. Regarding Step 1: Where did the small organic biopolymer building blocks for a self-replicating system originate? (§12.2) Regarding Step 2: How did these biopolymers organize into a self-replicating system to kick-start evolution? (§12.3) Regarding Step 3: Can replication and selection of mutation of a biopolymer generate macroevolution and cause all of life in the time available? Or ever? (§12.4;§16.7)

§12.2　　THE ORIGIN OF BIOPOLYMERS (STEP I)

Where did the small organic biopolymer building blocks for a self-replicating system come from? There are three principal theories.

§12.2.1　　The Darwin-Haldane-Oparin Primordial Soup

This idea was first developed by John B.S. Haldane and Alexander Oparin independently in the 1920's and is still standard textbook teaching.[21,22] It proposes the atmosphere of the early earth was reducing (i.e. rich in hydrogen) and that organic compounds accumulated under the action of energy sources like lightning in a "hot dilute soup" to use Haldane's now famous words.[21] The soup was a "vast chemical laboratory" from which life eventually arose. These ideas remained untested until 1953, the year the structure of DNA was deciphered by James Watson and Francis Crick. That same year a continent away, a graduate student named Stanley Miller at the University of Chicago working in the laboratory of Nobel laureate Harold Urey made an incredible discovery.

He was using the apparatus shown in Fig.12.1 to pass 60,000-volt sparks through an atmosphere of hydrogen, methane, ammonia and water vapor. His experiment was intended to simulate the Precambrian environment in a lightning storm. After several days a red precipitate formed on the sides of the flask. Subsequent analysis found this to be a 15% yield of carbon from the methane and to consist of organic compounds, most notably the amino acids glycine (a 2.1% yield), alanine (a 1.7% yield) and aspartic acid (a 0.026% yield). This impressive result was published in a modest 2-page report in *Science* the same year.[23] Analysis of samples withdrawn serially during the experiment demonstrated that hydrogen cyanide (HCN) and aldehydes were formed first, and the reaction products were predominantly α-substituted amino- and hydroxy- acids.[24] This indicated the Strecker synthesis reaction mechanism was being used as represented by the following formula:

$$RCHO + HCN \rightarrow \rightarrow RCH(OH)CN \rightarrow NH_3 \rightarrow RCH(NH_2)CN$$
$$\downarrow H_2O \qquad\qquad \downarrow H_2O$$
$$RC(OH)COOH \qquad RCH(NH_2)COOH$$

Subsequent experiments demonstrated various sugars could be also be made, as well as the nucleic acid base adenine and cyanoacetylene which is a potential precursor to the nucleic acid bases uracil and cytosine.[25] The results were not only informative about gas phase organic chemistry, they have great significance for cosmo-chemistry. For instance in explaining why organic molecules of similar identity and relative concentration the experiment are found in meteorites (§12.2.2).[24] What's the problem then? The problem is Miller's results were hailed as far more than this. They were, and in many quarters still are, taken as no less than the way the first biomolecules arose on the Earth. The question for us then, is whether the Miller-Urey experiments have this or anything to do with origin-of-life scenarios?

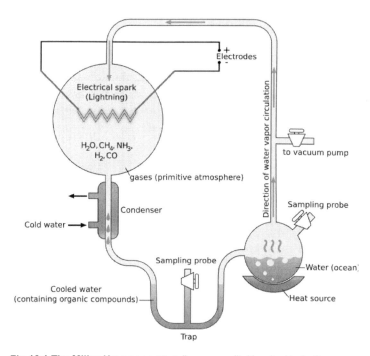

Fig.12.1 The Miller-Urey apparatus (Image credit: YassineMrabet)

They do not. The reason is Miller's model and the Haldane-Oparin theory that it tests both assume the early earth atmosphere was strongly reducing, with high concentrations of hydrogen, methane and ammonia. Instead the early earth atmosphere is believed to have been neutral or of slight reducing power[27] and "dominated by CO_2 [carbon dioxide]and N_2 [nitrogen], with traces of CO [carbon monoxide], H_2 [hydrogen] and reduced sulfur gases"[26] [specifically hydrogen sulphide (H_2S) and sulphur dioxide (SO_2)]. The Earth's gravitational field is unable to hold hydrogen to keep it reducing, and is lost into space precluding accumulation. Hydrogen is essential for reducing carbon dioxide and nitrogen to methane and ammonia in the experiments however. Without hydrogen neither could these compounds have been significant components of the early earth atmosphere either.[26,28] Another problem is that ultraviolet light is estimated to have been four to eight times greater than it is today. This would have dissociated any methane or ammonia photochemically preventing accumulation anyway.[27,29]

Another problem of the Miller-Urey simulation is that evidence indicates the early earth atmosphere contained oxygen.[26,29] The oxygen was from two sources, from ultraviolet light dissociating the water vapor (H_2O) in the atmosphere into oxygen and hydrogen,[29] and from volcanic outgassing of oxygen. A reducing atmosphere, and specifically a ratio of hydrogen to carbon monoxide (H_2/CO) of greater than one, and hydrogen to carbon dioxide (H_2/CO_2) of greater than two as well as the presence of methane, is essential for the success of the Miller-Urey experiment.[24] This is not just because the presence of oxygen in the apparatus will cause it to explode, but oxidization will break down the methane and ammonia into carbon dioxide, water and nitrogen. With the loss of methane and ammonia, spark discharges do not now make amino acids or any other significant organic molecules.[30-32] Salk Institute origin-of-life scientist Leslie Orgel sums up. "Recent studies have convinced most workers concerned with the atmosphere of the early earth that it could never have been strongly reducing. If this is true, Miller's experiments, and most other studies of prebiotic chemistry, are irrelevant".[33] We need another theory.

§12.2.2   The "Impact" Theory

This hypothesis offers a potential lifeline to rescue the primordial soup theory however. It contends that organic material was made in space, and delivered pre-made to the Earth's oceans by meteorites and interplanetary dust. Current consensus in the science is that meteorites struck the Earth with high intensity for about half a billion years after its formation 4.6 billion years ago. By some estimates the amount of material delivered was as much as 50,000 tons a year.[34] Even today about 300 tons of organic material reaches Earth from space annually.[35] As a result the study of carbon-rich meteorites called "carbonaceous chondrites" is of great interest. The Murchison meteorite that fell on Australia in 1969 is among the best characterized of these. It contains about 2% carbon and a variety of amino, sulphonic, phosphoric and hydroxy acids, as well as trace amounts of nucleic acid bases. All in all a very similar profile of compounds to those generated by the Miller-Urey apparatus.[36,37]

However there is still no proof for the impact theory, and many doubt that significant amounts of these compounds could survive the heat generated by meteorite entry into the Earth's atmosphere. Neither of the two hypotheses can explain a significant source of lipid amphiphiles needed for cellular and subcellular membranes either as the Miller-Urey experiments can only generate short chain fatty acids.[38] As a result "the formation of membrane structures of the kind needed for the earliest cells in this theory is an unsolved problem".[39]

A still bigger problem for both the impact and the primordial soup hypotheses, is there is no support from the fossil record of an amino acid-rich and nucleic acid-rich ocean by nitrogen-rich deposits in the Precambrian sedimentary rock.[40] The evidence indicates the opposite. Only nanomolar concentrations have been found, consisting mainly of glycine, and even that may even have been derived from the high concentrations of ammonium acetate used for the extraction process in the simulation experiment. The take-home point is this: "If there ever was a primitive soup, then we would expect to find at least somewhere on this planet either massive sediments containing enormous amounts of the various nitrogenous organic compounds, amino acids, purines, pyrimidines and the like, or alternatively in much metamorphosed sediments we should find vast amounts of nitrogenous cokes. In fact no such materials have been found anywhere on earth. Indeed to the contrary, the very oldest of sediments…are extremely short of nitrogen"[41]. We need another theory.

§12.2.3   The Hydrothermal Vent Theory

This theory was first proposed by organic chemist and German patent attorney Gunter Wächtershäuser. The main problem with his theory is that it has remained a theory. It holds there was no prebiotic soup and that organic molecules were formed on iron-sulfur surfaces, specifically iron pyrite, by submarine volcanism in reactions involving transition metal irons, heat, high pressure and hydrogen sulfide.[42] The location of these reactions in deep ocean is proposed to have offered shielding to protect emerging life from bombardment by asteroids and comets believed by the science as striking the earth back then. Hydrogen sulfide in the presence of iron(II) sulfide is a reducing agent, and the energy source for the reaction is from iron(II) sulfide and hydrogen sulfide, forming hydrogen and iron pyrite (FeS_2). Iron pyrite is the chemical term for fools' gold. This association that has not been lost on the skeptics. A notable strength of the theory is it can explain the formation of activated thiocarboxylic acids from which peptide synthesis by polymerization is possible, without the need for further activation. There are several difficulties however.

For while the reduction of carbon monoxide (CO) has been demonstrated, the all-important reduction of carbon dioxide (CO_2) to make methane has not. Furthermore when the theory is tested experimentally, "no amino acids, purines or pyrimidines are produced"[43] and thus no monomers are made at least as regard proteins and nucleic acids. The stability of organic molecules under the harsh conditions of volcanism (which might be expected to destroy or denature them) is also suspect as organic compounds (amino acids, proteins and nucleic acids in particular) are degraded at high temperatures. Modern day bacteria that live in hot springs have sophisticated methods to cope with this problem, and how exactly those developed all by themselves without design is another question begged. Wächtershäuser remains undeterred. Stanley Miller is among his critics. Miller says: "The high temperatures of the [deep ocean] vents would not allow the synthesis of organic compounds but would decompose them, unless the exposure time at vent temperatures was short. Even if the essential organic molecules were available in the hot hydrothermal waters, the subsequent steps of polymerization and the conversion of these polymers into the first organisms would not occur as the vent waters were quenched to the colder temperatures of the primitive oceans."[44] So much for the third theory. What can we say by way of summary?

There is no agreement about the source of prebiotic monomers and all the theories proposed so far have intractable problems. Any experimental support for a solution to the origin of life is absent, emphatically, and has been since before Pasteur's day. The claim that a naturalistic explanation exists for the origin of life is a lie (§25).

But what if biomonomers could arise spontaneously by an as yet undiscovered mechanism? This would be a big step forward, even though we would still be far from getting life. The next problem is how they could self-organize in a way to self-replicate. Let's look at this second step and inspect what evidence exists to account for that process, assuming the first step has got itself off the ground already somehow.

§12.3   THE ORIGIN OF A SELF-REPLICATING SYSTEM (STEP 2)

A fundamental feature of all life is an organization whereby genetic information for self-replication is stored in the form of nucleic acids. Apart from a few viruses that use ribonucleic acid (RNA), the rest of nature uses deoxyribonucleic acid (DNA). Nucleic acids specify the composition of the proteins in the cell by the sequence of their component bases that specify by this blueprint molecule the sequence of the amino acid constituents of a protein (§2.3.3). All proteins are derived from nucleic acids therefore. However proteins are required for nucleic acids to exist because the enzymes are proteins and enzymes are need to catalyze the replication and translation of DNA (§15.3.1-3). There is a "flow of biological information" then, a cycle of interdependence between nucleic acids and proteins, in all cells, whereby the one is required for the existence of the other. This creates a conundrum when a solution is sought as to how this chicken-and-egg started. Not only are these molecules highly complicated, they have the appearance of being "irreducibly" complex (§15.3). How do we solve this without admitting this obvious appearance is design?

A spontaneous concurrent emergence of both proteins and nucleic , functionally interdependent, would solve the problem. Spontaneously arising with that amount of complexity is a spectacular supernatural event in all but name however, and everywhere in science dismissed. A "protein-first" view is supported by experiments demonstrating abiotic polymerization of amino acids into molecules called proteinoids. Some of them have catalytic activity, including the formation of additional peptide bonds and nucleotide bonds (the chemical bonds between amino acids the nucleic acid monomers respectively). Sidney Fox (1912-1998) at the University of Miami has shown that proteinoids can even form vesicles called microspheres that he considers protocells.

The problem with the protein-first model is that proteins have no heritable mechanism. How do they replicate themselves given their typical 3-D globular shapes and concealed hydrophobic groups, without a known templating mechanism? How would descent with modification arise from a protein that serves as a "gene", and how does one get to nucleic acid genes and translation of nucleic acids? Nobody knows. Consider the alternative then. The "gene first" view.

§12.3.1   The RNA World Hypothesis

The "nucleic-acid first" theory is more promising in resolving this dilemma. Postulating RNA as the first biopolymer, this theory was first proposed in the 1960s by Leslie Orgel at the Salk Institute, Francis Crick at the Medical Research Council in England and Carl Woese at the University of Illinois. It was later dubbed the "RNA world" hypothesis by Harvard biologist Walter Gilbert, a name that still sticks.[45] Unlike proteins, RNA can serve both as a carrier of genetic information and as an enzyme, a property once thought restricted only to proteins. The RNA world hypothesis holds RNA was the original molecule of heredity and that it catalyzed all the reactions necessary for the first life to survive and replicate. Proteins and DNA evolved subsequently. It also represents the mainstream view of origin-of-life researchers today.[33,46]

This theory has appeal for several reasons. RNA is known to be the store of genetic information in certain viruses like HIV, it's more easily synthesized than DNA (and which itself is made from RNA), most cellular co-enzymes are nucleotides or compounds which plausibly could be derived from nucleotides, and certain RNA molecules can fold into complex three dimensional domains with enzyme activity. These latter molecules are called "ribozymes", a term derived from "ribonucleic acid" and "enzyme". Thomas Cech at the University of Colorado and Sidney Altman at Yale University were awarded the 1989 Nobel Prize for discovering these molecules.

Another chicken-and-egg predicament immediately arises however. Without evolution no mechanism exists for a self-replicating ribozyme to be spontaneously generated, but without some form of self-replication there is no way to evolve into the first self-replicating ribozyme. To accommodate this, the RNA world hypothesis of "nucleic acids before proteins" makes three crucial assumptions: 1) the RNA in the earliest cell could replicate without protein because of a so-called "autocatalytic replicase function"; 2) it could catalyze every step of protein synthesis, and 3) meaningful sequences of RNA spontaneously emerged, the so-called "template function".

The first assumption is accepted with the description of ribozymal enzyme activity. However in the more than half century since the RNA world theory was proposed, none of the necessary enzymatic properties of the second assumption have been shown for any RNA. Even peptide bond synthesis which would join amino acids up has not been demonstrated.[47] To be sure an increasing number of ribozyme activities are described, but they remain restricted (consisting of self-splicing activity of some group I and II introns, self-cleaving activity of hammerheads and related RNA structures, transfer RNA processing by an RNA component of a protein enzyme called RNAseP, and peptidyl transferase activity).[48] None of these actions can be related to a postulated prebiotic function. The reactions are also performed by rather large RNA molecules, raising questions how these could plausibly have spontaneously arisen. The crucial difference between these ribozyme activities and that of the polymerases they attempt to mimic, is the latter use an arbitrary external template (enabling them to replicate any sequence of bases), while ribozymes are sequence specific - they bind specific substrates and catalyze specific reactions. Theoretical biologist Stuart Kauffman of the Santa Fe Institute summarized the state of the art back in 1993 and it hasn't changed since: "In summary, to date no means have been found to achieve template-directed replication of arbitrary RNA sequences by sequential addition of monomers to an arbitrary external template."[49]

This has not deterred the optimism that such a function will be found. In a News and Views article written for *Science* in the same year, it was claimed "Betting odds are that a self replicating RNA molecule will be prepared in the next decade."[50] If you had responded with your wallet you could have made some money. What has been found since then though, by David Bartel and Jack Szostak at Harvard, is that ribozymes isolated from a large pool of random sequences were able to ligate RNA molecules with a 3',5' phosphodiester bond.[51] The reaction required extensive manipulation, including the use of pre-synthesized DNA sequences, a nucleotide template and immobilization of the RNA pool molecules on agarose beads to prevent their aggregation and precipitation out of solution given the high concentrations necessary. In view of these hurdles, a realistic scenario for a spontaneous origin is not remotely in view. As for their view, they concluded "an RNA replicase could only have arisen from primordial sequence pools that were not truly random", and from "a truly enormous amount of RNA", and that some other non-enzymatic mechanism is required to generate the template sequence.

An even bigger problem for the RNA world hypothesis relates to its third assumption, namely template function. Specifically how ribonucleotide monomers could spontaneously have self-organized themselves into ordered template base sequences with meaning for protein synthesis. The molecular structure of RNA is demonstrated in Fig.12.2. It consists of a long string typically folded, of purine (adenine and guanine) or pyrimidine (thymidine and uracil) nucleotide bases joined at the nitrogen 9 position to the carbon 1 on ribose (C1-N9 glycosidic bond). The backbone of the molecule is formed by ribose monomers joined by 3'-5' phosphodiester bonds. What is known about the abiotic synthesis of RNA?

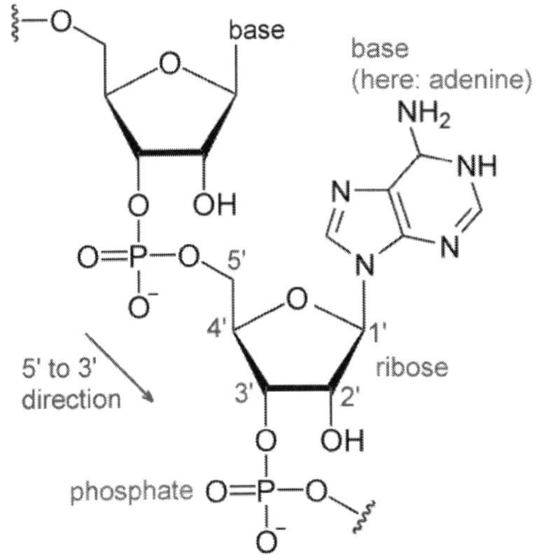

Fig.12.2 The structure of RNA. The backbone is formed by a long string of ribose monomers (only 2 shown) on the left joined together by 5'-3' phosphodiester bonds. The bases adenine, guanine, cytosine or uracil (adenine shown) attach to ribose through a C1-N9 glycosidic bond. (Image credit: Daniel Ramsköld).

As regards the purines (adenine and guanine), they can be made by self-condensation of hydrogen cyanide (HCN) using Miller-Urey conditions. The yield of adenine is low and its susceptibility to hydrolysis a concern for stability. As regards the pyrimidines (uracil and cytosine), uracil synthesis occurs either by acid hydrolysis of hydrogen cyanide to orotic acid which then undergoes photochemical decarboxylation, or by condensation of β-alanine and cyanate to dihydrouracil which then undergoes photochemical dehydrogenation. Cytosine synthesis is more problematic due to low yields "requiring concentrations of reactants [1M potassium cyanate and 0.1M cyanoacetylene] unlikely to have occurred on the primitive Earth" [52]. In a review in *Nature*, Scripps origin of life biologist Gerald Joyce concluded "Neither of these pathways for the prebiotic synthesis of pyrimidines is especially convincing".[52] These abiotic nucleotide synthetic pathways are not possible in conditions simulating the early earth atmosphere, and totally different from those used by living cells today. For example cells make purines though a sequence of intermediates requiring over 11 separate reaction steps! (§15.3.4)

As regards ribose, it is made autocatalytically by the formose reaction from formaldehyde (HCHO) and postulated to have been generated by solar ultraviolet photolysis of water vapor and carbon dioxide.[53] The formose reaction produces a complex mixture of more than 40 sugars including trioses and hexoses in addition to pentoses however, only one of which is ribose. Just among pentoses there are 12 stereoisomers, 8 aldoses and 4 ketoses. In other words there is a "dizzying array of related compounds"[52], in a racemic mix, all competing with D-ribose to bind to a nucleotide. How ribose and the correct pyrimidine or purine nucleotide might preferentially be selected from other potential reactants by abiotic RNA synthesis has no present explanation whatsoever.[54]

The stability of ribose is low also (73 minutes at 100°C and 44 years at 0°C), compared to hundreds of years for amino acids.[55] Even assuming the correct molecular coupling C1-N9 of pentose to nucleotide to generate mononucleosides notwithstanding these difficulties, and the problem of confounding side reactions, there are the further hurdles of generating multiple products with phosphorylation of nuclesides,[56] degradation, internal cyclization reactions, the tendency for nascent nucleotide chains to fold back onto themselves as a result of Watson-Crick complementary binding, and the difficulties of separating daughter from parental strand even if copying were to occur. And if a molecule were formed, it would need to associate with others like it before degrading again, like short-lived needles trying to finding each other in haystack in a race against time by an undirected random process.

Another problem with the notion of abiotic RNA synthesis, and for that matter DNA and protein and carbohydrates too, is the curious fact these essential life molecules are homochiral. The term chiral comes from the Greek *cheir*, or hand, and refers to their left or right hand mirror-image conformations or stereoisomers (Fig.8.1). With extremely rare exception biochemistry involves only the levo or left-handed form for amino acids, and the dextro-or right handed form for carbohydrates and nucleotides. By contrast all abiotic chemistries generate a racemic 50:50 mix of D and L forms. The homochirality attribute of biochemistry was first discovered by Louis Pasteur. In biochemistry if the "wrong" D-amino acid, L-nucleotide or L-sugar enantiomer adds to the growing end of a molecule, it prevents further chain elongation called enantiomeric cross-inhibition.[57]

Stuart Kauffman raises still further issues that need confronting.[49] Supposing somehow a replicating double stranded RNA molecule did form. It would represent a "nude gene". In other words genetic material and nothing else. How did it surround itself with the complex and interdependent web of chemical reactions that is cellular metabolism today? (§15.3) DNA is catalytically inert of course and it is proteins that perform these actions in the cell, not only metabolic pathways of carbohydrate and lipid biochemistry, but the translation, transcription and replication of nucleic acids also. This tight coupling of nucleic acids and proteins is as a result of the genetic code linking them. How did that arise, progressively, by evolution? As he says: "Its emergence seems to require its prior existence"[49]. He explains the crux of the problem:

"The coding problem, intense as it is, is really a fragment of a larger problem: How would a nude gene gather a coupled metabolism about itself?...It can be examined in a different form, a form which has received little attention. The simplest free-living entities, the pleuromona-like organisms, are on the order of 0.1 the size of a bacterium, have a very simplified cell boundary with a simple bilipid layer, and have a genome which encodes perhaps 1000 to 2000 proteins: these proteins play the usual structural and enzymatic roles in a coupled metabolism. Viruses can be as simple as RNA or DNA strands encoding perhaps a dozen proteins. But viruses are not free living entities; they are obligate parasites forced to usurp the machinery of their host cells in order to carry out their own replicative life cycle. Viruses are highly sophisticated parasites which have virtually certainly managed to simplify their metabolic system because of that present in the host. Thus it is an observed fact that all free-living organisms exhibit *a minimal and substantial level of complexity.* The deep question is, Why?"[58]

Indeed.

Of course the RNA or DNA first theory does not answer his question. The self-templating function of the theory is a simple mechanism that does not need this extra baggage of complexity. The standard explanation is progressive evolution took place, over eons, with the nude gene gradually accumulating metabolic pathways about itself and as it did, obtaining greater and greater competitive advantage over the more simple forms it overcame. Kauffman responds: "Yet it's only answer to why we now observe a given minimal level of complexity is historical accident. Entities having the complexity of pleuromona-like organisms happen to be the simplest free-living survivors. Why do we not observe systems as simple as viruses, or even simpler, living in Darwin's shallow pond? Because, we respond, such hapless forms would soon be outcompeted by present-day organisms. In short, the replicating RNA theory, ribozyme polymerase or otherwise, offers no *theory* on why free-living organisms "must" exhibit minimal complexity. We have another evolutionary just-so story. Like other just-so stories in evolution, of course it may be true."[59]

And so we conclude as Leslie Orgel does: "It is almost inconceivable that nucleic acid replication could have got started, unless there is a much simpler mechanism for the prebiotic synthesis of nucleotides"[33]. It's a sentiment shared by another origin of life scientist, Gerald Joyce. He said the same thing in *Nature* almost a decade before in an indication of the pace of this science in moving forward on its mission: "The most reasonable interpretation is that life did not start with RNA."[52] The RNA world concept remains very much alive in spite of the facts however. Why? Here is Nobel laureate Christian de Duve with his explanation: "Yet judging from the recent literature, the widespread faith in a primeval, protein-less RNA world remains unshaken and continues to be popularized. Perhaps, this attitude is to be explained by the lack of a plausible alternative, short of special creation. "It must have happened that way," the underlying reasoning seems to go, "because it could not have happened any otherwise."[60]

The difficulties identified above have focused attention on compounds of more simple chemistry which might have preceded RNA, like nucleic acid analogues including the puranosyl analogue of RNA (pyranose form replaces the furanose form of ribose). This is more stable as a helix than RNA and less likely to form competing multiple strand structures, but pyranosyl nucleotides are not expected to be any easier to make than the furanosyl form. Another suggestion is peptide nucleic acid (PNA), consisting of a backbone of amino acids rather than ribose. This is capable of forming stable double helices with RNA and DNA and is not a chiral molecule. Problems with PNA include the fact activated PNA monomers cyclize, a problem that does not have a solution in abiotic conditions. Enantiomeric cross-inhibition is also "just as serious in the polymerization of nucleotides on a PNA template as it is on a conventional RNA or DNA template".[61] We need another theory.

§12.3.2   The "Thioester World Theory"

First proposed by Christian de Duve, this theory holds thioester protoenzymes triggered an early metabolism in a primordial soup, out of which RNA was later generated.[10,62] A thioester (the chemical formula is R-S-CO-R') is formed when a thiol (chemical formula R-SH) joins with a carboxylic acid (R'-COOH), releasing a molecule of water. Carboxylic acids and amino acids are among the most abundant substances in Miller-Urey conditions and volcanic eruptions are rich in hydrogen sulfide (H_2S), providing substrates for the generation of thioesters. The thioester bond is also a high-energy bond which de Duve suggests could have supplied energy for early chemical reactions, just as the high energy phosphate bond of ATP now does for all extant life (§15.3.4). How that energy was harnessed or how ATP and all its complicated synthetic machinery could arise out of a thioester energy metabolism, or the metabolic reactions we find in cells today from those supposed origins, are questions still unanswered.

§12.3.3　　　　　The "Self-Replicating Clay Theory"

This theory was first proposed by Graham Cairns-Smith at the University of Glasgow and popularized by Richard Dawkins in *The Blind Watchmaker*. He suggests the first form of life was a self-replicating clay though there is no congruence of clay chemistry with metabolism. Cairns-Smith holds this remarkable inorganic matter was later overthrown by one of its progeny which then became organic in a "genetic takeover".[63] The details of how this coup d'état could ever have happened, even in theory, raise more questions than they attempt to solve but the clay theory's most serious problem is that it has no experimental support from biology at all. It didn't at the time Dr. Dawkins popularized it either.[33,64,65] It never has and so notions of a mineral catalyst making life can fairly be summed up as Leslie Orgel says: "In my opinion, there is no basis in known chemistry for the belief that long sequences of reactions can organize spontaneously – and every reason to believe that they cannot".[33]

§12.3.4　　　　　Kauffman's Self Organizing Hypothesis

In the face of the manifold difficulties, MacArthur Fellow Stuart Kauffman now at the University of Calgary offers a solution which not only takes care of explaining the origin of life but the complexity of nature as well. It's that the ability to spontaneously self-organize is an innate property of biological systems. On this view our problem is not difficult at all. It's easy! How you look is what makes all the difference. This is how he looks. See if you see:

"The central problem is this: How hard is it to obtain a self-reproducing system of complex organic molecules, capable of a metabolism coordinating the flow of small molecules and energy needed for reproduction and capable of further evolution? Contrary to all our expectations, the answer, I think, is that it may be surprisingly easy...I believe that the origin of life was not an enormously improbable event, but law-like and governed by new principles of self-organization in complex webs of catalysts."[66]

In other words life and all its complexity is a function of a law, a law of nature! This claim is fascinating, if only for the bravado. How does it happen exactly?

The idea is entirely theoretical and is the product of computer modeling. His system rests on ideas of "autocatalytic set theory". Specifically that in a collection of different polymers (a "set") that are autocatalytic (meaning they can catalyze one another's formation) above a critical threshold number of these different molecules, the entire set suddenly becomes able to self-reproduce. It becomes "alive"! How many does it take to produce the effect? Between 6,165 and 34 million he says.

Bringing this idea back to biology and the real world, the polymers in question in the simulation are peptides and small ribozymes. What molecule catalyzes exactly what molecule in the set, and which produces which other molecule, is beneath the level of scrutiny of the construct. Details will be needed to be filled in if we are to agree the computer simulation has any transference to reality then, and of course we will need to explain how a set of these molecules and with such remarkable properties can all arise spontaneously. He wrote these words more than two decades ago and still we wait...

Even if this feat did all happen by itself, we're still miles from getting life. The next hurdle is how the self-reproducing, stable set can now change itself, "evolve", from these short polymers into all the diversity of life. To give you some background to the problem, such modeling by Manfred Eigen back in 1971 showed that in the eventuality of an arising of a new "mutant" catalytic activity in a set of autocatalytic peptides, all the other members in the set must change also if it is to remain self-reproducing. Kauffman's solution is that change is produced when the set itself divides, randomly distributing the contents to the two daughter cells, which therefore have different abilities under different particular environmental conditions and so become selected, evolving as they do. These details are just as speculative as is the delivery and regulation of the energy for the reactions i.e. substrate supply, and the coordination needed to prevent inhibition of the self-sustaining activity knowing failure means extinction.

His ideas are extraordinary to say the least and they have been met with skepticism and more, even from evolution scientists. Reviewers in the pages of *Nature* called it a: "highly over-optimistic estimate...Any experimental evidence to the contrary would be most welcome"[52] and having a "statistical otherworldliness...The style is heady, sententious and dangerously seductive...There are times when the bracing walk through hyperspace seems unfazed by the nagging demands of reality."[68] We take a close look at this crucial idea biological complexity can be got for free, even in theory, much later (§16). It lies at the heart of a naturalistic paradigm of biology of course. At least for now, we know Dr. Kauffman's ideas are as empty empirically as they are devoid of biological reality. We are still left with nothing to support those claims the origin of life was spontaneous.

§12.4 THE ORIGIN OF CELLULAR COMPLEXITY FROM A SELF-REPLICATING SYSTEM OF BIOPOLYMERS (STEP 3)

We have no solution in view even in theory but what if we could get biomonomers to form spontaneously somehow (Step 1), and they could arrange themselves into a self-replicating system that would get around the need for genes before there was any metabolism to make them, or metabolism before there were any genes to specify them (Step 2), and it also fortuitously turned out also to be a system to permit mutation and natural selection, and allow it to evolve from there. Even then we would still need Step 3 that is a way to construct life from this, or some other primitive self-replicating system. We know that living things are purposeful and complex without exception (§1.2;§15.1). Could replication, mutation and natural selection generate the necessary biochemical complexity for the most basic cellular life? The question biology takes unequivocally in the affirmative. Here is Carl Sagan for instance:

> "It is evident that once a self-replicating, mutating molecular aggregate arose, Darwinian natural selection became possible and the origin of life can be dated from this event."[69] Absent any evidence it's merely wishing and masquerading here as science we can say nothing polite about it. But there is another way of tackling the question and thereby get a handle on whether we can believe what biologists say.

Do a feasibility study. You don't have to be a biologist to know a basic first step before setting out on any exacting task is to be sure it can be done! Not all science is possible. Nobody still tries to build a perpetual motion machine, or make gold from lead over a fire, although both occupied sound minds once.

To answer affirmatively for natural forces here we need to 1) know the time available for the task of constructing life and be sure there is enough, and 3) know what is required to construct even the simplest forms of life (setting the demands of Steps 1 and 2 aside) and be sure natural selection has the power. We perform a more formal analysis of this problem later (§16.7)

§12.4.1 How Much Time Was There For Life To Arise Spontaneously?

It turns out if we believe the fossils and believe the dating evidence, there wasn't much time at all. This is an extraordinary discovery, but this assumes the truth of the dating claims of course. Assuming that, the first fossil record of life is the 3.55 billion year old stromatolites from Warrawoona sediment in Australia and other fossils of dated age from Southern Africa (§5.5.2). It seems unnecessary say so, but just in case, these earliest fossils resemble extant cyanobacteria ("blue-green algae") and so represent sophisticated bacterial life probably capable of photosynthesis and all the complexity that entails (§15.3.6). The time line goes even further back than this date however.

Radioactive dating of meteorites, we are informed by the science, indicates the solar system and Earth is ±4.55 billion years old. The "late heavy bombardment" of the Earth is believed to have continued for a half billion years thereafter the science says, with impacts capable of generating enough heat to boil the oceans at their surface or even vaporize them, thus preventing the development of a terrestrial biosphere and sterilizing it of life.[33,70-72] The Moon is believed to have arisen as a result of the bombardment by a Mars sized body says the science, and some of the evidence of the bombardment is taken to be the craters on its surface today dated to 4 billion years old the science says. Significant numbers of large bodies (100 km diameter) continued to hit the Earth and the Moon until at least 3.8 billion years ago they say.[26] Most therefore conclude that unless life originated on the ocean floor and so was shielded from this onslaught (from ~4.0-4.2 billion years ago), or unless alien life landed here but which would only push the origins problem to some other solar system and leave it just as intractable (§12.5), it could only have emerged at 3.7 to 4.0 billion years when the battering dissipated.[72]

The oldest rocks found on earth are from Isua, Greenland and dated at 3.85 billion years old. Interestingly, they have a relative enrichment of the ^{12}C isotope over ^{13}C. This is indicative of biologic activity which is not explainable by abiotic processes, thus providing evidence although indirect and for a view not universally held, of the existence of life at the very bottom of the earth's sedimentary pile.[33,73] In a review of the subject in *Nature*, Scripps Institute biologist Gerald Joyce sums this up: "The oldest rocks that provide clues to life's distant past are 3.9 billion years old and by that time cellular life seems already to have been well established."[52].

At least on the basis of current data then, there is an indirect record of life from the near outset of the earth's documented existence, and arguably "as soon as environmental conditions allowed it"[74], and therefore with "breath taking rapidity",[75] overlapping the current estimate of the timing of the late heavy bombardment. This data does not leave much of a window to speak of - at the very most a few hundred million years,[33] and on the above data, perhaps only a very few million years.[10] Regardless, it is an instant in paleoevolutionary terms. Stanley Miller says: "in spite of the many uncertainties involved in the estimates of time for life to arise and evolve to cyanobacteria, we see no compelling reason to assume that this process, from the beginning of the primitive soup to cyanobacteria, took more than 10 million years." [76]

However life emerged, it did so remarkably quickly all agree. And yet for this quite brilliant purposeless spontaneity, we have not seen it happen again. In other words life emerged sometime within the first 0.25% of the Earth's habitable existence (using the 10 million yr estimate above and diving by the 4 billion year Earth existence), but not in the 99.75% rest. And you don't see design? What you see depends on what you use to look. Stephen Jay Gould, like academia generally and not even much different even from Stuart Kauffman far out in left field, see it completely the other way around however. Specifically "life on Earth evolved quickly and is as old as it could be. This fact alone seems to indicate an inevitability, or at least a predictability, for life's origin from the original chemical constituents of atmosphere and ocean."[5] In other words we know it was naturalistic, so must have been inevitable, as if that has now solved the problem (§12.2-4). The difference between the only two ways of seeing the evidence is determined by whether or not your mind is already made up before the evidence is inspected.

§12.4.2   Beyond Even Chance

Enough narrative. Let's run the numbers ourselves. It will also allow us to look at this problem from another angle. We know we need both protein and nucleic acid for any kind of life (§12.3). The probabilities of randomly constructing these polymers from their constituent monomers can be calculated very simply. Natural selection at least had to overcome those odds. What are they?

We know the genetic code is a language as real as English. There are twenty-six letters in English, four in the nucleic acid alphabet and twenty for that of proteins (§2.3). The odds of construction of a particular sequence are given by the reciprocal of the number of possible different sequences that can be generated from the total pool of potential monomers. For a nucleic acid of n bases length there are 4 equally probable bases and therefore 4^n different sequences possible. A typical gene has a sequence of 1000 bases. The number of possible such sequences is 4^{1000} or around 10^{60277}. We are already utterly beyond even twilight zone science fiction and this is to get one protein, but for the record the odds for the entire human genome are $4^{100,000,000}$ or $10^{600,000,000}$ and for the genome of the bacterium *E. coli* it is $4^{4,000,000}$ or $10^{2,400,000}$.[78]

Similar calculations can also be performed for the odds for the assembly of proteins (§8.7). For a small protein with a chain of 100 residues like cytochrome c the correct amino acid from the 20 possible must be found for each position and thus there are 20^{100} or about 10^{130} possible sequences. This combinatorial analysis does not account for the fact some amino acid substitutions can be tolerated without changing the overall protein conformation (so called synonymous amino acid substitutions §2.4.4), and that amino acids are found at different frequencies in nature. Allowing for these, Manhattan Project physicist Hubert Yockey at Berkeley has calculated the level of probability to be around one in 10^{65}.[79]

Remarkably similar results were found when this was tested by MIT molecular biologist Robert Sauer using cassette directed mutagenesis experiments. These calculations assume incorporation of L-amino acids only however. Incorporation of non-biological residues or amino acid analogues or an incorrect optical isomer into the lengthening chain will destroy it. Accounting for optical isomers alone the probabilities are an additional 2^{100}, which is around 10^{30}. But amino acid residues must be joined together by peptide bonds, not non-peptide linkages which have been estimated to occur about 50% of the time.[80] Thus another 10^{30} possibilities exist to generate a total odds of one in 10^{125} for a protein just 100 amino acids long. Typical proteins are much larger than this and some like the cystic fibrosis transmembrane receptor are over fourteen times longer at 1480 amino acids.

Speaking of cystic fibrosis, and as a measure of the precision demanded for life and for living, the loss of only one amino acid in that protein is the most common cause of cystic fibrosis. We need more than one protein for life of course. Thus the product of the probabilities of each of the individual proteins serves as the final answer. The total number of atoms in the universe is 10^{80} and the total volume of the Universe in milliliters is 10^{84}. Thus the odds for even one protein are so far beyond the power of undirected natural force they are, literally, impossible. The bottom line? Nucleic acids and proteins could not have arisen randomly and have a very obvious and exquisite appearance of design merely by their existence. This appearance of design is no illusion because the probabilities are demonstrable and mathematical and relentless. All of the discussion that is this chapter was unnecessary. The answer is given, emphatically, by a very simple elementary calculation on the back of a matchbox.

Stuart Kauffman calculates that since the universe began, assuming it to be around 13.7 billion years old, the maximum number of possible pairwise particle collisions is 10^{193} and that "The known universe has not had time since the big bang to create all possible proteins of length 200 once...It would take at least 10 to the 67th times the current lifetime of the universe for the universe to manage to make all possible proteins of length 200 at least once."[81] This is an interesting observation for a quite different reason, and it is also a question for you. Knowing the probabilities involved, why does he and the mainstream and elite science establishment speak the way they do about ultimate origins? (Table 3.1)

We saw earlier in the case of weasel natural selection that the problem is not solved by serial random modification of a sequence to a particular target either (§4.7). While that produces very successful results very quickly, it does so because it smuggles in design. Richard Dawkins offers another potential solution in *The Blind Watchmaker*. Clay crystals. Crystals are not infrequently used as examples of how order results from natural physical processes, and which is the general principle here. The issue to inspect in such examples is how much order and how much purpose they can really assemble. Dr. Dawkins equates crystals with DNA to show how they might have acted as simple replicators for the first evolutionary life forms using the Cairns-Smith crystal model (§12.3.3). He says: "DNA molecules, whose capacity for storing information has already impressed us, are something close to crystals themselves." Are they?

The trouble with this idea is that while crystals dazzle they are really simple iterative structures. Take sodium chloride or table salt for instance. After assembling only two monomers you already know the polymer. Here is the blueprint: "Take a sodium atom. Attach a chloride atom. Repeat". Making a diamond is easier still and it shines even more. DNA on the other hand is aperiodic. In other words in the reading frame used to code for a protein, the sequence of its bases changes so often that the shortest description of its construction can be the sequence itself. DNA is not remotely "something close" to a crystal.

To get such biochemical compounds by natural selection from a simple self-replicating system we need a mechanism to advance from simplicity to purposeful complexity. How does Dr. Dawkins propose clay crystals might store information and then replicate this to daughter crystals to do this? By flaws. But this requires both that a flaw be transmittable and that the flaw represent new and meaningful information. This would appear in contravention of the construction orders for a crystal as given for table salt (NaCl) above. This does not stand in the way of his reasoning. He explains: "Flaws can occur anywhere over the surface of a crystal. If you like thinking about capacity for information storage (I do), you can imagine the enormous number of different patterns of flaws that could be created over the surface of a crystal. All those calculations about packing the New Testament into the DNA of a single bacterium could be done just as impressively for almost any crystal...you could easily devise an arbitrary code whereby flaws in the atomic structure of the crystal denote binary numbers."[82]

He does not show how this could be done even for one verse or one word. And so this enterprise of trying to make solid mud seem like DNA is no longer just a dream, it's as ridiculous as it already sounds. All it takes to recognize this is to let the Naturalistic Premise go until the analysis is done.

There is still another little problem with his notion, and this one he does identify: "What DNA has over normal crystals is a means by which its information can be read." He does not offer a solution to this problem. He can hardly be faulted. Even assuming such an improbable system could arise, the odds against it generating a single DNA sequence of over 135 bases naturalistically is already more than all of the atoms in the universe. He still does not see design though. What is this mind blindness? (§19)

There is more that can be said but we need to move on because here is Francis Crick to tell us what this science actually knows about the matter, both by way of summary for us and by way of summarizing how science ends up when the science is already certain before the evidence is in: "...at the present time we can only say that we cannot decide whether the origin of life on earth was an extremely unlikely event or almost a certainty – or any possibility between these two extremes."[83] The uncertainty does not impugn the certainty he does have about a naturalistic origin however. Why does he have it? Actually he has even more interesting views than Dr. Dawkins does and we look them next.

§12.5   LIFE FROM OUTER SPACE?

Some have seen in this relentless logic and mathematics the need for a completely different solution and here it is. Life came to Earth from Space. The idea is an old one going back at least 24 centuries to Anaxagoras (c500 BC-428 BC) but popularized by English physicist Lord Kelvin and Swedish Nobel laureate Svante Arrhenius a century ago, and more recently revived by such notables as Cambridge astrophysicist Sir Fred Hoyle and the Salk Institute's Leslie Orgel. However probably the most famous of the believers in recent memory is Sir Francis Crick. As co-discoverer of DNA we can expect him to be familiar with the probabilities of making it naturalistically. And so his opting for this idea says two things to us. First, how much more crazy to him is getting DNA naturalistically than it is blindly reaching into blackness to pull out space aliens that merely shift the origins problem to another planet without changing it a whit. And second, given his indisputable credentials, we should take his idea seriously. Okay let's look at it.

The idea is formally called panspermia, meaning "seeds everywhere". It posits life originated somewhere else in the Universe then spread to the Earth. In the classical version this was as bacteria and spores on dust particles. Unfortunately the only evidence for the idea is the evidence of life on Earth. We might stop there but Dr. Crick adds an interesting twist because he rejects the idea of microbes traveling about in space in this way "because" he says, "it is difficult to see how viable spores could have arrived here, after such a long journey in space, undamaged by radiation". As a result he takes the idea further, as so-called "directed panspermia."[83-85] He explains the concept in his autobiography. The idea is quite simple really. It is that, in his words, "life on Earth originated from microorganisms sent here, on an unmanned spaceship, by a higher civilization elsewhere."[85]

To reject design on grounds of supposed implausibility in the face of what we now know to invoke space invaders is an extraordinary idea. The reason for believing in such a notion is far more interesting than even the idea itself. It is a serious proposal from a Nobel laureate and one with an echo in journals like *Scientific American*.[7] Dr. Crick said he came to this conclusion as a result of the fact that the genetic code is conserved across life which he concluded implied a "narrow bottleneck" of ancestry; and secondly that since the universe is over twice as old as the Earth so providing further time for evolution.[86] Thus the logic here is to buy time (§12.4.1) in hopes of circumventing the relentless improbabilities (§12.4.2), all the while never solving the question, just displacing it. Halving the odds does nothing to the probabilities of course and he went to his grave convinced life must have come about spontaneously, somewhere, somehow.

Speaking of space aliens, since we know the probabilities of making a molecule of DNA or protein, what are the probabilities of alien life arising? The way the idea of aliens is spoken of affirmatively by leading scientists like Richard Dawkins[87] and Stephen Jay Gould,[88] and the way SETI has captivated the popular imagination, the probabilities would seem at least reasonable scientifically even if low.

The odds are estimated by the so-called Drake equation, named after American astronomer Frank Drake who proposed an approach to this problem back in 1961. Here is his formula:

$$N = R_* \times f_p \times n_e \times f_l \times f_i \times f_c \times f_L,$$ where

N is the number of civilizations capable of communicating with Earth,
R_* is the rate of star formation suitable to support life,
f_p is the fraction of those stars with planets,
n_e is the number of planets that are habitable in a solar system with planets,
f_l is the fraction of those planets where live arises,
f_i is the fraction of planets with life that develop intelligence,
f_c is the fraction of planets with intelligent life with interstellar communication, and
L is the lifetime of communicating civilizations

So what's the answer?

Dr. Drake's is that there are about 10,000 communicating civilizations in our galaxy alone. Look at the equation again. There's no way to attach numbers to at least the last 5 terms without guessing and for the final result to be of reasonable probability every single one of the terms must be of reasonable probability because they are multiplicative. In other words if even one of the terms is wildly improbable, then so is the final answer. How can we estimate probability? Utterly useless is what it is!

This has not stopped the equation about little green men making a very credentialed tour through the scientific literature and the lecture halls of the world. The illogic of a spontaneous origin of life is compounded by a materialist worldview. On that view it's inescapable. If life arose so quickly on this planet then why not elsewhere in a Universe of billions and billions of galaxies each with billions and billions of suns and even more planets, after all the Earth isn't that special. Or is it? (§14.4)

While the equation cannot be tested directly it most certainly can be tested indirectly. The estimated probability of making protein and DNA without which life as we know it cannot exist, is a test right there. We know it is impossible (§12.4.2). Whether or not it's reasonable to infer there may be life out there but not-as-we-know-it is a lot harder to test of course. We can get some idea of the odds involved nevertheless, because here is another test and it also comes up with the same answer by the way. It is by physicist John Barrow at Cambridge University. He says:

"If t_{bio} [the typical time it takes for life to evolve in the universe] and t_* [the lifetime of main sequence stars] are independent, then the time that life takes to arise is random with respect to the stellar time scale t_*. Thus it is most likely that either $t_{bio} \gg t_*$ or $t_{bio} \ll t_*$. Now if $t_{bio} \ll t_*$, we must ask why it is that the first observed inhabited solar system (us) is most likely to have $t_{bio} \cong t_*$. This would be extraordinary unlikely. On the other hand, if $t_{bio} \gg t_*$, then the first observed inhabited solar system (us) is the most likely to have $t_{bio} \cong t_*$ since systems with $t_{bio} \gg t_*$ have yet to evolve. Thus we are a rarity, one of the first living systems to arrive on the scene. In this case, we are led to a conclusion, an extremely pessimistic one for the SETI enterprise, that $t_{bio} \gg t_*$."[89]

The fascination by this science with alien life, like the supposed certainty of the spontaneous origin of life from chemicals, remains unmoved by of the mathematics.

§12.8   IN CONCLUSION: HOW TO GET SOMETHING FROM NOTHING SCIENTIFICALLY

You have now reviewed the best evidence for the spontaneous origin of life. Origin of life scientist Klaus Dose reviewed the same evidence and said: "More than 30 years of experimentation on the origin of life in the fields of chemical and molecular evolution have led to a better perception of the immensity of the problem of the origin of life on Earth rather than to its solution. At present all discussions on principal theories and experiments in the field either end in stalemate or in a confession of ignorance." [90] To review the bankruptcy of ideas for how it happened spontaneously, and yet still believing that it did, here for one last time is Sir Francis Crick:

"An honest man, armed with all the knowledge available to us now, could only state that, in some sense, the origin of life appears at the moment to be almost a miracle, so many are the conditions which would have had to be satisfied to get it going. But this should not be taken to imply that there are good reasons to believe that it could not have started on the earth by a perfectly reasonable sequence of fairly ordinary chemical reactions. The plain fact is that the time available was too long, the many microenvironments on the earth's surface too diverse, the various chemical possibilities too numerous and our own knowledge and imagination too feeble to allow us to be able to unravel exactly how it might or might not have happened a long time ago, especially as we have no experimental evidence from that era to check our ideas against. Perhaps in the future we may know enough to make a considered guess, but at the present time we can only say that we cannot decide whether the origin of life on earth was an extremely unlikely event or almost a certainty – or any possibility between these two extremes."[83]

If we are to believe the word from science life is just a lucky accident. Even children know there's no such thing as a free lunch, and yet we are told by scientists that we can get something incomparably more complicated for nothing. Faced with the manifold evidence of purpose in living things that is the appearance of design, concluding it is an illusion is willful blindness. The real question in the origin of life literature is why they are so sure it is solved (Table 3.2), and if not solved why they are certain it was not by design. For instance Richard Dawkins says "We still don't know exactly how natural selection began on Earth",[91] yet he is certain life was not created. How is that grounded, scientifically?

In 2002 John Rennie, the editor of *Scientific American* tackled the origin of life problem in his article *15 Answers to Creationist Nonsense*. It was one of the questions he called "nonsense", namely the claim "Evolution cannot explain how life first appeared on earth". This was his answer:

"The origin of life remains very much a mystery, but biochemists have learned about how primitive nucleic acids, amino acids and other building blocks of life could have formed and organized themselves into self-replicating, self-sustaining units, laying the foundation for cellular chemistry."[7]

Stockbrokers who write like this get prosecuted. He admits this question has no answer, so it can hardly be "nonsense", and he goes on to imply there are models for how it happened. But there aren't (§12.2-4).

Nobel laureate and Harvard biochemist George Wald is perhaps the most frank because although his view is the same as his colleagues, he at least tells us why he believes what he believes. In doing so he also takes us a full circle back to Louis Pasteur:

"When it comes to the origin of life there are only two possibilities: creation or spontaneous generation. There is no third way. Spontaneous generation was disproved one hundred years ago, but that leads us to only one other conclusion, that of supernatural creation. We cannot accept that on philosophical grounds; therefore we choose to believe the impossible: that life arose spontaneously by chance. One has only to contemplate the magnitude of this task to concede that the spontaneous generation of a living organism is impossible. Yet here we are, as the result, I believe, of spontaneous generation."[92]

This sad display from science would only be a shame if it were not also a poison. Administering spontaneous generation and evolution to young minds in science class could not be more false or its consequences more dire (§11.5).

The National Academy of Sciences disagrees and so demonstrates a second conclusion we are forced to make in summary. In its document "*Science and Creationism*" which was addressed in their words, to a "broader audience, but particularly to "teachers, educators and policymakers who design, deliver and oversee classroom instruction in biology" they say:

> "The arguments of creationists are not driven by evidence that can be observed in the natural world. Special creation or supernatural intervention is not subjectable to meaningful tests, which require predicting plausible results and then checking these results through observation and experimentation. Indeed, claims of "special creation" reverse the scientific process. The explanation is seen as unalterable, and evidence is sought only to support a particular conclusion by whatever means possible."

And so we witness a problem far more pernicious than a problem of deluded minds, it's something pernicious being perpetrated by minds - deceit. Truth is suppressed and it is willful and consciously done. Evolution is kept alive by science telling lies just like this. There were four just in the passage above (§25). In staking a claim for the high ground of empiricism and "observation and experimentation," the Academy only condemns itself because "meaningful tests...observation and experimentation" deny spontaneous generation overwhelmingly. This removes ignorance as explanation for the behavior. This science already knows that life came about spontaneously you see. The conclusion is made for philosophical reasons, not because of the evidence. It's in spite of the evidence (§19). Emperor's New Clothes.

REFERENCES

1. National Academy of Sciences. *Science and Creationism: A View from the National Academy of Sciences*. 2nd Ed edn, (The National Academies Press http://books.nap.edu/html/creationism/, 1999).
2. Futuyma, D. J. *Science on trial*. 95 (Sinauer Associates Inc., 1995).
3. Ruse, M. *Darwinism Defended*. 168 (Addison-Wesley Publishing Company, 1982).
4. Ruse, M. *The evolution wars. A guide to the debates*. 168 (ABC-CLIO, Inc., 2000).
5. Gould, S. J. The evolution of life on earth. *Scientific American*, 85-91 (1994).
6. Ridley, M. *Evolution*. 544 (Blackwell Science, Inc., 1996).
7. Rennie, J. 15 Answers to creationist nonsense. *Scientific American*, 78-85 (2002).
8. Mayr, E. *What Evolution Is*. 43 (Basic Books, 2001).
9. Kauffman, S. in *Intelligent Thought* (ed J. Brockman) 169-178 (Vintage Books, 2006).
10. de Duve, C. The beginnings of life on earth. *American Scientist* 83, 428-437 (1995).
11. Cowen, R. *History of Life*. (Blackwell Science, Inc., 1995).
12. Cowen, R. *History of Life*. 1 (Blackwell Science, Inc., 1995).
13. Darwin, F. *The Life and Letters of Charles Darwin*. Vol. 3, 18 (John Murray, 1887).
14. Thaxton, C., Bradley, W. & Olsen, R. *The Mystery of Life's Origin*. (Philosophical Library of New York, 1984).
15. Tiner, J. H. *Louis Pasteur - Founder of Modern Medicine*. 90 (Mott Media, 1990).
16. Tiner, J. H. *Louis Pasteur - Founder of Modern Medicine*. 75 (Mott Media, 1990).
17. Bendall, D. S. Ed. *Evolution from Molecules to Men*. 128 (Cambridge University Press, 1983).
18. Farley, J. *The spontaneous generation controversy from Descartes to Oparin*. 2nd edn, 73 (The Johns Hopkins University Press, 1979).
19. Alberts, B. The cell as a collection of protein machines: preparing the next generation of molecular biologists. *Cell* 92, 291-294 (1998).
20. Pigliucci, M. Where do we come from? A humbling look at the biology of life's origin. *Skeptical Inquirer* 23, 21-27 (1999).
21. Haldane, J. B. S. in *The origin of life* (ed J.D. Bernal) 242-249. Originally published 1929 in Rationalist Annual 1923-1910 (Weidenfeld and Nicolson., 1967(1929)).
22. Oparin, A. I. in *The origin of life* (ed J.D. Bernal) 199-234. Originally published Moscow, 1924 (Weidenfeld and Nicolson, 1967(1924)).
23. Miller, S. L. A production of amino acids under possible primitive Earth conditions. *Science* 117, 528-529 (1953).
24. Mason, S. F. *Chemical evolution*. 236-7 (Oxford University Press, 1991).

25 Oro, J. Synthesis of adenine from ammonium cyanide. *Biochemical and biophysical research communications* 2, 407-412 (1960).
26 Kasting, J. F. Earth's early atmosphere. *Science* 259, 920-926 (1993).
27 Shapiro, R. *Origins. A skeptics guide to the creation of life on earth*. 111-2 (Summit Books, 1986).
28 Lahav, N. *Biogenesis*. 138 (Oxford University Press, 1999).
29 Towe, K. M. Environmental oxygen conditions during the origin and early evolution of life. *Advances in space research* 18, 7-15 (1996).
30 Ferris, J. P. & Chen, C. T. Photochemistry of methane, nitrogen and water mixture as a model for the atmosphere of the primitive earth. *J. American Chemical Society* 97, 2964 (1975).
31 Fox, S. W. & Dose, K. *Molecular evolution and the origin of life*. (Marcel Dekker, 1977).
32 Holland, H. D. *The chemical evolution of the atmosphere and oceans*. (Princeton University Press, 1984).
33 Orgel, L. E. The origin of life - a review of facts and speculations. *Trends in biochemical sciences* 23, 491-495 (1998).
34 Whittet, D. C. B. Is extraterrestrial organic matter relevant to the origin of life on Earth? *Origins of life and evolution of the biosphere* 27, 249-262 (1997).
35 Anders, E. Pre-biotic organic matter from comets and asteroids. *Nature* 342, 255-257 (1989).
36 Fry, I. *The Emergence of Life on Earth* 81 (Rutgers University Press, 2000).
37 Miller, S. L. Which organic compounds could have occurred on the prebiotic earth? *Cold Spring Harbor Symposia in Quantitative Biology* 52, 17-27 (1987).
38 Chyba, C. F. & Sagan, C. Endogenous production, exogenous delivery, and impact-shock synthesis of organic molecules: An inventory for the origins of life. *Nature* 355, 125-132 (1992).
39 Lahav, N. *Biogenesis*. 244 (Oxford University Press, 1999).
40 Lahav, N. *Biogenesis*. 139 (Oxford University Press, 1999).
41 Brooks, J. & Shaw, G. *Origin and development of living systems*. 359 (Academic Press Inc., 1973).
42 Wachtershauser, G. Groundwork for an evolutionary biochemistry: The iron-sulfur world. *Progress in biophysics and molecular biology* 58, 85-201 (1992).
43 Keefe, A. D., Miller, S. L., McDonald, G. & Bada, J. L. Investigation of the prebiotic synthesis of amino acids and RNA bases from CO_2 using FeS/H_2S as a reducint agent. *Proceedings of the National Academy of Sciences USA* 92, 11904-11906 (1995).
44 Miller, S. L. & Bada, J. L. Submarine hot springs and the origin of life. *Nature* 334, 609-611 (1988).
45 Gilbert, W. The RNA world. *Nature* 319, 618 (1986).
46 Shapiro, R. A simple origin for life. *Scientific American* 296, 46-53 (2007).
47 Lahav, N. *Biogenesis*. 213 (Oxford University Press, 1999).
48 Zhang, B. & Cech, T. R. Peptide bond formation in vivo selected ribozymes. *Nature* 390, 96-100 (1997).
49 Kauffman, S. A. *The Origins of Order*. 287-341 (Oxford University Press, 1993).

50 Benner, S. A. Catalysis:design versus selection. *Science* 261, 1402-1403 (1993).
51 Bartel, D. P. & Szostak, J. W. Isolation of new ribozymes from a large pool of random sequences. *Science* 261, 1411-1418 (1993).
52 Joyce, G. F. RNA evolution and the origins of life. *Nature* 338, 217-224 (1989).
53 Mason, S. F. Chemical Evolution 241 (Oxford University Press, New York, NY, 1991).
54 Lahav, N. *Biogenesis.* 201 (Oxford University Press, 1999).
55 Larralde, R., Robertson, M. P. & Miller, S. L. Rates of decomposition of ribose and other sugars: Implications for chemical evolution. *Proceedings of the National Academy of Sciences USA* 92, 8158-8160 (1995).
56 Ferris, J. P. *Cold Spring Harbor Symposia in QuantitativeBiology* LII, 29-39 (1987).
57 Joyce, G. F. *et al.* Chiral selection in poly(C)-directed synthesis of oligo(G). *Nature* 310, 602-604 (1984).
58 Kauffman, S. A. The Origins of Order. 294 (1993).
59 Kauffman, S. A. *The Origins of Order.* 294-5 (Oxford University Press, 1993).
60 de Duve, C. *Blueprint for a cell:the nature and origin of life.* (Neil Patterson Publishers, 1991).
61 Schmidt, J. G., Nielsen, P. E. & Orgel, L. E. Enantiomeric cross-inhibition in the synthesis of oligonucleotides on a nonchiral template. *Journal of the American Chemical Society* 119, 1494-1495 (1997).
62 de Duve, C. *Blueprint for a cell:the nature and origin of life.* (1991).
63 Cairns-Smith. *Genetic takeover and the mineral origins of life.* (Cambridge University Press, 1982).
64 Horgan, J. In the beginning... *Scientific American*, 117-125 (1991).
65 Lahav, N. *Biogenesis.* 259 (Oxford University Press, 1999).
66 Kauffman, S. A. *The Origins of Order.* xvi (Oxford University Press, 1993).
67 Dover, G. A. Observing development through evolutionary eyes: A practical approach. *BioEssays* 14, 281-287 (1992).
68 Dover, G. A. Self-organization and emergence in life sciences. *Nature* 365, 704-706 (1993).
69 Sagan, C. On the origin and planetary distribution of life. *Radiation Research* 15, 174-192 (1961).
70 Sleep, N. H., Zahnle, K. J., Kasting, J. F. & Morowitz, H. J. Annihilation of ecosystems by large asteroid impacts on the early Earth. *Nature* 342, 139-142 (1989).
71 Chyba, C. F. *Geochim. Cosmochim. Acta* 57, 3351-3358 (1993).
72 Maher, K. A. & Stevenson, D. J. Impact frustration of the origin of life. *Nature* 331, 612-614 (1988).
73 Mojzsis, S. J. *et al.* Evidence for life on Earth before 3,800 million years ago. *Nature* 384, 55-59 (1996).
74 Schopf, J. W. Microfossils in the early Archean apex chart: New evidence of the antiquity of life. *Science* 260, 640-646 (1993).
75 Hayes, J. M. The earliest memories of life on Earth. *Nature* 384, 21-22 (1996).
76 Lazcano, A. & Miller, S. L. How long did it take for life to begin and evolve to cyanobacteria? *Journal of Molecular Evolution* 1994, 546-554 (1994).

77 Eigen, M. *Steps toward life.* 9-11 (Oxford University Press, 1992).
78 Kuppers, B.-O. *Information and the origin of life.* 59-69 (The MIT Press, 1990).
79 Yockey, H. P. A calculation of the probability of spontaneous biogenesis by information theory. *Journal of Theoretical Biology* 67, 377-398 (1977).
80 Meyer, S. C. in *Mere creation. Science, faith and intelligent design* (ed W.A. Dembski) 113-147 (Intervarsity Press, 1998).
81 Kauffman. *Investigations.* 144 (Oxford University Press, 2000).
82 Dawkins, R. *The Blind Watchmaker.* 152 (W.W. Norton & Company, 1996).
83 Crick, F. R. *Life itself, its origin and nature.* 88 (Simon & Schuster, 1981).
84 Crick, F. & Orgel, L. E. Directed panspermia. *Icarus* 19, 341-346 (1973).
85 Crick, F. *What mad pursuit.* (Penguin Books, 1990).
86 Fry, I. *The Emergence of Life on Earth* 132 (Rutgers University Press, 2000).
87 Dawkins, R. *The Blind Watchmaker.* (W.W. Norton & Company, 1996).
88 Gould, S. J. *The Flamingo's Smile.* 403-416 (W.W. Norton, 1987).
89 Barrow, J. D. Cosmology, life, and the anthropic principle. *Annals of the New York Academy of Sciences* 950, 139-153 (2001).
90 Dose, K. The origin of life: more questions than answers. *Interdisciplinary Science Review* 13, 351 (1988).
91 Dawkins, R. *The Blind Watchmaker.* 162-3 (W.W. Norton & Company, 1996).
92 Wald, G. The Origin Of Life. *Scientific American* 191, 44-53 (1954).

ABOUT THE AUTHOR

Peter-Brian Andersson is a Rhodes Scholar with doctorates in medicine (MBChB, University of Cape Town *cum laude*) and philosophy (DPhil, University of Oxford) and an honors degree in medical biochemistry (BSc.Med(Hons) University of Cape Town). He was awarded a Junior Research Fellowship at Oxford University, a distinction afforded to the top 2% of its graduate researchers, and was elected to Alpha Omega Alpha honor medical society in the United States. He has published widely in neuroimmunology and neurology including in *Neuroscience, Trends in the Neurosciences, Journal of Experimental Medicine, The Lancet* and *Neurology*. He has held college lecturer or clinical instructor positions at the Universities of Oxford, University of California San Francisco and Stanford University, and currently serves as a Clinical Associate Professor at the University of California Los Angeles and as a neurologist in clinical private practice. He can be reached at pba@tlotenc.org. and www.bythethingsthataremade.org.

INDEX

Agassiz, Louis, 51, 183, 388, 397
Alberts, Bruce, 52, 61, 118, 397, 570
alleles, 77, 81, 83, 85, 87, 88, 89, 90, 92, 93, 94, 97, 98, 150, 153, 305, 365
American Association for the Advancement of Science, 103, 136, 371, 432
anagenesis, 75, 76, 277, 278, 279, 280, 284, 500
apparent design, 25, 37, 41, 42, 43, 44, 56, 58, 63, 64, 96, 97, 109, 149, 164, 167, 169, 266, 291, 299, 337, 342, 347, 353, 440, 588, 594; argument from, 107–25
appearance of design in nature: adaptiveness, 36; complexity, 34–36; diversity, 29–32; similarity, 36–37, 291–94
Archeopteryx, 251
Aristotle, 16, 70, 293, 294, 307, 308, 313, 335, 337, 343, 387, 389, 545
Arthur, Wallace, 119, 384, 411, 424, 436, 438, 440
artificial selection, 48, 128, 129, 130, 132, 140, 141, 153, 159, 162, 163, 166, 559
Ayala, Francisco, 88, 184, 215, 371, 373
Bauplan, 31
Behe, Michael, 25, 65
Benga, Ota, 552
Benton, Michael, 207, 209, 272, 273
Bible, 10; and evolution, 41; authority and theistic evolution, 40; evolution v. creation as a moral question, 40; on trial by evolution, 10, 17, 18, 456, 545
biodiversity, 31, 32, 65, 200, 202, 204, 207, 208, 270, 274, 555, 559
Biogenetic Law, 390, 392, 394, 395, 398, 405, 424, 427, 428, 438, 441; definition, 392
Biston betularia, 147, 148, 149, 177
blind watchmaker, 10, 42, 58, 63, 64, 80, 96, 133, 135, 142, 164, 200, 450
Bowring, Samuel, 205
C paradox, 84, 375
Cambrian explosion, 56, 204, 205, 206, 208, 214, 215, 216, 217, 254, 256, 259, 276, 277, 287; definition, 204
Carroll, Robert L, 184, 191, 212, 218, 236, 243, 252, 277

Carroll, Sean, 52, 383, 384, 420
chromosome, 84
chronospecies, 76
cladistics, 196, 227, 247, 315, 317, 324, 361, 393, 455, 461, 462, 464
cladogenesis, 76, 468
climbing Mount Improbable, 96, 139
coelacanth, 185, 189, 190, 193, 218
Collins, Francis, 4, 23, 40
Common Ancestry Inference, 109, 112, 131, 134, 386
Conundrum of Life, 21
Cowen, Richard, 183, 521, 568
Coyne, Jerry, 49, 119, 120, 121, 149, 153, 215, 390, 411, 432, 441
creation science, 20, 23
creationism, 4, 5, 9, 23, 25; definition, 4
Crick, Francis, 37, 83, 349, 571, 577, 590, 591, 593
crossing over (chromosomal), 87
Darwin, Charles: and Lyell's idea, 139; early life, 69; embryology evidence, 385; origin of life, 569; whale evolution, 225
Davies, Paul, 55, 108, 124, 127, 163, 354
Dawkins, Richard, 6, 7, 10, 11, 28, 34, 37, 42, 43, 45, 47, 52, 55, 56, 58, 60, 63, 76, 80, 96, 98, 105, 109, 122, 139, 144, 147, 149, 164, 169, 174, 214, 277, 280, 281, 319, 505, 549, 552, 583, 589, 591, 594
Dayhoff, Margaret, 351
de Duve, Christian, 124, 568, 582
Dembski, William, 131, 168
Dennett, Daniel, 19, 65, 133, 138
development. See embryology
Diamond, Jared, 133, 456, 546
Dobzhansky, Theodosius, 52, 74, 103, 104, 122, 139, 141, 319, 425, 438, 549
Dollo's law, 514
Drake equation, 592
Eldredge, Niles, 4, 44, 62, 76, 77, 81, 112, 120, 127, 168, 180, 189, 210, 211, 217, 251, 275, 278, 283, 323, 499
embryology: Biogenetic Law, 395, 396; Darwin's claims, 387; definition, 382; von Baer's laws, 390
evo-devo, 384

601

evolution: as fact, 103–7; biogeography, 555–59; descent with modification, 81; evidence of, 107–25; mechanism of, 81–97; negative evidence of, 121; of horses, 249; of whales, 220; subtheories of, 74–81; what's special about it, 52–62
evolutionary tree, 45, 82, 195, 212, 213, 216, 221, 357, 359, 362, 369, 454, 461, 468, 469, 482, 513, 514, 521, 526, 535
finch. See Galapagos finch
Flew, Anthony, 5
fossil: Burgess, 146, 189, 193, 206, 270, 277, 288; definition, 183; Ediacaran, 203; importance, 179; living, 141, 171, 190
fossil record: incompleteness, 189–91, 265–68, 273–78; incompletness testing for, 268–73; intermediate forms, 218–55; interpretation, 191–98, 454–74; pattern, 189–91; predicted by Darwinism, 189–91; pull of the recent, 207; stasis, 217–18; sudden appearance, 202–17
fossils: oldest, 203
Futuyma, Douglas, 61, 104, 112, 113, 114, 115, 116, 117, 119, 120, 121, 122, 149, 224, 276, 310, 349, 396, 431, 449, 452, 499, 510, 526, 551, 567
Galapagos finch, 61, 127, 128, 146, 556
Gee, Henry, 180, 461, 463, 478, 502
genetic code, 37, 84, 94
genetic drift, 93, 94, 98, 156, 364, 365, 524
genotype, 72
Georges Cuvier, 184
Gilbert, Scott, 337, 384, 387, 396, 438
Gingerich, Philip, 231, 237, 242, 245
Goldschmidt, Richard, 77, 169, 171, 397
Gould, Stephen Jay, 45, 62, 76, 77, 132, 145, 181, 189, 192, 205, 206, 209, 217, 218, 220, 227, 233, 236, 238, 241, 277, 278, 319, 328, 330, 332, 334, 351, 382, 388, 389, 390, 396, 410, 411, 420, 424, 429, 441, 468, 551, 587, 591
gradualism, 74, 79, 198, 516
Grant, Peter and Rosemary, 60, 61, 146, 153, 154,159
Grant, Verne, 107, 114, 116, 119, 120, 175, 181
Grassé, Pierre, 141, 171, 214

Haeckel, Ernst, 35, 45, 46, 61, 386, 392, 396, 398, 399, 408, 423, 425, 427, 428, 430, 439, 458, 484, 545, 570
Haldane, J.B.S., 56, 74, 147, 176, 254, 571, 573, 596
Hall, Brian, 306, 309, 311, 317, 322, 335, 341, 384, 394, 401, 406, 411, 430, 441
Hardy-Weinberg equilibrium, 90, 93
Hawking, Stephen, 123, 125, 138
heterochrony, 343, 403, 405, 407, 425, 426, 427, 428, 429, 430, 436, 439, 440; by Gould, 426, 427; by Haeckel, 393; by Smith, 427; significance of, 428
His, Wilhelm, 397, 406
Homo erectus, 453, 472, 481, 484, 485, 486, 497, 508, 509, 510, 517–28, 529, 543, 554, 561, 562, 563
homology, 291–342; as evolution, 294–300, 323–25; developmental approaches, 319–21; ear bones, 328–35; not from similar genes or development, 300–307; pentadactyl limb, 292, 325–27; phylogenetic approaches, 309–19
hopeful monsters, 77, 521
Hox genes, 413–21
Hoyle, Fred, 94, 370, 591
human evolution: arguments for, 451–54; cladistics failure, 464–68; erectus not a transitional species, 517–28; finding evolution, 468–74; fossil species, 483–93; habilis not a transitional species, 528–37; hominid vs hominin, 482; inspecting argument 1 (evolution a fact), 475–76; inspecting argument 2 (similarity with apes), 476–80; inspecting argument 3 (fossil record pattern), 499–537; inspecting argument 4 (human fossils all recent), 543; Laetoli footprints, 540–42; of brain, 503–9; Oldowan tools, 511, 512, 543; philosophical implications, 544–52; phylogenies, 466; reversal, 516; supposition of, 468–74
Hume, David, 42, 43, 66
Huxley, Julian, 74
Huxley, Thomas Henry, 79, 99, 250, 263, 457, 464, 550, 551
intelligent design: analysis, 28; vs. creationism, 25

602

Java man, 453, 485, 521, 526
Johanson, Donald, 489, 490, 526, 535, 536
Johnson, Paul, 552
Johnson, Phillip, 146, 338
Kant, Immanuel, 22, 40
Kauffman, Stuart, 167, 578, 580, 583, 587, 589
Kettlewell, Bernard, 60, 147, 153
Kimura, Motoo, 361, 365, 371
Klein, Richard, 490, 511, 512, 515, 518, 521, 523, 527
Lamarck, Jean Baptiste. See Lamarckism
Lamarckism, 48, 71, 72, 74, 146, 298, 299, 338, 395, 410, 425, 432, 439
Law of Non-Contradiction, 55
Law of The Emperor's New Clothes, 11, 51, 374
Le Gros Clark, Sir Wilfred, 182, 258, 485, 529
Leakey, Richard, 479, 485, 489, 491, 512, 533, 534, 535, 537
Lewin, Roger, 174, 283, 460, 530, 531, 533
Lewontin, Richard, 63, 127, 145, 470
Linnaeus, Carl, 16, 18, 21, 45, 58, 351, 456, 483, 497, 544
Lucy, 480, 489, 490, 492, 507, 513, 526, 535, 536, 542, 562, 563
Lyell, Charles, 79, 139, 184
macroevolution: definition, 76
Macroevolution Inference, 109, 112, 126, 131, 132, 134, 139, 141, 147, 158, 162, 386
Majerus, Michael, 149
Margulis, Lynn, 170, 178
Marsh, Frank Lewis, 139
Mayr, Ernst, 3, 17, 50, 52, 61, 74, 76, 78, 81, 104, 106, 114, 115, 116, 117, 118, 119, 120, 129, 134, 142, 160, 173, 181, 194, 195, 198, 251, 278, 310, 349, 381, 396, 419, 425, 473, 500, 545, 546, 555
Medawar, Sir Peter, 105, 173
meiosis, 85
mesonychid, 231, 232, 234, 238, 243, 244, 245, 246
methodological naturalism, 8, 18, 19, 21, 23, 41, 55, 57, 58, 122, 568, 570
Meyer, Steven, 215
microevolution: definition, 76
Miller, Kenneth, 4, 23, 112, 117

Miller, Stanley, 571, 575, 587
Miller-Urey apparatus, 571–73
Modern Synthesis, 74, 80, 99, 103, 113, 139, 143, 146, 149, 179, 196, 218, 226, 265, 278, 286, 287, 298, 300, 383, 437, 438, 456
molecular clock, 116, 358, 359, 360, 361, 362, 363, 364, 365, 366, 367, 368, 370, 371, 373, 374, 376, 377, 378, 380, 475; cannot keep time, 366–71; testing, 371–76
molecular evolution, 376; equidistance and separation, 353–59; neutral theory, 365; phylogenetics, 350–53
Monod, Jacques, 53, 548, 551
Morgan, Thomas Hunt, 437, 438
Morris, Simon Conway, 146, 207, 270
Mount Rushmore, 7, 24, 25, 26, 27
mutation, 90
National Academy of Sciences, 5, 20, 52, 61, 65, 66, 112, 114, 116, 118, 120, 137, 138, 159, 220, 231, 244, 246, 259, 260, 262, 263, 288, 339, 344, 380, 450, 467, 470, 494, 495, 497, 545, 559, 560, 564, 570, 595, 596, 597, 598
National Association of Biology Teachers, 20, 65, 112, 137
natural selection: adaptation, 73, 127, 339; as blind watchmaker, 107–25; as tautology, 143; Darwin's finches, 161; of weasels, 164; ring species, 130, 161, 162, 163; survival of the fittest, 80, 142, 143, 144, 147, 199, 438, 552
Naturalistic Premise, 108, 109, 112, 122, 126, 131, 134, 167, 387, 436, 557, 590
Neo-Darwinism. See Modern Synthesis
ontogeny recapitulates phylogeny, 61, 392, 395, 442
Orgel, Leslie, 573, 577, 581, 583, 591
origin of life: cell from self-replicating system, 585–90; from space, 590–93; hydrothermal vent theory, 574–76; impact theory, 574; Miller-Urey apparatus, 571–73; origin of biopolymers, 571–76; origin of self-replicating system, 576–85; primordial soup, 571–73; probability, 587–90; RNA world, 582; self-organizing hypothesis, 583–85; time available, 586–87
Orr, Allen, 49

Owen, Richard, 183, 228, 249, 294, 297, , 299, 306
Paley, William, 42, 43, 63
parallelism, 311, 337, 388, 392, 427, 429, 430, 444
Pasteur, Louis, 306, 569, 580, 594, 596
Paul, Chris, 211, 249
Peking man, 485, 526
Pennock, Robert T, 4, 20, 106, 276
pentadactyl limb, 44, 115, 252, 292, 294, 301, 325, 326, 327, 338, 408
phenotype, 72
phylotypic stage: definitions, 400; not conserved, 394–409
Pickford, Martin, 468, 479, 492
Pigliucci, Massimo, 97, 112, 143
Piltdown man, 56, 459
plesiomorphy, 196
polymorphism, 88, 89, 90, 153, 171, 373, 380
punctuated equilibrium, 62, 76, 77, 278–86, 286, 335, 347, 475, 519, 520, 554; definition, 278; problems with, 284–86
Raff, Rudolf, 32, 113, 312, 315, 319, 323, 327, 341, 384, 387, 397, 409, 410, 425, 514
recapitulation, 111, 119, 382, 389, 390, 392, 393, 394, 395, 398, 401, 402, 403, 406, 408, 409, 410, 423, 424, 425, 426, 427, 428, 429, 431, 436, 439, 441, 442
recombination, 80, 81, 85, 89, 90, 92, 131, 135, 142, 167, 373
Rennie, John, 46, 47, 104, 594
Richardson, Michael, 403, 407, 423, 428, 441, 442
Ridley, Mark, 41, 55, 105, 112, 113, 114, 115, 116, 122, 125, 127, 131, 143, 149, 159, 161, 162, 163, 174, 254, 255, 325, 338, 349, 366
Ruse, Michael, 47, 112, 117, 127, 133, 149, 197, 219, 251, 255, 396, 397, 400, 452, 568
Sagan, Carl, 18, 27, 65, 124, 585, 597, 598
Scala Naturae, 71, 388, 439
Schwartz, Jeffrey, 462, 468, 478, 480, 531
scientific creationism. See creation science
Seposki, John, 207, 272
SETI, 26, 591, 593

Shapiro, James, 5
Simpson, George Gaylord, 74, 80, 143, 179, 183, 191, 211, 214, 222, 229, 254, 276, 281, 283, 425, 454, 548
Slack, Jonathan, 398, 401, 402, 411
Smith, John Maynard, 76, 321, 341
Smith, Kathleen, 422, 427
speciation, 49, 61, 97, 159, 172, 279, 284, 556; definition, 76, 278
spontaneous generation, 568, 569, 594, 595, 596
Stanley, Steven, 76, 77, 97, 144, 179, 212, 217, 218, 219, 278
Stebbins, G. Ledyard, 74
stromalites, 586
symplesiomorphy, 196
synapomorphy, 196, 235, 237, 247, 314, 319, 323, 324, 335, 402, 433
Synthetic Theory. See Modern Synthesis
Tattersall, Ian, 458, 462, 468, 472, 520, 521, 530, 531, 533
taxonomy: Linnaean, 16, 31, 482; systematics, 45
The Everything From Nothing Premise, 123
theistic evolution, 4, 22, 38, 39, 40, 41, 295, 545
Thewissen, Hans, 220
Tobias, Phillip, 489, 525, 528, 536
ultra-Darwinist, 76, 168, 280, 282, 438
uniformitarianism, 49, 113, 131, 139, 174, 184, 440
Valentine, James, 118, 192, 204, 205, 207, 211, 217, 270, 273, 284, 285
vestigial organs, 111, 119, 229, 237, 238, 381, 382, 386, 431, 432, 433, 434, 435, 431–35, 439
von Baer, Karl Ernst, 119, 388, 389, 390, 391, 392, 394, 395, 400, 403, 407, 409, 411, 412, 421, 422, 424, 436, 439, 440; laws, 388
Waddington, Sir C.H., 146, 173
Wake, David, 299, 337, 339
Wald, George, 109, 124, 594
Walker, Alan, 468, 479, 485, 491, 492, 534, 537
Wallace, Alfred Russell, 70, 547, 558
Watson, James, 61, 83, 118, 397, 571
Weismann, August, 74
Wells, Jonathan, 154

White, Tim, 468, 490, 491, 492, 517, 520, 536, 542
Woese, Carl, 31, 577

Wood, Bernard, 464, 465, 468, 471, 472, 473, 490, 500, 503, 512, 513, 517, 521, 522, 525, 532, 535, 537, 543
Yockey, Hubert, 588

Made in the USA
San Bernardino, CA
28 December 2014